Interfacial Phenomena

A monolayer of hexadecanol spread on Loch Laggan, Scotland. Such a condensed monolayer can considerably reduce evaporation of the water (Chapter 7) and also effectively damps out the ripples (Chapter 5). Reproduced by permission of Price's (Bromborough) Limited.

[*Frontispiece*

Interfacial Phenomena

J. T. DAVIES

Professor of Chemical Engineering and Head of the Department, University of Birmingham

E. K. RIDEAL

Former Professor of Colloid Science, University of Cambridge

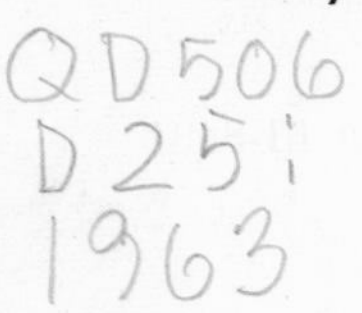

Second Edition

1963

Academic Press

New York and London

ACADEMIC PRESS INC.
111 Fifth Avenue
New York, 3, N.Y., U.S.A.

United Kingdom Edition
Published by
ACADEMIC PRESS INC. (London) Ltd.
Berkeley Square House, Berkeley Square, London W.1.

Second edition 1963

LIBRARY OF CONGRESS CATALOG CARD NUMBER 61-8494

PRINTED IN GREAT BRITAIN BY WILLMER BROTHERS & HARAM LTD.

Preface

Since the first edition of "Surface Chemistry" in 1926, the subject matter embraced by this title has grown both in importance and in complexity. There has appeared during the intervening years a number of excellent books on the subject, but with each new development the volumes become more compendious. The stage has now been reached when the adsorption of gases on to solids has merited special treatment, and specialized works have appeared on physical adsorption, chemisorption and heterogeneous catalysis.

It seems to us that while these properties of the solid-gas interface have received much attention, there has been a tendency for treatments of the various interfaces involving liquids to be confined to the more technical aspects of such subjects as detergency, flotation, foams and emulsions. In this volume we therefore examine particularly some of the more fundamental properties of the various liquid interfaces: we include a systematic presentation of the results of our studies together over a period of ten years, first at the Royal Institution and subsequently at King's College, London.

In Chapter 8 we discuss the more important characteristics of disperse systems and of adhesion, especially in so far as these follow from the fundamental interfacial properties described in the preceding chapters.

We wish to express our indebtedness to many friends and colleagues for stimulating discussions over the years.

This book was written while J.T.D. was a lecturer at the Department of Chemical Engineering, University of Cambridge, and while E.K.R. was at Imperial College, University of London. It will, we hope, be useful to chemical engineers, chemists and biologists, and prove stimulating in both industrial and academic laboratories.

For the second printing, we have taken the opportunity to correct several minor errors, and to bring up to date two sections. These concern the damping of waves and ripples (pp. 269–273), and the circulation within moving drops (pp. 335–336).

J.T.D.
E.K.R.

April 1963

CONTENTS

Chapter 5 PROPERTIES OF MONOLAYERS

Chapter I

The Physics of Surfaces

CONDITIONS AT A PHASE BOUNDARY

The boundary between two homogeneous phases is not to be regarded as a simple geometrical plane, upon either side of which extend the homogeneous phases, but rather as a lamina or film of a characteristic thickness: the material in this "surface phase" shows properties differing from those of the materials in the contiguous homogeneous phases. It is with the properties of matter in the surface layer that we are here concerned. Just as in bulk, the matter of the "surface phase" may exist in the solid, liquid, and gaseous states: i.e., there are various types of interfacial phase. We are here especially concerned with the boundaries between a liquid and a gas or vapour, and also with those brought into existence by the mutual contact of two immiscible or partly miscible liquids, or of a liquid against a solid.

It is a matter of common observation that a liquid behaves as if it were surrounded by an elastic skin with a tendency to contract. Drops of liquid, uninfluenced by external forces such as gravity, adopt a truly spherical shape: the determinations of the contractile behaviour and tension of a soap film stretched across a framework is a common laboratory experiment. Young[1] was the first to attempt an explanation of this "surface tension" in terms of the attractive and repulsive forces between the molecules constituting the liquid: the cohesion between the molecules of a liquid must surpass their tendency to separate under the influence of thermal motion. This net attraction between neighbouring atoms is fulfilled most completely in the interior of the phase, while those atoms or molecules at the surface are attracted less completely than they would have been in the bulk (Fig. 1-1). Consequently the energy of the latter is greater, and, since the free energy of a system tends to a minimum, the surface of such a pure phase will always tend to contract spontaneously.

In symbols, if γ_0 is the force per cm. tending to contract such a surface, and if S, T, P, V, A, μ, and n refer respectively to entropy, absolute temperature, pressure, volume, surface area, chemical potential, and number of molecules in the system, then

$$dF = -SdT - PdV + \gamma_0 dA + \mu dn \tag{1.1}$$

where F represents the total Helmholtz free energy of the system. At constant temperature and volume for a given number of moles of system, this reduces to

$$\gamma_0 = \left(\frac{\partial F}{\partial A}\right)_{T,V,n} \tag{1.2}$$

Under these conditions a spontaneous contraction of the surface area ($-\partial A$) will decrease F (∂F negative), provided γ_0 is positive. Since the surface of a stable liquid phase does in fact tend to decrease in area, γ_0 is

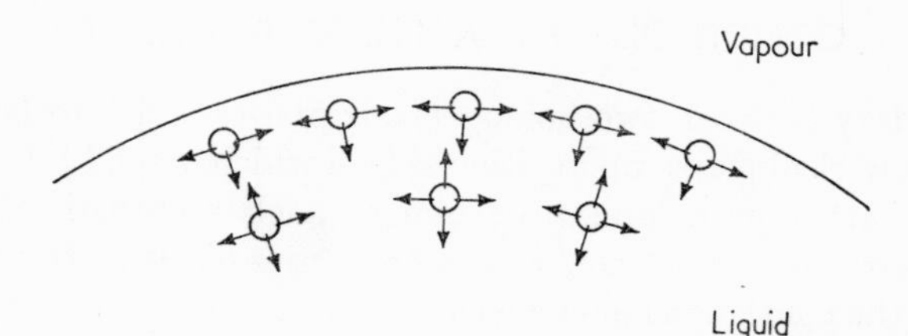

Fig. 1-1. Attractive forces (represented by arrows) between molecules (shown as spheres) at the surface and in the interior of a liquid.

always positive. It is called the surface tension, some values of which are quoted in Table 1-I. We shall now show that if F^s is the Helmholtz free energy per unit area of surface, F^s is equal to γ_0 in liquid systems. The argument is that, since under these conditions $d(AF^s) = dF$, we can substitute for F in eq. (1.2) to obtain:

$$\gamma_0 = F^s + A\left(\frac{\partial F^s}{\partial A}\right)_{T,V,n} \tag{1.3}$$

TABLE 1.I

Standard Surface Tensions of Pure Liquids against Air[2]

	t (°C)	γ_0 (dynes cm.$^{-1}$)
Water	20	72.8
Water	25	72.0
Bromobenzene	25	35.75
Benzene	20	28.88
Benzene	25	28.22
Toluene	20	28.43
n-Octanol	20	27.53
Chloroform	20	27.14
Carbon tetrachloride	20	26.9
n-Octane	20	21.8
Ethyl ether	20	17.01

But, in a one-component liquid, F^s depends only on the configuration of molecules in the surface, and not on area, so that $\partial F^s = 0$ at constant T and V. Hence, for such a system, eq. (1.3) reduces to:

$$\gamma_0 = F^s \tag{1.4}$$

Similarly, at constant T and P, $\gamma_0 = G^s$, the latter being the excess Gibbs free energy. Since in practice changes of pressure or volume accompanying surface changes are small, G^s and F^s are practically identical.

Thus the surface tension and the Helmholtz free energy per unit area of the surface are equal. In practice the former is measured in dynes cm.$^{-1}$, and the latter in ergs cm.$^{-2}$, though the dimensions of each are identical.

The relation (1.4) breaks down if the viscosity of the system is so high that the rearrangement of molecules when the surface is extended occurs more slowly than the relaxation of shear stress in the interior of the liquid or solid[3].

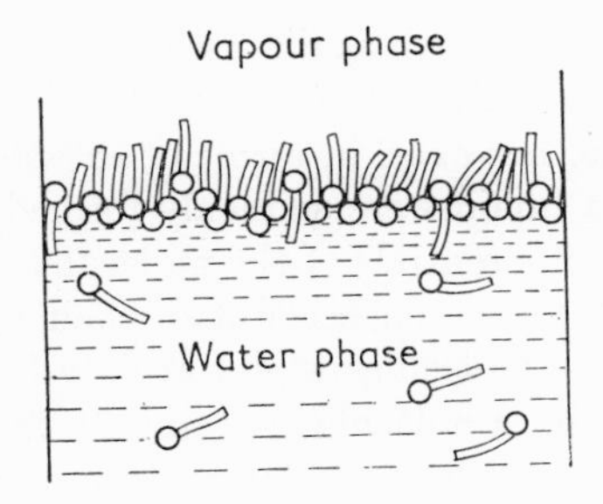

Fig. 1-2. Molecules of butanol adsorbing into a monolayer at the vapour-water surface, showing the hydrated polar "heads" (here hydroxyl), and the upward orientation of the hydrocarbon "tails". (Figure adapted from Harkins[2].)

The tendency of the surface area to decrease spontaneously as a result of the atoms or molecules in the surface entering the bulk of the phase can be altered by adding a second component. Suppose that to water ($\gamma_0 = 72$ dynes cm.$^{-1}$) we add a little butanol (C_4H_9OH). This will dissolve in the water because of the tendency of the hydroxyl group to be hydrated, in spite of the C_4H_9– chains' partial dislocation of the hydrogen-bonded water structure. If, however, the butanol molecules reach the surface, they can keep their hydroxyl groups in the water, while the hydrocarbon chains can escape into the vapour phase (Fig. 1-2), where they are energetically more welcome than in the water. Thus the molecules of butanol, unlike those of water, tend to accumulate in the surface rather than in the bulk of the liquid, forming an oriented monomolecular layer (or "*monolayer*"). This tendency for packing into the surface, or *adsorption* as it is called, must be considered in conjunction with the contractile tendency of the surface of the pure water, the net result being that, if Π is the repulsive pressure (positive in the sense that the film tends to expand by further spreading) of the adsorbed layer of butanol, the net surface tension is now lowered to γ by this amount:

$$\gamma = \gamma_0 - \Pi \tag{1.5}$$

Now Π is usually less than γ_0, so that γ remains positive, i.e. the phase is still coherent in that it tends to contract into as little volume as possible. If, however, Π exceeds γ_0, then $\gamma < 0$, and the net effect is that the surface tends to expand, leading to buckling of the surface or, in fluid liquid-liquid systems, to spontaneous emulsification (Chapter 8).

The Tension within a Liquid Surface

There is ample evidence that a free surface of a liquid behaves as if in a state of tensile stress—as if it contained a thin stretched rubber membrane within it. While this contractile tendency follows dimensionally from the positive free energy of the system as discussed above, the physical existence of a tensile stress in the surface has been disputed. The origin of a real stress is easily comprehensible, however, in terms of the following argument[3a].

If a fresh surface of a liquid were suddenly formed and were initially free from stress, the chemical potentials of the molecules in the surface would be higher than in the bulk phase, on account of their unsymmetrical environment. Consequently some molecules would rapidly leave the surface for the bulk, so increasing the intermolecular spacing in the plane of the surface. This increased molecular spacing must lead to an extra attractive tension between the molecules in the surface layer, reducing their "escaping tendency" or chemical potential. The net desorption of molecules from the surface would continue until this tension reached a value sufficient to reduce the chemical potential of the molecules in the surface to that of the bulk liquid. When this is achieved, the numbers of molecules entering and leaving the surface layer must be equal.

That the surface stress does not cause an overall movement of the liquid is explicable by the balancing effect of the difference of chemical potential between the surface and the bulk: consequently there is no reason to deny the existence of a physical stress or tension in the surface.

In mathematical terms, by eq. (4.49)

$$\left(\frac{\partial \mu}{\partial \gamma}\right)_T = -A_m$$

where A_m is the area occupied by 1 molecule in the surface. Hence the otherwise higher value of μ for the molecules in the surface is reduced by a positive tension stress γ, until the values of μ are the same in the surface and in the bulk.

Kinetics of Molecules in the Surface

Though the surface of water appears perfectly smooth and quiescent, the kinetic theory of gases and liquids shows that, on a molecular scale, it is in a state of violent agitation, with the surface molecules continually being replaced by others. Thus, from the kinetic theory of gases, the number of molecules of vapour (at a vapour pressure p dynes $cm.^{-2}$) striking unit area of surface per second is given by $p(2\pi m kT)^{-1/2}$ where kT is in ergs and m is the weight of one molecule. From this formula one calculates that, for saturated water vapour at 20°C, 8.5×10^{21} molecules of water vapour strike 1 $cm.^2$ of surface each second. If a fraction α of these enters the liquid surface, $8.5 \times 10^{21} \times \alpha$ molecules condense per $cm.^2$ per second, and, at equilibrium, this must be equal to the rate of evaporation of water. Now α is known to be between 0.034 and 1 for water, so that the lowest estimate of the number of water molecules leaving or condensing on each $cm.^2$ per second under these conditions is 2.9×10^{20}. The mean residence time in the surface of these molecules is designated $\bar{t}$, and is given[4] by $n \Big/ \left(-\dfrac{dn}{dt}\right)_{desorption}$ where n is the equilibrium number of molecules present per $cm.^2$ of surface, and $\left(-\dfrac{dn}{dt}\right)_{desorption}$ is the rate of desorption (and of adsorption). In the present example, putting $n = 10^{15}$ (from the dimensions of the water molecules) and $\left(\dfrac{dn}{dt}\right)_{desorption}$ equal to 2.9×10^{20} molecules $cm.^{-2}sec.^{-1}$, one finds $\bar{t}$ is 3.4 microseconds. This very short life of a molecule in the surface before it evaporates implies an extremely violent agitation[4,5], though, because of the strong cohesion in the liquid surface, the time-average position of the latter is definite to within a few molecular thicknesses. This rapid exchange is valid only for molecular exchange at equilibrium: if actual evaporation rates are measured they are much lower than those calculated in the above way, because stagnant layers of vapour and cooling of the water phase both considerably retard any overall mass-transfer process: this is discussed more fully in Chapter 7.

Desorption into the vapour may alternatively be expressed by the equation:

$$\left(-\frac{dn}{dt}\right)_{desorption} = K\theta e^{-q/RT} \tag{1.6}$$

where θ is the fractional surface coverage, and the energy barrier q equals the heat of vaporization. Putting $\theta = 1$, $\left(-\dfrac{dn}{dt}\right) = 2.9 \times 10^{20}$ molecules $cm.^{-2}sec.^{-1}$, and $q = 585 \times 18$ calories $mole^{-1}$, one obtains for K, the desorption constant for water at room temperature, a value of $1.2 \times 10^{28} cm.^{-2}$

sec.$^{-1}$ This figure is in excellent accord both with results for the physical and chemical adsorption of simple molecules on solid surfaces, and also with fundamental theory[6]. For example, according to the Eyring[7] theory, $K = \frac{nkT}{h} = 0.6 \times 10^{28}$ (where h is Planck's constant) assuming that the molecule does not undergo any change in its degrees of freedom during desorption. Quantum mechanical theory gives $K = 0.14 \times 10^{28}$, while the Polanyi-Wigner and Langmuir theories both give $K = 10^{28}$cm.$^{-2}$sec.$^{-1}$ at room temperature[6].

Exchange of the solvent molecules in the liquid surface with those in the immediately subjacent bulk is even more rapid on account of the very small distances involved and the rapid molecular motion. Quantitatively one may use the expression[7]

$$D = \lambda^2/\bar{t}$$

where D is the diffusion coefficient (for water in water this is 2×10^{-5}cm.^{2} sec.$^{-1}$)[8], λ is the distance between two successive equilibrium positions, and $\bar{t}$ is the time taken for a molecule to move from one position to the other. Now though D may be slightly altered by the lack of symmetry close to the surface, we may obtain at least the order of magnitude of $\bar{t}$: for liquid water (with $\lambda = 3.5 \times 10^{-8}$cm.)[7] $\bar{t}$ is 6×10^{-5}microsecond. The time required for a liquid surface to take up its equilibrium value of surface tension is somewhat greater than this, but is still very small: about 10^{-3} microsecond is required for the processes of re-orientation and re-arrangement of the molecules of a simple liquid following the exposure of new surface[9].

Again, one may alternatively calculate a K term to describe the exchange desorption of the surface molecules into the immediately subjacent bulk liquid. From $n = 10^{15}$ molecules cm.$^{-2}$ as before, and with $\bar{t} = 6 \times 10^{-11}$sec., $\left(-\frac{dn}{dt}\right)_{\text{desorption}}$ must be $10^{15}/6 \times 10^{-11} = 1.7 \times 10^{25}$ molecules cm.$^{-2}$sec.$^{-1}$ Hence, from eq. (1.6), with $\theta = 1$ and q the energy of activation for the self-diffusion of water (5300 cal.mole^{-1})[8], we find $K = 12 \times 10^{28}$cm.$^{-2}$sec.$^{-1}$, again in fair agreement with fundamental theory; that the value is rather high is in accord with the belief that the activated state for the diffusion of water occurs with an appreciable increase in entropy[7].

For an adsorbed monolayer of a surface-active agent in equilibrium with the solution, $\bar{t}$ for the agent may be considerably greater than for the solvent. Thus for butyric acid in water, the rates of desorption of acid from the monolayer (and so also of adsorption at equilibrium) are lower than for the water molecules by a factor $e^{-\lambda/RT}$ where λ is the energy of desorption, here about 2300 cal.mole.$^{-1}$; consequently $\bar{t}$ is of the order 3×10^{-9}sec.

Again, this equilibrium exchange rate of adsorption and desorption cannot be realized if there is a net flow: diffusion of the solute along the concentration gradients over a region of perhaps several millimetres below the surface will then greatly reduce the rate of adsorption or desorption, so that times ranging from a few milliseconds to several hours may be required before diffusion brings the adsorbed film and the bulk phase into equilibrium.

Vapour Pressures over Curved Surfaces

Imagine a very small water drop of radius a in equilibrium with vapour, and let μ_a be the chemical potential of this system. At the same temperature the equilibrium chemical potential of vapour over a plane water surface will be denoted by μ. We shall see that $\mu_a > \mu$, the difference between the two quantities depending on a and γ, as related by the Kelvin equation[10]. This is derived as follows.

Suppose that dn molecules of water are brought at constant temperature, constant total volume, and constant surface area from the interior of a large mass of liquid beneath the flat surface, and that these molecules are then added to the interior of the drop of radius a. The work dF done in this process is, by eq. (1.1), $(\mu_a - \mu)dn$, and this must be the same as if the dn molecules had been utilized in enlarging the surface of the drop, i.e. it is equal to γdA, where γ is the surface tension and A the area of the surface of the drop. The area A is related to a by the equation $A = 4\pi a^2$, and so

$$(\mu_a - \mu)dn = \gamma dA = 8\pi a\gamma da \tag{1.7}$$

We can eliminate dn and da from this equation by noting that the increase in volume of the drop when the dn molecules of water are added is $v dn$, v being the molecular volume of water, but is also dV where

$$V = \frac{4}{3}\pi a^3.$$

Thus
$$v dn = dV = 4\pi a^2 da$$

Hence, by substitution of da into eq. (1.7),

$$\mu_a - \mu = \frac{2\gamma v}{a} \tag{1.8}$$

which is the general form of the Kelvin equation. It is true for any liquid or solid. For ideal vapours one can substitute for μ_a and μ, using the relations

$$\mu = \mu_0 + \boldsymbol{k}T \ln p$$
$$\mu_a = \mu_0 + \boldsymbol{k}T \ln p_a$$

where p_a is the equilibrium vapour pressure over the curved surface and p that over the flat surface. These substitutions lead to the Kelvin equation for spherical drops:

$$\ln\left(\frac{p_a}{p}\right) = \frac{2\gamma v}{a\boldsymbol{k}T} \tag{1.9}$$

The calculated application of this equation to water droplets is shown in Table 1-II.

TABLE 1-II

radius of drop, a	p_a/p
1 micron	1.001
0.1 micron	1.01
0.01 micron	1.1
50 Å	2.0
10 Å	3.0
6.5 Å	4.2
6.0 Å	5.5

Calculation of the relative increase of vapour pressure of water over small spherical drops, using eq. (1.9). It is assumed that the value of γ is constant at γ_0.

The Kelvin equation can best be tested experimentally by measuring the growth to equilibrium of aerosol droplets[11]. A monodisperse aerosol, formed from a non-volatile material such as di-octylphthalate, is allowed to reach equilibrium over a mixture of di-octylphthalate and toluene having a flat surface: in the course of attaining equilibrium, toluene vapour condenses on to, and mixes with, the aerosol droplets of di-octylphthalate till the vapour pressure of the toluene over these droplets is equal to that over the flat surface of the mixture of di-octylphthalate and toluene, at which stage growth of the aerosol particles ceases. The vapour pressure over a curved surface being higher than over a flat surface, a rather smaller mole-fraction of toluene is found in the aerosol droplets at equilibrium than in the bulk phase. From the observed extent of growth of the aerosol droplets, therefore, relative to that which would give to the droplets the same composition as in the bulk mixture, the excess vapour pressure of toluene over surfaces of various curvatures can be found. For droplets of radii of the order 0.1 micron, experimental results agree with the Kelvin equation to within $\pm 2\%$.

Small drops of aqueous NaCl, in equilibrium with a plane sheet of water, also give agreement with theory to within a few per cent[12].

Over a cylinder of liquid, of radius a, the corresponding equation for the rise in vapour pressure is:

$$\ln(p_a/p) = \gamma v/a\boldsymbol{k}T \tag{1.9a}$$

and, over a trough of liquid, of radius a, the pressure is reduced according to the relation:

$$\ln(p/p_a) = \gamma v/a\boldsymbol{k}T \tag{1.9b}$$

Excess Pressures inside Bubbles

The work necessary to enlarge the gas bubble in Fig. 1-3 is, as given by eq. (1.7), $8\pi a\gamma da$ (i.e. γdA). But, instead of increasing the size of the bubble (or air-drop) by simply adding more molecules as above, let us apply an external pressure inside the bubble, e.g. with a fine capillary tube. Hence

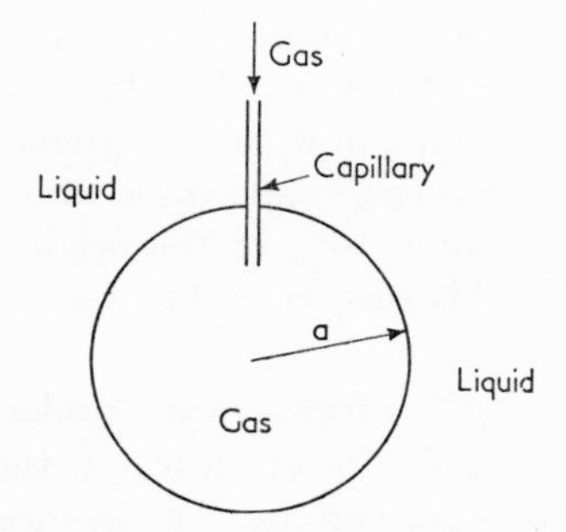

Fig. 1-3. Bubble of gas in liquid. The pressure excess inside is P_e (eq. (1.10)).

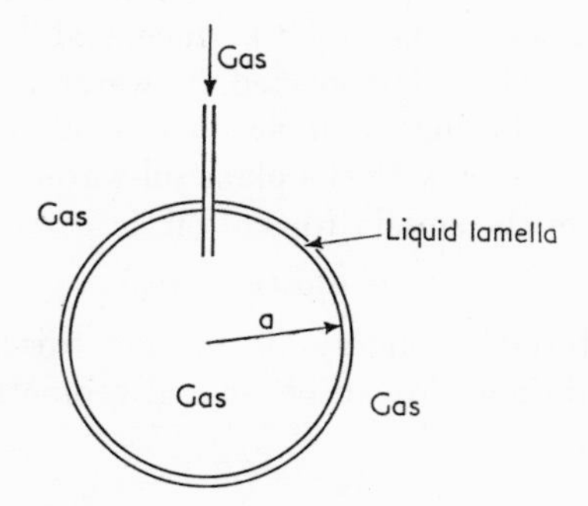

Fig. 1-4. Bubble of gas in gas. The excess pressure inside is again designated P_e (see eq. (1.11)).

the work done in increasing the size of the bubble is also given by $P_e \Delta V$, i.e. by $P_e \times 4\pi a^2 \times da$, where P_e is the excess pressure inside the bubble compared with that outside. Equating these two expressions for the work done in expanding the bubble, we have:

$$P_e \times 4\pi a^2 \times da = 8\pi a\gamma da$$

or

$$P_e = 2\gamma/a \tag{1.10}$$

If the bubble is blown in air (Fig. 1-4), there are two surfaces to take into account, and the work necessary to enlarge the bubble is now $2\gamma dA$, i.e.

$16\pi a\gamma da$. By similar reasoning to that above, we find that P_e, the excess pressure inside an air bubble of this type, is given by:

$$P_e = 4\gamma/a \tag{1.11}$$

Numerically we can illustrate the use of this equation by considering a soap bubble of 1 micron radius, having $\gamma = 25$ dynes cm.$^{-1}$ The excess pressure in such a bubble is thus 10^6 dynes cm.$^{-2}$, i.e. 1 atm. This enhanced pressure in small bubbles is of importance in the breakdown of foams, in that air will diffuse through the liquid lamellae from the small bubbles in the foam into the larger bubbles, in which the pressure is lower. In this way, in foams of rubber latex or Teepol solution the total number of bubbles may decrease in 15 minutes to only 10% of the original number, without any film rupture occurring[13]. This subject is discussed more fully in Chapter 8.

Solubility from Small Droplets

In an extremely fine emulsion of oil in water, the solubility of the oil in the water is slightly enhanced, because the chemical potential of the oil is raised by the appreciable effect of the interfacial energy in a system where the total interface is large. If μ be the chemical potential of the oil in contact with and dissolved in water over a flat surface, and if μ_a be the corresponding chemical potential of the oil dispersed in water in the form of drops of radius a, the work of taking dn molecules of oil from the interior of the aqueous solution in contact with the plane oil-water interface, and of placing these dn molecules on the small drops of oil, is:

$$(\mu_a - \mu)dn = 8\pi a\gamma_i da \tag{1.7a}$$

where γ_i is the interfacial energy or tension, discussed in detail below. Further, as before, dn and da can be related geometrically, and hence:

$$\mu_a - \mu = \frac{2\gamma_i v}{a} \tag{1.8a}$$

where v is now the molecular volume of the oil. For an ideal solution of oil in water we have also:

$$\mu_a - \mu = \boldsymbol{k}T \ln\left(\frac{c_a}{c}\right) \tag{1.12}$$

where c is the molar concentration of dissolved material in the water in contact with a plane interface, and c_a is the corresponding quantity when the interface is of radius a.

From equations (1.8a) and (1.12) one obtains

$$\ln\left(\frac{c_a}{c}\right) = \frac{2\gamma_i v}{a\boldsymbol{k}T} \tag{1.13}$$

which is similar in form to eq. (1.9). If a is small, c_a becomes greater than c, the solubility of the oil in water increasing as the size of the droplets is decreased. Applied to solids, this argument shows clearly why, if crystals have been formed so quickly that they are difficult to filter on account of their small sizes, standing can cause the larger crystals to grow at the expense of the smaller ones till finally only a few large crystals remain.

Surface Tension and Curvature

For liquid surfaces whose radius of curvature, a, is large compared with the thickness of the surface region, the surface tension γ is very closely equal to γ_0, the tension of a plane surface. If, however, we are dealing with drops of (say) water so small that a and the thickness of the surface layer are of the same order of magnitude, γ can differ appreciably from γ_0. This has been investigated quantitatively by three different approaches—the quasi-thermodynamical method of Tolman[14], the statistical mechanical treatment of Kirkwood and Buff[15], and the molecular interaction method of Benson and Shuttleworth[16]. Although all three methods predict that γ will become less than γ_0 when a is of a few molecular magnitudes, there is not precise agreement between the calculations. The first, for example, predicts a 50% fall in γ of a water droplet consisting of 13 molecules (a = 4.6 Å), while the last theory suggests that the fall in γ should be only 15% or less for such a droplet ($\gamma = 61$ dynes cm.$^{-1}$; $\gamma_0 = 72$ dynes cm.$^{-1}$). Since the quasi-thermodynamical argument is not rigorous at these very small values of a, we prefer the latter result, which is based on counting the number of bonds broken, using a scale molecular model. The conclusion is thus that γ does indeed decrease slightly in very small droplets (e.g. the embryos and nuclei responsible for phase transitions, Chapter 8), but that the change is not very large. In the ordinary determinations of surface tension by the drop-weight method, the effect of curvature in changing γ can always be completely ignored.

Total Surface Energy

We have seen that F^s, the excess *Helmhotz free energy* of unit area of a liquid, is equal at constant T and V to the tension γ_0. The *total surface energy* U^s, however, is greater than F^s, from which it may be calculated by its definition:

$$F^s = U^s - TS^s \tag{1.14}$$

where the thermodynamic quantities now refer to surface excesses. To find the entropy term S^s, it is convenient to use the relation:

$$S^s = -\left(\frac{\partial F^s}{\partial T}\right)_{n,V} = -\left(\frac{\partial \gamma_0}{\partial T}\right)_{n,V}$$

so that

$$U^s = \gamma_0 - T\left(\frac{\partial \gamma_0}{\partial T}\right)_{n,V} \tag{1.15}$$

For water at 25°C this becomes

$$U^s = 72 - 298\ (-0.154) = 118 \text{ ergs cm.}^{-2}$$

and for mercury at 25°C,

$$U^s = 475 - 298\ (-0.22) = 541 \text{ ergs cm.}^{-2}$$

In physical terms, the total surface energy U^s is the total potential energy of the molecules that form 1 cm.^{2} of the surface of the liquid in excess of that which the same molecules would possess in the interior of the liquid.

We may note that U^s is (unlike γ_0) virtually independent of temperature, which may be shown as follows.

If equation (1.15) is differentiated, we obtain

$$\left(\frac{\partial U^s}{\partial T}\right) = \left(\frac{\partial \gamma_0}{\partial T}\right) - T\left(\frac{\partial^2 \gamma_0}{\partial T^2}\right) - \left(\frac{\partial \gamma_0}{\partial T}\right)$$

i.e.

$$\left(\frac{\partial U^s}{\partial T}\right) = -T\left(\frac{\partial^2 \gamma_0}{\partial T^2}\right)$$

But, in practice, γ_0 decreases linearly with T to a very good approximation, so that $\left(\frac{\partial^2 \gamma_0}{\partial T^2}\right)$ is very small, and hence $\left(\frac{\partial U^s}{\partial T}\right)$ is also very small. For water, for example, it is -0.00048, and for benzene $+0.00012$ [2].

Surface Entropy

The molecules of a liquid in the bulk have other liquid molecules as the nearest neighbours. When a surface is formed, however, the molecules in the surface layer have a different environment on one side, so that compared with bulk liquid there is a new possibility of "randomness" in that a molecule may occupy a position either in the immediately subjacent bulk phase or in the surface. These two possibilities will give rise to an entropy increase of approximately Rln2, i.e. +1.4 e.u., this then being a standard entropy change associated with forming a surface[16a]. Consequently, the temperature coefficient of surface tension must be negative, just as bulk phases become more miscible with a rise of temperature.

From the temperature coefficient of the surface tension of water, one finds that the positive entropy of formation of the water surface of 0.154 erg cm.$^{-2}$(°K)$^{-1}$. This corresponds to an entropy change of +1.6 e.u., close to the figures of +1.4 e.u., estimated above for the standard entropy of formation of a surface. For mercury, the temperature coefficient of -0.22 corresponds to an entropy change of about +2 e.u.

Molecular Theories of Surface Energy

Since cohesive forces fall off very steeply with distance, one can consider as a first approximation interactions between neighbouring molecules only.

Here we shall consider how the total surface energy U^s is related to molecular cohesion in liquids. Both Frenkel[17] and Langmuir[18] suggested simple methods of establishing this relation, and we shall first discuss *Frenkel's method.*

Let the number of nearest neighbours of a molecule in the interior of the liquid be Z, and let the number of such neighbours for a molecule in the surface be Z'. Then, if ν is the number of molecules per unit volume, and if u_1 is the mutual cohesive energy of two neighbouring molecules, the surface excess energy per unit area is clearly given by:

$$U^s = u_1(Z - Z')\nu^{2/3} \tag{1.16}$$

To evaluate u_1 in this expression, consider the evaporation of the liquid mass. If the latent heat of evaporation of 1 mole of the liquid is L, then this is a close approximation to the total energy of vaporization of 1 mole of liquid, and hence

$$L = \tfrac{1}{2}\nu v Z u_1 \tag{1.17}$$

where v is the molar volume of the liquid. The factor $\frac{1}{2}$ is introduced to allow for the fact that, while the removal of one molecule from the bulk of the liquid breaks Z bonds, $\frac{Z}{2}$ new bonds will simultaneously re-form in the liquid. Elimination of u_1 between equations (1.16) and (1.17) gives:

$$U^s = \frac{2L}{v}\left(\frac{Z - Z'}{Z}\right)\nu^{-\frac{1}{3}} \tag{1.18}$$

To illustrate the use of this expression, let us assume throughout that $Z = 6$, $Z' = 5$. For water, $\frac{L}{v} = 582$ cal. cm.$^{-3}$ ($= 2.4 \times 10^{10}$ ergs cm.$^{-3}$), $\nu = 33.3 \times 10^{21}$, and hence U^s is calculated to be 240 ergs cm.$^{-2}$ (compared with 118 ergs cm.$^{-2}$ from experimental data). For n-octane calculation gives 56.7 ergs cm.$^{-2}$ (50.7 ergs cm^{-2}. from experiment), while for mercury the calculated U^s is 362 ergs cm.$^{-2}$, compared with 536 ergs cm.$^{-2}$ from experiment. These values are summarized in Table 1-III, where agreement to within about 50% is considered satisfactory for such an elementary theory.

Langmuir's method[18] consists in assuming that each individual molecule possesses a spherical surface of the usual surface tension: thus for water the volume of one molecule must be $\frac{18 \text{ cm.}^3}{N} = 30 \times 10^{-24}$cm.^{3}, and its surface area is consequently 47×10^{-16}cm.^{2} For 1 cm.^{3} water the surface energy is thus $47 \times 10^{-16} \times U^s \times (N/18)$ ergs, which should be equal to L/v. Hence $U^s = 150$ ergs cm.$^{-2}$ This and other values are shown in Table 1-III. Langmuir thus assumes that the forces between two molecules in contact are mainly

TABLE 1-III

Substance	U^s (expt.) (from eq. (1.15)) in ergs cm.$^{-2}$	U^s (calc. by Frenkel's method eq. (1.18)) in ergs cm.$^{-2}$	U^s (calc. by Langmuir's method) in ergs cm.$^{-2}$	U^s (calc. from exptl. distribution function) in ergs cm.$^{-2}$	U^s (calc. assuming face-centred cubic lattice) in ergs cm.$^{-2}$	U^s (Prigogine and Saraga[20]) in ergs cm.$^{-2}$
He	0.59	—	0.35	—	0.7	—
H_2	5.4	—	2.7	—	5.5	—
H_2O	118	240	150	—	—	—
N_2	26	—	—	—	28.1	26.2
Ne	15	—	—	—	11.4	18.5
A	35.3	—	18	27	32.9	35.4
Hg	541	362	230	490	—	—
n—C_8H_{18}	50.7	56.7	28	—	—	—
n—$C_8H_{17}OH$	50.7	74	41.5	—	—	—

dependent on the surface properties of these molecules: this is called "independent surface action".

To obtain better agreement with theory, however, a more elaborate picture of molecular interactions is required. If one can obtain experimentally (e.g. from x-ray analysis of the liquid) the function describing the distribution of molecules around any given molecule, one can then derive the intermolecular potential function from *quantum-mechanics*. In this way U^s for liquid argon and mercury[19] are found to be 27 ergs cm.$^{-2}$, and 490 ergs cm.$^{-2}$ respectively (column 5, Table 1-III): these are considerably more accurate estimates than are possible from Langmuir's theory. For other, less simple liquids the distribution function is not available, though as a first approximation it may be taken to be the same as in the solid state, allowing for the additional potential energy of the liquid molecules due to their movements about the positions constituting a solid lattice. Assuming a face-centred cubic structure in the liquid, one thus calculates the results in the sixth column of Table 1-III.

Statistical-mechanical calculation of the distribution and potential functions can lead to rigorous equations for U^s. Unfortunately, these are not suitable for numerical calculation[19,20], in consequence of which several simplified theories have been proposed. Prigogine and Saraga[19], for example, use a simple "cell" model for the distribution of molecules in the liquid, with a "well" type of potential distribution. Further, they assume that, in the surface, the motion of any molecule parallel to the surface of the liquid is the same as it is in the interior of the liquid, though perpendicular to the surface the molecules move in a larger effective free volume. This model gives U^s as 25.8 ergs cm.$^{-2}$ for argon (compared with the experimental value of 35.3): agreement is, however, much better if it is assumed that only 70% of the "cells" in the surface are occupied by molecules, the remainder being empty holes in the superficial layer. Results calculated by this modified theory are shown in the seventh column of Table 1-III. Other statistical theories give values of U^s of 19 and 27.2 ergs cm.$^{-2}$ for liquid argon (Hill; Kirkwood and Buff[20]). Statistical theories are particularly applicable to calculating the surface tensions of two component systems such as mixtures of ether and acetone, as Guggenheim and Prigogine and Defay have shown[20]: the shape of the γ vs.c curve even for an organic solute such as sebacic acid dissolved in water can be predicted. The films of long-chain molecules (e.g. stearic acid) often break up into "islands" of close-packed molecules, leaving the rest of the surface only sparsely covered. This "two-phase" formation can also be predicted from statistical theory (Saraga and Prigogine[20]) if the cohesion between the molecules in the surface is high compared with their attraction for water molecules.

Fused metals usually have rather high surface tensions, ranging from about 200 dynes cm.$^{-1}$ for K to about 1100 for Cu and 1819 for Pt. Both electrostatic and non-electrostatic forces must be allowed for in the theoretical treatment of such systems[21].

INTERFACIAL TENSION

If water is placed in contact with oil the interface between the two liquids has a contractile tendency. This is represented as γ_i, and is expressed in dynes cm.$^{-1}$, as is γ_0. This term must be included in eq. (1.1) for the total free energy of this whole system. For butanol against water γ_i is only 1.8 dyne cm.$^{-1}$, and such a low figure is characteristic of oils containing polar groups. This shows that the molecules of butanol ($\gamma_0 = 24$) must concentrate

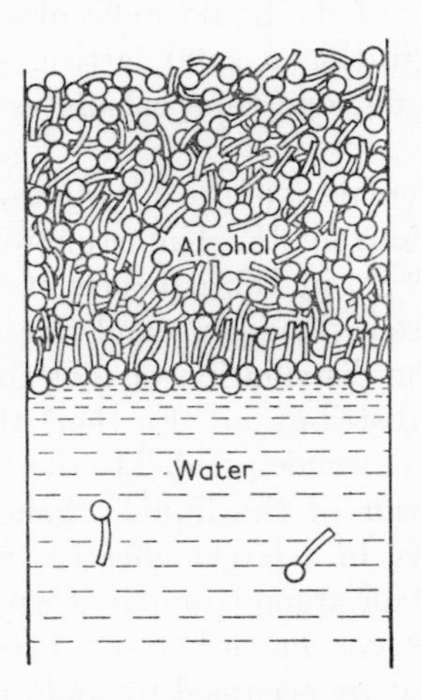

Fig. 1-5. Orientation of butanol molecules at the interface between butanol and water. The orientation of the water molecules at the interface is not represented. (After Harkins[2].)

at the oil-water interface, where the repulsion between the packed and oriented molecules (Fig. 1-5) offsets somewhat the usual contractile tendency of an interface. Interfacial packing occurs because the hydroxyl "heads" of the butanol molecules can escape from the oil into the water, while the chains remain in the oil, this process (Fig. 1-5) resulting in a state of low standard free energy.

Similarly, for nitrobenzene against water $\gamma_i = 25.1$, whereas for this oil $\gamma_0 = 43.9$ dynes cm.$^{-1}$. The difference indicates that considerable orientation of the dipolar molecules occurs at the interface. For hydrocarbon oils $\gamma_i \approx 50$, a figure much higher than γ_0 (≈ 22 dynes cm.$^{-1}$). Table 1-IV contains a summary of some accurately known interfacial tensions[2] for various organic liquids against water. As Bikerman has pointed out, for these hydrocarbon derivatives the mutual insolubility of the oil and water runs parallel to the interfacial tensions (Fig. 1-6), and several empirical relations have been suggested[2].

In three component systems the same types of relationships hold: Fig. 1-7 shows butanol, dissolved in benzene, adsorbing at the interface against water.

TABLE 1-IV

Standard Interfacial Tensions between Water and Pure Liquids[2]

	t (°C)	γ_i (dynes cm.$^{-1}$)
n-Hexane	20	51.0
n-Octane	20	50.8
Carbon disulphide	20	48.0
Carbon tetrachloride	20	45.1
Carbon tetrachloride	25	43.7
Bromobenzene	25	38.1
Benzene	20	35.0
Benzene	25	34.71
Nitrobenzene	20	26.0
Ethyl ether	20	10.7
n-Octanol	20	8.5
n-Hexanol	25	6.8
Aniline	20	5.85
n-Pentanol	25	4.4
Ethyl acetate	30	2.9
Isobutanol	20	2.1
n-Butanol	25	1.8
n-Butanol	20	1.6

Again the alcohol molecules, packing into the interface so as to immerse the hydroxyl "head groups" in the water, form an *orientated monolayer*, the repulsion within which reduces somewhat the contractile tendency of the interface. Quantitatively, resembling eq. (1.5), $\gamma = \gamma_i - \Pi$. Further, the greater the miscibility of the oil and water, in the presence of the third component, the lower is the interfacial tension. Thus ethanol, added in increasing amounts to a mixture of water and i-pentanol, reduces γ progressively from 4.4 dynes cm.$^{-1}$ to zero, the latter being the interfacial tension when 25%(wt.) ethanol is present: the whole system then becomes miscible and forms a single phase. With less of the third component than will produce complete miscibility of the phases, spontaneous emulsification may often occur by the "diffusion and stranding" mechanism, even though γ is still quite high. This process is explained in detail in Chapter 8. Although complete miscibility of the phases necessarily results in zero interfacial tension, the converse is not true: a monolayer of organic cations concentrated at the mercury-water interface by an electrical field will reduce γ through zero (with consequent spontaneous emulsification) without inducing any apparent miscibility of the phases.

Interfacial Entropy

At the interface between a hydrocarbon oil and water, some of the molecules of the latter will be oriented by the adjacent $-CH_2-$ groups into an ice-like form. Hydrogen bonds may be partly responsible, and a similar

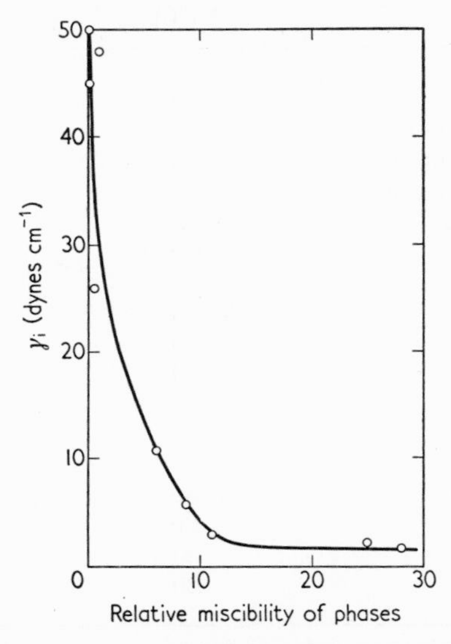

Fig. 1-6. Interfacial tensions in the two-phase systems of water and different organic liquids, as a function of the relative miscibility. (After Bikerman[2].)

tendency of water to form "ice-bergs" round dissolved molecules of both rare gases and hydrocarbon gases is well known, these layers of oriented water molecules having a density greater than that of the liquid water[21a].

From the temperature coefficient of the interfacial tension between paraffin oil and water, one calculates that ΔS for the formation of 8 Å^2 of interface is $+0.6$ e.u. The standard change of entropy on formation of an oil-water interface is, however, $+2.8$ e.u. per unit molecular area, since now both the water and also the oil molecules have more possible positions between the bulk phases and the interface. Consequently the entropy change is 2.2 e.u. less than expected, presumably due to the formation of an oriented layer of water against the hydrocarbon[16a]. This entropy of -2.2 e.u. for the orientation (and semi-solidification) of a molecule of water may be compared with the figure of about -5 e.u. for the formation of ordinary ice.

Cohesion and Adhesion

Dupré in 1869 formulated a relation defining the *work of adhesion*, $W_{O/W}$, between oil and water in the following terms[22]:

$$W_{O/W} = \gamma_{O/A} + \gamma_{W/A} - \gamma_{O/W} \tag{1.19}$$

where subscripts O, A, and W refer to oil, air, and water. This may be illustrated physically as follows. Suppose that the oil and water are initially in contact in a column of cross-section 1 cm². (Fig. 1-8a), the energy of the

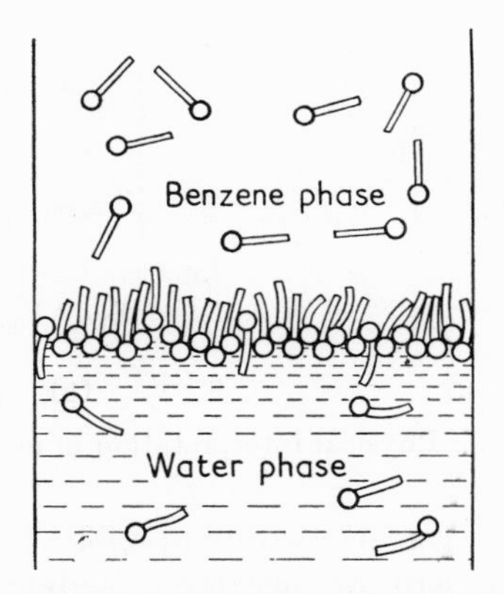

Fig. 1-7. Adsorption of butanol into an orientated monolayer at the benzene-water interface. (After Harkins[2].)

interfacial system being $\gamma_{O/W}$ ergs. They are now separated by a direct pull (with air allowed to enter between them), and the energy of the interfacial system is now $\gamma_{O/A} + \gamma_{W/A}$ (Fig. 1-8b). The work required to effect the separation of the liquids is defined as $W_{O/W}$ which is therefore given by eq. (1.19). A typical value for water and a paraffinic oil is 43 ergs cm.$^{-2}$ For a single liquid (e.g. oil) the same procedure gives

$$W_{oil} = 2\gamma_{O/A} \tag{1.20}$$

i.e. the *work of cohesion* of any single liquid is twice its surface tension. For water W is 144 ergs cm.$^{-2}$, and for paraffinic oils it is about 44 ergs cm.$^{-2}$

If in each liquid the molecules attract the molecules of the other liquid as much as or more than they are attracted by each other, then they become completely miscible since the free energy will decrease on mixing. This may also be seen in terms of adhesion and cohesion, since if

$$W_{O/W} > W_{oil}$$

and if $$W_{O/W} > W_{water}$$

then $$W_{O/W} > \frac{W_{oil}}{2} + \frac{W_{water}}{2}$$

i.e. $$W_{O/W} > \gamma_{O/A} + \gamma_{W/A}$$

and hence, by eq. (1.19), the necessary consequence of complete miscibility is

$$\gamma_{O/W} < 0 \tag{1.21}$$

In physical terms this means there is no stable interface, as the molecules of the two liquids tend to mix as completely as possible. This result is true only for a *two-component* system: the presence of a *third component*, spread as a

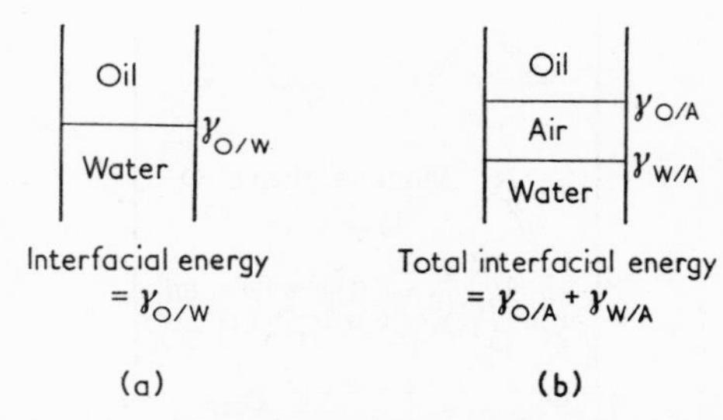

Fig. 1-8. Physical interpretation of eq. (1-19)

film at the interface between two immiscible liquids, can sometimes, momentarily, make γ_i zero or negative, leading now to *spontaneous emulsification* (Chapter 8).

SPREADING

Spreading of One Liquid on Another

If a small drop of high-boiling paraffin oil is placed on a water surface, it maintains its form as a drop, floating in a depression on the water surface as in Fig. 1-9. By equating the horizontal components of the tensions (and assuming as a first approximation that $\gamma_{W/A}$ acts exactly horizontally), we obtain

$$\gamma_{W/A} = \gamma_{O/A} \cos \theta_1 + \gamma_{O/W} \cos \theta_2 \tag{1.22}$$

Suppose now that we replace the paraffin oil by a drop of the lower paraffin octane. We then decrease $\gamma_{O/A}$ (Table 1-V) to such an extent that eq. (1.22) can only apply when $\gamma_{O/A}$ and $\gamma_{O/W}$ exert their full effect to balance $\gamma_{W/A}$, i.e. θ_1 and θ_2 must both become zero.

If a more polar oil such as n-octanol is used eq. (1.22) can never be satisfied because $\gamma_{W/A} > \gamma_{O/A} + \gamma_{O/W}$: momentarily the drop of octanol rests on the surface, then it spreads out till the whole surface is covered with a thin film of the oil. The tendency of the oil to spread is clearly positive, and since θ_1 and θ_2 must again approach zero as the drop thins out during spreading one can define an initial spreading coefficient[2] S of one liquid on another, here the oil on the water, as:

$$S = \gamma_{W/A} - (\gamma_{O/A} + \gamma_{O/W}) \tag{1.23}$$

where these quantities are measured before mutual saturation of the liquids has occurred and where, if the oil and water are ultimately miscible (e.g. ethanol and water), $\gamma_{O/W} = 0$. The general conditions for any oil to spread on

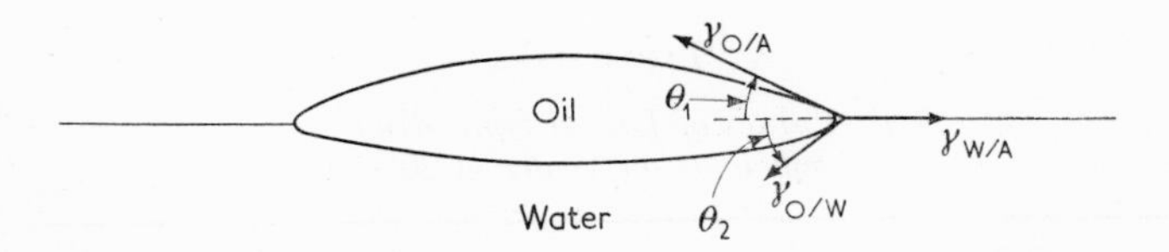

Fig. 1-9. Simplified representation of a drop of high-boiling paraffin oil lying on a water surface.

the water is that S be positive or zero: examples are shown in Tables 1-V and 1-VI. If S is appreciably positive, the oil will spread against contamination, i.e. $\gamma_{W/A}$ can be less than for pure water and spreading still occurs. This is of importance in the breaking of foams (Chapter 8).

As long ago as 1765 Benjamin Franklin observed that olive oil spreads over water to a thickness of 25 Å. This was, indeed, the first conclusive proof of the ultimate indivisibility of matter and of the atomic theory: the monomolecular layer cannot become thinner than this, since 25 Å is the length of the hydrocarbon chains. Attempts to expand the film further result in the approach of the tension to that for a clean water surface (the "islands" of monolayer which now float on the surface having little effect). Similar results, obtained in 1899 by Lord Rayleigh[23] with films of castor oil thinned to 14 Å,

TABLE 1-V

Derivation of Initial Spreading Coefficients for Spreading at the Air-Water Surface at 20°C [2]

Oil	$\gamma_{W/A} - \gamma_{O/A} - \gamma_{O/W} = S$	Conclusion
n-Hexadecane	72.8 − 30.0 − 52.1 = −9.3	Will not spread
n-Octane	72.8 − 21.8 − 50.8 = +0.2	Will just spread
n-Octanol	72.8 − 27.5 − 8.5 = +36.8	Will spread against contamination

are shown in Fig. 1-10. Here the total tension rises if the film is expanded so much that its thickness becomes less than 80 Å, while the tension tends to that of pure water if the spreading is allowed to expand the film to a thickness of 14 Å or less.

The coefficient of initial spreading of one liquid on another, S, can be simply related to the work of adhesion, since, by combining eqs. (1.19), (1.20), and (1.23),

$$S = W_{O/W} - 2\gamma_{O/A} \tag{1.24}$$

and

$$S = W_{O/W} - W_{oil} \tag{1.25}$$

The latter relation shows that spreading occurs on the clean water surface

TABLE 1-VI

Values of Initial Spreading Coefficients on Water at 20°C [2]

	S in dynes cm.$^{-1}$
Ethanol	+50.4
Methanol	+50.1
Propanol	+49.1
Ethyl ether	+45.5
Isoamylalcohol	+44.3
n-Octanol	+36.8
Undecylenic acid (at 25°C)	+32.0
Chloroform	+13.0
Benzene	+ 8.9
Toluene	+ 6.8
Nitrobenzene	+ 3.8
Hexane	+ 3.4
Carbon tetrachloride	+ 1.1
n-Octane	+ 0.22
Carbon disulphide	— 7.6
n-Hexadecane	— 9.3
Bromoform	— 9.6
Liquid petrolatum	—13.6
Methylene iodide	—26.5

(S zero or positive) only if the liquid forming the drop (e.g. a polar oil) adheres to the water more strongly than it coheres to itself: further, if S is positive, eq. (1.23) shows that $\gamma_{O/A} - \gamma_{W/A} + \gamma_{O/W} < 0$, and hence water cannot spread on this oil, since S' for this inverse system must be given by

$$S' = \gamma_{O/A} - \gamma_{W/A} - \gamma_{O/W} \tag{1.26}$$

which, being less than the expression in the above inequality, must be always negative. It is, however, possible for two liquids to be chosen neither of which will spread on the other, as may be seen from the data for n-hexadecane and water (Table 1-V).

Spreading at the oil-water interface is subject to exactly the same conditions. Suppose we have a benzene-water interface ($\gamma_i = 35$ dynes cm.$^{-1}$) in which we place a drop of ethanol. This is ultimately miscible with both benzene and water, so that S is now given by (35-0-0), i.e. 35 dynes cm.$^{-1}$ For ethanol at the n-hexane-water interface S is similarly 51 dynes cm.$^{-1}$ This high initial spreading power of alcohols at an oil-water interface is useful in assisting the spreading of monolayers of larger molecules (Chapter 5). It also causes interfacial turbulence (Chapter 7).

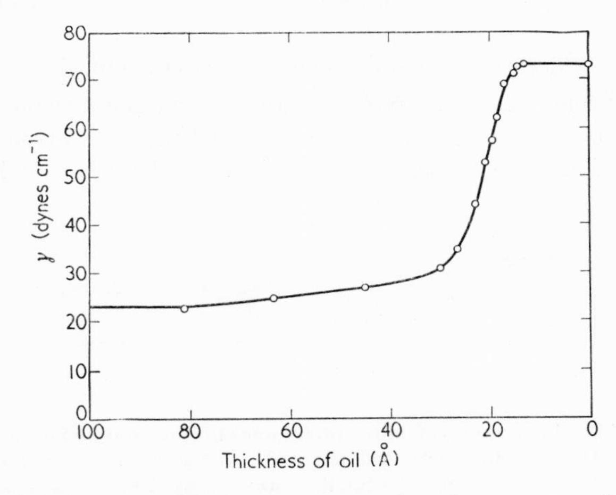

Fig. 1-10. Rayleigh's results[23] on the change of surface tension of water covered with a thin film of castor oil. If the oil film is thinned to less than about 25Å, the surface tension rises steeply towards the value for a clean water surface, showing that the oil film cannot be spread uniformly thinner than about 25Å.

Impurities may greatly affect the spreading of an oil on water. For example[2], if to the centre of a large lens of a high-boiling paraffinic oil (floating but not spreading on a water surface) some oleic acid is added, the lens suddenly breaks up into a number of smaller drops, resulting from the strong spreading action of the oleic acid. The latter may finally lie on the free surface as a monolayer, or, if enough oleic acid is present, a thicker film of a mixture of oleic acid and paraffinic oil covers the water surface between the lenses of oil. In mathematical terms, the cohesion between oil and water, $W_{O/W}$, has been locally increased enough to make S positive (eq. 1.25): this corresponds to decreasing $\gamma_{O/W}$ in eq. (1.23).

Impurities in the water usually reduce S, since $\gamma_{W/A}$ is reduced more than $\gamma_{O/W}$ by the impurity, especially if $\gamma_{O/W}$ is already low.

The spreading coefficients S, quoted in Tables 1-V and 1-VI, refer to the initial state of the system at the moment when a drop of the oil has been placed on a clean water surface and before any appreciable amount of spreading has occurred: the surface tension of water is still taken as 72.8 dynes cm.$^{-1}$ After the spreading of some of the material, however, this reduces the surface tension $\gamma_{W/A}$ of the water, and so S decreases (cf. eq. (1.23)). If for example, a drop of benzene is placed on water, its initial spreading coefficient is given by:

$$S = \gamma_{W/A} - (\gamma_{O/A} + \gamma_{O/W}) = 72.8 - (28.9 + 35.0) = 8.9 \text{ ergs cm.}^{-2}$$

However, when the benzene and water have had time to become mutually saturated,

$$S_{\text{final}} = 62.4 - (28.8 + 35.0) = -1.4 \text{ ergs cm.}^{-2}$$

the film of benzene on water having lowered the surface tension of the latter by 10.4 dynes cm.$^{-1}$, i.e. the film pressure of the benzene is given by $\Pi = 10.4$ dynes cm.$^{-1}$ The final state, therefore, is that the benzene has stopped

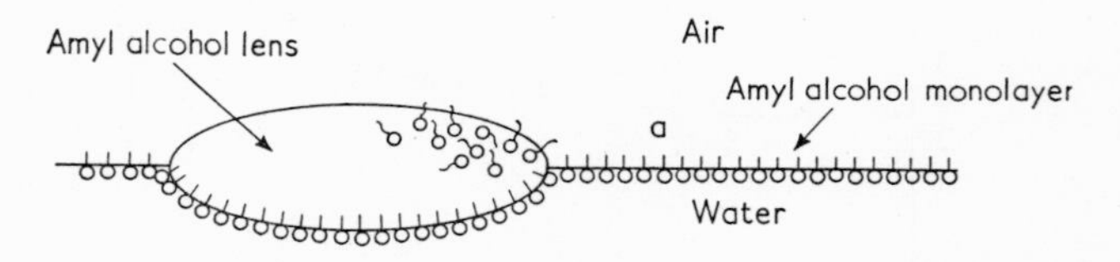

Fig. 1-11. Amyl alcohol does not finally spread over film-covered water at "a" because the non-polar CH_3— groups, oriented upwards, are slightly less polar than is the amyl alcohol: the latter coheres to itself more readily than it will adhere to the oriented film at "a".

spreading, or, if it has been spread right over the surface by the initial velocity of spreading, it retracts again into a very flat lens, in contact with the water surface covered with a monolayer of benzene: the slight orientation of the benzene rings against the water surface in this monolayer makes it slightly less attractive energetically for liquid benzene to cover it than for this liquid to form a lens. Rather similar conditions apply for amyl alcohol: S is 44.3, while $S_{\text{final}} = -2.0$. This is illustrated in Fig. 1-11, in which it is clear that final spreading would have to carry the polar amyl alcohol from the drop over the non-polar hydrocarbon surface of the oriented monolayer. Carbon disulphide, however, has a negative initial S (Table 1-VI):

$$S = 72.8 - (31.8 + 48.6) = -7.6 \text{ dynes cm.}^{-1}$$

Therefore a lens of this oil apparently does not spread, although an invisible

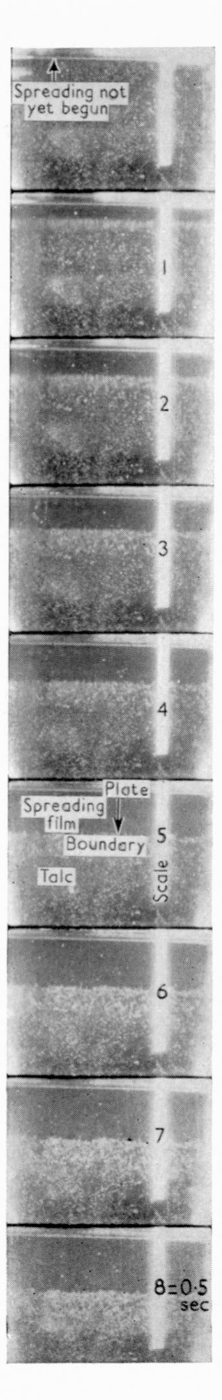

Fig. 1-13. The spreading of acetic acid at petrol-ether—water interface (filmed at 16 frames per sec.).

[*To face p. 25*

monolayer of carbon disulphide ($\Pi = 2.3$ dynes cm.$^{-1}$) does extend across the water surface, and hence S_{final} is given by:

$$S_{\text{final}} = 70.5 - (31.8 + 48.6) = -9.9 \text{ dynes cm.}^{-1}$$

By reducing $\gamma_{W/A}$ the carbon disulphide thus makes S still more negative. This is a general finding, i.e. saturation of the liquids makes S_{final} less favourable for spreading than S.

Kinetics of Spreading

The velocity of spreading of many polar oils on water is about 10 cm. sec.$^{-1}$ This figure is of interest in connection both with "kicking" droplets (see

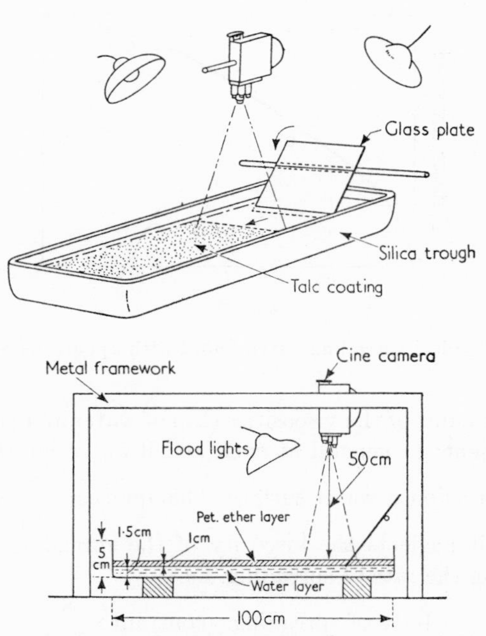

Fig. 1-12. Details of typical apparatus for measuring spreading rates [24].

Chapter 7) and with foam-breaking agents (Chapter 8). Fig. 1-12 shows a suitable apparatus for studying spreading-rates; a ciné-camera[24,25] at a speed of 16 frames sec.$^{-1}$ permits following of the initial stages of spreading by photographing talc particles pushed along the surface by the leading edge of the spreading material, as in Fig. 1-13. If an "oil" such as acetone is used, desorption of the film will cause the spreading rate to fall off after perhaps 1 sec., so that for such substances it is particularly important to observe the early stages of spreading with the ciné-camera. A silica trough $100 \times 30 \times 5$

cms. is convenient[24], and into one end of this is dipped a glass plate wet with the acetone or other liquid whose rate of spreading is to be studied.

The curves found are of the form of that in Fig. 1-14, the initial part "a" of the spreading possibly depending on the height of the lens of spreading liquid and other mechanical factors. The part "b" of the curve, where steady spreading occurs, persists until depletion of the spreading liquid, desorption, or evaporation becomes important, at which time the rate falls off ("c" in Fig. 1-14).

Experimental rate results are shown in Table 1-VII for several systems, together with the initial S divided by the viscosity of water (for an air-water

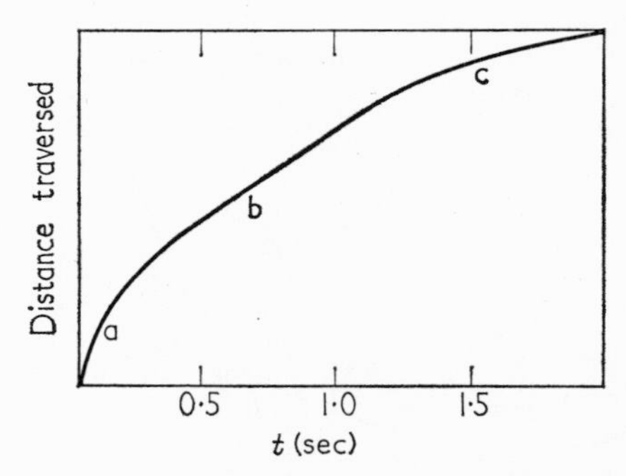

Fig. 1-14. Typical spreading curve found with apparatus of Fig. 1-12.

surface) or by the sums of the viscosities ($\Sigma\eta$) of water and petrol-ether where the latter is present: in general at such an oil-water interface spreading is about half as fast as on a water surface. This quotient $\frac{S}{\Sigma\eta}$ we may take as a rough measure (it neglects the viscosity of the spreading material) of the factors controlling the rate of spreading, i.e.

$$\text{Rate of spreading} = \text{constant} \times \frac{S}{\Sigma\eta} \tag{1.27}$$

As seen from Fig. 1-15, some correlation is indeed observed. Better agreement can hardly be expected without also taking the viscosity of the spreading liquid into account. It is clear from Table 1-VII that the presence of a surface-active agent, by lowering $\gamma_{W/A}$ and hence S, reduces the rate of spreading considerably. A silicone oil spreading over a variety of liquids moves a certain distance after times varying inversely with the viscosities of the liquids[27].

The reason that a film travels rather slowly relative to the kinetic molecular movements in the surface is that there is no slippage between it

TABLE 1-VII

Spreading Rates

Spreading oil	Interface	Measured initial rate of spreading for first $\frac{1}{16}$th sec. (in cm. sec.$^{-1}$) (Region "a" of Fig. 1-14)	Measured steady rate of spreading (in cm. sec.$^{-1}$) (Region "b" of Fig. 1-14)	$\frac{S}{\Sigma\eta}$ (calc. by eq. 1.23)
Ethanol	air-water	53.0	9.2	50
Ethanol	petrol-ether-water	27.2	6.4	25
Ethanol	petrol-ether against water containing 10^{-4} M-$C_{12}H_{25}N(C_3)_3^+$	12.8	3.4	21
Acetic acid	air-water	43.2	10.0	45
Acetic acid	petrol-ether-water	24.8	4.8	25
Acetone	air-water	33.6	10.4	50
Acetone	petrol-ether-water	24.6	8.2	25
Valeric acid	air-water	—	9.0	40
n-Octanol	air against water containing sodium lauryl sulphate	—	3.6	6
1:1 Dihydro-perfluoro-butanol	air against water containing sodium lauryl sulphate	—	4.6	23

Here η is expressed in centipoises, and is equal to unity for both water and petrol-ether. S is in dynes cm.$^{-1}$ All results are due to Davies, Boothroyd and Palmer[24] except that for valeric acid, which Burgers, Greup, and Korvezee[26] studied.

and the underlying or overlying liquid: a layer of the latter must, therefore, be carried along with the film, the resistance to motion depending on the viscosity of the water or petrol-ether. In an elegant study of this phenomenon Schulman and Teorell[28] found that a monolayer of oleic acid, flowing under a surface pressure gradient of 5 cm. sec.$^{-1}$, carried with it a water layer of mean thickness 10^{-3}cm. This drag effect must also be allowed for in determining surface viscosities (Chapter 5).

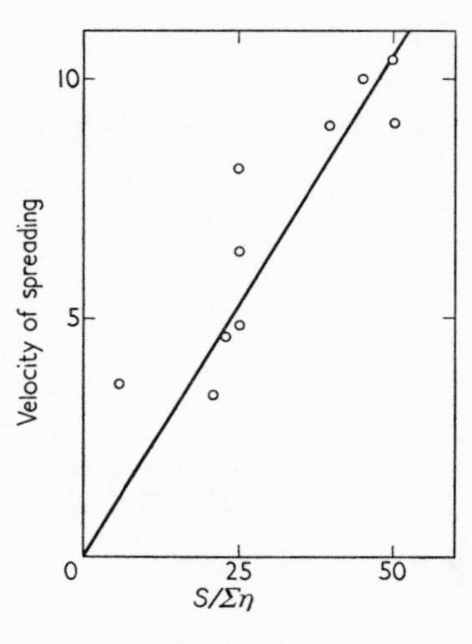

Fig. 1-15.

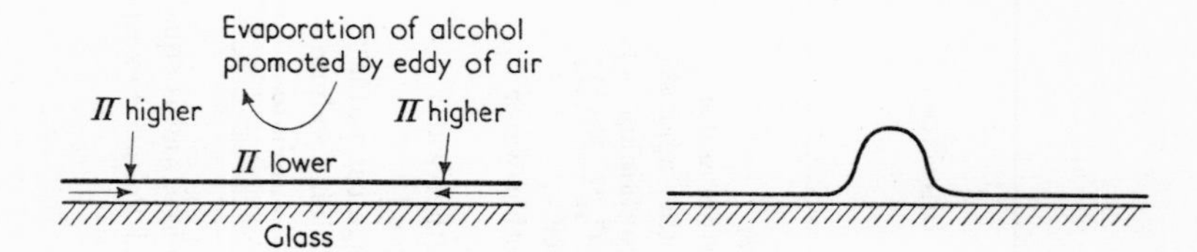

Fig. 1-16. Formation of drops by unequal evaporation of alcohol from an alcohol-water mixture. If the glass is inclined, the drop of liquid runs down as a "tear".

A well-known phenomenon is the formation of "tears" in the liquid wetting the side of a glass of liqueur. The reason for this lies in an eddy of air causing enhanced evaporation of alcohol at some point, which locally reduces the spreading pressure there. Spreading into this region then occurs from the adjacent film of liquid (Fig. 1-16), causing liquid to be pushed into a drop (cf. also Fig. 1-11) at the point where the original evaporation occurred. This phenomenon is sometimes of importance in distillation (Chapter 7).

Spreading from Solids

This is characterized by a very marked temperature dependence: Table 1-VIII shows the temperatures below which the spreading of various solids at the air-water surface becomes zero or extremely slow.[29] Above these temperatures, the energies at the triple interface solid-liquid-air and the energy of the molecules in the solid are of values such that the spreading coefficient becomes appreciably positive, and the molecules can leave the crystal lattice reasonably rapidly. Generally, detachment of molecules into the surface film occurs only at the periphery of the crystal in contact with

TABLE 1-VIII

Compound	Critical Temp. for rapid spreading on water[39] °K
C_{11} acid	233
C_{12} ,,	254
C_{14} ,,	273
C_{15} ,,	266
C_{16} ,,	278
C_{18} ,,	290
Oleic acid	221
Hexadecyl acetate	240
Octadecyl acetate	259
Ethyl palmitate	271

the air-water surface: the supply of material through the bulk of the water is negligible both because the energy barrier now includes not only the formation of a hole in the solid, but also the immersion of the hydrocarbon chain in the water, and also because diffusion through the bulk liquid is a rather slow process. A simple experiment demonstrates this clearly. A drop of water placed on the surface of a piece of solid stearic acid etches a shallow ring on the surface, though apparently no stearic acid dissolves directly into the water inside this ring. Spreading rates depend generally on the perimeter of the solid-gas-liquid interface[29].

Cetyl alcohol, of interest in retarding the evaporation of water, spreads from the solid at room temperature, forming a monolayer of pressure about 33 dynes cm.$^{-1}$ at 25°C. Higher alcohols give progressively reduced final film pressures and rates of spreading. The rate of spreading is between 0.8 and 1.0×10^{14} molecules cm.$^{-1}$ sec.$^{-1}$ on a clean surface[29], and, in general, obeys the relation:

Number of molecules spreading sec.$^{-1}$ = constant × perimeter × $(\Pi_e - \Pi)$ where Π_e is the equilibrium spreading pressure. The constant is therefore $2.4 - 3.0 \times 10^{12}$ molecules dyne^{-1} sec.$^{-1}$

Hughes and Rideal[30] have shown that protein films spread rapidly over a water surface if this is touched with a crystal of the solid protein. Quantitatively, one may coat a fine, hydrophobic fibre with the protein, weighing the fibre both before and after bringing it into contact with the water. In this way the spread film of protein may be shown to be a monolayer. Again, a rise in temperature greatly accelerates spreading.

RELATIONS BETWEEN SURFACE TENSIONS AND INTERFACIAL TENSION

Treatment of Gibbs

Gibbs considered the interfacial tension of mercury in equilibrium with water, in relation to the surface tensions of the separate components of the system in equilibrium with each other. He deduced[31] that if $\gamma_{Hg/W}$ represents the interfacial tension between mercury and water, then this is given by

$$\gamma_{Hg/W} = \gamma_{Hg(H_2O\ sat.)/A} - \gamma_{W(Hg\ sat.)/A} \tag{1.28}$$

and subscripts sat. and A in the other terms denote that the surface tensions of the mutually saturated liquids are measured against air. His reasoning in deriving this equation was as follows. On a clean mercury surface (of high surface tension) water will be adsorbed (lowering the surface tension); this film of water corresponds at equilibrium to adsorption from saturated water vapour, and may reach such a thickness that its interior behaves as if it were bulk liquid water (Fig. 1-17). If this is true, the total surface tension (or

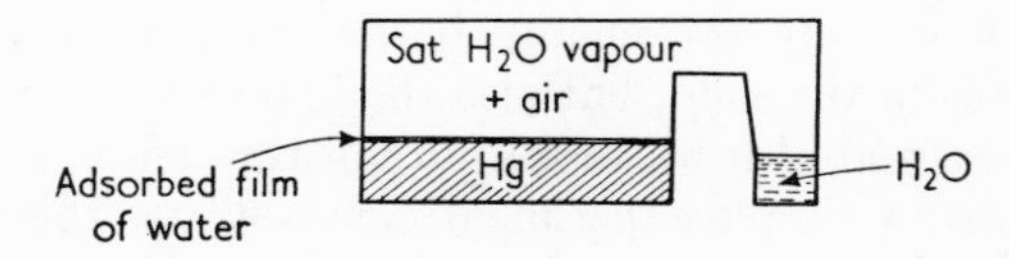

Fig. 1-17. Water adsorbing on to the mercury surface from saturated water vapour to form an equilibrium layer of water.

energy) should thus be the sum of the surface tension of water and the interfacial tension between mercury and water, i.e.

$$\gamma_{Hg(H_2O\ sat.)/A} = \gamma_{W(Hg\ sat.)/A} + \gamma_{Hg/W}$$

which is identical with equation (1.28). In practice the effect of mercury on the surface tension of water is negligible, as the surface tension of water is much lower than that of mercury.

The relation (1.28) has been tested[32] with the following results. At 25°C, $\gamma_{Hg/W}$ is 374.0 dynes cm.$^{-1}$, while $\gamma_{Hg(H_2O\ sat.)/A} - \gamma_{W(Hg\ sat.)/A} = (447.6 - 72.0)$ dynes cm.$^{-1}$, i.e. 375.6 dynes cm.$^{-1}$ The assumption behind eq. (1.28) that the water layer absorbed on the mercury surface behaves as bulk water is thus substantiated. Similar confirmation comes from studies of organic liquids in equilibrium with mercury.

Antonoff's relationship

This states that the interfacial tension of two mutually saturated liquids is equal to the difference between their surface tensions, the latter being measured when each liquid has become saturated with the other[33,34]. It is thus a generalization of Gibbs's treatment of the mercury-water interface (eq. (1.28)). For water and organic liquids the mutual saturation often occurs slowly; as long as 10 days is sometimes required to reach true equilibrium.

If B(W) and W(B) refer to two typical mutually saturated phases (e.g. benzene and water each saturated with the other), and if the surface tension against air of W is higher than that of B, Antonoff's relation may be written in general terms as:

$$\gamma_{B/W} = \gamma_{W(B)/A} - \gamma_{B(W)/A} \qquad (1.29)$$

where $\gamma_{B/W}$ is the interfacial tension when equilibrium has been attained. Some of the experimental data used to test this are shown in Table 1-IX. The phenol-water system also obeys Antonoff's relation after careful equilibration for about a week[33,34].

Adam[5] has pointed out that if eq. (1.29) be combined with the Dupré equation (1.19) in the form:

$$W_{B/W} = \gamma_{B(W)/A} + \gamma_{W(B)/A} - \gamma_{B/W}$$

one obtains finally:

$$W_{B/W} = 2\gamma_{B(W)/A}$$

In words, this means that the work of adhesion between the two mutually saturated liquids is equal to the work of cohesion of the liquid of lower surface tension after saturation with that of higher surface tension, or, for the benzene-water system, it is equal to the work of separating two identical surfaces each consisting of air against benzene saturated with water. This implies that, after saturation with benzene, the water-air surface formed in the separation defining W (Fig. 1-8b) must be very similar in its external field of force to that of liquid benzene, through becoming covered with an adsorbed monolayer of benzene. The significant lowering of surface tension of the water from 72 dynes cm.$^{-1}$ to 56.5 dynes cm.$^{-1}$ suggests that such a benzene monolayer has a pressure of 16.5 dynes cm.$^{-1}$, which is not surprising since benzene vapour is, indeed, known to adsorb on a water surface (Chapter 4) giving monolayers of about this pressure. This approach corresponds exactly to Gibbs's suggestion that the adsorbed film of water on mercury behaves as normal liquid water.

TABLE 1-IX

Oil (B)	Reference	t°C	$\gamma_{W(B)/A}$	$\gamma_{B(W)/A}$	$\gamma_{B/W}$	$\gamma_{W(B)/A}-\gamma_{B(W)/A}$	$\gamma_{W(B)/A}-\gamma_{B(W)/A}-\gamma_{B/W}$ ($=S_{final}$) Cf. eq. (1.29)
Benzene	32	25	62.1	28.2	33.9	33.9	0.0
	34	25	56.5	29.3	27.2	27.2	0.0
	2	20	62.4	28.8	35.0	33.6	− 1.4
	36	25	61.8	28.4	33.6	33.4	− 0.2
Chloroform	34	25	51.7	27.4	23.0	24.3	+ 1.3
Carbon tetrachloride	32	25	69.7	26.2	43.5	43.5	0.0
Ether	34	25	26.8	17.4	8.1	9.4	+ 1.3
Toluene	32	25	63.7	28.0	35.7	35.7	0.0
n-Propylbenzene	32	25	68.0	28.5	39.1	39.5	+ 0.4
n-Butylbenzene	32	25	69.1	28.7	40.6	40.4	− 0.2
Nitrobenzene	32	25	67.7	42.8	25.1	24.9	− 0.2
	36	25	68.0	43.1	24.8	24.9	+ 0.1
Isoamylalcohol	2	20	25.9	23.6	5.0	2.3	− 2.7
	36	25	27.6	24.6	4.7	3.0	− 1.7
n-Heptyl alcohol	2	20	28.5	26.5	8.0	2.0	− 6.0
	36	25	29.0	26.9	7.7	2.1	− 5.6
Carbon disulphide	2	20	70.5	31.8	48.6	38.7	− 9.9
Methylene iodide	36	25	71.9	52.3	40.5	19.6	−20.9
	2	20	72.2	50.5	45.9	21.7	−24.2

All values are in dynes cm.$^{-1}$ and refer to mutually saturated systems consisting of water and various oils.

The liquid of lower surface tension, however, according to the Gibbs adsorption relation (Chapter 4), should have its surface tension raised only slightly by the presence of solutes. This is found in practice, since water dissolved to saturation in the benzene raises its surface tension only slightly, from 28.2 dynes cm.$^{-1}$ to 29.3 dynes cm.$^{-1}$; similarly water raises the surface tension of chloroform and ether by as little as 0.8 dynes cm.$^{-1}$ and 1.0 dynes cm.$^{-1}$ Antonoff's relation is thus in effect a statement of the known facts that, while a liquid of high surface tension tends readily to adsorb a monolayer of a less polar substance, the less polar substance will have its surface tension raised only slightly by the presence of more polar molecules.

This explanation breaks down if at equilibrium neither liquid spreads on the other, not even to the extent of a monolayer. This occurs with the higher paraffins and water, and also with methylene iodide and water. It follows, therefore, that the Antonoff relation should not be valid for such systems. In fact we now have the non-spreading conditions:

$$\gamma_{W(B)/A} + \gamma_{B/W} > \gamma_{B(W)/A}$$

and

$$\gamma_{B(W)/A} + \gamma_{B/W} > \gamma_{W(B)/A}$$

which, combined with the Dupré equation (1.19), give

$$W_{B/W} < 2\gamma_{W(B)/A}$$

and

$$W_{B/W} < 2\gamma_{B(W)/A}$$

showing that the adhesion between the saturated liquids is less than the cohesion of either. If Antonoff's relation were to hold quite generally, we should expect to find that spreading must *always* just occur on one of the two liquids once mutual saturation is established, since $S_{\text{final}} = \gamma_W - (\gamma_B + \gamma_{B/W})$, which, by eq. (1.29), should be zero. The last column of Table 1-IX shows that such a conclusion is not in accord with experiment. For slightly polar oils, such as chloroform or benzene, it is approximately true, but cannot be correct for more polar oils (e.g. n-heptyl alcohol), nor for initially non-spreading oils (such as carbon disulphide or methylene iodide).

Drops of Oils on Water

The form of a drop of oleic acid, floating on water and in equilibrium with the monolayer-covered surface, has been extensively studied[37,38,39]. Since the angles θ_1 and θ_2 of Fig. 1-9 must be adjusted so that the forces balance, the drop of oil must be limited in size if it is to be in equilibrium. One may demonstrate this by the following experiment. A drop of a spreading oil is placed on a clean water surface. Spreading is generally rapid enough to carry out enough liquid from the drop to form a multimolecular layer on the water surface, and interference colours can often be seen at this stage. After a few seconds the excess oil retracts into drops, leaving the rest of the surface covered with a monolayer. The energy of the system comprising the monolayer and drops will be a minimum for a certain drop size: simple theory predicts that coalescence should not occur after this size is attained[40]. In practice, even thin layers of oleic acid on water do not obey the theoretical equations, because the small oil drops initially formed on the water have enhanced spreading pressures compared with large drops (a phenomenon similar to the enhanced solubility of particles of high curvature).

The film pressure of the monolayer in equilibrium with the floating drops of oil has received much attention[38]. We mention particularly various indicator oils (which are useful in determining the surface pressure on ponds and lakes) and piston oils (which maintain a given surface pressure as long as drops are present). *Indicator oils*[41] are conveniently made by diluting a filtered sample of old crankcase oil from a car with a non-spreading oil such

TABLE 1.X

Piston oil	Pressure exerted (dynes cm.$^{-1}$ at 20°C)[41]
Tricresyl phosphate	9.5
Rape seed oil	10.5
Castor oils	about 16.5
Neatsfoot oil	19.0
Oleic acid	29.5

as Nujol, a viscous white paraffin oil. Such indicator oils are so called because their tendency to spread, and hence their thickness on the surface, is determined by the surface pressure of the surface on which they are placed. They do not form visible lenses and, after previous calibration, the interference colour of the multimolecular patch of indicator oil gives an accurate measure of the pressure obtaining on any given surface such as a pond or lake[42]. *Piston oils*[41] spread only to a monolayer, the excess oil remaining as a lens which acts as a reservoir. This produces a constant spreading pressure on anything in the surface (Table 1-X). Naturally, such piston oils will spread as a monolayer only if the surface pressure around them does not exceed the spreading pressure of the oil, and this enables one to obtain a rough estimate of the surface pressure by using a range of piston oils.

CONTACT ANGLES

Theory

If a small drop of liquid is placed on a uniform, perfectly flat, solid surface, it will, possibly, not spread completely over this surface, but its edge may make an angle θ with the solid, as shown in Fig. 1-18. This is exactly analogous to the behaviour of a drop of one liquid on another, with θ being equal to $(\theta_1+\theta_2)$ (Fig. 1-9). At its simplest, the theory of the contact angle allows us to resolve horizontally the equilibrium tensions acting in Fig. 1-18, to give

$$\gamma_{S/A}=\gamma_{S/L}+\gamma_{L/A}\cos\theta \tag{1.30}$$

where the subscripts have their usual significance. This relation applies for contact angles θ less than, equal to, or greater than 90°.

Equation (1.30) can be combined with the appropriate form of the Dupré relation:

$$W_{S/L}=\gamma_{S/A}+\gamma_{L/A}-\gamma_{S/L} \tag{1.31}$$

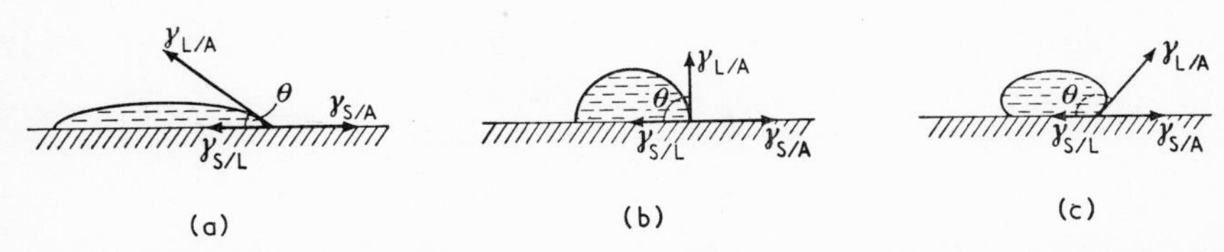

Fig. 1-18. Drop of liquid resting on solids, each having a smaller molecular attraction for the liquid than the preceding one.

(derived in the same was as was eq. (1.19)) to give what is known[1,43] as Young's equation:

$$W_{S/L}=\gamma_{L/A}(1+\cos\theta) \tag{1.32}$$

which is a more useful relation than eq. (1.30), in which $\gamma_{S/A}$ and $\gamma_{S/L}$ cannot easily or accurately be measured.

The concept of a tension acting along a solid surface is, however, a difficult one to interpret precisely, and it is preferable to deduce eqs. (1.30) and (1.32) in terms of surface energies of the solid as follows. Suppose[44] that in a virtual displacement of the equilibrium (i.e. $\theta>0$) 3-phase boundary, constituted as in Fig. 1-19 so that the liquid surface is horizontal on the left side, a length δl of the upper solid-gas interface is invaded by the liquid, by immersion of the solid sheet (1 cm. in width) in the direction shown by the arrow. During this immersion the tension of the liquid-air interface must help to pull the plate down, its work on the plate being given by $\delta l\gamma_{L/A}\cos\theta$. Since the process is an infinitesimal change in an equilibrium system, the total work done is zero, so that if the F^s terms for the solid represent surface energies per unit area, and the area change is $\delta l\times 1$ cm.^{2}, we have

$$\delta l(F^s_{S/L}-F^s_{S/A})+\delta l\gamma_{L/A}\cos\theta=0 \tag{1.33}$$

i.e.

$$F^s_{S/A}-F^s_{S/L}=\gamma_{L/A}\cos\theta \tag{1.34}$$

or

$$F^s_{S/A}=F^s_{S/L}+\gamma_{L/A}\cos\theta \tag{1.35}$$

Although the last relation may be obtained directly by resolving the tensions in Fig. 1-18 (if F and γ can be equated), it is not easy to visualize a solid in a state of tension. By the more rigorous method of surface energies, however, and a similar treatment in defining $W_{S/L}$, Young's equation (1.32) remains unaffected.

The action of the vertical component, $\gamma_{L/A} \sin \theta$, on the solid is of interest, and can be directly studied for a drop of mercury placed on a sheet of mica about 1 micron thick[45]. The rather high surface tension of mercury deforms the mica slightly all round the edge of the drop (Fig. 1-20). In general, however, solids are so little deformable that ϕ_1 and ϕ_2 are close to zero.

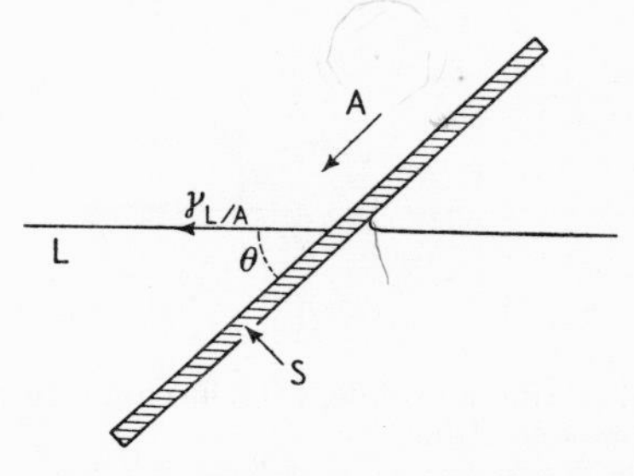

Fig. 1-19. Plate of solid immersed in liquid so that the meniscus on the left is horizontal.

Magnitude of Contact angles of Liquids on Solids

Young's equation (1.32) shows that the angle θ is determined by the relative magnitudes of the adhesion of the liquid for the solid ($W_{S/L}$) and the self-cohesion of the liquid ($2\gamma_{L/A}$). For a given value of $\gamma_{L/A}$, the contact angle will increase as the adhesion between the liquid and the solid decreases, and an angle of 180° would be indicative of zero adhesion. This is never reached in practice, though mercury on steel gives $\theta = 154°$. Water on paraffin wax gives angles up to 110°, and water on polyethylene gives $\theta = 94°$. The former angle indicates that $W_{S/L} = 48$ ergs cm.$^{-2}$, close to the value between a liquid paraffin and water (about 43 ergs cm.$^{-2}$).

A high θ can also be achieved on normally hydrophilic surfaces. Whereas glass, carefully cleaned in chromic acid, has a contact angle of zero for water, a thin coating of a silicone or dimethyl-silane derivative will completely alter the wetting properties: even a monolayer of the dimethyl-silane will render glass or steel lipophilic (oil-wetted), a fact often useful in preparing suitable rings and dropping tips for the measurement of surface tension (see below).

For example, dimethyldichlorosilane, $(CH_3)_2SiCl_2$ [structure: CH_3, CH_3 above Si; Cl, Cl below], attaches itself chemically to the $-OH$ groups on the outside of the silicate lattice of glass, eliminating HCl to give

```
CH3     CH3  CH3     CH3
   \   /        \   /
    Si           Si
   /  \         /  \
  O    O       O    O
  |    |       |    |
—O—Si—O—Si—O—Si—O—Si—O—
```

Under the names Teddol, Drifilm, and M441 (I.C.I.), this material is readily available. It may be applied from solution in CCl_4[46]. Heavy-metal soaps, long-chain fatty acids and amines produce the same effects, though silicone and silane treatment is preferable where it can be used, as these materials are bound so tightly to the surface that they do not contaminate the adjacent liquids.

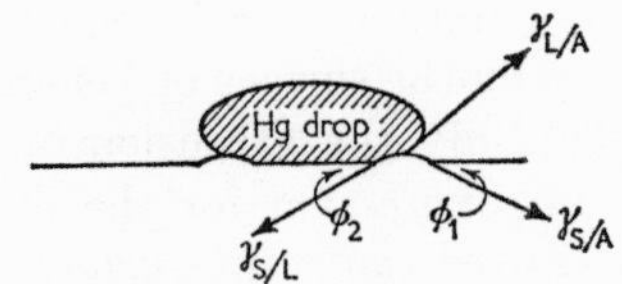

Fig. 1-20. Mercury drop resting on thin sheet of mica[45].

Polytetrafluoroethylene (Fluon or Teflon) is strongly water-repellent, having θ greater than for polyethylene by 14°, i.e. 108° instead of 94°: it can be wetted by aqueous solutions only if these contain a strong wetting agent such as sodium di-octylsulphosuccinate.

All the contact angles quoted above refer to smooth surfaces, soon after formation. The contact angle depends, however, on the roughness of the surface, on its absorption of impurities and of water, and on a possible molecular re-orientation of the solid surface in the presence of the water.

Roughness of a surface has the effect of making the contact angle further from 90°; if the smooth material gives an angle greater than 90°, roughness increases this angle still further, but if θ is less than 90°, roughness decreases the angle. For example, on rough paraffin wax the contact angle is 132°, compared with about 110° on smooth wax[47]. Roughening is used in the Wilhelmy plate apparatus to ensure that the contact angle of a thin mica plate in contact with surface-active solutions is always close to zero. Quantitatively, roughening is measured by the ratio of the real to the apparent surface, this ratio being called r. Wenzel's relation[48]

$$\cos\bar{\theta} = r\cos\theta \tag{1.36}$$

relates r to the angle θ on a smooth surface and the average angle $\bar{\theta}$ on a rough surface. This may be proved as follows. In equation (1.33) above, the total solid surface energy term $\delta l(F^{s}_{S/L} - F^{s}_{S/A})$ should be replaced, if the surface is rough, by $r\delta l(F^s_{S/L} - F^s_{S/A})$, since $r.\delta l \times 1$ cm.^{2} is the real change in area involved. Hence, if the *roughening is small-scale*, so that the method of virtual displacement remains valid,

$$r\delta l(F^s_{S/L} - F^s_{S/A}) + \delta l \gamma_{L/A} \cos\bar{\theta} = 0$$

where $\bar{\theta}$ is the new contact angle for the roughened surface. Thus

$$r(F^s_{S/A} - F^s_{S/L}) = \gamma_{L/A} \cos\bar{\theta}$$

or, since

$$F^s_{S/A} - F^s_{S/L} = \gamma_{L/A} \cos\theta \qquad (1.34)$$

$$\cos\bar{\theta} = r \cos\theta \qquad (1.36)$$

This relation can be valid only for sub-microscopic roughening, since *coarse roughening* implies that close to the solid the liquid boundary must become ragged, with the further consequence that $\bar{\theta}$ is not a constant independent of δl as assumed in deriving eq. (1.36). Detailed knowledge of the pattern of the roughness and any "grooving" is then required[49], though, with only occasional grooves on the surface of Teflon, the mean contact angle is still given[50] by eq. (1.36). By the extension of the method of deriving eq. (1.36) one can find[51] an expression for the apparent contact angle θ' observed for a finely constituted *composite surface* made up of a fraction f_1 having a contact angle θ_1 and a fraction $(1-f_1)$ of contact angle θ_2:

$$\cos\theta' = f_1 \cos\theta_1 + (1-f_1) \cos\theta_2 \qquad (1.37)$$

For rough surfaces f_1 is replaced by $r_1 f_1$, and $(1-f_1)$ by $r_2(1-f_1)$: if $f_1 = 1$, Wenzel's relation (1.36) results. When part of the surface is composed of air spaces, as on the surface of a porous solid, $\theta_2 = 180°$ (since small water drops remain spherical in air): i.e.

$$\cos\theta' = f_1(\cos\theta_1 + 1) - 1 \qquad (1.38)$$

If most of the surface consists of air-spaces, as on certain textiles and on certain plants and insects, θ' may be of the order of 150° when θ_1 is only 90°.

In experimental measurements it is frequently difficult to obtain reproducible contact angle values: in particular, the angles measured when the solid surface is advancing into the liquid may be greater than when the plate is being withdrawn. It appears, however, that the cleaner the surface, the smaller is the *hysteresis of the contact angle*. The large advancing contact angle may be due to a film of some material which prevents the liquid adhering to the solid; after contact with the liquid, this film may be wholly or partially removed, so that contact between the liquid and the solid becomes more complete, giving the smaller receding angle. This implies that $F^s_{S/L}$ becomes reduced by removal of the film on the solid, and $\cos\theta$ is therefore greater to balance eq. (1.34), i.e. θ must be less. If the film consists of greasy material adsorbed in minute traces from the air (as is found experimentally by potential measurements after a few minutes' exposure of a cleaned air-water surface) it may leave the solid to be preferentially adsorbed at the air-liquid surface. If this occurs, $\gamma_{L/A}$ is now reduced and, again by eq. (1.34), $\cos\theta$ must be increased, i.e. θ is reduced. It is known from independent studies on flotation that only 10% coverage of a solid surface by a monolayer of a long-chain compound may drastically increase θ, so that this explanation of the hysteresis seems reasonable. Needless to say, the liquid surface must be cleaned as scrupulously as possible (Chapter 5) before measurements of θ are made, and the common practice of taking the mean of the advancing and receding contact angles has little to recommend it. Even if all traces of contamination by greasy material are eliminated, the advancing contact angle may be too large if the solid is covered partly or wholly with an adsorbed film of air. This increases the apparent contact angle (eq. (1.37)) and is specially important on rough solids. It decreases $W_{S/L}$ in eq. (1.32), and hence makes θ too great; significantly, very clean and smooth[44,47] solid surfaces show no "hysteresis" of the contact angle.

Adsorption of water by the solid, either as a monolayer[52] or in small cracks or scratches in the surface, will decrease the receding angle, since $F^s_{S/A}$ will be reduced by the monolayer. This phenomenon, therefore, also produces hysteresis effects, although they may be eliminated by working at high humidity[47]. A drop of water resting on the surface of solid stearic acid gradually spreads out, but whether this is a consequence of absorption of water into cracks or of a re-orientation of the molecules of the stearic acid in the surface (to present $-COOH$ groups to the overlying water) is not yet established. The idea has been generalized[51], however, that for such organic solids as stearic acid or long-chain alcohols, θ is given by eq. (1.37), where $\theta_1 = 0$ for that part of the surface with exposed polar groups, and where $\theta_2 (\approx 110°)$ refers to that part of the surface having exposed hydrocarbon chains, i.e.

$$\cos\theta' = f_1 - 0.34(1 - f_1)$$

For the solid C_{16} acid and alcohol, θ can be anywhere between 100° and 50°, depending on the way the crystal has been cut[5]: the former figure corresponds closely to only the ends of the hydrocarbon chains being exposed, while the latter implies that about 70% of the surface consists of exposed carboxyl groups. Zisman et al.[53], in an extended investigation on the wetting of various solid surfaces with a range of liquids, find linear relations between $\cos\theta$ and $\gamma_{L/A}$; $\cos\theta$ decreases linearly from 1 to a low value as $\gamma_{L/A}$ increases above some critical value, though the reason for this is still not clear.

Spreading Coefficients of Liquids on Solids

By analogy with the spreading coefficient of one liquid on another (eq. (1.23)), we may write for the spreading coefficient of a liquid on a solid

$$S_{S/A} = F^s_{S/A} - (\gamma_{L/A} + F^s_{S/L}) \tag{1.39}$$

where S, A, and L refer to solid, air, and liquid. If an equilibrium contact angle θ can be maintained in the liquid, this relation may be combined with eq. (1.35) to give:

$$S=\gamma_{L/A}(\cos\theta-1) \tag{1.40}$$

which is accordingly valid for $\theta>0$. From this expression S may be determined experimentally by measurements of $\gamma_{L/A}$ and either θ or $\gamma_{L/A}\cos\theta$ (see pp. 46 and 49 below). It may also be found directly by the sessile drop method (see below).

The spreading coefficient of water on clean paraffin wax has been frequently determined, with good agreement between these three different methods to give $S=-99\pm1$ ergs cm.$^{-2}$ It is interesting to note that this figure agrees with that calculated directly (by an equation similar to eq. (1.23)), taking the free energy terms for solid paraffin wax as equal to the corresponding tension terms measured for liquified paraffin wax:

$$S=30-56-72=-98.$$

For oils spreading on chromium S is positive, and the method of restraining the spreading with a monolayer at an air-water interface (see below) must be used. For 7-butyltridecane S is 25 dynes cm.$^{-1}$, though with much chain branching S may be reduced to about 17 dynes cm.$^{-1}$ The history of the metal surface appears important, however, and chromium surfaces cleaned by ion-bombardment and immediately wetted with water are oleophobic, i.e. the oil does not spread, and S must be negative[54].

Lubricating oils spread readily over steel, especially if they are somewhat oxidized to polar compounds[55]. Indeed, the spreading pressures of such oils on steel are only slightly less than for the same oils on water. That some oils do not spread on metals is, indeed, an indication that the metal surface is not clean, for, if it were, $F^{s}_{S/A}$ in eq. (1.39) would be so high (several hundred to several thousand ergs cm.$^{-2}$) that S would always be positive. Many metal surfaces, however, adsorb a monolayer of the organic substance which may so reduce $F^{s}_{S/A}$ that S may be negative; chlorinated hydrocarbons and phosphate esters are examples of this. Volatile liquids also cover the metal surface in the region of measurement with an adsorbed monolayer, and θ as normally measured will be the angle of contact of the liquid against the monolayer-coated solid. The value of S in eq. (1.40) will consequently refer to this system[53,54,56]. To obtain a value of S which would apply to the clean solid, we must apply the procedure described below for correcting the work of adhesion.

Adhesion of Liquids to Solids

By a derivation similar to that of eq. (1.19) the work of adhesion of a liquid L to a solid is given by the Dupré equation:

$$W_{S/L} = F^s_{S/A} + \gamma_{L/A} - F^s_{S/L} \tag{1.31}$$

But
$$S = F^s_{S/A} - \gamma_{L/A} - F^s_{S/L} \tag{1.39}$$

hence
$$W_{S/L} = S + 2\gamma_{L/A} \tag{1.41}$$

For example, the adhesion of water to paraffin wax is calculated by eq. (1.41) to be $-98+2\times72=46$ ergs cm.$^{-2}$, compared with the value of about 43 ergs cm.$^{-2}$ for water and liquid paraffins. $W_{S/L}$ may also be obtained from eq. (1.32). These values all refer to S and θ as measured, and sometimes θ will be anomalously high due to adsorption of vapour at the air-solid interface, and $W_{S/L}$ will then refer to the adhesion of the liquid to the monolayer-covered solid surface.

To obtain values of $W_{S/L}$ for clean surfaces one must therefore correct for this[56], adding to $W_{S/L}$, as obtained above, the change in energy of the solid surface due to adsorption of the vapour. This is possible by combining the measured adsorption isotherm with the Gibbs equation (4.63).

De-wetting by Surface-Active Agents

By eq. (1.32)

$$\cos\theta = \frac{W_{S/L}}{\gamma_{L/A}} - 1$$

Hence, when surface-active agents, such as long-chain quaternary ammonium ions, are added to water in which is immersed a sheet of glass, θ may change with concentration in a rather complicated manner[57]. For pure water $W_{S/L}$ is high, and the measured θ is zero, though this is not an equilibrium contact angle. Adding a little of the quaternary salt reduces $W_{S/L}$ to such an extent that it is now approximately equal to the new value of $\gamma_{L/A}$, and $\cos\theta$ is small, i.e. the glass is not wetted by the solution. The principal effect of the addition of further quaternary salt is to continue to reduce $\gamma_{L/A}$, and $\cos\theta$ increases again, so that the solid again becomes wetted by the solution. This subject is treated more fully on pp. 435-438.

Contact between Two Liquids and a Solid

If in Fig. 1-19 the air is replaced by a second liquid M, and θ is measured as shown in this figure, the modified form of eq. (1.35) is:

$$F^s_{S/M} = F^s_{S/L} + \gamma_{L/M}\cos\theta \tag{1.42}$$

Further, by the Dupré equation (1.19),

$$W_{S/L} = \gamma_{L/A} + F^s_{S/A} - F^s_{S/L}$$

and
$$W_{S/M} = \gamma_{M/A} + F^s_{S/A} - F^s_{S/M}$$

where as usual the subscript A denotes air. Hence

$$W_{S/L} - W_{S/M} = \gamma_{L/A} - \gamma_{M/A} + \gamma_{L/M}\cos\theta \tag{1.43}$$

Experimentally the displacement of a film of one liquid from the solid surface by the other liquid may be very slow, and the hysteresis of the contact angle may therefore be large. The small receding angle of the water favours the hanging plate method in studying monolayers at the oil-water interface (Chapter 5).

Liquid L will displace liquid M from a solid powder or capillary system if the advancing contact angle made by the liquid L is acute: the free energy of the system tends to be reduced by the larger $F^s_{S/M}$ being replaced by the smaller $F^s_{S/L}$ (cf. eq. (1.42) with $\cos\theta > 0$). Eventually the whole powder or capillary system becomes spontaneously wet with L. In practice an applied pressure difference is usually necessary to facilitate this process, as in oil-well flooding (Chapter 8).

The two-liquid contact angle could be useful in finding the adhesion of a liquid L to the solid when, with this liquid alone, the solid is wetted completely ($\theta = 0$) and eq. (1.32) cannot be used. One must now find a liquid M such that this alone does not wet the solid completely, i.e. $W_{S/M}$ can be found from eq. (1.32). Hence, from eq. (1.43), $W_{S/L}$ can be computed. The liquids should be mutually saturated before determining the surface tensions and angles[5].

MEASUREMENTS OF SURFACE AND INTERFACIAL TENSION

Before any measurements are carried out, it is essential to clean the surface thoroughly, to remove traces of contamination which may otherwise greatly lower the tension. Impurities on the air-liquid surface are best removed by blowing them to one side with a fine jet of air, and then sucking this surface off with a fine capillary attached to a water-pump. The surface may conveniently be sprinkled with a little ignited talc to facilitate the cleaning process, as described more fully on p. 220. Interfaces between two liquids are cleaned by applying the above procedure to the surface of the heavier liquid before the upper liquid is poured on, followed by sucking from different points on the interface (see p. 224).

The Ring Method

The surface tension determines the force required to detach a metal ring from the surface of a liquid. This has been used for many years as a method of measuring the surface tension, with the wire ring connected either to one arm of a balance (in place of the pan) or to a light beam carried on a horizontal torsion wire whose constants are known. The simplest theoretical interpretation of the results equates the pull required to detach the ring from the surface to the total perimeter of the ring multiplied by the surface tension: the total perimeter of the ring is twice the length, since the meniscus pulls on each side of the wire, as in Fig. 1-21. Equating the vertical pull f on the ring, at the moment of the ring breaking away from the surface, to the radius r of the ring, we have:

$$f = 4\pi r\gamma$$

or

$$\gamma = \frac{f}{4\pi r} \tag{1.44}$$

This assumes that the surface tension acts vertically and that the contact angle is zero, i.e. that the liquid completely wets the ring; to ensure this, platinum rings may be readily cleaned by gentle flaming. With such a platinum ring attached to a torsion wire, the instrument is known as the du Noüy tensiometer[58]. Its use is fully described by Reilly and Rae[59].

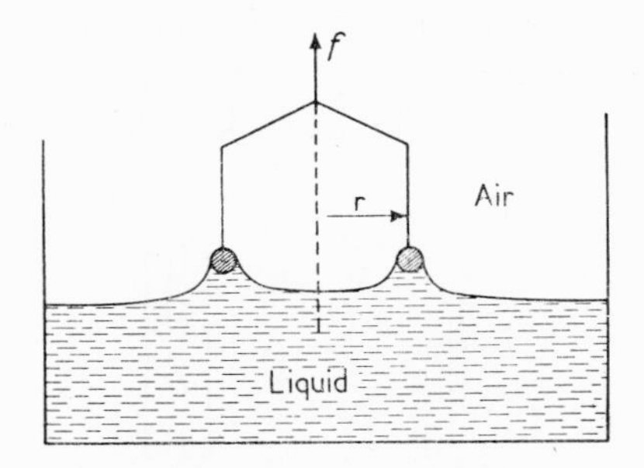

Fig. 1-21. Du Noüy tensiometer (ring method).

Since the surface tension does not in general act exactly vertically (Fig. 1-21), a correction factor β is necessary in eq. (1.44), which becomes:

$$\gamma = \frac{\beta f}{4\pi r}$$

where β depends[60] on r, on the radius of the wire used for the ring, and on the density of the liquid. For a 1 cm. radius ring of wire of radius 0.02 cms. detaching from pure water at 25°C, β is 0.93, and for a 1.85 cm. radius ring of wire of radius 0.03 cm., β is 0.88. Standard surface tensions, which may be used to check the procedure or to calibrate the ring, are listed in Table 1-I.

The ring method can also be applied to determining interfacial tensions[60] provided that the lower liquid preferentially wets the platinium ring. Water, with overlying benzene or other oil, satisfies this condition, but if the ring is wetted by the upper liquid (e.g. water overlying CCl_4), the method requires modification. Possibly a stainless-steel ring coated with silicone to make it preferentially oil-wetted would be satisfactory here. Best results may be obtained by using large rings made of fine wire, such as 3 cm. diameter rings of wire of 0.02 cm. radius. Standard interfacial tensions are listed in Table 1-IV.

The Drop-weight Method

The weight (or volume) of each liquid drop which detaches itself from the tip of a vertical tube is determined largely by the surface tension of the liquid. Assuming that the drops are formed extremely slowly, they detach themselves completely from the tip when (Fig. 1-22a) the gravitational pull just reaches the restraining force of surface tension:

$$Mg = V\rho_l g = 2\pi a\gamma$$

or

$$\gamma = \frac{Mg}{2\pi a} = \frac{V\rho_l g}{2\pi a}$$

where g is the gravitational acceleration, M and V are the mass and volume of each drop that falls from the tip, ρ_l is the density of the liquid, and a is the radius of the tip of the tube. These relations, however, require correction,

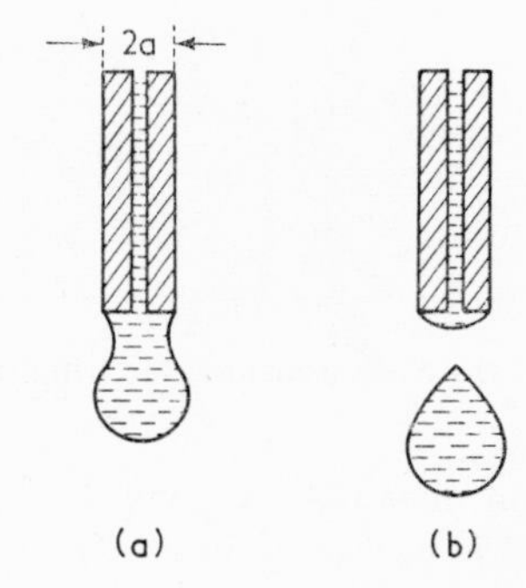

Fig. 1-22. Drop-weight or drop-volume method for surface tension measurements. In (a) the drop is about to fall from the tip, and in (b) it has just fallen.

because the liquid forming the drop does not completely leave the tip (Fig. 1-22b), and the surface tension seldom acts exactly vertically. The correction factors are rather important in this method, as drops are often as much as 40% smaller than predicted by the above relations.

The correction factor φ is known from the work of Harkins and Brown[61], who found that φ is a function of a and of the volume of the drop, where

$$\gamma = \frac{\varphi Mg}{2\pi a} = \frac{\varphi V\rho_l g}{2\pi a} \tag{1.45}$$

In practice the factor φ is readily obtained from Fig. 1-23. All liquids that have been tested give the same value of φ, showing that viscosity and other differences need not be allowed for in this method, provided that the drops are formed extremely slowly. It is convenient in practice to choose tips of such a size that $a/V^{1/3}$ lies between 0.75 and 0.95, where φ is least sensitive to variations in $a/V^{1/3}$.

For interfacial work the theory of the method is again directly applicable, M now being the apparent mass of the drop and ρ_l the density difference between the two phases. For example, in one experiment at 25°C with a tip of radius a of 0.045 cm., the limiting volume of a drop of CCl_4 formed in water was 20.8×10^{-3}cm.^{3} Hence $a/V^{1/3} = 0.14$, and from Fig. 1-23 we find φ is

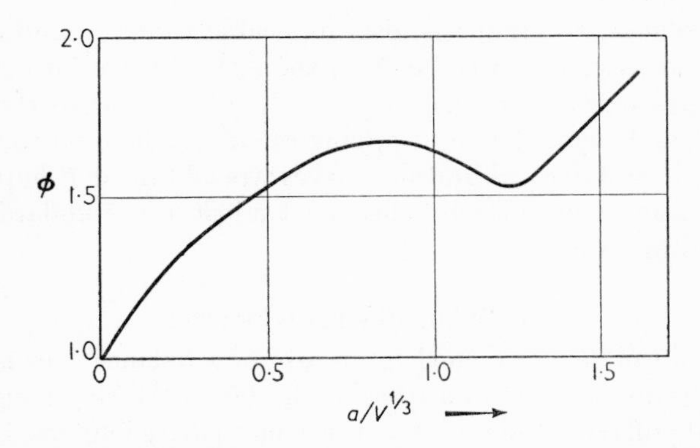

Fig. 1-23. Correction factor φ of eq. (1.45) as a function [2,61] of $a/V^{1/3}$.

about 1.1. We can now calculate γ_i using eq. (1.45) modified for density differences:

$$\gamma_i = \frac{1.1 \times 20.8 \times 10^{-3} \times (1.58 - 1.00) \times 981}{2\pi \times 0.045} = 45.8 \text{ dynes cm.}^{-1}$$

Experimentally, one must ensure that the outside of the tip from which the drops are formed is either completely wetted or completely non-wetted by the liquid, otherwise the liquid drop will form with an uncertain value of a. On the other hand, the interior of the tube must not be too hydrophobic, or the drops break off inside the tube instead of right at the end. Thus Teflon cannot be used, though very lightly siliconized stainless steel is quite reliable. The tips should be made on a watchmaker's lathe, so that the ends are sharp and regular (Fig. 1-22). They may be made hydrophobic with silicone and they must be handled with great care so that they do not become contaminated in any way.

Drops can be counted (not faster than one every three minutes) into a weighing bottle, which is weighed before and after a certain number has been formed, or, more conveniently for pure liquids, the steel tip may be joined (preferably by a ground joint) to an Agla micrometer syringe, by which the volume of a single drop may easily be read directly. The initial formation of the drop can then be performed rapidly, while the final stages leading to detachment can be carried out very slowly (over a period of a minute or two). In this way single drop studies can yield surface tensions with an accuracy of a few tenths of one per cent. If the drop is formed too rapidly it may be too heavy, because the tail of the detaching drop gets "blown up" with more liquid coming down the tube. This can, however, be studied separately if experiments on the ageing of surfaces are to be undertaken (Chapter 4).

The experimental technique is also applicable to determining interfacial tensions. The heavier liquid is fed from the Agla syringe into the steel tip. Again the drops must be formed very slowly, at least when they approach the size for detachment. Faster dropping rates may be used to study interfacial ageing if suitable calibration curves are obtained (Chapter 4). The method and apparatus may be checked against the standard interfacial tensions listed in Table 1-IV.

The Wilhelmy Plate Method

If a very thin plate is attached to an arm of a balance[62], or is suspended from a light beam attached to a torsion wire (as in the ring method above), the additional pull on the plate when it becomes partly immersed (Fig. 1-24a)

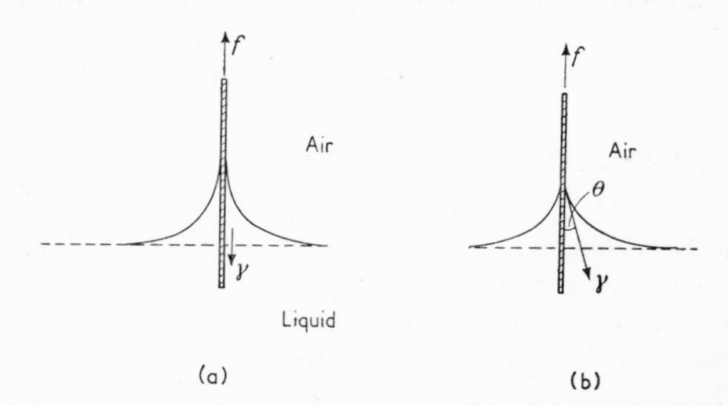

Fig. 1-24. Pull on a vertical plate of unit perimeter if (a) it is completely wetted and $\theta = 0$; and if (b) the contact angle in the liquid is θ. In (a) the pull is γ; in (b) it is $\gamma \cos \theta$, where γ refers to the liquid-air tension.

is equal to the product of the perimeter and the surface tension. To obtain accurate results, it is preferable not to detach the plate from the surface, but to keep the plate partly immersed and to correct for buoyancy. The latter effect is very small if a thin mica plate be used: convenient dimensions are 10 cm. wide, 5 cm. high, and 0.02 cm. thick. The method is valid only when the contact angle is zero, the plate being completely wetted by the liquid. To promote wetting the mica plate may be roughened by rubbing with a very fine emery paper, moving this in small circles over the mica surface to make the roughening as uniform as possible in all directions[62].

In general the meniscus takes the typical form shown in Fig. 1-24b; it is the weight of the liquid pulled up above the mean level of the liquid surface that one measures. This weight is $\gamma \cos \theta$ per unit perimeter of the plate[63], or, if $\theta = 0$, it is equal to γ.

Given suitable wetting characteristics, so that the plate is completely wetted by one liquid, the method may also be used for interfacial tensions. To obtain a plate strongly water-wetted ($\theta = 0$ in the water), one may use a glass cover-slide (cleaned in hot chromic acid), or roughened mica. Mica coated with lamp-black from a smoky Bunsen-burner flame is very strongly preferentially oil-wetted, i.e. $\theta = 0°$ in the oil and 180° in the water. Fuller details of the method are given in Chapter 5.

The Pendant Drop Method

If a pendant drop of liquid is photographed or projected on to squared paper, one may compute the surface tension from the shape of the drop[64].

Other Methods

Historical details of the above and of other methods, which include the ripple method, the sessile drop method, the capillary rise method, and the maximum bubble pressure method, are to be found in the literature[2,5,39,59,65].

Differential Measurements

The lowering of surface tension due to a monolayer is best studied by a "surface balance" (Chapter 5). With suitable instruments changes of only a few millidynes cm.$^{-1}$ can be accurately recorded. The drop volume method and the Wilhelmy plate method are, however, often used to study adsorbed films, particularly at the oil-water interface.

EXPERIMENTAL STUDIES OF SOLID-LIQUID INTERFACES

The Plate Method for obtaining θ

When a solid surface can be obtained in the form of a relatively flat plate a few centimetres across, the angle at which this plate dips into a liquid can conveniently be altered till the liquid surface remains planar right up to the solid surface (Fig. 1-19). The solid is held in an adjustable holder capable of being tilted to any angle and of being raised and lowered, so that both the advancing and receding angles may be measured. The liquid is contained in a Langmuir trough filled to the brim, the surface being repeatedly swept with barriers[44] so that no contamination remains to lower $\gamma_{L/A}$ below the value for the pure liquid. The Langmuir trough and barriers, which are discussed in detail in Chapter 5, may be conveniently made of polytetrafluoroethylene (Fluon or Teflon)[66]. For accurate work it is necessary to study the liquid surface where it meets the solid by an optical technique: when the plate is adjusted so that the liquid surface is flat up to the line of contact, the reflection of an illuminated slit, held at an oblique angle to the line of contact, is straight, whereas the reflection is curved if the surface is not quite flat. A microscope may also be used to study the liquid surface near the solid.

The plate method can also be used to find the contact angle between a solid and two liquids, the air in Fig. 1-19 being replaced by (say) benzene. The liquid-liquid interface must, like the air-liquid interface, be carefully cleaned before the measurements are made, the cleaning procedure again being described in Chapter 5. As pointed out above, the difference between advancing and receding angles may be very considerable, indicating that if a film of one liquid is formed on the solid, it can be only slowly displaced by the other liquid.

If only a small specimen of solid is available, the plate method may be modified by partly immersing the solid in a flat-sided cell which has been scrupulously cleaned, followed by blowing gently across the interface and sucking off the contamination (p. 220). The angle of the solid is again adjusted till the meniscus remains horizontal right up to the solid surface, as may be seen conveniently in a low-power binocular microscope.

The solids should be clean and smooth; freshly cut materials may, for example, be polished with increasingly fine emery cloths, followed by a leather hone, and finally polished with a fine grade of filter paper. The solid is then washed in redistilled ethanol and then in redistilled water, and is dried under vacuum in a desiccator[44]. Paraffin wax may be prepared smooth by dipping a clean microscope slide quickly in and out of molten paraffin-wax; and talc by simple cleavage[44].

Other Methods for obtaining θ

If the surface shows a difference between advancing and receding contact angles, this may be conveniently studied by the technique of Macdougall and Ockrent[52], by which a drop of the liquid is observed on an inclined plane of the solid. For single filaments and individual films Derjaguin's calculations[52] may be used, or the fibres may be chopped, and treated as a bed of powder (Chapter 8).

Exact temperature control is not important: the contact angle varies by only about 0.06° per °C.

The "Wetting Balance" Method for finding θ, W, and Negative S

The "wetting balance" of Guastalla[67] uses the Wilhelmy hanging plate to obtain the work of adhesion (or work of wetting) without measuring the contact angle directly. The work of adhesion is, as usual, relative to the solid surface containing an adsorbed monolayer of water or surface-active agent. A thin slide (perimeter 1 cm.) of the solid material is initially immersed in liquid, and is then raised slowly through the surface, till the external pull f becomes constant: after correction for buoyancy this force is $\gamma_{L/A}\cos\theta$ (p. 47 and Fig. 1-24b). The work done in withdrawing the slide a little further is now the work of de-wetting the slide and is given by the product of $\gamma_{L/A}\cos\theta$ and the change in the height of the slide. If the latter is hydrophobic $\left(\text{i.e. } \theta > \frac{\pi}{2}\right)$, cos θ will be negative; work will then be gained when the slide is withdrawn, and work will be done when the slide is immersed.

Since advancing and receding contact angles are equal only for smooth and perfectly clean surfaces, some hysteresis on raising or lowering the slide is often found with practical systems: the area of the hysteresis loop corresponds to work lost in removing adsorbed gas, for example, from any surface cracks or scratches, and sometimes also in leaving a thin film of water in these after withdrawal.

For a paraffin wax slide (1 cm. perimeter) in pure water, $f = -24$ dynes, so that $\gamma_{L/A}\cos\theta = -24$ dynes cm.$^{-1}$ This is in accord with measured values of θ of about $-110°$ for this system. If now we use a solution of a surface-active agent such as myristyltrimethyl ammonium bromide, adsorption of monolayers at the interfaces makes the wax surface wettable, with $f = +15$ dynes, and hence $\gamma_{L/A}\cos\theta = +15$ dynes cm.$^{-1}$ That f is positive implies that the adsorption of the monolayers will cause the waxed slide to be immersed spontaneously unless now restrained by an external pull of this amount. The wetting will be complete if f reaches the value of $\gamma_{L/A}$ for the solution (i.e. if $\theta \leqslant 0$): this is attained for paraffin wax at sufficiently high concentrations (above 10^{-3} N) of myristyltrimethyl ammonium bromide. It is interesting to observe that the same solution de-wets glass, the latter adsorbing a monolayer of the long-chain cations and emerging dry. By measuring $\gamma_{L/A}$ as well as $f(=\gamma_{L/A}\cos\theta)$, one can also obtain θ by this method. Experimentally one first obtains f for the slide under test, and then replaces the latter by a platinum or mica slide, roughened to make it completely wetted ($\theta = 0$) as in the Wilhelmy plate method (p. 47 this chapter, and Chapter 5). From f and $\gamma_{L/A}$ one may also calculate $W_{S/L}$ by eq. (1.32), and S by eq. (1.40).

The Sessile Drop Method for Negative S Values

This method[68] depends on measuring the height h of a large sessile drop, lying in equilibrium on a smooth surface as in Fig. 1-25. The relation of h and

S may be shown as follows. Let ρ_l and r be respectively the density of the liquid and the radius of the (circular) drop, the volume V being constant. Now imagine this drop to spread a little over the surface, with an increase in r by an amount dr, and a decrease in h of dh. The total change in energy in the equilibrium system must be zero, so

$$2\pi r dr(F^s_{S/A} - F^s_{S/L} - \gamma_{L/A}) - \tfrac{1}{2}\rho_l g V \, dh = 0 \qquad (1.46)$$

where the first term allows for the energy change in the spreading over the

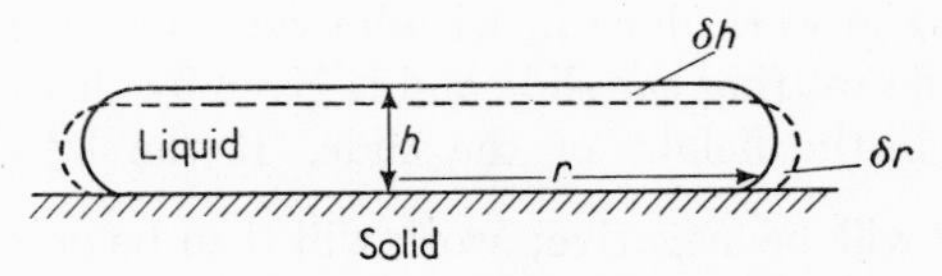

Fig. 1-25. Cross section of a large sessile drop.

additional area $2\pi r dr$, and the second allows for the loss in potential energy due to the lowering of the centre of gravity of the drop from $\frac{h}{2}$ to $\frac{h-dh}{2}$. The acceleration due to gravity is represented by g. Now by the geometry of the drop, V is given approximately (since the drop is large) by

$$V = \pi r^2 h$$

In differential form, since V is a constant, this becomes

$$2\pi r dr = -\frac{V}{h^2} dh$$

and this, substituted into eq. (1.46), gives:

$$F^s_{S/A} - F^s_{S/L} - \gamma_{L/A} = -\tfrac{1}{2}\rho_l g h^2$$

or, using eq. (1.39),

$$S = -\tfrac{1}{2}\rho_l g h^2 \qquad (1.47)$$

Experimentally one places the drop of liquid on to a horizontal, clean solid surface, and then waits for equilibrium to be established. The height of the drop is then measured either with a spherometer or with a long-focus cathetometer. To ensure that the approximations in above derivation are valid, a must be large compared with h: a drop of volume 1 $cm.^3$ usually ensures that this is so. Clean, smooth, solid surfaces are required if results are to be reproducible. Water on paraffin wax is found to have an S of -100 dynes $cm.^{-1}$ by this method[63], -99 to -100 by the "wetting balance" method, and -99 by direct observation of θ combined with eq. (1.40)[44,48].

Direct Measurement of Positive S Values

If S is positive, one may measure it by just restraining the drop of liquid (e.g. oil) from spreading, by the pressure of a monolayer at the air-water interface. For this purpose the solid surface is placed just below the surface of the water in a Langmuir trough: the oil drop can then just be prevented from spreading over the solid by a monolayer of stearic acid on the water surface, as in Fig. 1-26a. If monolayers other than those of stearic acid are used, the experimental results are unchanged: this effectively answers the obvious criticism of the method that the film material may contaminate the metal surface and so alter $F^s_{S/L}$.

The theory of the method[69] is based on a combination of the following equations. Firstly, the definition of S for the liquid L on the solid is

$$S = F^s_{S/A} - (\gamma_{L/A} + F^s_{S/L}) \tag{1.39}$$

Next, for the two-liquid system (Fig. 1-26a),

$$G + F^s_{S/L} + \gamma_{L/A} = \gamma \cos\theta_1 + F^s_{S/W} \tag{1.48}$$

where subscripts S and W refer to the solid and to water, L refers to the liquid (e.g. oil) under test, and A refers, as usual, to air, G represents the

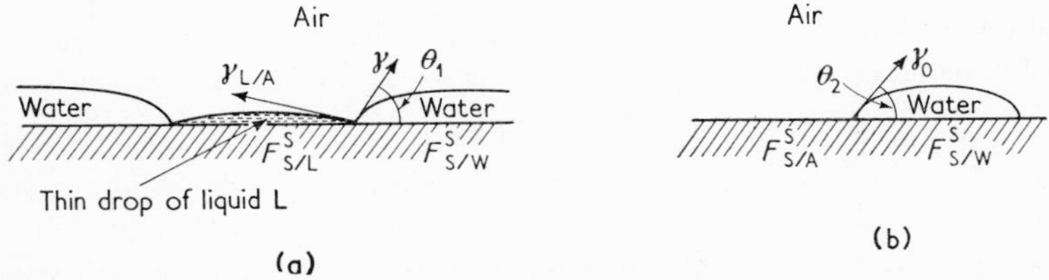

Fig. 1-26. Oil or other liquid (L) just spreading against thin layer of water, on the surface of which is a monolayer of film pressure Π. (b) Water with no monolayer rests on the same solid with angle θ_2.

hydrostatic spreading force of the water layer, and $\gamma(=\gamma_0 - \Pi)$ is the surface tension of the air-water surface when, as in Fig. 1-26a, this is covered with an insoluble monolayer of film pressure Π.

Finally, for a drop of pure water lying at equilibrium on the solid (Fig. 1-26b):

$$G + F^s_{S/A} = \gamma_0 \cos\theta_2 + F^s_{S/W} \tag{1.49}$$

with γ_0 the surface tension of pure water.

Substitution into eq. (1.39) of the values of $F^s_{S/L}$ from eq. (1.48) and of $F^s_{S/A}$ from eq. (1.49) gives

$$S = \Pi \cos\theta_1 + \gamma_0(\cos\theta_2 - \cos\theta_1) \tag{1.50}$$

Fortunately the experimental use of eq. (1.50) is simplified by the fact that $\cos\theta_1$ is found to be nearly constant (independent of Π), being 0.525 for different oils on chromium-plated steel. For clean water on steel, $\cos\theta_2 = 0.707$ and $\gamma_0 = 73$, so for this system eq. (1.50) reduces to

$$S = 0.525\Pi + 13.3$$

Experimentally it is convenient to use a water layer 0.05 cm. or 0.1 cm. thick, and an oil drop of 0.02 ml. dropped from a height of 3 cm. to the water surface on which a monolayer of cetyl or other long-chain alcohol, or stearic acid, is spread. Typical values of S for an oxidized lubricating oil on steel are +10 to +20 dynes $cm.^{-1}$, and for hydrocarbons on chromium-plated steel S is +17 to +25 dynes $cm.^{-1}$

Indirect Measurement of Positive S Values

Due to the adsorption of a monolayer of the vapour, organic liquids may not in practice spread over a metal surface though, could the surface be kept clean, the liquid would spread. With the measured θ one finds from eq. (1.40) an S value appropriate to the contaminated solid surface, and then adds the change in energy of the solid surface due to the adsorbed vapour, from separate experiments [54,56] (p. 41).

Adhesion Energies

These may be determined from the "wetting balance" or from determination of θ, using eq. (1.32), or by any other method of finding S, since

$$W_{S/L} = S + 2\gamma_{L/A} \tag{1.41}$$

It must again be emphasized that $W_{S/L}$ will refer to the solid surface on which may be adsorbed vapour of water or other liquid: though this effect is small for water on a surface of paraffin wax, metals may adsorb vapours quite strongly. The $W_{S/L}$ appropriate to the clean surface may be obtained as described on p. 41.

REFERENCES

1. Young, *Phil. Trans.* **95**, 65, 82 (1805).
2. Harkins, "The Physical Chemistry of Surface Films", Reinhold, New York (1952);
   Bikerman, "Surface Chemistry", Academic Press Inc., New York (1958)
3. Shuttleworth, *Proc. phys. Soc. Lond.* **A63**, 444 (1950).

3a. Gurney, *Nature, Lond.* **160**, 166 (1947).

4. de Boer, "The Dynamical Character of Adsorption" pp. 3–8, Clarendon Press, Oxford (1953).
5. Adam, "The Physics and Chemistry of Surfaces", Oxford University Press (1941).
6. Trapnell, "Chemisorption" pp. 90–96, Butterworths, London (1955).
7. Glasstone, Laidler, and Eyring, "The Theory of Rate Processes" pp. 519–21, McGraw-Hill, New York (1941).
8. Orr, W. J. C. and Butler, J. A. V., *J. chem. Soc.* 1273 (1935).
9. Bond, W. N. and Puls, *Phil. Mag.* **24**, 864 (1937).
10. Thomson (Lord Kelvin), *Phil. Mag.* **42**, 448 (1871).
11. La Mer and Gruen, *Trans. Faraday Soc.* **48**, 410 (1952).
12. Orr, C., Hurd, and Corbett, *J. Colloid Sci.* **13**, 472 (1958).

13. de Vries, *Proc. 2nd Internat. Congr. Surface Activity* **1**, 256, Butterworths, London (1957).
14. Tolman, *J. chem. Phys.* **17**, 333 (1949).
15. Kirkwood and Buff, *J. chem. Phys.* **17**, 338 (1949).
16. Benson and Shuttleworth, *J. chem. Phys.* **19**, 130 (1951).
16a. Davies, unpublished calculations.
17. Frenkel, "Kinetic Theory of Liquids", Clarendon Press, Oxford (1947).
18. Langmuir, *Chem. Rev.* **6**, 451 (1929); **13**, 147 (1933).
19. Kassel and Muskat, *Phys. Rev.* **40**, 627 (1932).
Fowler, *Physica's Grav.* **5**, 39 (1938).
Eisenstein and Gingrich, *Phys. Rev.* **58**, 307 (1940).
Prigogine and Saraga, *J. Chim. phys.* **49**, 399 (1952).
Harasima, *Advanc. chem. Phys.* **1**, 203 (1958).
20. Guggenheim, *Trans. Faraday Soc.* **41**, 150 (1945); "Mixtures" Chapter 9, Oxford University Press, London (1952).
Prigogine and Defay, *J. Chim. phys.* **46**, 367 (1949).
Kirkwood and Buff, *J. chem. Phys.* **17**, 338 (1949).
Hill, *J. chem. Phys.* **19**, 261 (1951); **20**, 141 (1952); *J. phys. Chem.* **56**, 526 (1952).
Saraga and Prigogine, "Changements de phases", Réunion Soc. Chim. Phys., Paris (1952).
Defay, *J. chim. Phys.* **51**, 299 (1954);
Englert-Chwoles, and Prigogine, *J. chim. Phys.* **55**, 16 (1958).
21. Samoilovich, *Z. exp. theor. Phys.* **16**, 135 (1946); *J. phys. Chem., Moscou,* 1127 (1949).
Glauberman, *J. phys. Chem., Moscou* **23**, 115 (1949).
21a. Frank and Evans, M. W., *J. chem. Phys.* **13**, 507 (1945).
Claussen and Polglase, *J. Amer. chem. Soc.* **74**, 4817 (1952).
Franks and Ives, *J. chem. Soc.* 741 (1960).
Bernal, *Trans. Instn. chem. Engrs., Lond.* (1960).
22. Dupré, "Théorie Méchanique de la Chaleur" p. 369 (1869).
23. Rayleigh, *Proc. roy. Soc.* **47**, 364 (1890); *Phil. Mag.* **48**, 321 (1899).
24. Davies, Boothroyd, and Palmer, Research Project in Department of Chemical Engineering, Cambridge (1956).
25. von Guttenberg, *Z. Phys.* **118**, 22 (1941).
26. Burgers, Greup, and Korvezee, *Rec. Trav. chim. Pays-Bas* **69**, 921 (1950).
27. Banks, *Proc. 2nd Internat. Congr. Surface Activity* **1**, 16, Butterworths, London (1957).
28. Schulman, and Teorell, *Trans. Faraday Soc.* **34**, 1337 (1938).
29. Cary and Rideal, *Proc. roy. Soc.* **A109**, 301 (1925);
Roylance and Jones T. G., *J. appl. Chem.* **9**, 621 (1959);
Mansfield, *Aust. J. Chem.* **12**, 382 (1959).
30. Hughes and Rideal, *Proc. roy. Soc.* **A137**, 62 (1932).
31. Gibbs, "Collected Works" **1**, 235, 258, Longmans, New York (1928).
32. Bartell, Case, and Brown, H., *J. Amer. chem. Soc.* **55**, 2769 (1933).
33. Antonoff, *J. Chim. phys.* **5**, 364, 372 (1907).
34. Antonoff, in "Colloid Chemistry" **7**, 38, (ed. J. Alexander) Reinhold, New York (1950).
35. Goard and Rideal, *J. chem. Soc.* **127**, 780 (1925).
36. Carter and Jones, D. C., *Trans. Faraday Soc.* **30**, 1027 (1934).
37. Quincke, *Phil. Mag.* **41**, 454 (1871);

Hardy, *Proc. roy. soc.* **A86**, 610 (1911);
Bancroft and Tucker, *J. phys. Chem.* **31**, 1681 (1927).
38. Langmuir, *J. chem. Phys.* **1**, 756 (1933);
Sawyer and Fowkes, *J. phys. Chem.* **60**, 1235 (1956).
39. Rideal "Surface Chemistry", Cambridge University Press (1930).
40. Feachem and Rideal, *Trans. Faraday Soc.* **29**, 409 (1933).
41. Blodgett, *J. opt. Soc. Amer.* **24**, 313 (1934);
Blodgett, *J. Amer. chem. Soc.* **57**, 1007 (1935);
Langmuir and Schaefer V. J., *Chem. Rev.* **24**, 181 (1939).
42. Adam, *Proc. roy. Soc.* **B122**, 34 (1937);
Goldacre, *J. Anim. Ecol.* **18**, 36 (1949).
43. Adam, *Endeavour* **17**, 37 (1958); *Nature, Lond.* **180**, 809 (1957).
44. Fowkes and Harkins, *J. Amer. chem. Soc.* **62**, 3377 (1940);
Derjaguin, *Proc. 2nd Internat. Congr. Surface Activity*, **3**, 446, Butterworths, London (1957).
45. Bailey, *Proc. 2nd Internat Congr. Surface Activity* **3**, 189, Butterworths, London (1957).
46. Davies and Wiggill, *Proc. roy. Soc.* **A255**, 277 (1960).
47. Ray and Bartell, *J. Colloid Sci.* **8**, 214 (1953).
48. Wenzel, *Industr. Engng. Chem.* (*Anal.*) **28**, 988 (1936).
49. Shuttleworth and Bailey, *Disc. Faraday Soc.* **3**, 16 (1948).
50. Allan and Roberts, *J. Polym. Sci.* **39**, 1 (1959).
51. Doss and Rao, *Proc. Indian Acad. Sci.* **7A**, 113 (1938);
Cassie and Baxter, *Trans. Faraday Soc.* **40**, 546 (1944); *Nature, Lond.* **155**, 21 (1945).
52. Macdougall and Ockrent, *Proc. roy. Soc.* **A180**, 151 (1942);
Derjaguin, *C. R. Acad. Sci. U.R.S.S.* **51**, 519 (1946).
53. Zisman et al., *J. Colloid Sci.* **1**, 513 (1946); **2**, 563 (1947); **7**, 166, 428, 465 (1952); **8**, 194 (1953); *J. phys. Chem.* **57**, 622 (1953); **58**, 236 (1954); **59**, 335 (1955); U.S. Navy Rep. NRL 4569 (1955).
Fowkes, *J. phys. Chem.* **57**, 98 (1953).
54. Harkins and Loeser, *J. chem. Phys.* **18**, 556 (1950).
55. Washburn and Anderson, E. A., *J. phys. Chem.* **50**, 401 (1946).
56. Harkins and Livingston, H. K., *J. chem. Phys.* **10**, 342 (1942);
Adam and Livingston, H. K., *Nature, Lond.* **182**, 128 (1958).
57. Ténèbre, *Mémor. Serv. chim. Etat.* **40**, 77 (1955).
58. du Noüy, *J. gen. Physiol.* **1**, 521 (1919); **5**, 429 (1923).
59. Reilly and Rae, "Physicochemical Methods" Vol. I, Methuen, London (1943).
60. Freud, B. B. and Freud H. Z., *J. Amer. chem. Soc.* **52**, 1772 (1930);
Harkins and Jordan, H. F., *J. Amer. chem. Soc.* **52**, 1751 (1930);
Macy, *J. chem. Educ.* **12**, 573 (1935);
Fox, H. W., and Chrisman, *J. phys. Chem.* **56**, 284 (1952);
Zuidema, and Waters, *Industr. Engng. Chem.* (*Anal.*), **13**, 312 (1941).
61. Harkins and Brown, F. E., *J. Amer. chem. Soc.* **41**, 499 (1919); *Int. crit. Tab.* **4**, 435 (1928);
Ward and Tordai, *J. Sci. Instrum.* **21**, 143 (1944).
62. Wilhelmy, *Ann. Phys.* **119**, 177 (1863);
Abribat and Dognon, *C. R. Acad. Sci., Paris* 208 1881 (1939);
Cheesman, *Ark. Kemi. Min. Geol.* **22B**, 1 (1946).
63. Padday, *Proc. 2nd Internat. Congr. Surface Activity*, **3**, 187, Butterworths, London (1957);

Allan, *J. Colloid Sci.* **13**, 273 (1958).
64. Andreas, Hauser, and Tucker, *J. phys. Chem.* **42**, 1001 (1938); Fordham, *Proc. roy. Soc.* **A194**, 1 (1948).
65. Partington, "An Advanced Treatise on Physical Chemistry" **2**, 174–192, Longmans, London (1951).
66. Fox, H. W. and Zisman, *Rev. sci. Instrum.* **19**, 274 (1948).
67. Guastalla, *Proc. 2nd Internat. Congr. Surface Activity*, **3**, 143, Butterworths, London (1957).
68. Padday, *Proc. 2nd Internat. Congr. Surface Activity*, **3**, 136 London (1957).
69. Washburn and Anderson, E. A., *J. phys. Chem.* **50**, 401 (1946); Clayfield, Dear, Matthews, and Whittam, *Proc. 2nd Internat. Congr. Surface Activity*, **3**, 165, 199–200, Butterworths, London (1957).

Chapter 2

Electrostatic Phenomena

INTRODUCTION

The origin of interfacial potentials was for many years a matter of considerable controversy. Beutner[1,2] and Baur[3], in the early years of this century, initiated a series of experiments with polar oils, such as salicylaldehyde and o-toluidine, placed in a tube between two aqueous phases. The oil thus formed a rather thick membrane between the two aqueous phases, and it was hoped that the behaviour of such artificial membranes would be useful in elucidating the behaviour of the more complex membranes

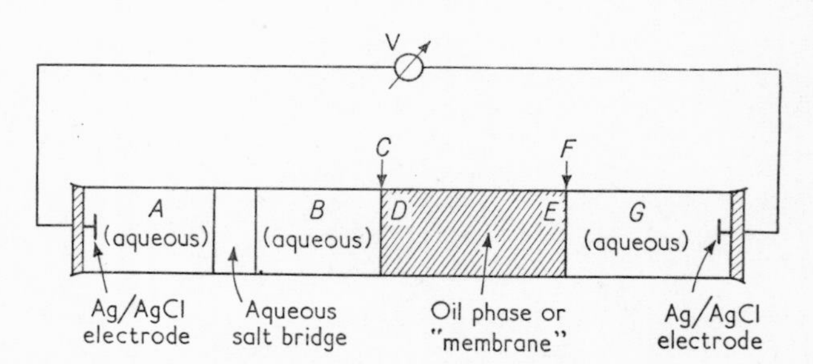

Fig. 2-1. Diagrammatic representation of cell used to measure potential differences across an oil "membrane". The organic salt is usually added to the aqueous phase B, and the only potential changed in consequence is that across the interface C: in general this potential change is called $\Delta\Phi$.

surrounding living cells. When organic salts (such as tetramethylammonium salts or picrates) are added to one of the aqueous phases, a difference of potential is set up between two electrodes in the aqueous phases, as illustrated in Fig. 2-1.

According to Beutner, the observed potential differences are caused by an unequal distribution of the various anions and cations across the interface, this leading to the phases on each side of the interface becoming electrically charged (Fig. 2-2). Lange has called such distribution potentials the ψ or outer potentials. According to Baur, however, the potentials arise from selective adsorption of a monolayer of the organic ions at the oil-water interface; such potentials (Fig. 2-3) Lange calls $\delta\chi$ potentials, though in England and America they are known as changes in contact potential, ΔV.

Here we shall define a further potential, Φ, the total potential drop across the interface. It is given at equilibrium by the relation $\Phi = \psi + V$, or, in Lange's notation, $\Phi = \psi + \chi$. The potential Φ is called the "inner" or Galvani potential of the oil phase relative to the water (Fig. 2-4).

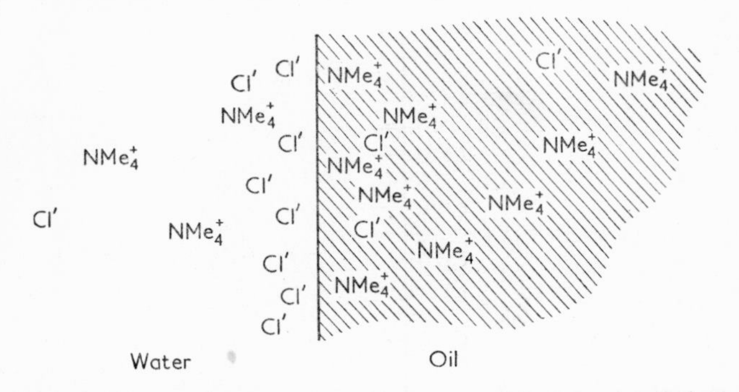

Fig. 2-2. Anions and cations may have different solubilities in the oil and water phases, setting up a distribution potential called ($\Delta\psi$) across the interface. The mean distances of the charges on each side of the interface are given by the Debye-Hückel function $1/\varkappa$. This usually has values of tens or hundreds of ångstrom units in water, but may reach 10^{-2}cm. in an oil.

There was heated argument for many years between Baur and Beutner about the origin of the measured oil-water potentials, but finally Baur accepted Beutner's explanation. The problem has now been studied in considerable detail[4,5], and it is accepted that if one or more types of ions can dissolve in both oil and water, no stable interfacial potential due to adsorption can be set up. In other words, $\Delta\Phi$, though temporarily non-zero when a monolayer is formed, decreases with time and finally becomes zero. The proof of this[4] depends on the fact that at equilibrium the diffusible ions must always distribute themselves between the phases on each side of the interface so that their electrochemical potentials $\bar{\mu}$ are restored to the initial values in each phase. In addition, the thickness of the monolayer is so small compared with that of each of the bulk phases, that the presence of the monolayer does not alter the chemical potentials, μ, of these phases. Also unaltered, therefore, is the potential Φ between the phases, since in the equation

$$\bar{\mu} = \mu + \Phi \boldsymbol{F}$$

the changes in both $\bar{\mu}$ and μ are zero ($\boldsymbol{F}$ being the Faraday unit).

This process of ionic redistribution corresponds to the system of Fig. 2.2 being already set up, when a monolayer as in Fig. 2-3 is subsequently spread at the interface. The ionic distribution of Fig. 2-2 is then automatically readjusted so that the total potential drop across the interface is unaltered by the presence of the monolayer. The potentials about which Beutner and Baur argued so long are, therefore, not monolayer adsorption potentials, since, in the highly polar oils Beutner and Baur used, fluctuations of V are unstable, and must decay to zero: this process is usually extremely rapid, as

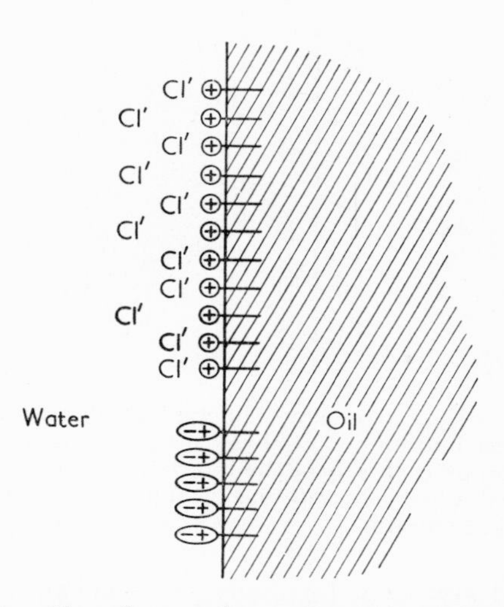

Fig. 2-3. Selective adsorption of a monolayer of organic ions at the oil-water interface changes the potential across the interface by ΔV (or $\Delta\chi$). For generality, the adsorbable organic ions are represented by a charged "head" and a hydrocarbon "tail". In the lower part of the figure the potential ΔV is due only to a monolayer of permanent dipoles, as would occur in the adsorption of a long-chain alcohol.

is shown below. We shall see also that any time-variations observed in these systems over intervals of several hours cannot be due to changes in V.

The origin of the various potentials is best illustrated by a careful choice of systems; with different non-aqueous phases we can now measure *changes* in potential ascribable wholly to ψ or to V at an oil-water interface, or to changes in V at an air-water interface. The changes in ψ may be brought about by adding organic salts (including surface-active agents) to either phase, while adsorption potentials ΔV are measured between water and air, or between water and a non-polar oil, when any surface-active material is spread or allowed to adsorb at the interface. We shall adhere to custom in referring to the changes in V as "interfacial potentials" or "surface potentials" according to whether paraffinic oil or air constitutes the non-aqueous phase: changes in the interfacial distribution potential ψ will be called "distribution potentials".

DISTRIBUTION POTENTIALS

According to Nernst's classical theory[6], distribution potentials arise from the different solubilities of the positive and negative ions in the two phases. These ionic distributions are defined by the following relations:

$$RT\ln B_+ = {}_w\mu^o_+ - {}_o\mu^o_+$$
$$RT\ln B_- = {}_w\mu^o_- - {}_o\mu^o_-$$

where B_+ and B_- refer to the cation and anion distribution coefficients, μ^o

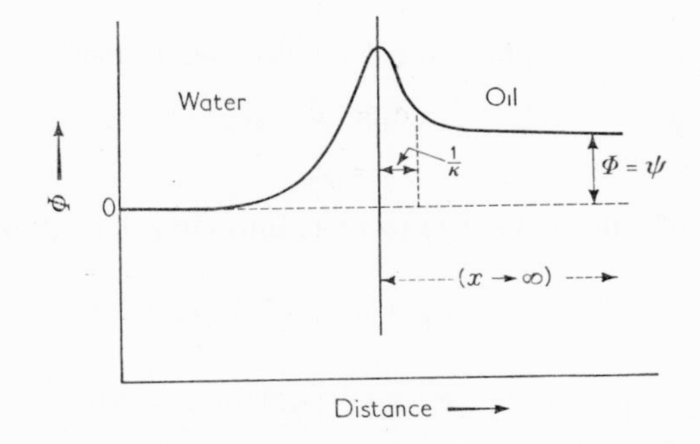

Fig. 2-4. Variation of Φ (relative to the interior of the water phase) near an oil-water interface. The oil is assumed polar. The distance from the interface is represented by x, and measurements are made where $x \rightarrow \infty$ compared with the ionic double layer thickness $1/\kappa$.

is the standard chemical potential, and subscripts o and w refer to the oil and water phases respectively.

The bulk aqueous phase is generally assigned zero potential (since absolute potentials across the interface cannot be measured); the oil phase has a potential Φ, determined in general by both the ψ and V potentials.

As an example, suppose that water is in contact with a polar oil such as amyl alcohol, nitrobenzene, or salicylaldehyde. To this system tetramethylammonium chloride is added. The organic cations tend to dissolve preferentially in the oil, while the chloride ions tend to remain in the water. This must cause the oil phase to be positive with respect to the water, as illustrated in Fig. 2-2. In such a system where a polar oil is used, we know that ΔV must decay (quite rapidly) to zero, so that when we measure $\Delta\Phi$ we shall also measure $\Delta\psi$. Electrokinetic potentials measured on emulsion drops confirm this view. Most of the potential change across the interface occurs within the phase of lower dielectric constant[7].

We now relate Φ to B_+ and B_- by means of their electro-chemical potentials $\bar{\mu}$, defined as above:

$${}_w\bar{\mu}_+ = {}_w\mu^o_+ + RT \ln {}_wc_+ \tag{2.1}$$

$${}_o\bar{\mu}_+ = {}_o\mu^o_+ + RT \ln {}_oc_+ + \Phi \boldsymbol{F} \tag{2.2}$$

$${}_w\bar{\mu}_- = {}_w\mu^o_- + RT \ln {}_wc_- \tag{2.3}$$

$${}_o\bar{\mu}_- = {}_o\mu^o_- + RT \ln {}_oc_- - \Phi \boldsymbol{F} \tag{2.4}$$

At equilibrium,

$${}_w\bar{\mu}_+ = {}_o\bar{\mu}_+ \tag{2.5}$$

and

$${}_w\bar{\mu}_- = {}_o\bar{\mu}_- \tag{2.6}$$

Further, the bulk of each phase must be electrically neutral, and therefore:

$${}_wc_+ = {}_wc_- \tag{2.7}$$

$${}_oc_+ = {}_oc_- \tag{2.8}$$

By substitution of equations (2.1) to (2.4) into equations (2.5) and (2.6), we find that

$${}_w\mu^o_+ + RT \ln {}_wc_+ = {}_o\mu^o_+ + RT \ln {}_oc_+ + \Phi \boldsymbol{F}$$

and

$${}_w\mu^o_- + RT \ln {}_wc_- = {}_o\mu^o_- + RT \ln {}_oc_- - \Phi \boldsymbol{F}.$$

If these relations are subtracted,

$$({}_w\mu^o_+ - {}_o\mu^o_+) - ({}_w\mu^o_- - {}_o\mu^o_-) = 2\Phi \boldsymbol{F}$$

by virtue also of eqs. (2.7) and (2.8). Further simplification is possible from the above definitions of B_+ and B_-, to give

$$RT \ln B_+ - RT \ln B_- = 2\Phi \boldsymbol{F}$$

or

$$\Phi = \left(\frac{RT}{2\boldsymbol{F}}\right) \ln \left(\frac{B_+}{B_-}\right) \tag{2.9}$$

This shows that, since the tetramethylammonium ion is more soluble in the oil phase than is chloride ion ($B_+ > B_-$), Φ is positive, as predicted above.

Further, it is easy to show that, if an overall distribution coefficient B is defined as (concn. of salt in oil)/(concn. of salt in water),

$$B = (B_+ . B_-)^{1/2} \tag{2.10}$$

and hence, from eqs. (2.9) and (2.10),

$$\Phi = (RT/2\boldsymbol{F}) \ln(B^2/B_-^2).$$

This equation cannot be used directly, as an absolute potential cannot be measured. The difference in Φ for the system nitrobenzene-water containing respectively added NaI and KI, however, is measurable, and according to the above theory should be independent of salt concentration and given by

$$\Delta\Psi = \Delta\Phi = (RT/\boldsymbol{F})\ln(B_{NaI}/B_{KI}) \tag{2.11}$$

once now B_- is the same for NaI and KI. Indeed, $\Delta\Phi$ is itself independent if the common ion, since from eq. (2.9)

$$\Delta\Psi = \Delta\Phi = (RT/2\boldsymbol{F})\ln(B_{Na^+}/B_{K^+}) \tag{2.12}$$

Experimental evidence to test eq. (2.11) may be obtained using the precision cell shown in Fig. 2-5. Karpfen and Randles[8] made direct current measurements of the changes in potential $\Delta\Phi$ across this cell when different

Hg, Hg_2Cl_2	KCl 3*M* water	M^+X^- water	M^+X^- nitrobenzene	$(Et)_4NPi$ di-isopropyl ketone	$(Et)_4NPi$	HgPi, Hg
	Φ_1	Φ	Φ_2	Φ_3		

Fig. 2-5. Cell for measuring $\Delta\Phi$ at interface of water and a polar oil such as nitrobenzene. Salt bridges are chosen so that when the salt MX is changed, the measured potential change is $\Delta\Phi$, i.e. Φ_1, Φ_2, and Φ_3 remain in constant. In such an experiment $\Delta\Phi$ is measured across thicknesses of solvent great compared with the thickness of the interfacial electrical double layers. Pi is picrate.

salts MX were used. These potentials may be compared (Table 2-I) with those calculated by substituting distribution coefficients[9] into eq. (2.11): by virtue of eq. (2.12) the common ion is of no account in determining $\Delta\Phi$. The agreement in the table is strong support for the validity of eqs. (2.11) and (2.12) if the oil is polar.

In the cell of Fig. 2-1, measurements[10] with complex organic salts added at the point B give potentials which are not even approximately constant, but vary markedly with concentration of additive (Table 2-II), indicating severe non-ideality. The oil is quinoline and the added salt is 2-phenyl-

TABLE 2-I

System (see Fig. 2-5)	$\Delta\Phi$ calculated[8] from distribution experiments[9] using eq. (2.11)	$\Delta\Phi$ observed
K^+ replaced by $N(Et)_4^+$	+124 mV	+126 mV
K^+ replaced by Na^+	+ 50 mV	+ 53 mV
Cl′ replaced by I′	— 95 mV	—102 mV

Changes in potential with nitrobenzene as the oil, using the precision cell shown in Fig. 2-5.

quinoline-4-carboxylic acid potassium salt (P.Q.K.) or quinine hydrochloride. From the potential changes (in the most dilute solutions, which are the most nearly ideal) the partition coefficients can be calculated for different pairs of ions, using equations of the type of (2.11). Comparison with the appropriate analytical figures at high dilutions is shown in Table 2.III; the agreement confirms that the potential changes are $\Delta\psi$ values, and that the theory above is valid at low concentrations, when the system tends to behave ideally.

TABLE 2-II

Moles of organic salt distributed between 100 cc. each of water and quinoline	$\Delta\Phi$ in mV[10]	
	P.Q.K.	Quinine hydrochloride
0.02	—77	+111
0.001	—50	+ 92
0.0001	—37	+ 69

TABLE 2-III

Salt	B calc. from $\Delta\Phi$	B determined analytically
KCl	0.024 ± 0.006	0.020 ± 0.002
Quinine salt of 2-phenyl-quinoline-4-carboxylic acid	1.7 ± 0.8	2.0 ± 1.3

Comparisons of partition coefficients B, in dilute solutions, calculated from potentials and measured analytically, taken from the results of Kahlweit and Strehlow[10].

DIFFUSION POTENTIALS

If a layer of polar oil separates two aqueous phases of different salt concentrations, a transient potential $\Delta\Phi$ will be measured across the electrodes in contact with the aqueous phases. The origin of the potential lies in the different transport numbers of anions and cations as they diffuse through the oil. The two opposing phase boundary potentials are equal by eq. (2.11) (if the system is ideal), and are established quickly since the oil is polar (see below). The measured diffusion potentials across the oil layer are usually small, perhaps 15 mV. Charged monolayers at the interfaces bounding a very thin film (perhaps 1 micron thick) of a polar oil may so influence the relative ionic mobilities as to change the diffusion potential[4,11].

MISCELLANEOUS EXPERIMENTAL RESULTS

Many other workers have carried out experiments with cells of similar type to that shown in Fig. 2-1. Ehrensvärd and Cheesman[12], for example, constructed an apparatus with a very long diffusion path of salicylaldehyde, so that additives injected at the point D can affect the interface C but not the interface F. They found that, when the compartments B and G contain saturated KCl, alcohols added to the oil at D do not affect Φ, but if 0.0001 N-KCl is used in B and G, the potential changes by amounts of 50 mV or more, the KCl solution in B becoming more negative. Although Ehrensvärd and Cheesman state that the potential they measure is a ΔV rather than a $\Delta\psi$, we believe that in such a polar oil as salicylaldehyde ($D=16$) the ΔV or adsorption potential must decrease to zero extremely rapidly, in perhaps 0.001 sec. However, the added alcohols would alter the ionic distribution coefficients at the interface C, thus setting up a diffusion potential in the dilute salt cell, also changing Φ according to eq. (2.12). That a change in Φ is involved is confirmed by the findings that cyclohexane, added at D, has a similar effect to the alcohols, changing the potential by up to 100 mV; the change in potential is, however, only small for added n-heptane. The general lack of effect when saturated KCl is used still awaits explanation. Osterhout[13] used nitrobenzene as the oil phase, with different inorganic salts in aqueous solution on either side (compartments B and G, Fig. 2-1). The resultant potential arises both from the non-ideality of the system and from a diffusion potential. McMullen[14] finds that cetyltrimethylammonium bromide, placed at B (Fig. 2-1) in contact with octanol, gives an initially high potential $\Delta\Phi$ (of about 300 mV), which falls again quite rapidly, the rate depending on the thickness of oil. If the system is ideal, the potentials at the interfaces C and F should ultimately balance.

More recently, Mme. Dupeyrat[15], using nitrobenzene as the oil, finds that $\Delta\Phi$ is constant at the high figure of 300 to 400 mV when myristyltrimethylammonium bromide is added to compartment B (Fig. 2-1). This is certainly a change in ψ, as such a polar oil could not maintain an altered V potential for more than about a second. Further, since the free energy of transferring each $-CH_2-$ from water to nitrobenzene is calculated from partition data[9] to be 585 cal.mole^{-1}, we have the relation:

$$RT \ln B_{+} = 585m + \text{constant} \tag{2.13}$$

where m is the number of $-CH_2-$ groups involved; this relation may be combined with the potential result for $N(Et)_4^+$ in Table 2-I, so that a relation of the type of eq. (2.12) can be applied to the myristyl-trimethyl- and tetraethyl-ammonium ions. The result is that the potentials in Mme. Dupeyrat's system are calculated to be about 265 mV greater than for

$N(Et)_4^+ Br'$, i.e. about 365 mV relative to KCl. The agreement with the experimental figure of about 300 to 400 mV confirms that the measured potential is $\Delta\psi$.

INTERFACIAL AND SURFACE POTENTIALS

If a non-aqueous phase is chosen in which practically no salt can dissolve (e.g. a paraffinic oil such as decane, or air), an ionic double layer cannot build up in the thickness of oil (say 1 mm.) available. Hence eq. (2.8) will no longer apply, although, since the system is at equilibrium, equations (2.1) to (2.7) still hold.

Dividing the potential on the oil side of the interface into V (due to either a dipole array or to a charged film, either adsorbed or spread at the interface) and ψ (see Fig. 2-6), we shall show that, if the oil solubilities of the interfacial film and of the salt present are very small, only ΔV contributes to the measured change in $\Delta\Phi$ when a film is spread at the interface[5].

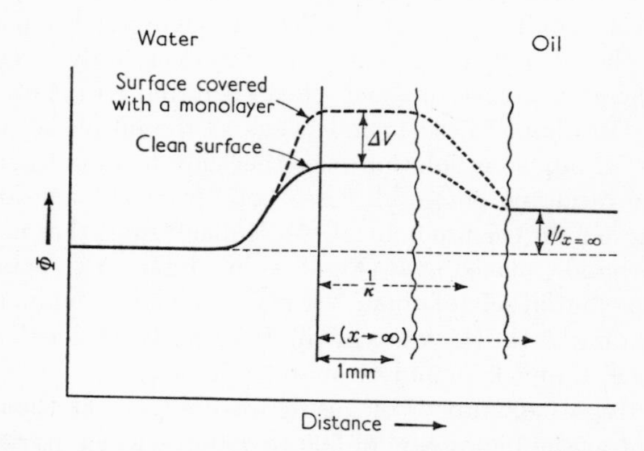

Fig. 2-6. Variation of Φ (relative to the interior of the water phase) at the interface with a paraffinic oil. Measurements are made with $x \sim 1$ mm., i.e. less than $1/\kappa$.

If, then, eq. (2.8) is no longer true, eq. (2.9) must be generalized to:

$$\Phi = (RT/2\boldsymbol{F})\ln(B_+/B_-) - (RT/2\boldsymbol{F})\ln({}_oc_+/{}_oc_-) \tag{2.14}$$

in which ${}_oc_+$ and ${}_oc_-$ refer to ionic concentrations in the oil at the point where Φ is measured. To express these concentrations in terms of measurable quantities we apply the equations of Boltzmann in the form applicable near the interface, where there is a contact potential V:

$$ {}_oc_+ = ({}_oc_+)_{x=\infty} \cdot \exp(-\varepsilon(V - \psi_{x=\infty})/\boldsymbol{k}T) \tag{2.15}$$

and

$$ {}_oc_- = ({}_oc_-) . \exp(+\varepsilon(V - \psi_{x=\infty})/\boldsymbol{k}T) \tag{2.16}$$

and further, in the hypothetical region of the non-aqueous phase where the electrical double layer is complete, at $x = \infty$,

$$({}_0c_+)_{x=\infty} = ({}_0c_-)_{x=\infty} \tag{2.17}$$

With these relations eq. (2.14) becomes:

$$\Phi = (RT/2\boldsymbol{F})\ln(B_+/B_-) - (RT/2\boldsymbol{F})\ln\exp(-2\varepsilon(V - \psi_{x=\infty})/\boldsymbol{k}T).$$

The double layer is complete at $x = \infty$, so $\Phi_{x=\infty} = \psi_{x=\infty}$ (Fig. 2-6) and hence, according to eq. (2.9),

$$\psi_{x=\infty} = (RT/2\boldsymbol{F})\ln(B_+/B_-).$$

Therefore

$$\Phi = \psi_{x=\infty} + (2RT\varepsilon/2\boldsymbol{k}T\boldsymbol{F})(V - \psi_{x=\infty})$$

or

$$\Phi = \psi_{x=\infty} + (V - \psi_{x=\infty})$$

or

$$\Phi = V.$$

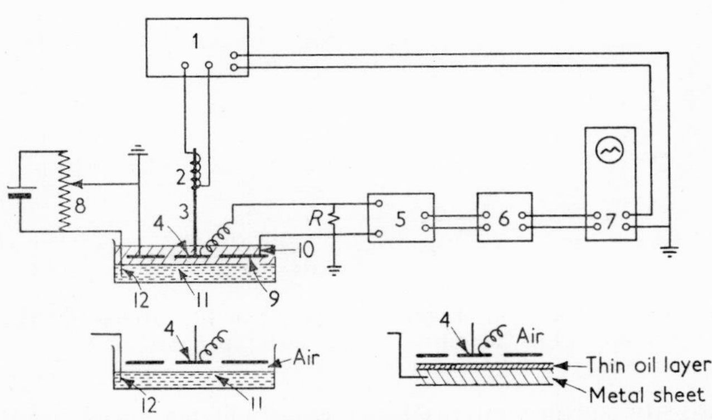

Fig. 2-7. Measurement of interfacial potentials. 1, oscillator; 2, vibrator coil; 3, Bakelite driving rod; 4, vibrating gold plate; 5, amplifier; 6, wave-filter; 7, oscilloscope; 8, potentiometer; 9, guard discs; 10, non-aqueous phase; 11, aqueous phase; 12, reversible electrode in aqueous phase. The resistance R may conveniently be 50 megohms[16,17]. Below are modifications applicable to the air-water surface, and the oil-metal interface.

Hence for non-polar oils or for air, in contact with water, Φ is independent of both B_+ and B_- (in contrast to eq. (2.11)), and

$$\Delta\Phi = \Delta V (= \delta\chi, \text{ Lange nomenclature}) \tag{2.18}$$

Whereas eq. (2.11) could be simply tested using direct current, this is no longer possible if the oil is paraffinic; non-polar oils do not conduct electricity. A capacity method[16] is therefore necessary for testing or using eq. (2.18); the alternating current is generated by the small-amplitude vibration of a gold plate immersed in the oil, being amplified and balanced to zero with a potentiometer connected to the electrode in the water (Fig. 2-7): the potentials of the gold plate and aqueous surface are then equal. To avoid disturbing the oil close to the monolayer at the interface, it is necessary to keep the amplitude of vibration of the gold plate to about 0.01 cm., the frequency to a few hundred cycles per second, and not to bring the gold plate closer than 0.5 mm. to the interface. All these factors decrease the sensitivity of the apparatus, and in consequence very high amplification of the signal is required. This has to be effected with a tuned circuit, feeding through a wave-filter to a cathode ray oscilloscope. The resultant accuracy is 1 mV. The amplifier circuit recommended is that of Kinloch and McMullen[17]. The method may also be modified to measure ΔV at the oil-metal or gas-metal interfaces[18].

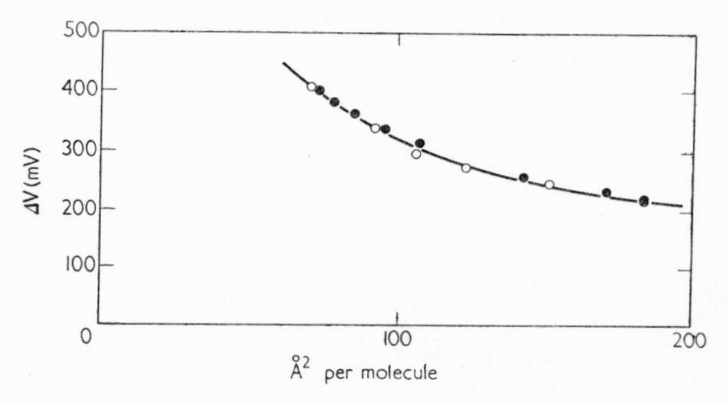

Fig. 2-8. ΔV for long-chain quaternary salts spread on 10^{-2} M-NaCl at the air-water surface (filled circles), and at the oil-water interface (open circles)[16].

At the air-water surface potentials can readily be measured with this same apparatus, or with a less sensitive version. Yamins and Zisman[18] were the first to measure surface potentials in this way. However, since stray ions are particularly unstable near a water surface because of the image forces, one may alternatively render the air and CO_2 above the water slightly conducting with α-particles from a little mesothorium bromide or polonium in the vicinity: the gas-ions so formed reduce the resistance of the narrow air gap to about 10^{14} ohms, and potentials can be measured[19] to better than 1 mV with a Lindemann electrometer or by using a special electrometer valve and D.C. amplifier[20]. As before, these are usually balanced with a potentiometer, this being done before the film is adsorbed or spread at the surface, and afterwards. The difference in potentiometer reading is then ΔV. This method has proved very valuable in following the course of chemical reactions and penetration phenomena in surface films, as explained in Chapter 6.

From eq. (2.18) we see that, if the non-aqueous phase is paraffinic or consists of air, the measured change in potential Φ should depend on V but not on B, i.e. only on the properties of the monolayer and not on those of the non-aqueous phase. This is indeed true for spread monolayers of $C_{18}H_{37}N(CH_3)_3^+$ ions, as shown in Fig. 2-8. Further, such potentials are completely stable with time (Fig. 2-9). In the measurement of the potentials of monolayers at the oil-water interface[16,21,22,23], a suitable oil is the high-boiling fraction of petrol-ether, made up of decane and near homologues.

If, however, one uses a fairly polar oil, there should, according to eq. (2.11), be no change in potential between the system with a clean interface and one with a monolayer spread at the interface; only if the distribution coefficients are changed[4] will there be a permanent potential $\Delta\Phi$. This is in full accord with experiment.

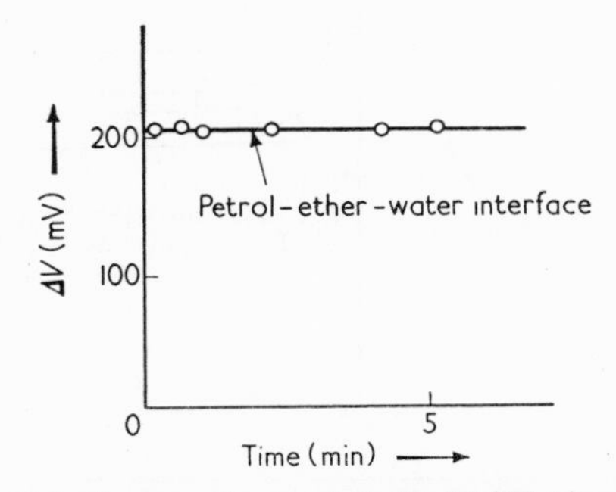

Fig. 2-9. Surface potential of a monolayer of $C_{18}H_{37}N(CH_3)_3^+$ spread to 92 $Å^2$ per chain between petrol-ether and water. The latter contained 1 N-NaCl. The potential does not change with time.

Between these two extremes lie the potential differences measured against very slightly polar oils: a monolayer spread at the oil-water interface now gives a measured potential difference $\Delta\Phi$ which tends to decrease with time towards zero. The rate and the extent of the decay of $\Delta\Phi$ are of some interest, as with very slightly polar oils one passes (Fig. 2-10) from the region of applicability of eq. (2.18) (where $\Delta\Phi = \Delta V$ and the potential is stable indefinitely) to the region of eq. (2.11) (where ions from the water will redistribute themselves after the film has been spread, and set up a new and compensating double layer in the oil, making ΔV zero). Further, for the latter system $\Delta\Phi = 0$, since a monolayer will not appreciably affect the distribution coefficients (i.e. $\Delta\psi = 0$)[4]. Thus the potential difference $\Delta\Phi$ should decrease with time and, if a complete compensating layer can be set up in the thickness of oil available, will eventually become zero. This is born out by experiment (Fig. 2-11).

The measurement of $\Delta\Phi$ in these systems is of some interest, in that with a very slightly polar oil such as benzene, $\Delta\Phi$ may be determined with either the vibrating plate, immersed as usual in the liquid[5,22], or with a radioactive source attached to a stationary electrode suspended in the air above a thin oil layer[5,22,24], 1 to 2 mm. in depth. For more polar oils still, an ordinary D.C. circuit may be used. The vibrating plate apparatus is less sensitive than with petrol-ether or air, however, because of the leak resistance through the benzene across the condenser formed by the gold plate and the aqueous surface[22]. If this resistance becomes too low, not only is sensitivity lost but spurious values of ΔV may be obtained. With benzene as the oil phase,

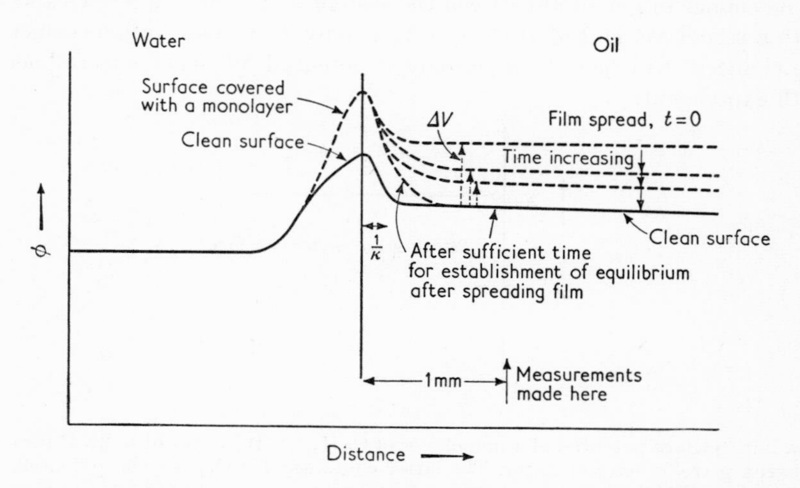

Fig. 2-10. Potential redistribution on oil side of interface after spreading a film if the oil is slightly polar. The (negative charged) ionic atmosphere slowly builds up in the oil till ΔV has decreased to zero, provided that the thickness of oil used is sufficient to dissolve the necessary ions.

however, the radioactive source method and the vibrating plate method give concordant results (Fig. 2-11).

If now a little highly polar oil such as bromobenzene is added to the benzene[5,22,24], results can no longer be obtained with the vibrating plate apparatus, and the potential difference of the monolayer (measured by the radioactive source method) soon vanishes (Fig. 2-11). With more polar oils, potentials can be measured directly with a normal electrode and electronic amplifier. If the times for the potential to decay by one-half are found, a rough correlation with the rates of diffusion of the counter-ions from the water to the oil (to build up the new double layer) might be expected. That this is so is seen from Table 2-IV, in which the calculated times are functions of the polarity of the oil, i.e. of the equilibrium concentration of chloride ions immediately on the oil side of the interface. From here the ions must diffuse through the oil to a mean distance of $1/\varkappa$ (the Debye-Hückel function) away

TABLE 2-IV[5,11,22]

Non-aqueous phase	Specific conductance ($ohm.^{-1} cm.^{-1}$)	Calculated concentration of Cl^- in non-aqueous phase	Depth of ionic atmosphere, i.e. $1/\varkappa$	Time calculated for Cl^- to reach equilibrium positions after film is spread	Observed time of decay of ΔV to 1/e of original value
Air	0	10^{-65}N	10^{24}cm.	10^{54} sec.	∞
Petrol-ether (decane)	10^{-15}	10^{-33}N	10^{8}cm.	10^{21} sec.	∞
Benzene	10^{-10}	10^{-11}N	5×10^{-3}cm.	$\sim$1 sec.	60 sec.
Isopropyl ether	1.4×10^{-10}	10^{-11}N	5×10^{-3}cm.	$\sim$1 sec.	1800 sec.
Amyl butyrate	10^{-9}	10^{-10}N	10^{-3}cm.	$\sim$0.3 sec.	600 sec.
Benzene + bromobenzene	10^{-8}	10^{-8}N	5×10^{-4}cm.	$\sim$0.01 sec.	$\sim$1 sec.

from the interface. Although benzene is only slightly more polar than paraffinic oils, it dissolves more water, so that ions in benzene can be at least partly hydrated, with the consequence that more ions can dissolve in benzene than in petrol-ether. Interfacial potentials at the benzene-water interface will therefore decrease with time, though only slowly since the number of chloride counter-ions is limited by their low solubility in benzene. In petrol-ether this solubility is extremely small, and the potentials are stable indefinitely.

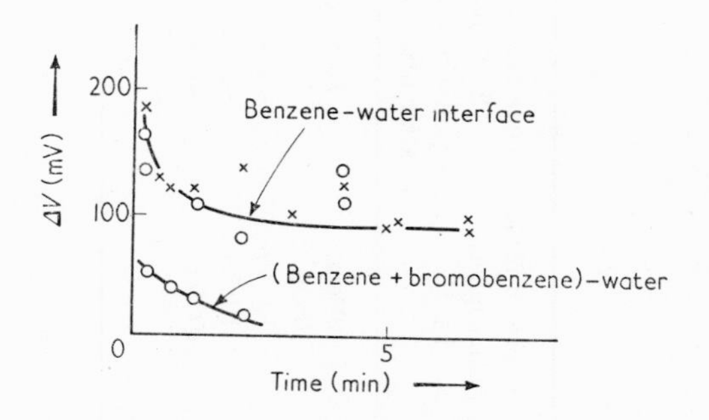

Fig. 2-11. Interfacial potentials of film of $C_{18}H_{37}N(CH_3)_3^+$ spread to 92 Å² per chain between water and slightly polar oils. The decrease in ΔV with time reflects the change from the region of applicability of equation (2.18) to that of equation (2.11). Crosses denote potentials measured with vibrating plate apparatus, and circles those measured with the radioactive source method.

Components of ΔV due to a Monolayer

At a clean surface of pure water, the potential ψ due to free charge may be taken as zero. Consequently, the *absolute potential* Φ_{H_2O} of the water is V or χ, the potential of the dipoles at the water surface. We shall avoid such absolute potentials here, except to point out that this absolute potential V has been estimated by different workers. Their values, however, range from -0.5 volts to $+0.4$ volts, though several are in agreement on a value between -0.1 and -0.2 volts.

If, however, a monolayer is spread or adsorbed on a clean water surface, the water dipoles will generally be re-orientated about the film-forming molecules, because of the new dipoles introduced into the surface. This change, for each molecule in the monolayer, will be denoted μ_1. The dipoles of the film-forming molecules (e.g. $\rangle$C$-NH_2$ in a long-chain amine) will also contribute to ΔV by an amount depending on the group dipole moment μ_2. A third component of ΔV for such electrically neutral films is μ_3, the moment of the bond (e.g. $\rangle$C$-$H) at the upper limit of the monolayer (Fig. 2-12).

If we apply the Helmholtz formula for an array of n dipoles per cm.^{2}, and if they are all vectorially additive in the vertical direction (Eucken), we obtain:

$$\Delta V = 4\pi n \mu_1 + 4\pi n \mu_2 + 4\pi n \mu_3 \tag{2.19}$$

Unfortunately, μ_1 cannot be measured, so it is usual to combine this term with μ_2; this allows for the fact that the re-orientation of the water dipoles, as expressed by μ_1 may well depend on μ_2, as illustrated in Fig. 2-12.

The term μ_3 is normally constant since paraffinic chains are usual in surface-active molecules, so then (2.19) simplifies to:

$$\Delta V = 4\pi n \mu_D \tag{2.20}$$

where $\mu_D = \mu_1 + \mu_2 + \mu_3$, characteristic of the dipole moment of the head-group (e.g. $>C{=}O$, $>C{-}NH_2$) of the molecules of the monolayer. This is true for any electrically neutral film, comprising only dipoles. As the film is compressed, μ_D will change if the dipolar head is unsymmetrical and is forced to

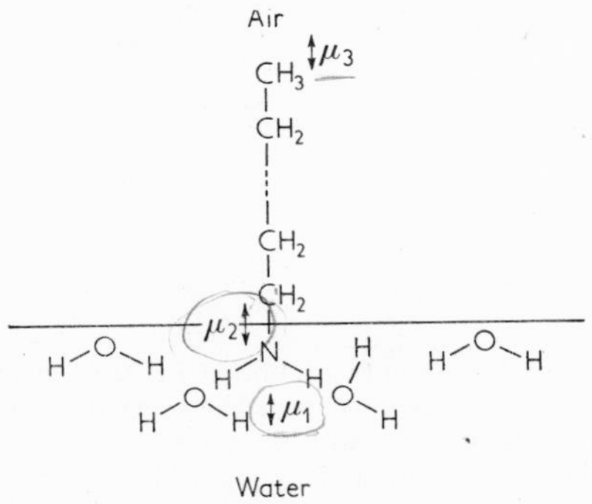

Fig. 2-12. Components μ_1, μ_2, and μ_3 of the surface potential ΔV for an electrically neutral monolayer.

change its orientation, as is shown in Fig. 2-13 for long-chain esters. Any re-arrangement of the dipoles upon compression of the film can consequently be deduced from changes in μ_D, if one can estimate the effective moment of each separate bond[25]. Some typical values of bond moments are shown in the figure: they are vectorially additive in any composite head-group, and are expressed as millidebyes (mD.). Many polymer monolayers are folded differently upon compression at the air-water surface, though μ_D at the oil-water interface is independent of the compression of the film[25].

In recent years there has been added interest in substituted hydrocarbon chains, and to investigate these one uses eq. (2.19) in the form:

$$\Delta(\Delta V) = 4\pi n(\Delta \mu_3) \tag{2.21}$$

Thus, by comparing films with the same numbers of molecules cm.$^{-2}$ spread at the same interface and having the same polar head-groups, we measure the difference in potential due to substitution for the $-CH_3$ group at the upper limit of the film of (say) $-CH_2Br$ or $-CF_3$, as shown in Fig. 2-14. Some results are shown in Table 2-V. That $\Delta\mu_3$ is less than the moment of the free carbon-halogen link suggests that there is mutual polarization of the dipoles in condensed films. Further, the dipoles of such ω-halogen links are unlikely to be orientated vertically, but rather at half the tetrahedral angle.

Consider now the potential due to a monolayer of long-chain ions such as $C_{18}H_{37}N(CH_3)_3^+$: the monolayer now contributes an electrostatic term as well as the ordinary dipole terms to $\Delta\Phi$ (Fig. 2-15). This electrostatic potential

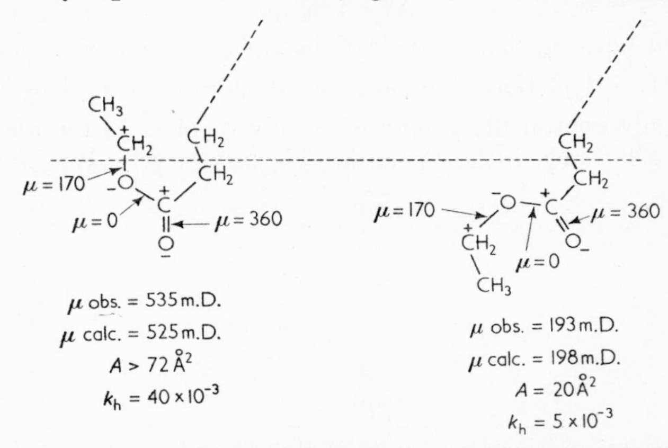

Fig. 2-13. Configuration and moments of bonds in head-groups of ethyl palmitate spread at the air-water surface[25]. At high areas the vertical component of the moments is high, while at low areas the ethyl chain is forced below the water surface, and the vertical component of the total moment is lower. The k_h values refer to the velocity constants for the hydrolysis when the ester is spread on 1 N-alkali, in units of litres.gm.mole^{-1} min.$^{-1}$ The hydrolysis is much slower when, at low areas, the ethyl groups form a protective sheath beneath the carbonyl groups.

arises from the unequal distribution of ions (Figs. 2-16, and 2-23, A and B) in the vicinity of the adsorbed monolayer, and though it is therefore strictly a ψ potential, it is convenient to include this in the ΔV term because it arises directly from the presence of the monolayer. To distinguish this monolayer electrostatic potential from $\Delta\psi$, however, it is generally termed ψ_0; it represents the electrostatic potential in the interface (at zero distance from it—hence the subscript zero) relative to the subjacent aqueous phase. With these conventions eq. (2.19) and (2.20) may be generalized[5,23,27,28] to apply to such charged monolayers:

$$\Delta V = 4\pi n\mu_1 + 4\pi n\mu_2 + 4\pi n\mu_3 + \psi_0 \tag{2.22}$$

$$\Delta V = 4\pi n\mu_D + \psi_0 \tag{2.23}$$

$$\Delta V = 4\pi n\mu_{ov} \tag{2.24}$$

TABLE 2-V

Variations in μ_3 for different films at the Air-Water and Oil-Water Interfaces[5,22,26]

Film	ΔV	Δ(ΔV)	$\Delta\mu_3$ (vertical component of dipole differences in ω-bonds)	Difference in μ for ω-bond and C—H bond, from bulk measurements
Myristic acid with carboxyl group ionized (25 Å²). Air-water	−50 mV	−900 mV	−600 mD.	−1800 mD.
Perfluorodecanoic acid with carboxyl group ionized (25 Å²). Air-water	−950 mV			
Stearic acid (pH 8.2). Air-water	0 mV	−1190 mV	−800 mD.	−1800 mD.
ω-trifluorostearic acid (pH 8.2). Air-water	−1190 mV			
Hexadecanoic acid on 1 N-NaOH (at 66 Å²). Air-water	−28 mV	0	0 mD.	−1900 mD.
ω-Bromohexadecanoic acid on 1 N-NaOH (at 66 Å²). Air-water	−28 mV	−132 mV	−230 mD.	−1900 mD.
ω-Bromohexadecanoic acid on 1 N-NaOH (at 66 Å²). Oil-water	−160 mV			
Hexadecanoic acid (pH4). Air-water (20 Å²)	+390 mV	−1260 mV	−660 mD.	−1900 mD.
ω-Bromohexadecanoic acid (pH4). Air-water (20 Å²)	−870 mV			

where μ_{ov} is the overall dipole contribution from the dipoles on the surface and from the counter-ions in the water below. In monolayers carrying no net electric charge (e.g. of long-chain alcohols) ψ_0 is always zero.

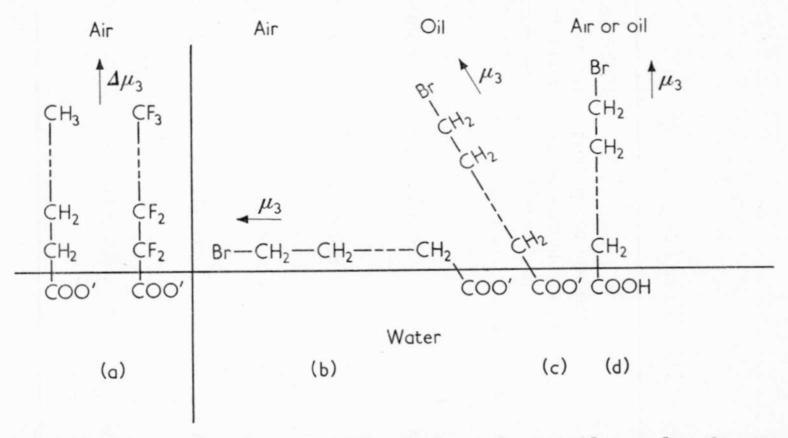

Fig. 2-14. (a) Anions of an aliphatic acid and of a perfluoro acid spread at the air-water interface. The difference in ΔV is due to the change in μ_3, the vertical dipole moment of the uppermost bond. (b) Anion of ω-bromohexadacanoic acid at air-water interface. The vertical component of μ_3 is zero. (c) Anions of ω-bromohexadecanoic acid at decane-water interface. Vertical component of μ_3 now appreciable, as the ω-bromo group is easily lifted into the oil. (d) Molecule of ω-bromohexadecanoic acid in close-packed film at pH4 at air-water surface. The dipole μ_3 is now relatively large.

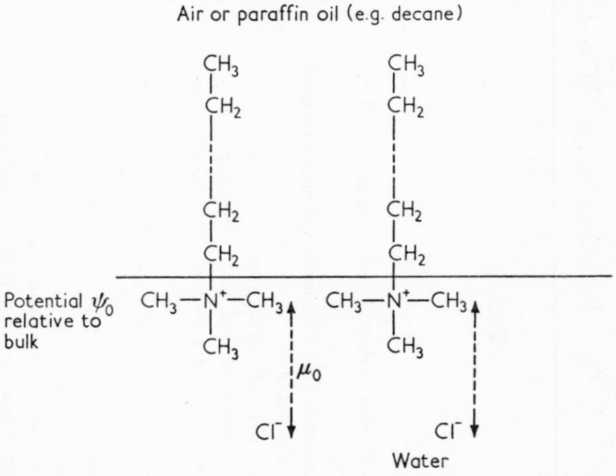

Fig. 2-15. A monolayer of $C_{18}H_{37}N(CH_3)_3^+$ at air-water or decane-water interface. The mean depth of the Cl^- counter-ions is $1/\varkappa$ and this makes possible very large dipole contributions if the ionic strength is low (Table 2-VI). The electrostatic potential at the interface is ψ_0 relative to the bulk of the aqueous phase.

Calculation of ψ at the Surface from the Equations of Gouy

In 1910 Gouy[29] put forward his theory of the diffuse double layer. The basic assumptions are that the charged surface is impenetrable, that the charge (σ per cm.^{2}) is uniformly spread over it, and that the counter-ions behave as point charges, being able to approach right up to the plane of the charges. These conditions are summarized in Fig. 2-16. On the basis of

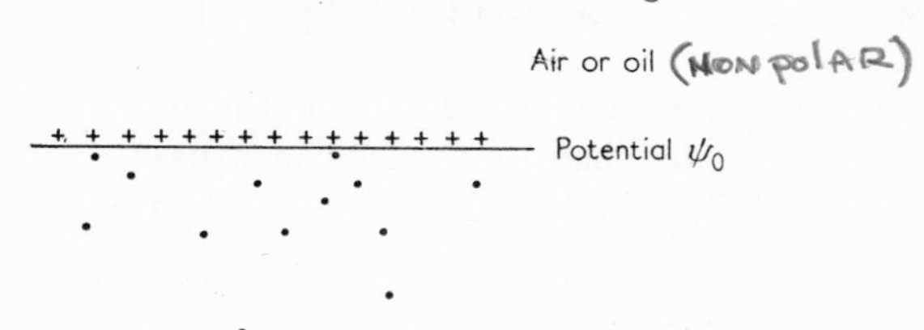

Fig. 2-16. Idealized double layer, dots representing point negative charges.

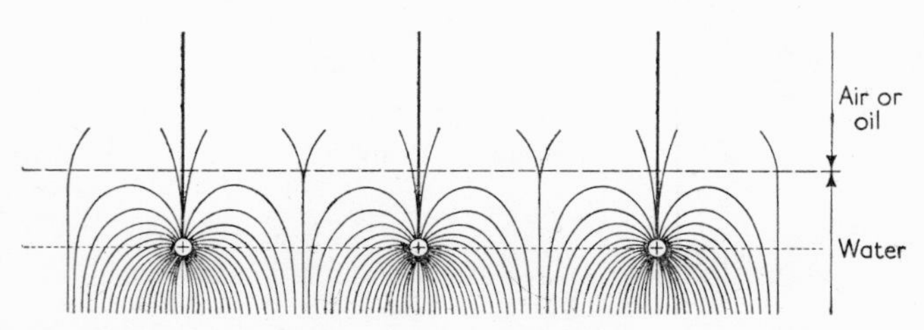

Fig. 2-17. Arrangement of long-chain ions at aqueous interface, showing lines of force. The depth of immersion is about 3 Å. The three vertical lines represent hydrocarbon chains, and $\oplus$, the ionic group.

these assumptions Gouy solved the Boltzmann equations for the distribution of cations and anions in terms of a potential ψ near the positive surface relative to the bulk of the liquid. The Boltzmann equations are:

$$_{s}c_{+} = c.e^{-\varepsilon\psi/kT} \tag{2.25}$$

and

$$_{s}c_{-} = c.e^{+\varepsilon\psi/kT} \tag{2.26}$$

which, on the basis of the above assumptions, lead to a calculated value of ψ_0 (called ψ_G) *at* the surface, given by:

$$\psi_G = \left(\frac{2\boldsymbol{k}T}{\varepsilon}\right) \sinh^{-1}\left\{\frac{\sigma}{c_i^{1/2}}\left(\frac{500\pi}{DRT}\right)^{1/2}\right\} \tag{2.27}$$

In these equations c refers to ionic concentrations (moles l.$^{-1}$), subscript s refers to the surface, c_i is the total (uni-univalent) electrolyte concentration (in moles l.$^{-1}$), $\boldsymbol{k}$ is the Boltzmann constant, T is the absolute temperature,

ε is the electronic charge, D is the dielectric constant of water (80 at 20°C; 78 at 25°C), and R is the gas constant.

At 20°C eq. (2.27) simplifies to

$$\psi_G = \left(\frac{2\boldsymbol{k}T}{\varepsilon}\right) \sinh^{-1}\left(\frac{134}{Ac_i^{1/2}}\right) \tag{2.28}$$

or

$$\psi_G = 50.4 \sinh^{-1}\left(\frac{134}{Ac_i^{1/2}}\right) \tag{2.29}$$

with ψ_G expressed in millivolts and with A (the area available in the surface to each ionogenic long-chain) expressed in square angstroms per charged group. Thus $A = 10^{16}/n$ if all the ionizable groups are dissociated.

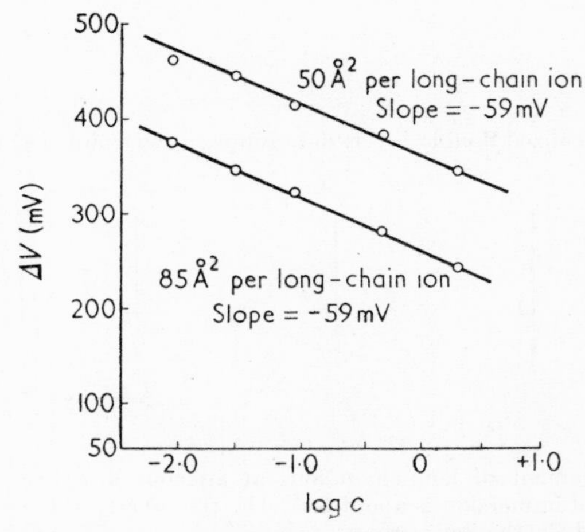

Fig. 2-18. Test of eq. (2.32) for monolayers of $C_{18}H_{37}N(CH_3)_3^+$ spread between air and aqueous NaCl[23].

At 50°C, (2.27) becomes

$$\psi_G = 55.3 \sinh^{-1}\left(\frac{139}{Ac_i^{1/2}}\right) \tag{2.30}$$

At high potentials ($\psi_G > 100$ mV) eq. (2.30) may be written

$$\psi_G = (\boldsymbol{k}T/\varepsilon)\ln(139^2 \times 4/A^2c_i) \tag{2.31}$$

which gives the relation

$$(\partial\psi_G/\partial \log c)_A = -2.303\boldsymbol{k}T/\varepsilon (= -58 \text{ mV at } 20°\text{C or } -63 \text{ mV at } 50°\text{C}) \tag{2.32}$$

This differential equation, however, is not as selective as eq. (2.27) since the Stern equations or even simple thermodynamic reasoning lead to the same differential form.

The question now arises of how far the Gouy potential, ψ_G, can be identified with ψ_0 in real monolayers, whether adsorbed or insoluble. Monolayers do not strictly fulfil the requisite conditions for the use of (2.27), but in practice the fact that the charge is *discretely* distributed in a monolayer consisting of long-chain ions should not greatly affect the validity of the Gouy theory as applied to charged films. The reason for this is that at relatively short

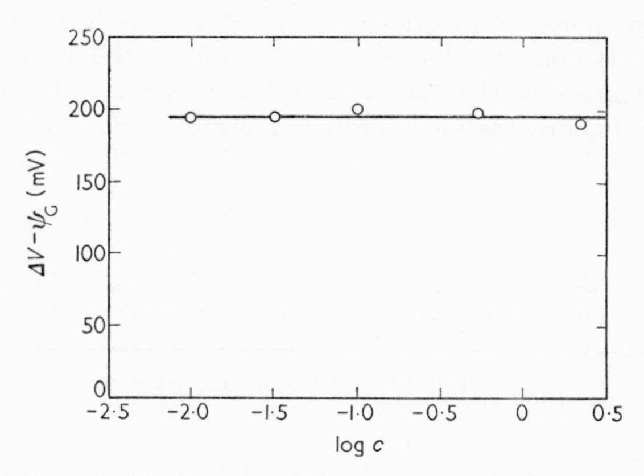

Fig. 2-19. Test of eq. (2.33) for same system as in previous figure, with A = 85 Å[23].

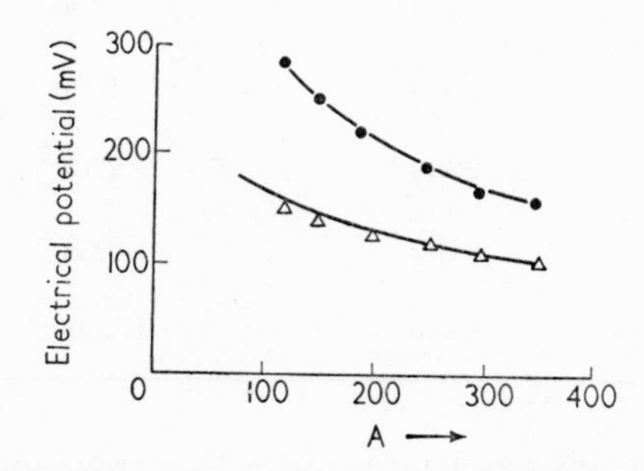

Fig. 2-20. Test of eq. (2.33)[28]. Points refer to ΔV; triangles to the derived values of ψ_0 (eq. (2.23)). The lower line represents ψ_G calculated by eq. (2.29).

distances below the charged interface the lines of force from the charged head-groups (e.g. $-N(CH_3)_3^+$) are all nearly equidistant and perpendicular to the interface, as required by the Gouy theory. This is illustrated in Fig. 2-17.

Agreement with eqs. (2.23) and (2.32), assuming $\psi_G = \psi_0$, is shown by the data of Fig. 2-18, but numerical comparison of ψ_0 and ψ_G cannot be made directly unless μ_D in eq. (2.23) is known. However, if μ_1 is unaffected by the electrical field (see p. 141), μ_D should be constant at constant area since in neutral films μ_D is practically independent of ionic strength[30]. Hence, by eq. (2.23), if $\psi_0 = \psi_G$ we should have:

$$\Delta V - \left(\frac{2kT}{\varepsilon}\right) \sinh^{-1}\left(\frac{134}{Ac_i^{1/2}}\right) = 4\pi n\mu_D \tag{2.33}$$

where the product $4\pi n\mu_D$ is constant at a given value of A or n. That this is indeed so is clear from Fig. 2-19 and supports the assumptions that $\Delta\mu_1$ is small and that ψ_G may be identified with ψ_0. Surprisingly, the Gouy equation seems to give the correct value of ψ_0 even at quite high ionic strengths, when the mean thickness of the ionic double layer, $1/\varkappa$, is only about 2 Å, i.e. less than the ionic radius for hydrated chloride ions[23]. The reasons for this are discussed below. Fig. 2-20 shows that eq. (2.33) applies to a film of

TABLE 2-VI

Calculated and measured Dipole Moments $\mu_{ov.}$ for completely ionized films of $C_{18}H_{37}N(CH_3)_3^+$ at the Air-Water or Decane-Water Interfaces[5,16]

NaCl concentration, c	$\mu_{ov.} = 450 + \frac{(1/\varkappa)\times 4.8 \times 1000}{80}$ (mD.)	$\mu_{ov.}$ (measured) as $n \to 0$, (mD.)	Potential ΔV at $n = 10^{13}$ (1000 Å² per chain)
2 N ($1/\varkappa$ = 2.1 Å)	578	500	22.5 mV
10^{-1} N ($1/\varkappa$ = 9.5 Å)	1020	1026	43 mV
10^{-2} N ($1/\varkappa$ = 30 Å)	2250	2700	86 mV
10^{-3} N ($1/\varkappa$ = 95 Å)	6150	6170	124 mV
10^{-4} N ($1/\varkappa$ = 300 Å)	18,450	19,300	220 mV

$C_{16}H_{33}N(CH_3)_3^+$ at the oil-water interface with a constant value for μ_D of 450 millidebyes, independent of A. The same value of μ_D is found at the air-water surface for various long-chain quaternary ions, the constancy of μ_D at different molecular orientations being a consequence of the symmetrical structure of the quaternary ammonium ion. Long-chain sulphates show no such constancy of μ_D, and this restricts the application of eq. (2.33) to their monolayers at a given orientation (i.e. at a given area, chain-length, and interface).

In the limit of very high areas per charged group, each charge may be regarded as acting separately on its counter-ion, the latter being at a mean distance $1/\varkappa$ away (Fig. 2-24a). This distance is equal to $3/c_i^{1/2}$ for uni-univalent electrolytes in water at room temperature, so that the ionic separations may be considerable in dilute solutions. Table 2-VI shows typical values of $1/\varkappa$ together with the total μ values calculated: these are made up of μ_D and of the dipoles constituted by the separation of unit electronic charges by the distance $1/\varkappa$, converting the latter values to millidebyes by the factor 4.8×10^3 with 80 as the dielectric constant of water. These calculated values are in good agreement with those obtained by extrapolation to $n = 0$, from the data for films of very low charge densities. Though the dipole moments are enormous under these conditions, the potentials remain moderate at $n = 10^{13}$ long-chain ions cm.$^{-2}$, as shown in the last column.

Heats of formation of the electrical double layer may be calculated for an ideal diffuse system: both these values and those measured are small, and agree rather poorly with one another[31]. It is likely that changes in hydration near the interface may be responsible for the discrepancies.

Corrections to the Gouy Equation

Even in an ideal system, consisting of a plane, impenetrable, uniformly charged surface, deviations from eq. (2.27) are to be expected[32], since:

(i) the dielectric constant of water may fall in the intense electrical field near the surface;

(ii) polarization of the water around each counter-ion should also be allowed for;

(iii) corrections are required for image forces, and for the coulombic energy of interaction between different counter-ions, increasing the ionic concentration near the surface;

(iv) the non-zero sizes of the counter-ions decrease the ionic concentration near the surface.

Detailed mathematical study of these points shows that, if the field at the interface is not in excess of 5×10^6V.cm.$^{-1}$, factor (i) is small (cf. p. 142). Secondly, for small counter-ions (of radius 2 to 3 Å in the hydrated state), correction (ii) is also small, and almost exactly cancels correction (i). Thirdly, for these small counter-ions, corrections (iii) and (iv) partly cancel each other, particularly if c_i is small ($< 10^{-3}$ N): for very small or very large counter-ions the surface potentials ψ_0 may, however, be altered by as much as 40 mV. This latter correction, incidentally, can explain the ion-exchange phenomena between Na^+, K^+, and other alkali metal ions on clays and on biological membranes, the relative surface concentrations differing by a factor of up to 5 times. The error due to factor (iii) may become quite large (e.g. 20% of ψ_0) if $c_i = 10^{-2}$ N, and at higher values of c_i and with high values of ψ_0, factors (i) and (iv) predominate.

The exact position of the long-chain ions in the surface is still uncertain: it is now agreed[23,33,34] that the ionic head-groups must be slightly immersed below the non-aqueous phase, as in Fig. 2-17. Calculation[23] gives about 3 Å as a typical depth of immersion of the head-groups. They may also be "staggered" at different depths of immersion to reduce their mutual repulsion.

To understand why eq. (2.27) is often quite satisfactory, even though the charges do not lie on a smooth plane, we must remember that up till now the possibility of penetration of some counter-ions between the charged groups in the surface has been neglected, as has also the appreciable size of the real counter-ions. In practice, at least with the long-chain quaternary ion monolayers, errors due to inapplicability of these two assumptions individually may balance out: penetration (position C, Fig. 2-23) will increase the number of counter-ions at the surface and so reduce ψ_0 below the calculated ψ_G, while the non-zero size of the counter-ions, in preventing their very close approach to the charged surface (B in Fig. 2-23), will give the result that $\psi_0 > \psi_G$. The approximate cancellation[23,35] of these two effects is apparently responsible for the agreement of the Gouy theory with experiment, at least to a first order of approximation. The process just described may also be illustrated by reference to Fig. 2-17 where penetration of the counter-ions between the charged heads, not allowed for in deriving ψ_G, offsets the non-zero size of the counter-ions. Measurements[35] of surface viscosity (p. 262), surface pressure and of second-order potential effects strongly support this idea of penetration (see p. 90); the more polarizable counter-ions such as iodide penetrate more readily than fluoride ions.

In brief, the Gouy equations hold surprisingly well in many systems at ionic strengths of NaCl up to 2 N (Fig. 2-18), due to a fortunate compensation of factors (i) and (ii) above, and of factors (iii) and (iv) with the penetration of counter-ions into the plane of the charged head-groups. Such cancellation of errors cannot be general; with very large or very small counter-ions, or with films packed so tightly as to prevent penetration, we cannot expect eq. (2.27) to hold accurately.

Calculation of ψ near the Surface from the Equations of Donnan

An alternative and often valuable approach to the properties of insoluble charged monolayers lies in considering the large, long-chain ions to be restrained in the surface by a membrane permeable to small inorganic ions only: this is a mathematical fiction equivalent to a very high energy of desorption of the hydrocarbon chains. We shall first derive the general expression for membrane potentials following Donnan[36], and then consider its application to ionized monolayers.

Suppose that, as in Fig. 2-21, a membrane separates two bulk aqueous phases. The right-hand phase contains non-diffusible p-valent ions which provide a fixed charge in this region: the membrane prevents these charged groups from being distributed into the aqueous solution on the left. Small cations (such as those of sodium) from a uni-univalent electrolyte of concentration c ($=c_+=c_-$) can pass through the membrane: their chemical potentials must therefore be the same on both sides of the membrane. This may be expressed as

$$\boldsymbol{k}T \ln({}_1c_+) = \boldsymbol{k}T \ln({}_2c_+) + \varepsilon\psi_{Don.} \tag{2.34}$$

where the subscripts 1 and 2 refer to the aqueous phases without and with the colloidal ions, and $\psi_{Don.}$ is the potential of aqueous phase 2 with respect to the aqueous phase 1. It is assumed that there are no specific effects: the μ^0 terms are identical on each side of eq. (2.34), and they are therefore omitted. In this respect eq. (2.34) is simpler than eqs. (2.1) and (2.2) referring to an oil "membrane", since the μ^0 terms in the latter are different from those in water.

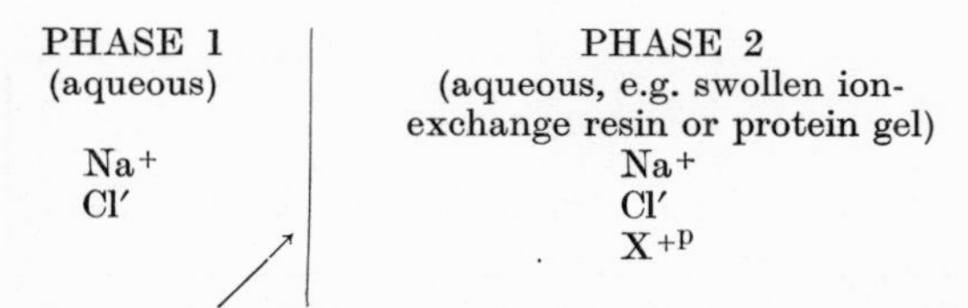

Fig. 2-21. The origin of the Donnan potential, $\psi_{Don.}$, of phase 2 relative to phase 1 lies in the inability of the charges X^+ in phase 2 to enter phase 1.

For the anions (e.g. Cl′) the equation is similar:

$$\boldsymbol{k}T \ln({}_1c_-) = \boldsymbol{k}T \ln({}_2c_-) - \varepsilon\psi_{Don.} \tag{2.35}$$

By adding eqs. (2.34) and (2.35) we obtain

$${}_1c_+ . {}_1c_- = {}_2c_+ . {}_2c_- \tag{2.36}$$

and, by subtracting and using eq. (2.36),

$$\psi_{Don.} = (\boldsymbol{k}T/\varepsilon)\ln({}_2c_-/{}_1c_-) = (\boldsymbol{k}T/\varepsilon)\ln({}_1c_+/{}_2c_+) \tag{2.37}$$

This shows the physical meaning of the potential $\psi_{Don.}$: if phase 2 carries fixed positive charges (as does a protein in slightly acid solution, or a cationic ion exchange resin), this phase will be positive with respect to the aqueous sodium chloride and it will attract chloride ions into itself, so that ${}_2c_- > {}_1c_-$. Similarly it will repel sodium ions, and so ${}_2c_+ < {}_1c_+$.

Before we can calculate $\psi_{Don.}$, the concentrations of diffusible ions in phase 2 must be obtained from the condition for overall electrical neutrality in phase 2:

$${}_2c_+ + pc_{x+} = {}_2c_- \tag{2.38}$$

where c_{x+} refers to the concentration of the non-diffusible p-valent ions X in phase 2. By solving eqs. (2.36) and (2.38), we obtain an expression for ${}_2c_+$ in terms of known quantities:

$$_2c_+ = \left(\frac{^2c_{x+}^2}{4} + c^2\right)^{1/2} - \frac{pc_{x+}}{2} \tag{2.39}$$

and, similarly,

$$_2c_- = \left(\frac{p^2c_{x+}^2}{4} + c^2\right)^{1/2} + \frac{pc_{x+}}{2} \tag{2.40}$$

Hence $\psi_{Don.}$ can be calculated directly from eqs. (2.37) and (2.39):

$$\psi_{Don.} = (\boldsymbol{k}T/\varepsilon)\ln\left(\frac{pc_{x+}}{2c} + \left[\frac{p^2c_{x+}^2}{4c^2} + 1\right]^{1/2}\right) \tag{2.41}$$

For example, if in a certain ion-exchange resin (which constitutes its own membrane) pc_{x+} is 0.3 moles litre^{-1}, and if the resin is in equilibrium at 25°C

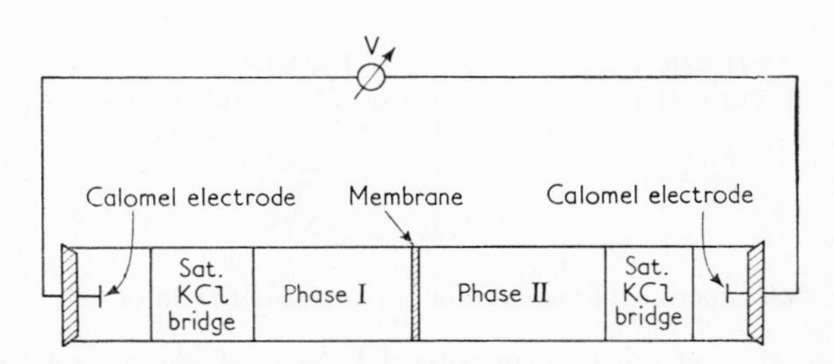

Fig. 2-22. Measurement of $\psi_{Don.}$.

with 10^{-2} N-NaCl, then, since $\boldsymbol{k}T/\varepsilon$ is 25.6 mV at 25°C, $\psi_{Don.}$ is 87 mV. If the resin is negatively charged (e.g. a phenyl-formaldehyde sulphonic acid condensate) $\psi_{Don.} = -87$ mV.

Experimentally, the Donnan potential may be measured directly across an uncharged membrane separating solutions of (say) protein and of a salt, as in Fig. 2-22: the potential $\psi_{Don.}$ is measured across two KCl salt bridges joined to the two aqueous solutions[36]. Strictly, activity coefficients should be taken into account, though in practice this is difficult to achieve in the phase containing fixed charges. The potential between the KCl salt bridge and phase 2 may also be affected by the different mobilities of the diffusible ions in the presence of the fixed charges. In consequence, experimental agreement with eq. (2.41) is considered satisfactory if within 1 or 2 mV in concentrated solutions, though deviations of 10 mV are usual in dilute solutions (where the changes in ionic mobilities become important).

Overbeek[36] describes another correction to the simple Donnan treatment: since the charge and potential are not uniformly distributed through phase 2 (as is assumed in eqs. (2.34 and 2.37) above), the very high potential close to certain ionic sites within phase 2 results in a calculated average accumulation of counter-ions higher than that given by eq. (2.37). This correction for $_2c_+$

and ${}_2c_-$ is more important at low salt concentrations (when ψ is high), and may reach 50%.

In spite of these limitations, the Donnan treatment is often usefully extended to describe the properties of *surfaces bearing a fixed charge*[37,38], the forces fixing the charged groups to the surface corresponding to a membrane preventing these charges entering the bulk phase (phase 1 in the above nomenclature). One serious difficulty, however, lies in defining the term c_{x+} of eq. (2.38), for which purpose the surface phase (2) must be assigned some particular thickness. This is at best a compromise for, according to the idea of a diffuse double layer, a few of the excess counter-ions will be found at any distance, however great, from the surface. However, if the thickness of the "surface phase" is taken to be the Debye-Hückel term $1/\varkappa$ (i.e. $3/c_i^{1/2}$Å for uni-univalent electrolytes in water at 25°C), the surface phase within this distance is so nearly electrically neutral that no great error is introduced[37,39]. The counter-ion concentration calculated from $\psi_{\text{Don.}}$ is then a mean of the values calculated using the potentials from the Gouy theory, the latter concentrations being averaged from the surface to a distance $1/\varkappa$ from it[39]. Fig. 2-24a illustrates this point.

The potential $\psi_{\text{Don.}}$ calculated for the "surface phase" will necessarily be less than ψ_G, since the former is calculated assuming the surface is of thickness $1/\varkappa$ while the latter assumes that all the fixed charge is concentrated on a plane. The exact relation between the two is best shown[39,40] by expressing $\psi_{\text{Don.}}$ of eq. (2.41) in a mathematically equivalent form and substituting numerically for $1/\varkappa$ at 20°C:

$$\psi_{\text{Don.}} = \left(\frac{\boldsymbol{k}T}{\varepsilon}\right) \sinh^{-1}\left(\frac{2 \times 134}{Ac_i^{1/2}}\right) \tag{2.42}$$

This relation, only valid if the "surface phase" is taken to be of thickness $1/\varkappa$, may be compared with eq. (2.28). When $Ac_i^{1/2} = 134$, $\psi_{\text{Don.}}$ is 81% of ψ_G, the divergence being larger when $1/\varkappa$ is large (c_i small), and being lower when the "surface phase" is very thin (high c_i). The use of eq. (2.42) is advisable at high charge densities and ionic strengths, as penetration of the counter-ions can cause the calculated ψ_G to be too high in these circumstances: the "surface phase" concept requires no separate mention of penetration nor of ionic sizes. It has been found applicable to the titration of proteins and the evaluation of the ionization of certain groups close to charges on a composite surface[37,38].

Sometimes it is preferable to take a value of the thickness of the "surface phase" other than $1/\varkappa$, as in the analysis of the kinetics of alkaline hydrolysis of monocetylsuccinate ions[38]. Here the ester group is situated 10Å from the negative carboxyl group, and the rate results (p. 291) fit well the assumption that the hydroxyl ions are present at a concentration given by eq. (2.37),

$\psi_{Don.}$ now being calculated from (2.41) with $c_{x'}$ obtained assuming this constant thickness of the "surface phase".

Specific Adsorption—Reversal of Charge

Certain counter-ions may be held in the surface by forces additional to those of purely electrostatic origin: on a negatively charged monolayer (as from $C_{22}H_{45}SO_4'$), for example, counter-ions of Ag^+ will be held also by the energy of polarization λ_p, while $C_2H_5NH_3^+$ ions are adsorbed partly by the van der Waals energy W associated with the ethyl group. On a phosphate surface[41], the negative charge is just compensated by Ag^+ ions adsorbing from 0.15 N-Ag^+; similarly 0.3 N-Tl^+, 1 N-Li^+, or 2.8 N-Na^+ also just compensate the charge, this sequence reflecting the highest polarization energy for Ag^+ and the lowest for Na^+. Higher concentrations than these impart a net positive charge to the surface, though for K^+ the polarization is so weak that no reversal occurs. For anions at positively charged surfaces, the order of polarization is $CNS' > I' > Br' > Cl'$. The sequence and values of the concentrations are functions not only of the counter-ions, but also of the surface ionogenic groups, the latter dependence making this "reversal of charge spectrum" a valuable tool in characterizing unknown surfaces, both of colloidal particles and of living cells. Since the "reversal of charge" concentration is most readily determined by electrokinetic rather than by electrostatic measurements, further discussion of these "spectra" is in the next chapter. Specific adsorption is of great importance in ion-exchange.

Since, according to the Gouy equation (2.27), ψ does not change sign however high c_i may be but merely tends towards zero, it is clearly necessary to introduce an additional energy term, allowing for the specific interaction. Considering HCl bound on to sites on the surface of wool, Gilbert and Rideal[42] calculate adsorption of H^+ from acid of concentration c according to:

$$\log\left(\frac{\theta}{1-\theta}\right) = \log c + \frac{(\lambda_p + W) - \varepsilon\psi}{2.303\,\boldsymbol{k}T} \qquad (2.43)$$

where $(\lambda_p + W)$ is the total specific adsorption energy of the positive counter-ions, and θ is the fraction of the possible "sites" on to which adsorption of H^+ has occurred. Adding this equation to a similar equation for Cl' and putting $\theta_{H^+} = \theta_{Cl'}$, gives an experimentally useful equation not involving ψ: this latter equation is not only in accord with the experimental variation of θ with c, but the treatment applies equally well to the adsorption of dyes[43]. Typical values of the derived specific interactions are shown in Table 2-VII, with an estimated separation of the total specific adsorption energy into its components.

Specific Adsorption—The Stern Theory

In his classical theory of 1924 Stern[45] allowed not only for the specific adsorption energy of an ion at a plane surface, but also for the non-zero sizes of counter-ions. His equation also includes ψ, and it tends in the appropriate limits to eq. (2.43) or to eq. (2.25). If n_c is the number of counter-ions of valency z_2, adsorbed per cm.^{2} from a bulk concentration c gm. ions. l.$^{-1}$, then Stern's relation is:

$$n_c = c\left\{\frac{n_s - n_c}{(1000/M)}\right\} e^{(-z_2\varepsilon\psi_\delta + \lambda_p + W)/kT} \tag{2.44}$$

where n_s is the number of possible adsorption "spots" or "sites" on the surface, M is the molecular weight of the solvent (usually water), and ψ_δ is

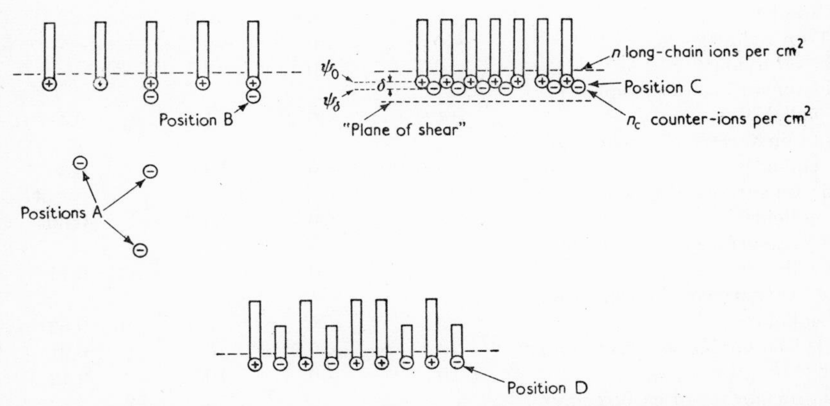

Fig. 2-23. Possible positions of negative counter-ions beneath a positively charged monolayer.

the potential at the plane of the first layer of adsorbed counter-ions, which are at a mean distance δ from the original charged surface (Figs. 2-23 and 2-24b). The latter is again represented as a uniformly charged, impenetrable plane. In this equation the term in $(n_s - n_c)$ allows for the non-zero size of the counter-ions, placing a limit on n_c, and the terms λ_p and W allow for the specific adsorption due to polarization and van der Waals forces.

For a single ionic species and with $\theta = n_c/n_s$, eq. (2.44) becomes

$$\frac{\theta}{1-\theta} = \frac{c}{(1000/M)} e^{((\lambda_p + W) - z_2\varepsilon\psi_\delta)/kT}$$

or

$$\log\left(\frac{\theta}{1-\theta}\right) = \log c + \log\left(\frac{M}{1000}\right) + \frac{(\lambda_p + W) - z_2\varepsilon\psi_\delta}{2.303kT} \tag{2.45}$$

TABLE 2-VII

System	Total specific interaction energy, relative to kT	W/kT	λ_p/kT	Reciprocal of difference of electronegativity between metal and oxygen[46]
Cl′ on wool[43]	~0	~0	~0	—
H^+ on wool[42,43]	10.2	~0	10.2	0.7
Orange II dye anion on wool[43]	7.3	7.0	0.3	—
Naphthalene-β-sulphonate ion on wool[43]	3.3	3.0	0.3	—
Metanil Yellow YK anion on wool[43]	8.9	8.6	0.3	—
Cl′ on nylon[43]	~0	~0	~0	—
H^+ on nylon[43]	16.7	~0	16.7	—
K^+ on surface of phosphate colloid[41]	<1.8	~0	<1.8	0.4
Na^+ on surface of phosphate colloid[41]	2.6	~0	2.6	0.5
Li^+ on surface of phosphate colloid[41]	3.6	~0	3.6	0.69
Tl^+ on surface of phosphate colloid[41]	4.8	~0	4.8	0.71
Ag^+ on surface of phosphate colloid[41]	5.5	~0	5.5	0.83
Cu^{++} on bovine serum albumin[53]	10	~0	10	0.91
Th^{+4} on glass	10	~0	10	0.42
Li^+ on surface of carboxylate colloid[41]	1.9	~0	1.9	0.69
Na^+ on surface of sulphate colloid[41]	0.7	~0	0.7	0.5
Na^+ on carboxylate surface[53a]	~1.0	~0	~1.0	0.5
Mg^{++} on carboxylate surface[53b]	~2.1	~0	~2.1	0.6
Ca^{++} on carboxylate surface[53b]	~4.4	~0	~4.4	0.7
Cu^{++} on carboxylate surface[53b]	9.7	~0	9.7	0.91
K^+ on to 1:1 mixed monolayer of long-chain sulphate and quaternary ammonium ions[48]	1.0 (arbitrary)	~0	1.0	0.4
Na^+ on to 1:1 mixed monolayer of long-chain sulphate and quaternary ammonium ions[48]	~1.0	~0	1.0	0.5
Cl′ on to 1:1 mixed monolayer of long-chain sulphate and quaternary ammonium ions[48]	3.0	~0	3.0	—

TABLE 2-VII *Cont.*

System	Total specific interaction energy, relative to kT	W/kT	λ_p/kT	Reciprocal of difference of electronegativity between metal and oxygen[46]
SO_4'' on to 1:1 mixed monolayer of long-chain sulphate and quaternary ammonium ions[48]	2.7	~0	2.7	—
F′ on to monolayer of $C_{18}H_{37}N(CH_3)_3^+$ (at 85 Å^2)[35]	2.1*	~0	2.1	—
Cl′ on to monolayer of $C_{18}H_{37}N(CH_3)_3^+$ (at 85 Å^2)[35]	3.0*	~0	3.0	—
I′ on to monolayer of $C_{18}H_{37}N(CH_3)_3^+$ (at 85 Å^2)[35]	3.9*	~0	3.9	—
Cl′ on to 1:1 mixed monolayer of long-chain sulphate and pyridinium ions[49]	3.5	~0	3.5	—
Ethylamine ions on to monolayer of long-chain carboxylate ions[44]	3.6	3.6	~0	—
n-Propylamine ions on to monolayer of long-chain carboxylate ions[44]	5.8	5.8	~0	—
Benzylamine ions on to monolayer of long-chain carboxylate ions[44]	8	8	~0	—
Octylamine ions on to monolayer of long-chain carboxylate ions[44]	13.5	13.5	~0	—
Quinolinium ions into monolayer of long-chain sulphate ions[54]	7.5	7.5	~0	—
Cl′ from water on to Hg at zero potential[65]	4.6	~0	4.6	—
Cl′ from water on to Hg at positive potentials[65]	up to 10	~0	up to 10	—
I′ from water on to Hg at zero potential[70]	~9	~0	~9	—

Many of the values have been calculated from the experimental data in the literature. The asterisk denotes values calculated from surface viscosities. The values of λ_p are only approximate, and probably vary with charge density.

Apart from the small constant term, this is identical with the Gilbert-Rideal equation (2.43) above.

The limit of eq. (2.44), for dilute solutions and only moderate potentials, is illustrated by first writing it in the form:

$$n_c = n_s\left(1 + \frac{1000}{Mc}\, e^{(z_2\varepsilon\psi_\delta - (\lambda_p + W))/kT}\right) \tag{2.46}$$

If now $z_2\varepsilon\psi_\delta - (\lambda_p + W)$ is numerically smaller than about $8kT$, and if the salt concentration does not exceed about 10^{-3} normal, the second term in the large brackets is great compared with unity, and hence

$$n_c = \frac{n_s Mc}{1000}\, e^{(-z_2\varepsilon\psi_\delta + (\lambda_p + W))/kT} \tag{2.47}$$

where both terms in the exponential brackets become positive upon numerical substitution. If, further, the specific adsorption energy is small,

$$n_c = \frac{n_s Mc}{1000}\, e^{-z_2\varepsilon\psi_\delta/kT}$$

and, hence, by assuming a "surface thickness" of (say) 2×10^{-8}cm., one can convert n_c to a surface concentration $_sc$, giving, for an aqueous system with $n_s = 3 \times 10^{14}$ negative sites,

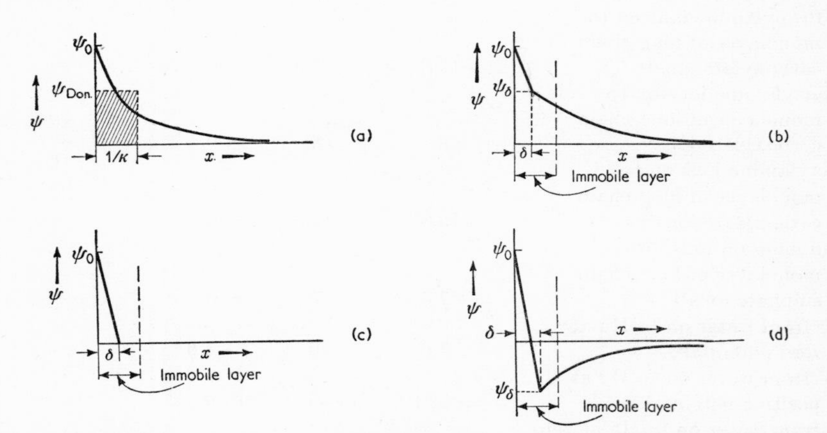

Fig. 2-24. (a) Typical potential variation with distance from surface with no specific interaction; the double layer is entirely diffuse (Gouy). The shaded region shows the "surface phase". (b) Typical potential variation with slight formation of Stern layer: n_c is small compared with n (usually implying that the specific adsorption energy is small). (c) By eqs. (2.48 and 2.49) when $n_c = n$, $\psi_0 = 0$; this corresponds to the "reversal of charge" concentration and there is no diffuse double layer. Hence also $\zeta = 0$. The condition that $n_c = n$ is reached at concentrations of counter-ions which are lower when the specific interaction is high. (d) When $n_c > n$, the potential is reversed in the diffuse double layer, by eqs. (2.48 and 2.49). Hence though ψ_0 may be but little affected by the specific adsorption, ζ has changed sign.

$$_s c_+ = 0.45 c e^{-\varepsilon\psi_\delta/kT}$$

This differs only slightly from eq. (2.25).

Reversal of the overall surface charge occurs (Fig. 2-24) when n_c is so high that εn_c (called σ_c) exceeds σ (the charge density on the interface, which, if the fixed charges are not close-packed, may well be less than εn_s). Alternatively, one may write that $n_c > n$, the latter being the number of fixed charges on 1 cm.^{2} of surface. The larger the standard free energy of adsorption ($\lambda_p + W$), the lower is the value of c required by eq. (2.46) or (2.47) to reach this critical value of n_c. Polyvalent ions similarly reduce ψ_δ to zero when $\sigma_c (= z_2 \varepsilon n_c)$ reaches σ for the fixed charge, i.e. $n_c = \sigma/z_2\varepsilon = n/z_2$: consequently, if these ions are strongly adsorbed, c in eq. (2.47) may be very low, because the "reversal of charge concentration" is then proportional to $(1/z_2)e^{(\lambda_p + W)/kT}$. Higher concentrations of the polyvalent ions will cause the surface to become oppositely charged, a condition well-known for glass and other surfaces in very dilute solutions of Th^{+4}; these cations must be strongly polarizing with λ_p of the order $10kT$. The polarizing tendency of an ion is also a measure of its tendency to form co-valent bonds, as is clear by considering H^+ or Ag^+. Consequently, it should be related to the electronegativity as defined by Pauling[46]: the difference of the electronegativity values between (say) oxygen in the head-groups of the monolayer and the metal ions in solution is a measure of the tendency for the bond to be purely ionic. Hence, the reciprocal of this electronegativity difference should correlate with λ_p, as is shown in Table 2-VII. The exact relation[38a] is complex, though useful in explaining the toxicity of heavy metal ions[38a,47].

Outside the Stern layer there is a diffuse double layer, the charge density σ_D of which is related to ψ_δ by the usual Gouy equation (cf. eq. (2.27)):

$$\psi_\delta = \left(\frac{2kT}{\varepsilon}\right) \sinh^{-1} \left\{\frac{\sigma_D}{c_i^{1/2}}\left(\frac{500\pi}{DRT}\right)^{1/2}\right\} \tag{2.48}$$

where D is the dielectric constant of the bulk liquid.

This equation, together with eq. (2.44), the relation $\sigma_c = n_c z_2 \varepsilon$, and the equation of conservation of charge,

$$\sigma_c + \sigma_D = \sigma \tag{2.49}$$

permit evaluation of σ_c, σ_D and ψ_δ (if σ is known). One may further compute ψ_0, the total potential drop across the surface, using the capacity of the condenser formed (Figs. 2-23 and 2-24) by the fixed charges of the surface and the Stern layer: this is

$$\sigma = \frac{D}{4\pi\delta}(\psi_0 - \psi_\delta) \tag{2.50}$$

in which D is now the dielectric constant within the fixed part of the double layer: a reasonable assumed value (see below) for $\frac{D}{4\pi\delta}$ is 10^7cm., or 11 μf.

The specific adsorption energy, $\lambda_p + W$, must be found independently, e.g. from studies of adsorption at a solid or mercury surface, from reversal of charge, or, for an ionic monolayer, from excess surface pressures (see p. 92) or surface viscosities (p. 262).

The total measured potential drop ψ_0 is not suitable for determining the extent of the specific interaction at liquid surfaces, because, as explained above, the compensation of errors in the simple Gouy formulation often allows eq. (2.27) to be used, even when, as is clear from other studies, a Stern layer is present. Even with a specific interaction of about $2\boldsymbol{k}T$, ψ_0 is insensitive to the model assumed, though ψ_δ and σ_D may be greatly affected.

Position of Polarized Counter-ions at a Liquid Surface

The experimental evidence for the inorganic counter-ions such as Cl′ being in the position C of Fig. 2-23 is as follows. The penetrability of the film of long-chain ions follows from the observed applicability, to a first approximation, of the Gouy equations when $1/\varkappa$ is so small as to be physically impossible at an impenetrable interface. Second-order electrical effects, however, show a sharp break when the NaCl concentration reaches 0.5 N (Fig. 2-25). As expected also, the surface pressure reflects, as a first-order effect, the presence of penetrated counter-ions, both in charged films[35] and in 1:1 films of long-chain cations and anions[23,35,48,49] (Fig. 2-26). These counter-ions must be held with an energy at least comparable to $\boldsymbol{k}T$, since they are not expelled during the kinetic movement of the film, but remain in position and increase Π. For the same reason the surface viscosity increases sharply above a certain NaCl concentration.

These effects are all shown graphically in Fig. 2-25, together with the calculated fraction of the counter-ions in the Stern layer if $\psi_0 = 200$ mV and $\lambda_p = 2\boldsymbol{k}T$. The effects of the penetrating Stern layer for these liquid films in which each long-chain ion occupies 85 to 100 Å^2 become apparent when about 60% of the counter-ions are adsorbed into this layer. From this conclusion, and from the changes in surface viscosity for other anions (Fig. 5-38), one may compute the λ_p values for the latter. These are shown in Table 2-VII. The physical meaning of this penetration is that the counter-ions in position C are able to satisfy their polarization tendencies with several neighbouring fixed ions simultaneously; that they do not penetrate further between the fixed head-groups is shown by the validity of the Gouy or Stern treatments with the normal value of σ in the plane of the head-groups. Nor would we expect the hydrated inorganic ions to enter this region of lower dielectric constant.

There are many undetermined factors in applying the Stern treatment to ionized monolayers. In particular, there is no reason to believe that λ_p should remain constant at different charge densities: the appreciable variations between the reversal of charge concentrations for different phosphate colloids[42], and also the increasing λ_p for anions adsorbing on to mercury surfaces of increasing positive potential (p. 99), constitute strong evidence that λ_p may be variable[50], especially as the arrangement of polarizing groups in the surface is also a function of charge density. Among other undetermined factors are the numbers and precise locations of the adsorption sites; the consequent diversity of explanations[33,40,41,51] of the behaviour of charged monolayers is not unexpected.

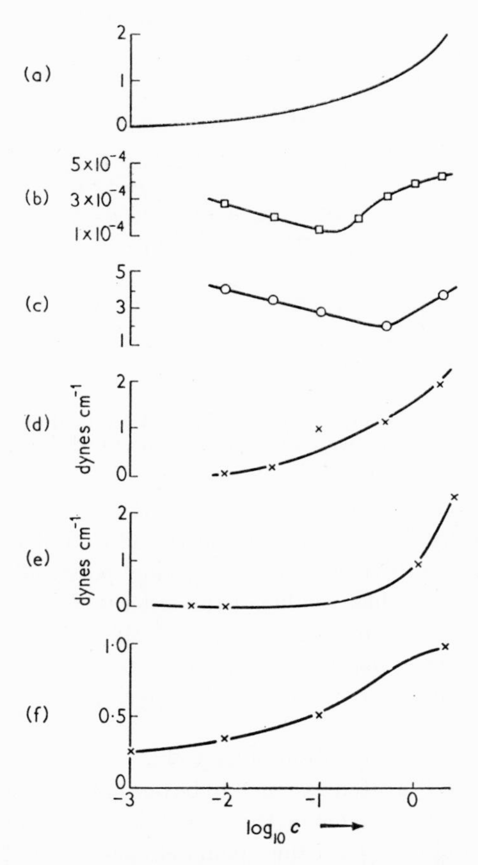

Fig. 2-25. Evidence for penetration of counter-ions into position C. (a) Relative sizes of Cl′ counter-ions and $1/\varkappa$. (b) Surface viscosity of monolayer of $C_{18}H_{37}N(CH_3)_3^+$ spread between air and aqueous NaCl, at 85 Å² per long-chain ion[35]. (c) Slope of μ_{OV}.vs. A for same film[35]. (d) Pressure increase for same film above that calculated by eq. (2.51)[35]. (e) Pressure increase for 1:1 mixed film of cationic and anionic long-chain ions, at 100 Å² per long-chain, on KCl solutions[28,48]. (f) Calculated fraction of counter-ions in Stern layer, with $\psi_0 = 200$ mV and a specific attraction o 2kT, recalculated from Verwey and Overbeek[45].

In calculating the λ_p values of Table 2-VII we have necessarily assumed a reasonable value for n_s, which renders the results only approximate. Further difficulties arise in finding the relative contributions from the inorganic anions and cations adsorbing into 1:1 films of long-chain anions and cations. Further, that ions from the bulk can penetrate into the lower part surface film (position C in Fig. 2-23) implies that their adsorption energy is more than enough to overcome the work of forcing their way into the monolayer

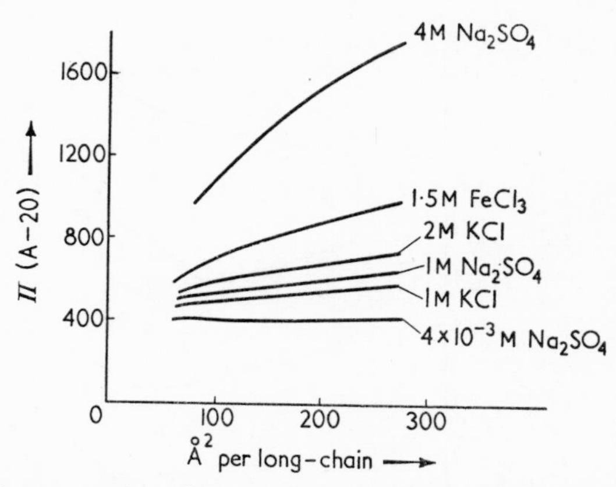

Fig. 2-26. Specific penetration of counter-ions into a 1:1 mixed film of $C_{18}H_{37}N(CH_3)_3^+$ and $C_{22}H_{45}SO_4'$ increases the pressure at high salt concentrations[48].

against the kinetic surface pressure Π_k, i.e. that $\lambda_p > \Pi_k A^*$, where A^* is the actual area of the (hydrated) ion entering the film. Calculations from the data of Fig. 2-26 will give an overall specific potential which must be $\lambda_p - \Pi_k A^*$. The $\Pi_k A$ term has a value of the order $0.3\boldsymbol{k}T$ in the films studied here, and allowance has been made for this in Table 2-VII. These results are consistent with the known high polarization of bromide and iodide ions, and with the high polarizing power of quaternary pyridinium ions[52].

Relation of ψ_δ and ζ

The relation of ψ_δ to ζ is often assumed to be one of equality[51]: consideration of Fig. 2-23 shows, however, that if the immobile, specifically adsorbed ions (position C) are of radius r, then the "plane of shear" is not at a distance δ from the plane of the fixed charges, but at a distance $\delta + r$. For this reason, as well as because of the enhanced viscosity to be expected in the electrical field of the surface (p. 141), we should expect that $\zeta < \psi_\delta$. This is indirectly confirmed by calculating the apparent specific interaction term from ζ instead of from ψ_δ: the consequent values are as high as $10\boldsymbol{k}T$ for Na^+ on a sulphate film, compared with the correct value of about $1\boldsymbol{k}T$ (Table 2-VII).

It is clear from experience, however, that ζ and ψ_δ are always of the same sign, and of the same order of magnitude. This is also seen from Figs. 2-23 and 2-24. The relation of ζ to ψ is discussed further on p. 140.

Position of Counter-ions held in a Liquid Surface by van der Waals Forces

At a liquid interface organic counter-ions will be held in the surface partly by van der Waals forces between the organic part of the molecule and the hydrocarbon chains across the interface (Fig. 2-23, position D). An extreme example is furnished by the 1:1 mixed spread films of (for example) $C_{18}H_{37}N(CH_3)_3^+$ and $C_{22}H_{45}SO_4'$[28,48]. Here either species may be considered the counter-ion but, as the surface potential is now virtually independent of salt concentration, $\psi_0 = 0$. Consequently the cations and anions are now situated in the same plane, or, in terms of the picture of Fig. 2-23, $\delta = 0$. That the counter-ions are pulled into position D by the van der Waals forces is shown for charged monolayers generally by an increase in the concentration of suitable organic counter-ions. The addition of an ionized amine below an anionic monolayer caused the ψ_0 potential to fall much more sharply than with inorganic ions, the values of $\partial\psi/\partial \log c$ being respectively -101 mV and -59 mV[34]. The explanation is that when counter-ions penetrate into position D they reduce σ, so that, by eq. (2.27), ψ_0 changes steeply. Such effects cannot be interpreted by the Stern equations, since δ is now zero: instead, one must modify σ in eq. (2.27)[54]. The ions of quinine, quinoline, and guanidine, as well as the straight-chain amines[27,42,54], behave in this way and, for the latter, W increases by about 1.3 $\boldsymbol{k}$T for each additional $-CH_2-$ group. Similar results are found for counter-ions of organic sulphates adsorbing on to positively charged surfaces[42].

FILM PROPERTIES AND ψ_0

The presence of an electric charge on a monolayer greatly alters its properties and, in particular, if a fatty acid film becomes ionized by formation of its alkali soaps, there is a great diminution of the net lateral adhesion between the film-forming molecules. This considerable loss of net adhesion is obviously due to the repulsion between the similar electric charges developed on adjacent end-groups, when electrolytic dissociation takes place[55].

Films of the C_{20} and C_{22} alkylpyridinium ions and alkyltrimethylammonium ions are virtually insoluble: they are, however, "gaseous" in spite of the length of the alkyl chain, and, again, repulsion between the charged head groups is responsible.

The repulsive energy, Π_r, due to the interaction between the charged head-groups in a monolayer, can be obtained, on certain assumptions[23,28], from the Gouy equation (2.28), calculating the electrical free energy of the ionic double layer. Hence follows the Davies equation:

$$\Pi_r = 6.1\, c_i^{1/2}\left\{\cosh \sinh^{-1}\left(\frac{134}{Ac_i^{1/2}}\right) - 1\right\} \tag{2.51}$$

at 20°C. This repulsive pressure Π_r is additional to that which the film would have if uncharged. If $Ac_i^{1/2} < 38$, eq. (2.51) reduces to the simple form

$$\Pi_r = \frac{2kT}{A} - 6.1\, c_i^{1/2} \tag{2.52}$$

That the Π_r values found experimentally at high values of c_i exceed those calculated from these equations is evidence (see above) of penetration of the counter-ions: this is a first-order effect on the measured surface pressure (Fig. 2-25).

Not only does ψ_0 influence the *surface pressure*, but it also affects the *extent and rate of adsorption* of a soluble film, the kinetics of surface reactions, and the unfolding of polymers in surface films. Of wider interest are the influences of ψ_0 on the stability of oil-in-water emulsions, and on the permeability of the walls of living cells. These problems are all approached through the principle that the electrical energy of a long-chain ion is reduced when the ion desorbs from the surface into the bulk solution: it passes from a region of potential ψ_0 to one of zero potential, and so loses energy of $z_1\varepsilon\psi_0$, where z_1 is the valency of the long-chain ion. This energy assists desorption, increasing both its rate and extent relative to a film of un-ionized molecules of the same chain length: the electrical work lowers the energy required to desorb the entire long-chain ion from W to $W - z_1\varepsilon\psi_0$. In the succeeding chapters there are numerous applications of this relation.

The pH near a Charged Surface

Another effect of ψ_0 is to alter the pH value at the surface: this is reflected in the pH at which occurs the change of interfacial tension indicative of half-ionization. Thus, if one measures the interfacial tension of a long-chain acid adsorbed at the benzene-water interface, one finds[56] that the change occurs at about 3 pH units to the alkaline side of the point at which such materials are half-ionized in bulk aqueous solution. Similarly, amines in the surface are apparently half-ionized at about 3 or 4 pH units to the acid side of the point at which they are in the bulk of the solution.

If pK, the ionization property of the head-group, is now assumed not to change (see below—the head-group always has an aqueous environment), then these experiments show that the pH in the surface, as indicated by the ionization, must be different from the measured value in the bulk phase. Quantitatively, since by eq. (2.25)

$$_s c_{H^+} = c e^{-\varepsilon\psi/kT}$$

the surface and bulk pH values must be related by

$$pH_s = pH_b + \frac{\varepsilon\psi}{2.3\, kT} \qquad (2.53)$$

where pH is defined as $-\log c_{H^+}$.

Only when $\psi = 0$ will the surface and bulk pH values be equal, and, if ψ is negative, $pH_s < pH_b$ because the charge attracts hydrogen ions into the vicinity of the surface. The potentials may arise either from ionization of adsorbed groups at the surface or from fixed charges.

The only difficulty in applying this equation is to know the value of ψ: for the interpretation of the ionization properties of the films, quoted above, ψ_0 should be used, since the measured repulsion is in the plane of the surface. The ψ_0 potentials for such systems may be calculated by eqs. (2.27)-(2.30); a typical value around 200 mV would account for the measured changes of 3-4 pH units.

In the adsorption of dyes at a surface, pH changes are also observed. Thus acid (sulfonated) dyes in aqueous solution, when shaken with benzene to give an emulsion, show colour changes indicative of a more acid pH, while the reverse is true of basic dyes[57]. The addition of a small amount of dye into an emulsion stabilized with long-chain ions constitutes an ingenious method of finding directly the pH of the interfacial region into which the dye is adsorbed. In this way it has been shown[58] that the electrokinetic potential ζ (Chapter 3) is sometimes satisfactory in eq. (2.53). The same conclusion is valid for the surface of the ovalbumin molecule in solution, for which the use of ζ is in agreement with titration results[59]. Since ζ is the potential at an appreciable distance from the plane of the fixed charges, it would seem reasonable to employ ζ in eq. (2.53) when there is evidence that the surface pH with which we are concerned is to be measured close to, but not at, the plane of these fixed charges.

For certain systems $\psi_{Don.}$ may be used in eq. (2.53), calculating this potential by eq. (2.41) or eq. (2.42). For proteins[38,59] the pH_s values thus calculated are in good agreement with the titration curves, and also with the values calculated from ζ. In the hydrolysis of monolayers of monocetyl succinate ions by OH′ in the underlying liquid, the kinetics can be explained[39] only by considering surface, and not bulk, concentrations of OH′ (Chapter 6).

Weakly Ionized Monolayers

If the monolayer is made of a weakly ionized long-chain compound, the dissociation constant of the head-group may possibly be affected by the proximity of the strong heterogeneous electrical field at the surface: up to this point we have assumed such effects unimportant. To examine more closely this assumption, one calculates Π_r from eq. (2.51), remembering that the charge density will now be reduced by a factor α representing the degree of dissociation: the terms in brackets in eqs. (2.28) and (2.51) will now become $\left(\frac{134\alpha}{Ac_i^{1/2}}\right)$. Comparison with experiment for films spread on substrates of differing salt concentrations (up to 0.1 N) shows appreciable disagreements for stearylphosphoric acid, which Payens[60] attributes to a change in the dissociation constant (by a factor of about 30, i.e. $\Delta pK = 1.3$). However, other explanations, in terms of deviation from the simple Gouy model, are also possible[34]. Betts and Pethica[40] report much smaller shifts in pK (by about 0.7 units) for both nonadecylamine and stearic acid, after extrapolating their results to zero α to lessen errors in using the Gouy treatment. The question is complicated further by the appreciable penetration of the counter-ions into the film when $c_i > 0.1$ N, this effect increasing Π_r (Fig. 2-25). We believe that any changes of pK at the surface must be small, and that, within the accuracy of the mathematical models, one may calculate surface properties, including pH_s, using the normal bulk value of pK for the head-groups.

ELECTROCAPILLARY CURVES

Because of the difficulty of discharging ions on a mercury surface, one may impart various electrical potentials to a mercury interface against an aqueous electrolyte. Just as with the air-water and oil-water surfaces, the interfacial tension depends on the electrical potential, E_{Hg/H_2O}, which with mercury may be altered by an external potentiometer (Fig. 2-27). Fig. 2-28 shows typical plots of this dependence: if the reference electrode is a normal

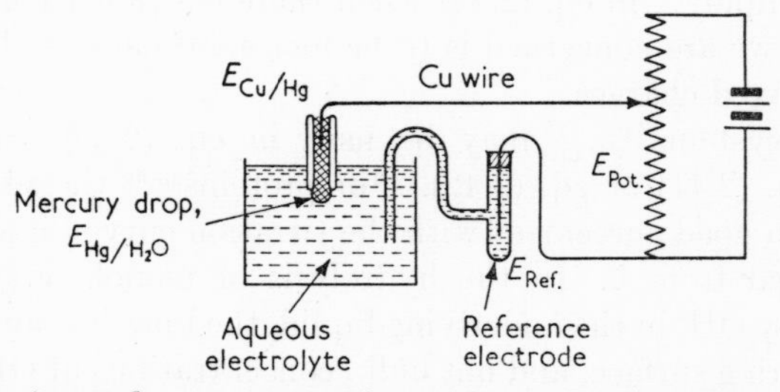

Fig. 2-27. Experimental arrangement for studying electrocapillary curves, with the potentials set up at the different junctions.

calomel electrode, concentrated solutions of various ions which are not specifically adsorbed give maximum values of γ when the potential across the potentiometer is close to -0.488 volts, the negative sign indicating that the negative terminal from the potentiometer is joined to the mercury. Unless otherwise stated, therefore, it is convenient to add 0.488 volts to the potential differences measured, so that all potentials (Figs. 2-28) are quoted relative to this value for the maximum for ions not strongly adsorbed and present in fairly high concentrations. Liquid junction potentials are ignored. On this scale, the position of the maximum shifts to slightly positive potentials at low electrolyte concentrations, even in the absence of specific adsorption—an effect pointed out by Esin and Markov[61].

The interfacial tension may be measured by contact angles (cf. eq. (1.22)) or by the drop weight method, allowing mercury drops to form on a tip immersed in the aqueous phase[62]: other methods have also been described[63,64,65].

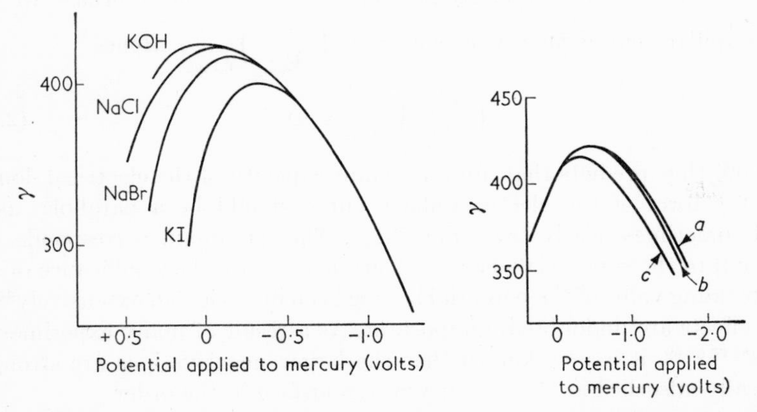

Fig. 2-28. Interfacial tensions (dynes cm.$^{-1}$) between mercury and 1 N-aqueous solutions, at various potentials relative to that of the maximum in the absence of specific adsorption. Curve (a) refers to 0.5 M-Na_2SO_4, with no surface-active agents curve (b) to the same solution, but containing 0.1 M-tetramethyl ammonium sulphate; curve (c) with 0.75 M-tetraethyl ammonium sulphate added[64,65].

The simple theory of the change in interfacial tension with potential (electrocapillarity) is as follows. The superficial electronic charge density on the mercury interface is denoted by σ, and the corresponding potential is E_{Hg/H_2O}. The latter is constituted by $E_{Pot.} + E_{Ref.} + E_{Cu/Hg}$ (see Fig. 2-27) or, since the latter term is constant,

$$dE_{Hg/H_2O} = dE_{Pot.} + dE_{Ref.} \tag{2.54}$$

The associated change in free energy of the mercury-water interface, constant temperature being understood throughout, follows from the theory of electrical condensers. It is

$$dF^s = \sigma dE_{Hg/H_2O} \tag{2.55}$$

whence

$$dF^s = \sigma dE_{Pot.} + \sigma dE_{Ref.}. \tag{2.56}$$

If the reference electrode be the commonly used normal calomel, $E_{Ref.}$ is constant, and hence

$$dF^s = \sigma dE_{Pot.}.$$

This electrical free energy increases the repulsion within the interface, owing to the mutual repulsion of the electrons within the mercury, and γ is reduced to $\gamma - d\gamma$. Hence

$$-d\gamma = \sigma dE_{Pot.} \tag{2.57}$$

or

$$\left(\frac{\partial \gamma}{\partial E_{Pot.}}\right)_{T,P,\mu_i} = -\sigma \tag{2.58}$$

This is known as the Lippmann[66] equation: it may be differentiated further to relate the differential electrical capacity C of the interface to the electrocapillary curve, since C is defined as $\left(\frac{\partial \sigma}{\partial E_{Pot.}}\right)_{T,P,\mu_i}$. Hence

$$\left(\frac{\partial^2 \gamma}{\partial E^2_{Pot.}}\right)_{T,P,\mu_i} = -C \tag{2.59}$$

Eq. (2.59) thus predicts that, for a constant capacity of the electrical double layer, the form of the electrocapillary curve should be a parabola, as is indeed sometimes nearly true (Fig. 2-28). The maximum corresponds (by eq. (2.58)) to the superficial charge density being zero, the significance of the corresponding value of the potential having been investigated extensively[64,67]; dipole effects are undoubtedly important. As a result of many experimental studies[65,66,68,69], it is now known that the larger negative ions are strongly specifically adsorbed at the mercury-water surface in the order

$$S'' > I' > CNS' > Br' > Cl' > OH' > F'$$

The electrocapillary curves of Fig. 2-28 show this clearly, the adsorption effects of these anions naturally being enhanced when the applied potential renders the mercury positive, and vice-versa.

Similar effects of the sequence of cations

$$N(C_2H_5)_4^+ > N(CH_3)_4^+ > Tl^+ > Cs^+ > Na^+$$

are visible (Fig. 2-28) in the region of negative potentials applied to the surface: the larger ions are adsorbed in this region by both electrostatic potentials and also by a specific adsorption potential $\lambda_p + W$. The quaternary ammonium ions are of particular interest in that they are resistant to discharge at the negative mercury electrode, and potentials high enough to cause the electrocapillary curve to cut the axis of $\gamma = 0$ can be maintained across the interface. The consequent spontaneous emulsification is discussed in Chapter 8 (p. 363).

Neutral molecules, such as n-heptanol, adsorb appreciably near the maximum when the electrical effects are small (Fig. 2-29a) but are displaced by the ions which are preferentially attracted to the mercury when the latter bears a sufficient electrical potential, either positive or negative. Aromatic molecules are particularly strongly adsorbed on the positive branch of the curve, due to interaction between the π-electrons of the former with the positive charges on the mercury surface[69].

The overall effect of adsorption on γ (or Π) is complicated, since both σ on the mercury surface and also the repulsion within the aqueous surface film are affected. Further, both are dependent on the same external variable,

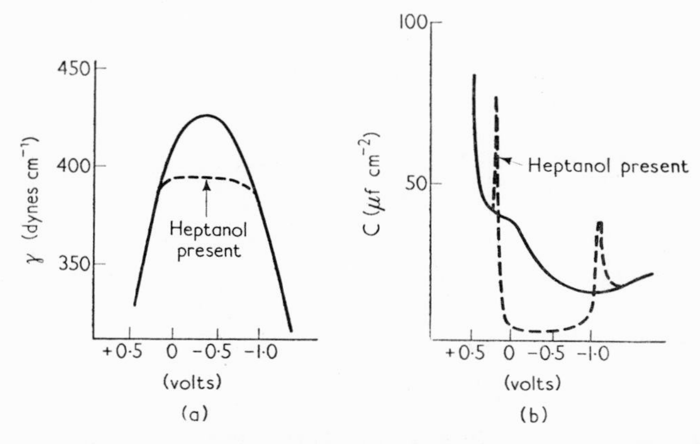

Fig. 2-29. Interfacial tension (a), and differential capacity (b) with 0.5 M-Na_2SO_4 alone (full curves). When n-heptanol is present, the curve becomes that shown by the broken line[65]. All voltages (abscissae) relative to standard E.C. maximum.

$E_{Pot.}$, as well as on the specific adsorption effects. Accordingly, one is restricted to studying differential effects rather than predicting the lowering of γ. The differential effect of adsorption on γ at any particular value of $E_{Pot.}$ follows from the Gibbs equation (4.54); while Na^+ and certain other univalent cations are negligibly adsorbed even at zero potential, chloride ion is specifically adsorbed under these conditions with $\lambda_p = 4.6\boldsymbol{k}T$. On positively charged mercury, moreover, λ_p for Cl′ increases with the positive charge, reaching $10\boldsymbol{k}T$ or more: this is because the positive charge enhances the deformation, polarization, and dehydration of the ion. The specific adsorption potential of F′ is too small ($<1.5\boldsymbol{k}T$) to be detected in this system, but the heavier anions are adsorbed in excess of the amounts calculated, so that the Na^+ counter-ions are adsorbed in a diffuse double layer. Fig. 2-30, obtained from adsorption equations, shows this clearly. Specific adsorption is also reflected in the divergence of the curves for positive potentials in Fig. 2-28. Inorganic univalent cations are not specifically adsorbed, but the quaternary ammonium ions have a relatively constant and rather small specific adsorption to mercury: most of the lowering of interfacial tension (Fig. 2-28) appears to be due to electrical effects.

An interesting analogy to the mercury interface, with its imposed potential, is the *oil-water interface* containing an insoluble monolayer of long-chain ions. These ions are unable to escape if the oil is paraffinic and the chain long enough to prevent desorption into the water, and one can vary the charge density by spreading films of various known ratios of long-chain cations and anions, the total number of long-chain ions per unit area being kept constant.

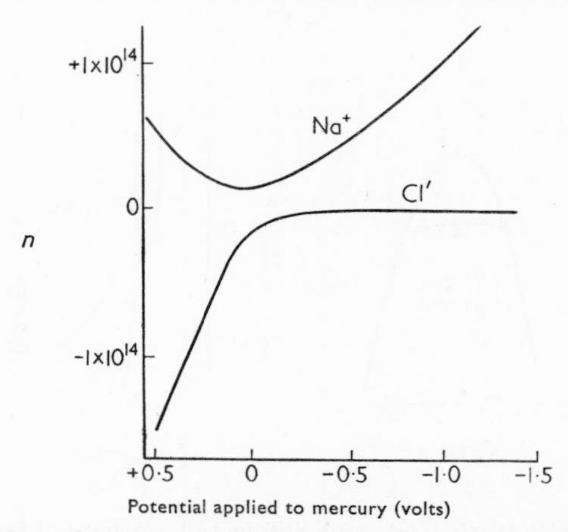

Fig. 2-30. Adsorption of ions on to mercury at various potentials relative to the standard electrocapillary maximum from 0.3 MNaCl at 25°C[65]. Even at zero potential, some Cl′ is adsorbed, and compensating Na^+ ions enter the diffuse layer.

The interfacial tension then depends on the net charge density, and hence on ψ_0. The formal expression of this effect follows from eq. (5.10a), since $\Pi = \gamma_0 - \gamma$

$$\gamma = \gamma_0 - \left(\frac{\boldsymbol{k}T}{A - A_0}\right) - \Pi_r$$

or, using (2.28) and (2.51),

$$\gamma = \gamma_0 - \left(\frac{\boldsymbol{k}T}{A - A_0}\right) - 6.10 c_i^{1/2}\left\{\cosh\left(\frac{\varepsilon\psi_G}{2\boldsymbol{k}T}\right) - 1\right\} \tag{2.60}$$

Hence, using the known surface charge densities to obtain ψ_0 (by eq. (2.27)), one can predict the form of the curve (Fig. 2-31), and compare this with experiment[67]. Though the lack of constancy in the capacity of a diffuse double layer (see below) prevents the curve being a true parabola, an expansion of the hyperbolic cosine for small values of ψ_G shows that the curve is parabolic in this region. By comparison with Fig. 2-28 it is clear that there is considerable specific anion adsorption: this is comprehensible in that the fixed cations are here cetylpyridinium ions, the positive charge on which is unshielded by organic groups and hence is always found to be strongly polarizing, promoting specific interactions. Even in the 1:1 mixed film with cetylsulphate ions Π is 2.2 dynes cm.$^{-1}$ higher than calculated by eq. (2.60) with $\psi_G=0$. Further, from the λ_P value of Na^+ we may calculate that λ_P is $3.5\boldsymbol{k}T$ for chloride ions in contact with pyridinium ions (Table 2-VII). This explains the shift of the maximum to net negative potentials of the fixed ions (Fig. 2-31). Though ΔV can be measured for these systems, the uncertainty about the values of μ_D and their constancy reduces the value of such experiments.

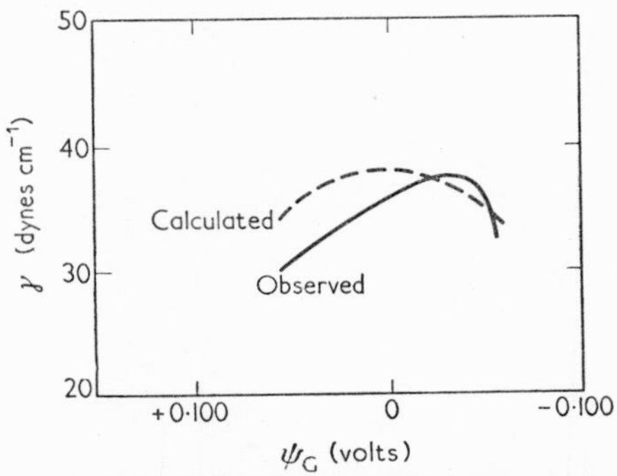

Fig. 2-31. Mixed films of cetylpyridinium ions and cetylsulphate ions spread at the oil-water interface, against 1 N-NaCl. The potentials are calculated from the net composition of the film[67].

ELECTRICAL CAPACITY OF THE DOUBLE LAYER

Direct measurements (with an impedance bridge or otherwise) of the differential capacities[65,70,71,72,73] are more accurate than double differentiation of the electrocapillary curve, though within the limits of accuracy of the latter procedure eq. (2.59) is confirmed[65]. Typical results are shown in Fig. 2-32, these always being interpreted theoretically as being due to the capacitance effects of the Stern and Gouy layers acting in series (Fig. 2-33), so that the measured capacity C should be given by

$$\frac{1}{C}=\frac{1}{C_{\text{Stern}}}+\frac{1}{C_{\text{Gouy}}} \tag{2.61}$$

The capacity of a Stern layer (Figs. 2-23 and 2-24) may be calculated from the formula for a condenser of unit area:

$$C_{\text{Stern}}=\frac{D}{4\pi\delta}$$

where δ is here (since the surface is impenetrable) the radius of the counter-ion, possibly in its hydrated state. Numerically, this expression reduces to

$$C_{\text{Stern}} \approx 9D/\delta \quad \mu\text{f. cm.}^{-2} \tag{2.62}$$

when δ is in Å. The values of D and δ in such a layer of cations are not known exactly, but reasonable values (assuming hydration and nearly complete dielectric saturation) are 9 and 4 Å respectively: these give a calculated capacity in accord with the nearly constant figure of 20 μf. cm.$^{-2}$ commonly found at appreciably negative potentials, when there are reasons (see below) for believing that the Stern capacity is dominant. For hydrated H^+ ions, slightly smaller capacities of the order 15 μf. cm.$^{-2}$ are found[65].

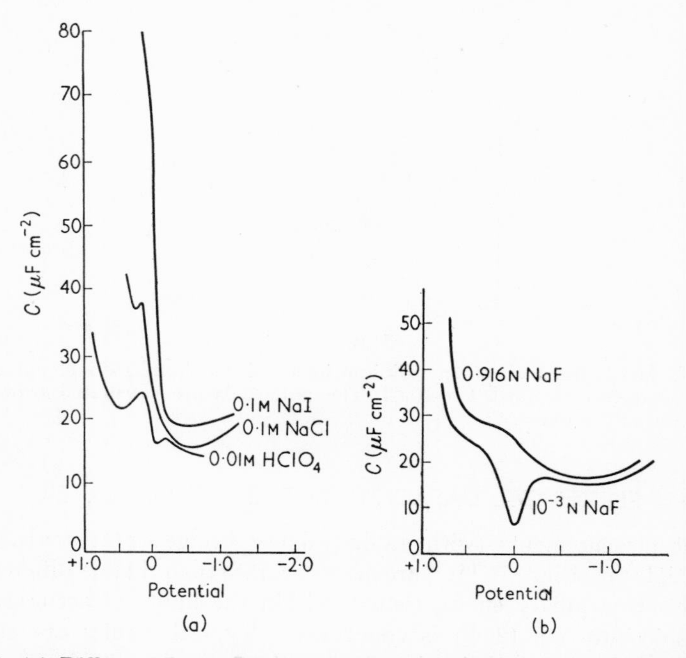

Fig. 2-32. (a) Differential capacities at different potentials relative to the electro-capillary maximum for each salt, at 25°C[65]. (b) The same for fluoride of different concentrations[70]. Potentials are measured in volts throughout.

At positive potentials, when anions are adsorbed, the capacities are much higher, and increase with increasing potential. This is compatible with the high specific interactions between the anions and the mercury discussed above, the tendency being to dehydrate the anions, whose larger size and surplus electrons renders them very liable to this process. The larger anions may also be strongly deformed and polarized. Limiting values of δ, for the completely dehydrated ions, are 1.4 Å for fluoride (giving a calculated C of 60 μf. cm.$^{-2}$), and 2.1 Å for I′ (giving a calculated C of 40 μf. cm.$^{-2}$). As seen from Fig. 2-32, these values are of the right order of magnitude, but it is useless to press the calculations further at present.

For a diffuse double layer the capacity is given from the Gouy equation (2.48): and at 25°C numerical substitution gives for uni-univalent electrolytes:

$$C_{Gouy} = 228.5\, c_i^{1/2} \cosh\left(\frac{\varepsilon\psi_\delta}{2\boldsymbol{k}T}\right)\ \mu\text{f. cm.}^{-2} \tag{2.63}$$

At ionic concentrations in excess of 0.1 N, C_{Gouy} exceeds 70 μf. cm.$^{-2}$ even when $\psi_\delta = 0$: as eq. (2.61) shows, such large capacities have a negligible influence on C, and only C_{Stern} is measured. The physical significance of these large capacities lies in the implicit use of 80 as the dielectric constant within

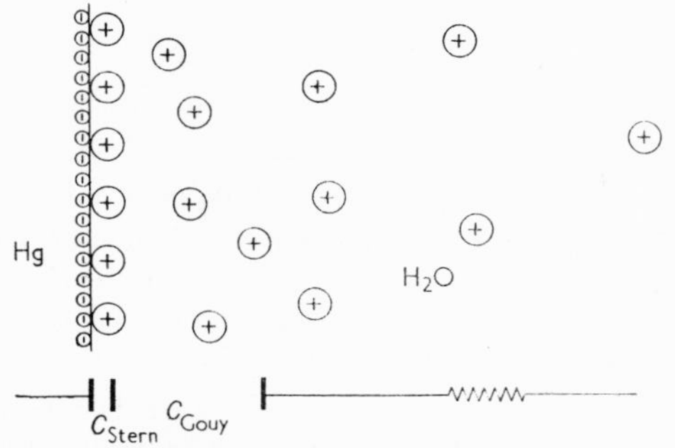

Fig. 2-33. Counter-ions at a negatively charged mercury interface. Those in the Stern layer may be less hydrated than those in the diffuse part of the double layer.

the diffuse part of the double layer. At lower ionic strengths C_{Gouy} rapidly decreases, becoming only 7 μf. cm.$^{-2}$ if $\psi_\delta \to 0$ and $c_i = 10^{-3}$ N. The "trough" in the capacity plot for 10^{-3} N NaF (Fig. 2-32b) at the potential of the electrocapillary maximum is determined largely by C_{Gouy} according to eq. (2.61), since this capacity is now the lower of the two in series: by eq. (2.62) it is 7.2 μf. cm.$^{-2}$ when $\psi_\delta = 0$. Further, the shape[70,72] of this trough is roughly that predicted by eq. (2.63).

The presence of heptanol or octanol in the solution affects the electrocapillary curve, as in Fig. 2-29a above, and the rather sharp discontinuities should be very steeply reflected in the second differential of this curve (i.e. in the differential capacity) according to eq. (2.59). Direct study of the capacity confirms this[65,70,71] (Fig. 2-29b), the capacity across the adsorbed monolayer of heptanol being only of the order 3.64 μf. cm.$^{-2}$ With an assumed dielectric constant of the oriented alcohol of 4, this capacity is consistent with the counter-ions being separated from the mercury by a monolayer of heptanol: this effect increases considerably the term δ in the equation for C_{Stern}.

For longer-chain additives, the effective dielectric constant of a close-packed monolayer must approach that of a hydrocarbon (about 2.1): the measured capacity when hexadecanol is present supports this figure, the value of δ found from eq. (2.62) being 19 Å, equal to the length of the molecule. The latter equation, however, gives too small values of δ for oleic acid, suggesting that these molecules, which cannot pack so closely because of the double bond, are partially penetrated by water or even by counter-ions[71].

If only a small amount of surface-active material is present the Stern part of the double layer may be regarded as two condensers in parallel, one containing the surface-active material as dielectric, and the other water[75]. Both the capacities are measured at the same electrode potential, and one may write

$$C_{\text{Stern}} = \theta C_{\text{S.A}} + (1-\theta)C_{\text{W}}$$

where θ is the fraction of the interface covered, and the subscripts refer to the regions covered with surface-active material and with water. By working in turn with completely bare and completely covered interfaces, one may use this equation to find the $\theta - c$ relationship for the surface-active additive.

The peaks in Fig. 2-29b are an exaggerated reflection of the suddenly changing thickness of the Stern layer, due to the desorption of the alcohol monolayer from the interface: the height of the peak is determined by the sharpness of the transition between the two Stern layers, and for polymers tends to increase markedly with molecular weight[74]. The differential capacity may also be frequency-dependent if there is an organic monolayer adsorbed on the mercury[70,74], suggesting a "leak" resistance due to passage of ions through this film.

Without added organic materials, the inflection is much slighter, though in certain systems (Fig. 2-32) one or two "humps" are visible. These must reflect rather discontinuous changes in the adsorption layers, in which may be found, according to the applied potential, strongly polarized anions, water dipoles only, or hydrated cations. The "hump" observed at 5°C with NaF, when the mercury is slightly positive, becomes smoother at 25°C and disappears at higher temperatures[70], these effects being independent of the concentration of NaF. The "hump" is also absent in methanolic solutions, suggesting that a weakly adsorbed layer of water dipoles may be replaced by adsorbed counter-ions outside a narrow potential region around the electrocapillary maximum. The unspecific "humps" generally found on the negative branches if the salt concentration is 0.01 N or lower[65] are a consequence of the forms of eqs. (2.61) and (2.63): if C_{Gouy} and C_{Stern} are nearly equal, C passes through a slight maximum.

REFERENCES

1. Beutner, *Z. Elektrochem.* **19**, 319, 467 (1913); *J. Amer. chem. Soc.* **36**, 2040, 2045 (1914).
2. Beutner and Barnes, *Biodynamica* **5**, 117 (1945);
   Beutner, *Science* **104**, 569 (1946).
3. Baur, *Z. Elektrochem.* **19**, 590, (1913); **32**, 547 (1926).
4. Craxford, Gatty, and Rothschild, *Nature, Lond.* **141**, 1098 (1938);
   Dean, Gatty, and Rideal, *Trans. Faraday Soc.* **36**, 161 (1940).
5. Davies, and Rideal, *Canad. J. Chem.* **33**, 947 (1955).
6. Nernst, *Z. phys. Chem.* **9**, 140 (1892).
7. Cheesman, and King, *Trans. Faraday Soc.* **36**, 241 (1940);
   Verwey, *Trans. Faraday Soc.* **36**, 192 (1940); *Phil Mag.* **28**, 435 (1939).
8. Karpfen and Randles, *Trans. Faraday Soc.* **49**, 823 (1953).
9. Davies, *J. phys. Chem.* **54**, 185 (1950).
10. Kahlweit, and Strehlow, *Z. Elektrochem.* **58**, 658 (1954).
11. Dean, *J. exp. Biol.* **16**, 134 (1939); *Cold Spr. Harb. Symp. quant. Biol.* **8**, 62 (1940).
12. Ehrensvärd, and Cheesman, *Tidskr. Särtr. Svensk Kem.* **53**, 126 (1941).
13. Osterhout, *Cold Spr. Harb. Symp. quant. Biol.* **8**, 51 (1940).
14. McMullen, PhD. thesis, Cambridge (1948).
15. Dupeyrat, *Mémor. Serv. chim. Etat* **40**, 51 (1955).
16. Davies, *Z. Elektrochem.* **55**, 559 (1951); *Nature, Lond.* **167**, 193 (1951).
17. Kinloch and McMullen, *J. sci. Instrum.* **36**, 347 (1959).
18. Yamins and Zisman, *J. chem. Phys.* **1**, 656 (1933);
    Frost, *Trans. electrochem. Soc.* **82**, 259 (1942);
    Kemball, *Proc. roy. Soc.* **A201**, 377 (1950).
19. Guyot, *Ann. Phys. Paris* **2**, 506 (1924);
    Frumkin, *Z. phys. Chem.* **116**, 485 (1925);
    Schulman and Rideal, *Proc. roy. Soc.* **A130**, 259 (1931).
20. Compton and Haring, *Trans. electrochem. Soc.* **62**, 345 (1932).
21. Davies, *Nature, Lond.* **167**, 193 (1951); *Biochim. biophys. Acta* **11**, 165 (1953); *Biochem. J.* **56**, 509 (1954).
22. Davies, *Trans. Faraday Soc.* **49**, 683, 949 (1953).
23. Davies, *Proc. roy. Soc.* **A208**, 224 (1951).
24. Alexander A. E. and Teorell, *Trans. Faraday, Soc.* **35**, 727, 733 (1939); **37**, 117 (1941);
    Dean, *Nature, Lond.* **144**, 32 (1939); *Trans. Faraday Soc.* **36**, 166 (1940).
25. Alexander A. E. and Schulman, *Proc. roy. Soc.* **A161**, 115 (1937);
    Davies, *Biochim. biophys. Acta* **11**, 165 (1953).
26. Gerovich and Frumkin, *J. chem. Phys.* **4**, 624 (1936);
    Gerovich and Vargin, *Acta phys.-chim. U.R.S.S.* **8**, 63 (1938);
    Klevens and Davies, *Proc. 2nd. Internat. Congr. Surface Activity* **1**, 31, Butterworths, London (1957);
    Fox, H. W., *J. phys. Chem.* **61**, 1058 (1957).
27. Schulman and Rideal, *Proc. roy. Soc.* **A130**, 259, 284 (1931);
    Schulman and Hughes, *Proc. roy. Soc.* **A138**, 430 (1932);
    Cassie and Palmer, *Trans. Faraday Soc.* **37**, 156, (1941);
    Crisp, in "Surface Chemistry" p. 65, Butterworths, London (1949).
28. Davies, *Trans. Faraday Soc.* **48**, 1052 (1952); in "Surface Phenomena in Chemistry and Biology" p. 55, Pergamon Press, London (1958).

29. Gouy, *J. Phys. Radium* **9**, 457 (1910).
30. Frumkin and Pankratov, *Acta. phys.-chim. U.R.S.S.* **10**, 45, 55 (1939).
31. Matijevic and Pethica, *Trans. Faraday Soc.* **54**, 1390, 1400 (1958).
Betts and Pethica, *Trans. Faraday Soc.* **56**, 1515 (1960).
32. Müller, *Kolloidchem. Beih.* **26**, 257 (1928);
Bikerman, *Phil. Mag.* **33**, 384 (1942);
Grahame, *J. chem. Phys.* **18**, 903 (1950);
Buckingham, *J. chem. Phys.* **18**, 903 (1950);
Conway, Bockris and Ammar, *Trans. Faraday Soc.* **47**, 756 (1951);
Freise, *Z. Elektrochem.* **56**, 822 (1952);
Bolt, *J. Colloid Sci.* **10**, 206 (1955);
Sparnaay, *Rec. Trav. chim. Pays-Bas* **77**, 872 (1958);
Brodowsky and Strehlow, *Z. Elektrochem.* **63**, 262 (1959).
Levine and Bell, G. M., *J. Phys. Chem.* **64**, 1188, 1195 (1960).
33. Samis and Hartley, *Trans. Faraday Soc.* **34**, 1288 (1938);
Aickin and Palmer, *Trans. Faraday Soc.* **40**, 116 (1944).
34. Haydon and Taylor, F. H., *Phil. Trans.* **252A**, 225 (1960).
35. Davies, *Trans. Faraday Soc.* **47**, 414 (1951);
Davies and Rideal, *J. Colloid Sci. Suppl.* **1**, 1 (1954).
36. Donnan, *Z. Elektrochem.* **17**, 572 (1911); *Chem. Rev.* **1**, 73 (1924);
Tiselius and Svensson, *Trans. Faraday Soc.* **36**, 16 (1940);
Overbeek, *Progr. Biophys.* **6**, 57 (1956).
37. Wilson, J. A., *J. Amer. chem. Soc.* **38**, 1982 (1916);
Danielli, *Biochem. J.* **35**, 470 (1941).
38. Davies, in "Surface Chemistry" p. 95, Butterworths, London (1949);
Davies and Rideal, *Proc. roy. Soc.* **A194**, 417 (1948);
38a. Davies, calculations quoted in Danielli and Davies, *Advanc. Enzymol.* **11**, 35 (1951).
39. Davies and Rideal, *J. Colloid Sci.* **3**, 313 (1948).
40. Betts and Pethica, *Trans. Faraday Soc.* **52**, 1581 (1956).
41. Kruyt, "Colloid Science" I and II, Elsevier, Amsterdam (1952, 1949).
42. Gilbert and Rideal, *Proc. roy. Soc.* **A182**, 335 (1944).
43. Meggy, *Trans. Faraday Soc.* **43**, 502 (1947).
Vickerstaff, "The Physical Chemistry of Dyeing", Oliver and Boyd, London (1950).
Alexander P. and Kitchener, *Text. Res. J.* **20**, 203 (1950).
44. Crisp, as in ref. 27.
45. Stern, *Z. Elektrochem.* **30**, 508 (1924).
Verwey and Overbeek, "Theory of the Stability of Lyophobic Colloids", Elsevier, Amsterdam (1948).
46. Pauling, "The Nature of the Chemical Bond", Cornell University Press (1944).
Chapman, *Nature, Lond.* **174**, 887 (1954); *Research Correspondence* in *Research* **8**, March (1955).
47. Somers, *Nature, Lond.* **184**, 475 (1959); **187**, 427 (1960).
48. Phillips and Rideal, *Proc. roy. Soc.* **A232**, 149, 159 (1955).
49. Strehlow, *Z. Elektrochem.* **59**, 744 (1955).
50. Pethica and Few, *Disc. Faraday Soc.* **18**, 258, 307 (1954).
51. Anderson, P. J., *Trans. Faraday Soc.* **55**, 1421 (1959).
52. Richter, *Z. phys. Chem.* **12**, 247 (1957);
Hartley, *Kolloidzschr.* 88, 22, (1939).
53. Klotz and Curme, *J. Amer. chem. Soc.* **70**, 939 (1948).

53a. Davies, unpublished calculations.
53b. Webb and Danielli, *Nature, Lond.* **146**, 197 (1940);
Danielli, *J. exp. Biol.* **20**, 167 (1944);
Havinga, *Rec. Trav. chim. Pays-Bas* **71**, 72 (1952);
Porter, *Proc. 2nd. Internat. Congr. Surface Activity* **4**, 103, Butterworths, London (1957).
54. Phillips, *Trans. Faraday Soc.* **51**, 1726 (1955).
55. Adam. *Proc. roy. Soc.* **A101**, 516 (1922);
Adam, "The Physics and Chemistry of Surfaces", Clarendon Press, Oxford (1941).
56. Peters, *Proc. roy. Soc.* **A133**, 147, (1931); *Trans. Faraday Soc.* **26**, 197 (1930).
57. Deutsch, *Z. phys. Chem.* **60**, 353 (1928).
58. Hartley and Roe, *Trans. Faraday Soc.* **36**, 101 (1940).
59. Danielli, *Proc. roy. Soc.* **B122**, 155, (1937); *Biochem. J.* **35**, 470 (1941).
60. Payens, "Ionized Monolayers" **10**, 425, Philips Research Reports, (1955).
61. Esin and Markov, *Acta phys.-chim. U.R.S.S.* **10**, 353 (1939);
Parsons, *Proc. 2nd Internat. Congr. Surface Activity* **3**, 38, Butterworths, London (1957).
62. Burdon, "Surface Tension and Spreading of Liquids", Cambridge University Press (1949).
Craxford and McKay, *J. phys. Chem.* **39**, 545 (1935).
63. Rideal "An Introduction to Surface Chemistry", Cambridge University Press (1930).
64. Adam "The Physics and Chemistry of Surfaces", Clarendon Press, Oxford (1941);
Gouy, *Ann. Chim.* (*Phys.*) **9**, 75 (1906).
65. Grahame, *Chem. Rev.* **41**, 441 (1947).
66. Lippmann, *Ann. Chim.* (*Phys.*) **5**, 494 (1875).
67. Oel and Strehlow, *Z. phys. Chem.* **4**, 89 (1955); *Z. Elektrochem.* **58**, 665 (1954); **59**, 744 (1955).
68. Gouy, *Ann. Phys. Paris* **6**, 5 (1916); **7**, 129 (1917).
69. Frumkin, *J. phys. Chem.* **34**, 74 (1930); *Phys. Z. Sowjet.* **1**, 255 (1932); "Surface Phenomena in Chemistry and Biology", p. 189, Pergamon Press, London (1958); *Proc. 2nd. Internat. Congr. Surface Activity* **1**, 58, Butterworths, London (1957).
70. Grahame, *J. Amer. chem. Soc.* **63**, 1207 (1941); **68**, 301 (1946); **71**, 2978 (1949); **74**, 1207 (1952); **76**, 4819 (1954); *J. chem. Phys.* **22**, 449 (1954); **23**, 1725 (1955); *J. Amer. chem. Soc.* **80**, 4201 (1958).
71. Gorodezkaya and Frumkin, *C. R. Acad. Sci. U.R.S.S.* **18**, 639 (1938);
Proskurnin and Frumkin, *Trans. Faraday Soc.* **31**, 110 (1935).
72. Brodowsky and Strehlow, *Z. Elektrochem.* **63**, 262 (1959).
73. Breyer, *Proc. 2nd Internat. Congr. Surface Activity* **3**, 34, Butterworths, London (1957);
Watanabe, *ibid.* p. 94.
74. Miller and Grahame, *J. Amer. chem. Soc.* **79**, 3006 (1957);
Miller, *Proc. 2nd Internat. Congr. Surface Activity* **3**, 53, Butterworths, London (1957).
75. Frumkin, *Z. Phys.* **35**, 792 (1926);
Parsons, *Trans. Faraday Soc.* **55**, 998 (1959).

Chapter 3

Electrokinetic Phenomena

SURFACE CONDUCTANCE

If two parallel thin wires of length l be placed in a surface at a distance b cm. apart (Fig. 3-1), the conductance of the surface between them (corrected for bulk effects) will be given by $K_s l/b$ where K_s is the specific surface conductance and has the dimensions of ohm^{-1}.

Suppose now that above the aqueous electrolyte we replace the air by an ionogenic material such as glass. The surface of the glass contains silicic acid groups from which some cations (e.g. H^+) will split off into the water, while the silicate groups remain part of the solid sheet. The latter will now bear a

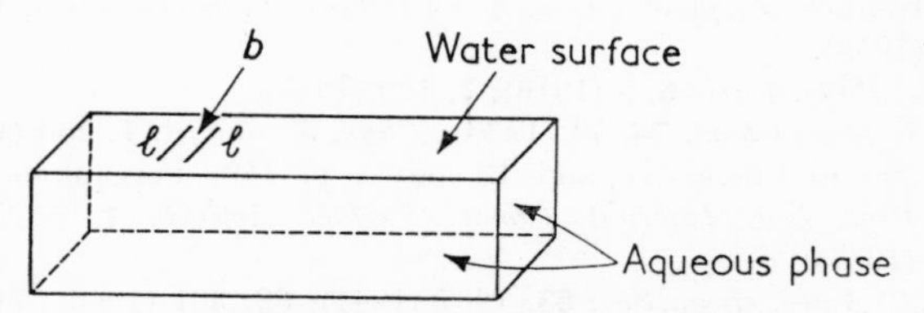

Fig. 3-1. Definition of surface conductance.

negative charge which, although causing a redistribution of the ions of the electrolyte near the surface, will not alter its actual conductivity: only the mobile cations (forming a diffuse double layer), added to the system by the ionization of the silicate groups, will alter the conductivity. This latter increment will be the surface conductivity. It was observed[1] as early as 1893, and was shown to be consistent with the idea of an electrical double layer by Smoluchowski. This conductance, associated with the high concentration of ions at a charged surface, may significantly reduce the values of the streaming potential, anomalous viscosity, and electrophoretic velocity, partially "short-circuiting" these electrical phenomena (see below).

Smoluchowski's theoretical treatment[1] has now been superseded by more sophisticated approaches[2,3,4]. The following calculation is sufficiently exact for most purposes. Suppose that a negatively charged surface, as of glass, is in contact with an aqueous solution of HCl whose bulk concentration is c gm.ions l^{-1}. The concentration of H^+ near the surface will be relatively high because of the negative potential of the surface, and, expressed in electronic charges per cm.^{2}, the charge density σ_D of the univalent counter-ions in the diffuse (Gouy) part of the double layer must, if electrical neutrality is to be preserved, be equal to the net surface charge density, and hence by eq. (2.27),

$$\sigma_D = -\left(\frac{D\mathrm{RTc}}{500\pi}\right)^{1/2} \sinh\left(\frac{\varepsilon\zeta}{2\boldsymbol{k}\mathrm{T}}\right) \tag{3.1}$$

Here ζ is the potential, relative to the bulk, of the innermost part of the diffuse ionic layer; in general, due perhaps to specific adsorption of some of the counter-ions on to the surface, to surface roughness, to variations of D or of viscosity near the interface, ζ will be rather less than ψ_0. Conversion of σ_D to gram equivalents per cm.^{2} and multiplication by 1_+, the ionic conductance (in ohm^{-1}cm.^{2}) of the counter-ions (e.g. H^+), gives the surface conductance K_s in ohm^{-1}:

$$K_s = \frac{\sigma_D 1_+}{\boldsymbol{F}} \tag{3.2}$$

or

$$K_s = \left(\frac{D\mathrm{RTc}}{500\pi \boldsymbol{F}^2}\right)^{1/2} 1_+ \sinh\left(\frac{\varepsilon\zeta}{2\boldsymbol{k}\mathrm{T}}\right) \tag{3.3}$$

Here we must remember that the movement of the counter-ions in the double layer will also cause a movement of the water (electro-osmosis): this increases the ionic conductance of the counter-ions[3,4] by the term $D\mathrm{RT}/2\pi\eta$ which for water at 20°C is 34 ohm^{-1} cm.^{2} equiv.$^{-1}$ Thus 1_+ in surface conductance is, for H^+, $330+34=364$, for Na^+, $45+34=89$, and for K^+, $68+34=102$ ohm^{-1}cm.^{2} equiv.$^{-1}$ This electro-osmotic augmentation of l will occur at all frequencies of the A.C. used in the conductance studies[4], up to about 10^5 cycles per second.

As an example, consider a glass surface in contact with 10^{-5} N-HCl. The ζ potential is found (if $D=80$, see below) to be about -80 mV, and application of eq. (3.3) with $1_{H^+}=364$, gives $K_s=0.3\times10^{-9}$ohm^{-1}. In 5×10^{-4} N-KCl, ζ is about -95 mV, and with $l_{K^+}=102$, $K_s=0.9\times10^{-9}$ ohm^{-1}. In 10^{-5} N-KCl, $\zeta=-140$ mV, and K_s is reduced slightly, to a calculated value of 0.3×10^{-9}ohm^{-1}, and in 1.3×10^{-5} $AgNO_3$, K_s is 0.34×10^{-9} ohm^{-1}. Comparison with recent experimental data in Table 3-1 shows that agreement is quite good. It may be noted that even if the counter-ions originally associated with the glass are (say) Na^+, the preliminary washing which always precedes measurements will exchange these for the cations of the aqueous electrolyte solution. Further, since ζ decreases as the salt concentration c is raised, eq. (3.3) shows that K_s will vary only slightly with concentration: this is confirmed experimentally (Table 3-I).

For a glass capillary containing 10^{-5} N-KI in methanol, ζ is -92 mV[4]. However, the lower dielectric constant makes the calculated K_s rather small, about 0.04×10^{-9} ohm^{-1}: this is in accord with experiment (Table 3-I). Agreement of calculation and experiment for clay is also satisfactory.[4a]

In calculating ζ from electrokinetic phenomena (eq. (3.8) below) it is

TABLE 3-I

Experimental Surface Conductances

System	Method	Experimental K_s (ohm^{-1})	Reference
Jena glass single capillary containing water	2	0.64×10^{-9}	7
Jena glass single capillary containing 10^{-5} N-KCl	2	0.5×10^{-9} to 1.0×10^{-9}	7 4
Jena glass single capillary containing 2.5×10^{-5} N-KI	2	0.9×10^{-9}	8
Jena glass single capillary containing 1.3×10^{-5} N-$AgNO_3$	2	0.6×10^{-9}	8
Pyrex glass spheres in 0.9×10^{-4} N-HCl	3	0.42×10^{-9}	9
Pyrex glass spheres in 1.7×10^{-4} N-KCl	3	0.45×10^{-9}	9
Pyrex glass spheres in 0.9×10^{-3} N-KCl	3	0.58×10^{-9}	9
Pyrex glass fibres in 4.2×10^{-3} N-KCl	4	4.5×10^{-9}	12
Pyrex glass spheres in 7.9×10^{-3} N-KCl	3	1.8×10^{-9}	9
Jena glass capillary containing 10^{-5} N-KI in methanol	2	0.05 to 0.12×10^{-9}	4
Collodion membrane in 10^{-2} N or 10^{-3} N-KCl	4	0.16×10^{-9}	10
Collodion membrane dyed with nigrosin	4	0.07×10^{-9}	10
Collodion membranes in distilled water and dilute KCl	4	order of 10^{-9}	Earlier results recalc. in 10
Nylon in 4.4×10^{-4} N-KCl	4	0.5×10^{-9}	12
Nylon in 3.6×10^{-4} N-HCl	4	2.2×10^{-9}	12
Dacron in 4.7×10^{-4} N-KCl	4	1.9×10^{-9}	12
Dacron in 2.3×10^{-4} N-HCl	4	1.6×10^{-9}	12
Orlon in 4.9×10^{-4} N-KCl	4	3.5×10^{-9}	12
Kaolinite in $CaCl_2$	3	0.7×10^{-9}	9
Kaolinite in HCl	3	0.6×10^{-9}	9
Ceramic filter in 10^{-3} N-KCl	4	1.3×10^{-9}	10
Monolayer of stearic acid on distilled water (pH 5.7)	1	35×10^{-9} to 38×10^{-9}	6
One face of Na soap lamella	1	7×10^{-9} to 16×10^{-9}	recalc. from 1

assumed that $D=80$ in the double layer. But if the mean value, $\bar{D}$, in the double layer is, say, 64 for a 10^{-3} N-KCl solution, this would increase ζ numerically from -81 mV to -100 mV, and the calculated surface conductance (eq. 3.3) would now be 1.1×10^{-9} ohm^{-1} (instead of 0.9×10^{-9} ohm^{-1} if $D=80$). Although this difference is, unfortunately, less than experimental error, surface conductance does suggest a method of checking directly the order of magnitude of ζ as calculated from the electrokinetic phenomena described below[4]: most other methods merely give values of $D\zeta/\eta$ from which expression ζ is determined, assuming that D and η have their normal values within the electrical double layer. If, however, the variation of η within the double layer is important, eq. (3.2) must be replaced by:

$$K_s = \frac{\eta_0 l_+}{\boldsymbol{F}} \int \frac{\rho dx}{\eta} \tag{3.2a}$$

where η_0 is the viscosity of the liquid in zero electric field, and η that in a field ψ' $(=d\psi/dx)$ near the surface: the volume electrical charge density at any distance x from the surface is ρ. This integral may be written:

$$K_s = \frac{\eta_0 l_+}{\boldsymbol{F}} \int_{\psi=0}^{\psi=\psi_0} \rho \frac{dx}{d\psi'} \cdot \frac{1}{\eta} \cdot d\psi' \tag{3.2b}$$

and integrated numerically since ρ and η (see p. 141) are functions of ψ'.

For most purposes, however, the simplified model of the viscosity remaining at a value of η_0 up to a "plane of shear" and then increasing discontinuously to infinity when ψ reaches the value of ζ, is sufficiently accurate.

Experimental Methods

The first and most obvious method of studying surface conductivity simply measures the increased conductivity of a liquid phase when a monolayer is adsorbed on the plane surface. Historically this method is of interest, in that it gave the first surface conductance measurements[1]: it is still important[5].

A liquid lamella, such as constitutes a bubble wall in a foam, is formed between two wires, which act as electrodes. The lamella drains till a black film (see Chapter 8) is produced, several hundred Ångstrom units thick. From the dimensions of this thin sheet of liquid the conductance expected from the bulk phase is calculated, and any excess conductance which is measured is the surface conductance of the two monolayers bounding the lamella. This method is very sensitive in that the bulk phase is so thin that the surface conductance terms are relatively quite important: in a film 2960 Å thick formed from sodium soap, the ratio of the surface conductance to the bulk conductance is 0.98, rising to as much as 4.8 in a film 277 Å thick[1]. To obtain these large ratios it is important that no electrolyte other than the soap be present; thus 3% of KNO_3 in the solution so increases the bulk conductivity that the effect is masked. From these results we may calculate K_s as follows. The bulk conductance of a thin sheet (1 cm.^{2} in area) of the solution of the sodium soap is given by the equivalent conductance of soap at this concentration multiplied by its concentration (in equiv. cm.$^{-3}$) and by the thickness of the sheet, i.e. $20 \times 0.0555 \times 10^{-3} \times 2960 \times 10^{-8}$, which is 33×10^{-9} ohm^{-1}. The conductance from the *two* faces of the liquid lamella is 0.98 times this, hence $K_s = 16 \times 10^{-9}$ ohm^{-1}. For the thinnest film $K_s = 7 \times 10^{-9}$ ohm^{-1}. A value of K_s of 10×10^{-9} ohm^{-1} corresponds, according to eq. (3.2), to a charge density of 8.5×10^{13} mobile K^+ ions per cm.^{2} at the surface of the lamella—a very reasonable figure.

Studies of the surface conductance of insoluble monolayers, as of stearic acid, may be carried out by direct spreading of the monolayer on the surface between two electrodes[6]. Spreading a monolayer of stearic acid on water between two cylindrical electrodes of platinum (Fig. 3-2) gives an enhanced conductance of 0.18×10^{-5} ohm^{-1}, from which $K_s = 38 \times 10^{-9}$ ohm^{-1}. This

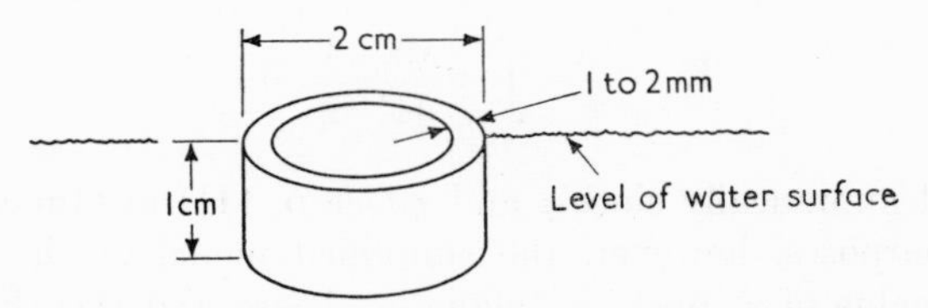

Fig. 3-2. Direct measurement of surface conductance.

corresponds to a fairly high charge density: about 11% of the molecules of stearic acid in the film must contribute a hydrogen ion to the diffuse part of the double layer. This is consistent with Fig. 5-18, in which the ΔV potential curve suggests a total dissociation (in the fixed and diffuse double layers) of the order 20% in a monolayer of a long-chain acid.

The second method consists of measuring the conductance of a capillary tube filled with electrolyte, and comparing this directly with that calculated for the bulk electrolyte in such a tube[7,8]. A capillary of Jena 16^{III} glass, 5 cm. long and of radius a of the order 10^{-2} cm., is suitable for such experiments, and K_s is calculated from the following relations, in which K is the specific conductance of the solution:

Total conductance for unit length $= 2\pi a K_s + \pi a^2 K$

Conductance relative to that for bulk liquid only

$$= \frac{2\pi a K_s + \pi a^2 K}{\pi a^2 K} = \frac{2K_s}{Ka} + 1 \tag{3.4}$$

Results of such measurements are shown in Table 3-I.

McBain[6] and his colleagues used a bank of glass capillary tubes in parallel to facilitate the conductance measurements. The method quoted by Zhukov[8] for making a bank of uniform capillaries in a block of polystyrene could also be used for measuring the surface conductance of this material.

Slit methods[6], in which the conductance of liquid in a very narrow slit is measured, are rather unreliable and often give high results for K_s.

The third method consists in placing a suspension of spheres between electrodes: if the spheres are of the order 1 micron diameter, the surface conductance through the suspension can equal or exceed the bulk conductance through the liquid between the spheres. Glass spheres may be conveniently prepared by blowing powdered glass through a blow-torch, and many polymers can be obtained directly as uniform small spheres. From the conductance K of the liquid phase (after centrifuging off the particles), the conductance K_1 of the suspension, the volume fraction φ of the dispersed phase, and the radius a of the spheres, one may calculate K_s from the relation[9]

$$K_s = \left[\frac{\left(\frac{K_1}{K} - 1\right) + \frac{\varphi}{2}\left(\frac{K_1}{K} + 2\right)}{\varphi\left(\frac{K_1}{K} + 2\right) - \left(\frac{K_1}{K} - 1\right)} \right] Ka \tag{3.5}$$

In practice it is often simplest to prepare the suspension first, and to find φ subsequently. This may be achieved either by density measurements, or by increasing K_1 by adding so much KCl to the system that $\frac{K_s}{K_1}$ is insignificant: hence, from the above relation

$$\frac{\varphi}{2} = \left(1 - \frac{K_1}{K}\right) \Big/ \left(\frac{K_1}{K} + 2\right)$$

Typical results of this method for spheres of Pyrex glass of 1.7 microns diameter are shown in Table 3-I. For particles other than spheres, a geometrical correction must be applied[9].

One should strictly apply to K_1 the Henry correction[9] for the amount of electricity carried by the movement of the charged particles in the electric field (electrophoresis), although in practice such a correction is important only if the glass spheres are smaller than 1 micron.

The fourth method consists in using a membrane of known pore size: collodion is a very suitable material. The conductance of the solution (e.g. of KCl) is measured before and after inserting the membrane between the electrodes. If a fraction $1/\beta$ of the cross-section of membrane consists of conducting pores, and if the solution in these pores has a total conductance α times that expected in the absence of surface conductance, then the conductance of the membrane is α/β times that of the free solution which was displaced by the membrane. In a typical collodion membrane[10], of pore radius 40 Å, $\alpha = 7.6$ and $\beta = 9$ in the presence of 10^{-3} N-KCl, while in a ceramic diaphragm, also containing 10^{-3} N-KCl and of pore radius 1160 Å $\alpha = 2.6$. From α, β, and the pore diameter, the surface conductance can be calculated. Thus, if $\alpha = 7.6$, eq. (3.4) applied to unit length of collodion capillary gives: $2K_s = 6.6$, $Ka = 6.6 \times 140 \times 10^{-6} \times 40 \times 10^{-8}$, or $K^s = 0.19 \times 10^{-9}$ ohm^{-1}.

McBain demonstrated that glass filters also show surface conductance effects[11], but no quantitative results are available. In all membrane experiments air-bubbles, bacteria, and algae are liable to lead to erroneous results, and these must therefore be carefully eliminated. Further, the pores will generally not all be of the same radius, nor will they even remain of the same diameter throughout their length[4,7]. They may also form an inter-connecting lattice. Though Klinkenberg[10] has shown that distribution of mean pore radii can be readily found, the latter complications make the results on membranes less acceptable than those on single capillaries. A further complication lies in the possible swelling of the walls of fine glass pores: this could lead to anomalously high surface conductances if ions can pass along the swollen region, as Bikerman points out[11].

For fibres, this method has proved useful; chopped fibres may be compressed into a "pad" across which the conductance can be measured. Interpretation of the results depends on a model of straight cylindrical capillaries inclined at various angles[12]. This model may give rather high values of K_s, as seen by the value for glass in Table 3-I.

ELECTRO-OSMOSIS

If a potential gradient is applied across the ends of an insulating tube containing a liquid, the latter will move (electro-osmosis). The liquid velocity v will depend on the potential ζ measured between the liquid immediately adjacent to the solid surface and a point in the liquid far away from the solid surface (Fig. 3-3). The relation between v and ζ may be found as follows. When a steady liquid movement has been set up under the applied electrical potential gradient of $E'_{app.}$ volts cm.$^{-1}$, each layer of liquid of thickness dx (Fig. 3-3) moves with a uniform velocity parallel to the wall, the total force on such a layer being zero. This means that the electrical forces on the counter-ions in this liquid layer must be balanced by frictional drag on neighbouring layers. If the flow is non-turbulent, if ρ is the volume charge density, η_0 is the viscosity of the liquid, and x is the distance of the layer from the plane of shear, then

$$E'_{app.}\rho\, dx = \eta_0\left(\frac{dv}{dx}\right)_{x+dx} - \eta_0\left(\frac{dv}{dx}\right)_x$$

which may be written

$$E'_{app.}\rho=\eta_0\left(\frac{d^2v}{dx^2}\right) \tag{3.6}$$

where the viscosity is assumed constant at η_0 through the electrical double layer.

To solve eq. (3.6) for v it is combined with the Poisson electrostatic equation relating ρ and the electrical potential ψ; which, if the dielectric constant retains its usual value of D_0, is:

$$\frac{d^2\psi}{dx^2}=-\frac{4\pi\rho}{D_0} \tag{3.7}$$

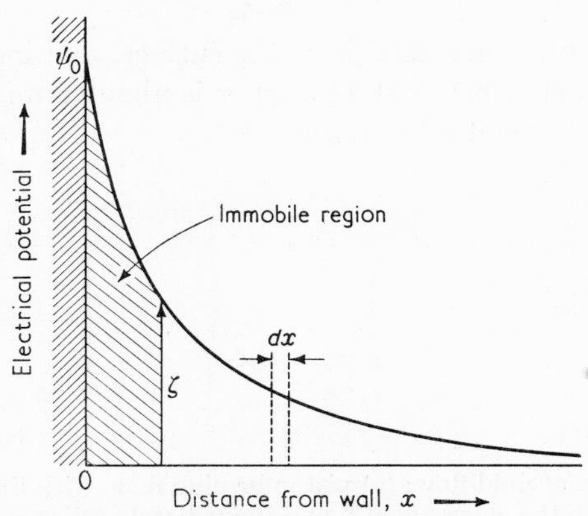

Fig. 3-3. Simplified representation of potential fall away from a charged surface.

This substitution gives:

$$\frac{d^2v}{dx^2}=-\frac{E'_{app.}D_0}{4\pi\eta_0}\cdot\frac{d^2\psi}{dx^2}$$

which on integration becomes:

$$\left\{\frac{dv}{dx}\right\}_x^a=-\frac{E'_{app.}D_0}{4\pi\eta_0}\left\{\frac{d\psi}{dx}\right\}_x^a$$

where a is the radius of the tube. Now if the tube is wide compared with $1/\varkappa$, i.e. if $\varkappa a\gg1$, conditions at $x=a$ are that

$$\frac{dv}{dx}=0,\quad \frac{d\psi}{dx}=0$$

and hence the flow velocity profile is given by:

$$\frac{dv}{dx}=-\frac{E'_{app.}D_0}{4\pi\eta_0}\cdot\frac{d\psi}{dx}$$

Since ψ becomes virtually zero at a very small fractional distance from the wall of the tube, this corresponds to almost perfect "piston" or "plug" flow (Fig. 3-4 (a) and (c)).

Further integration gives:

$$\{v\}_{x=0}^{x=a} = -\frac{E'_{app.} D_0}{4\pi\eta_0} \{\psi\}_{x=0}^{x=a}$$

Since at the "plane of shear", from which distances are measured, $\psi = \zeta$ and $v=0$ (no slippage), and since also at $x=a$, $\psi=0$ and $v=v$, the final relation between v and ζ (assuming D and η are constant) is:

$$v = \frac{E'_{app.} D_0 \zeta}{4\pi\eta_0} \tag{3.8}$$

Here ζ is a constant, characteristic of the surface, and independent of $E'_{app.}$ For example, if $\zeta = 100$ mV and $D_0 = 80$, v is about 7 microns sec.$^{-1}$ for an applied potential gradient of 1 volt cm.$^{-1}$

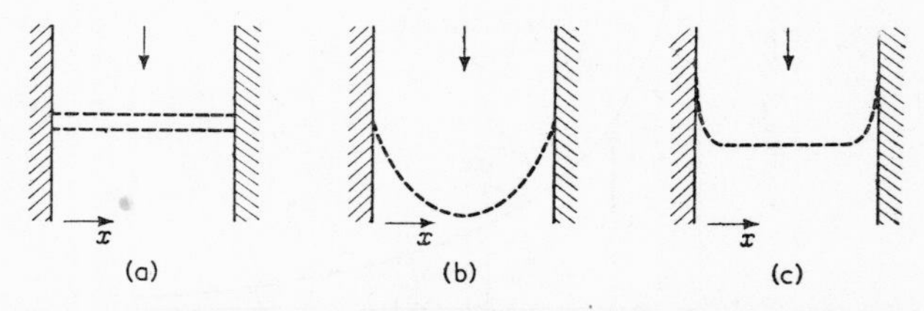

Fig. 3-4. Three types of fluid flow: (a) piston or plug flow, with linear velocity profile; (b) laminar flow, with the elements of liquid immediately adjacent to the wall moving infinitely slowly, and the resultant velocity profile being parabolic; (c) steep velocity gradients near walls, but fairly uniform velocity over most of tube.

Now within the double layer the local electrical field may be very high, since the potential may fall as much as 100 mV in a distance of 20Å from the surface. In such a field η may well be increased by electrical saturation, since this potential drop corresponds to a local field of 5×10^5 volts cm.$^{-1}$, i.e. 1.7×10^3 e.s.u. cm.$^{-1}$; this field also reduces D to 45 (see below). With greater regard for reality, therefore, let us assume that η and D are not constant, but vary with $\frac{d\psi}{dx}$, the local field strength at any point in the double layer, i.e. that

$$D = f_1(\psi'), \ \eta = f_2(\psi') \tag{3.9}$$

where ψ' is $\frac{d\psi}{dx}$. Fox[2] has shown that the equation of electro-osmotic flow is now:

$$v = \frac{E'_{app.}}{4\pi} \int^{\psi_0} \frac{D}{\eta} d\psi \tag{3.10}$$

where the integration is carried out to ψ_0 since there is now no sharply defined "plane of shear". Overbeek[2] gives the same relation without proof. The velocity should still be strictly proportional to $E'_{app.}$.

Experimentally, one measures in a tube the linear flow-rate v (though this can be found accurately only if there is no gassing at the electrodes for which Ag−AgCl is a suitably reversible system). Further, to permit evaluation of ζ, the capillary must have a pore size considerably greater than the thickness of the electrical double layer, otherwise the limits of $\psi=0$, $d\psi/dx=0$, at $x=a$ will not be fulfilled. Usually the pore radius will be of the order 10^{-3} cm., and for aqueous systems $1/\varkappa$ will be of the order 10^{-5} to 10^{-6} cm., so this condition is satisfied.

The volume flow rate u of liquid in the capillary is often conveniently measured. It is given by:

$$u=\pi a^2 v \tag{3.11}$$

If now Ohm's law applies to 1 cm. of capillary, with K the specific conductance of the solution, and i the current flowing in the capillary,

$$i=E'_{app.}\pi a^2 K \tag{3.12}$$

whence, by combination with eq. (3.11), we may eliminate the cross-sectional area (i.e. πa^2):

$$u=\frac{iv}{KE'_{app.}}$$

or, by eq. (3.8),

$$u=\frac{iD_0\zeta}{4\pi K\eta_0} \tag{3.13}$$

The quantities u, i, and K can be measured, and hence ζ can be found from this equation. Some values of ζ are shown in Table 3-II.

Often, however, Ohm's law does not hold for these systems because of surface conductance, and eq. (3.12) must now be written as

$$i=E'_{app.}\pi a^2 K+E'_{app.}2\pi a K_s \tag{3.14}$$

where K_s is the specific surface conductance. Thus, in general, eq. (3.13) becomes

$$u=\frac{i}{4\pi\left(K+\frac{2K_s}{a}\right)}\cdot\frac{D_0\zeta}{\eta_0} \tag{3.15}$$

To measure ζ by electro-osmosis one can therefore either use eqs. (3.8) or (3.11) directly, or use eq. (3.15) after finding the surface conductance. One can also work under conditions such that K and a are high, so that eq. (3.15) reduces to eq. (3.13); this is then also valid for electro-osmosis through a porous plug of arbitrary internal geometry. However, if surface conductance is important in the plug it is more difficult to allow for it precisely. The method of Zhukov and Fridrikhsberg[10] could be used, however, if the pores are assumed to be non-interconnecting.

TABLE 3-II

Electrokinetic Potentials of Glass in contact with Aqueous Electrolytes*

Aqueous phase	(mV)	Reference
Distilled water (J)	—157 ± 10	13
	—180 ± 5	8
Distilled water (P)	—139	14
10^{-4} N-KOH (J)	—148	13
10^{-5} N-KOH (P)	—155	14
10^{-4} N-KOH (P)	—143	14
10^{-5} N-HCl (J)	—80	13
10^{-4} N-KNO_3 (P)	—118	14
10^{-5} N-KCl (J)	—140	13
10^{-4} N-KCl (J)	—121	13
10^{-3} N-KCl (J)	—81	13
4.2×10^{-3} N-KCl (P)	—60	12
10^{-2} N-KCl (J)	—32	13
10^{-4} N-$CaCl_2$ (J)	—76	13

*P denotes Pyrex glass, J denotes Jena 16 III glass

STREAMING POTENTIALS

Suppose now we apply no potential gradient but, instead, force liquid through the tube and measure the potential set up thereby between electrodes placed at the ends of the tube. This potential is called the streaming potential, and arises from the liquid stream carrying with it the mobile part of the electrical double layer near the walls of the capillary. When steady state flow has been reached, with no external current flowing, the electrical current, which is proportional to the pressure difference P, is entirely dissipated by back-conduction through the liquid. This latter current is therefore proportional to the streaming potential.

If the liquid flow is slow enough to be laminar (Fig. 3-4), the capillary large compared with the term $1/\varkappa$, and surface conductance negligible, the streaming potential $E_{str.}$ must be proportional to ζ. The argument is as follows.

The potential $E_{str.}$ volts is generated by the imposed laminar liquid flow in the electrical double layer near the wall of the tube. The flow profile is given by the classical equation for laminar flow (the velocity profile being parabolic, as in Fig. 3-4b):

$$v = \frac{P}{4\eta_0 l}(a^2 - r^2) \tag{3.16}$$

where in the tube of length l and radius a, at any distance r from the centre, the linear velocity of flow of liquid (of viscosity η_0) is v, under the pressure P. Hence, by differentiation, the velocity gradient at (and very close to) the wall is:

$$\left(\frac{dv}{dr}\right)_{r=a} = -\frac{Pa}{2\eta_0 l}$$

If, as is usual, the ionic atmosphere at the wall is extremely thin in comparison with the diameter of the tube, the velocity v at any distance x ($x = a - r$) from the wall is given by:

$$v = \int_0^x \left(\frac{dv}{dx}\right)_{x=0} dx = \frac{Pax}{2\eta_0 l} \tag{3.17}$$

where we assume that η_0 is constant within the double layer. A similar equation $\left(\text{with a different value of } \left(\frac{dv}{dx}\right)\right)$ will be true of flow in the laminar sublayer near the wall even if the flow is so fast as to be turbulent elsewhere: the theory, therefore, remains valid if the ionic double layer lies within this laminar layer. Since also, by definition, the charge $I_{str.}$ carried per second by the liquid (the streaming current) is

$$2\pi a \int_0^a \rho v dx$$

substitution of ρ from eq. (3.7) and of v from eq. (3.17) gives:

$$I_{str.} = -\frac{Pa^2}{4\eta_0 l}\int_0^a D_0 \frac{d^2\psi}{dx^2} . x . dx$$

If D_0 is also independent of x (i.e. independent of the presence of the electrical double layer), integration by parts gives:

$$I_{str.} = -\frac{PD_0 a^2}{4\eta_0 l}\left\{\left[\frac{x d\psi}{dx}\right]_{x=0}^{x=a} - \{\psi\}_{x=0}^{x=a}\right\}$$

and, since $d\psi/dx = 0$ and $\psi = 0$ when $x = a$ (the radius of the tube being much greater than the thickness of the ionic atmosphere), and since also $\psi = \zeta$ when $x = 0$, this reduces to:

$$I_{str.} = -\frac{PD_0 a^2 \zeta}{4\eta_0 l} \tag{3.18}$$

or, as can be shown (Fox[2]) if D and η vary with field-strength in the double layer:

$$I_{str.} = -\frac{Pa^2}{4xl}\int_0^{\psi_0} \frac{D}{\eta}\,d\psi \tag{3.19}$$

This current $I_{str.}$ is responsible for the streaming potential $E_{str.}$ At steady state it must be balanced by a current I_2 conducting the charge back through the liquid and at the surface:

$$I_2 = \frac{E_{str.}\pi a^2 K}{l} + \frac{E_{str.}2\pi a K_s}{l}$$

where K and K_s are the specific bulk and surface conductivities. At the steady state the total flow of electricity is zero, so $I_{str.} + I_2 = 0$, i.e.

$$E_{str.} = \frac{PD_0\zeta}{4\pi\eta_0\left(K + \frac{2K_s}{a}\right)} \tag{3.20}$$

The streaming potential is simply related to the electro-osmotic flow of liquid: comparison of eqs. (3.15) and (3.20) shows that $E_{str.}i = uP$. This relation also holds if $\frac{D_0\zeta}{\eta_0}$ is replaced by $\int_0^{\psi_0}\frac{D}{\eta}d\psi$, and is a consequence of the reciprocal nature of the processes[2], provided only that the velocity gradient near the wall is non-turbulent in both.

Equation (3.20) is valid only if the following conditions are fulfilled[13].

(a) The radius of the capillary must considerably exceed the thickness of the electrical double layer (i.e. $\varkappa a \gg 1$), otherwise the condition $\psi = 0$ at $r = a$ will not be satisfied, and the streaming current or potential will be anomalously low.

(b) The length of the capillary must considerably exceed the "characteristic flow length of the liquid" given by $\bar{v}D_0/4\pi K$, where $\bar{v}$ is $u/\pi a^2$ and K is the specific conductance of the solution. If not, the effects are again anomalously low.

(c) The flow must be laminar: if pressures and rates of flow are so high as to lead to turbulent flow, the streaming potential (or current) increases non-linearly with P. While this effect is not important for aqueous solutions under practical conditions, it is significant for non-aqueous solutions[13,15] where the ionic double layer is much more diffuse and may extend into the region of turbulent flow.

Streaming potential measurements give satisfactory results for aqueous KI and $AgNO_3$ solutions in glass capillaries. If the term $2K_s/a$ in eq. (3.20) is neglected, the ζ potential apparently passes through a maximum, at a concentration of $AgNO_3$ of 2×10^{-5} N (Fig. 3-5): correctly calculated by the inclusion of the term 2Ks/a in eq. (3.20), ζ shows no such maximum[8].

For a porous plug it is more difficult to apply an exact surface conductance correction[2,4,13,14], but the conclusion is the same in that the measured

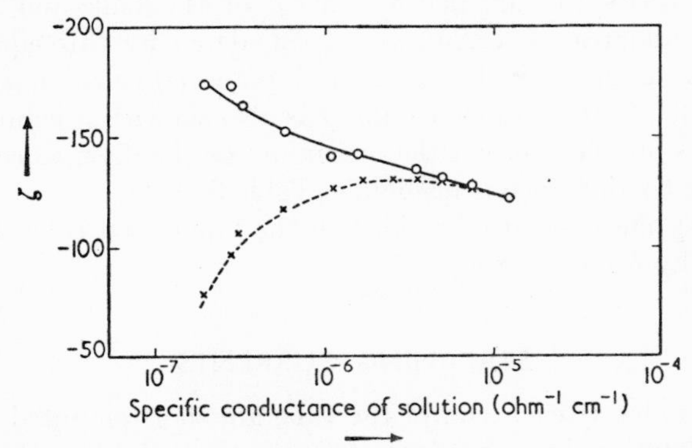

Fig. 3-5. Electrokinetic potential (mV) for glass against aqueous $AgNO_3$. The broken line shows the spurious maximum by neglecting the surface conductance term in eq. (3.20), and the full line shows the potentials correctly calculated[8].

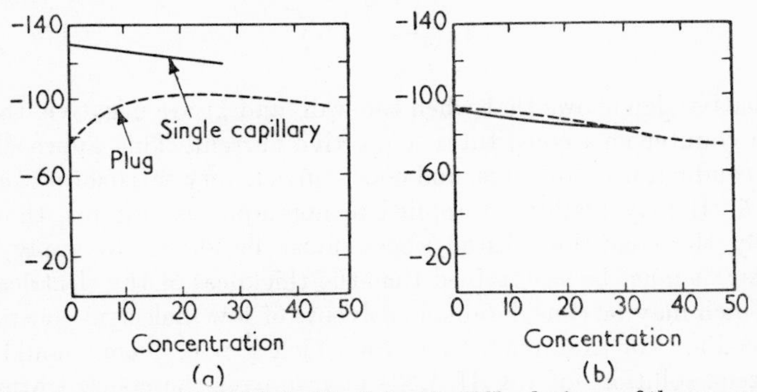

Fig. 3-6. Electrokinetic potentials (mV) of glass in KCl solutions, of the concentrations shown (in μ equivs. $1.^{-1}$). (a) For aqueous solutions, the surface conductance term cannot be allowed for exactly in a porous plug, and the results using eq. (3.20) are anomalously low. Data for the single capillary can be interpreted exactly. (b) For methanolic solutions, the surface conductance is so small that results are practically identical[4].

streaming potential will decrease in fine pores at low electrolyte concentrations, due to short circuiting by the surface conductance. Henniker[16], for example, found that there was no detectable streaming potential when an aqueous solution of 2×10^{-5} N-KCl was allowed to flow through a Chamberland 5.L.13 bacterial filter of pore radius 0.15 micron, though a 10^{-3} N solution of KCl gave a potential of 7 mV.

The surface conductance term in eq. (3.20) will, according to the results in Table 3-I, reduce the streaming potential of 2×10^{-5} N-KCl in a pore of this size by 98 or 99%: however, when 10^{-3} N-KCl is used, the calculated reduction is only about 38%. If the porous plug consists of pores of different diameters in series the unequal weighting of the bulk and surface conductances invalidates eq. (3.20), and ζ calculated by this equation passes through a maximum, as in Fig. 3-6: the ζ potential calculated by correctly applying eq. (3.20) to a simple capillary is also shown[4]. A solution of KI in methanol does not show either the maximum or the divergence of ζ values, as the surface conductance is negligible (Table 3-I).

By imposing mechanical vibrations on the liquid in a tube, $E_{str.}$ can also be measured by A.C. devices[17].

STREAMING CURRENTS

If the electrodes at each end of the tube are short-circuited, the current $I_{str.}$ produced by an imposed flow of liquid through the tube will return through this short-circuit rather than through the liquid. For laminar flow, the current $I_{str.}$ is related to ζ by:

$$I_{str.} = -\frac{P D_0 a^2 \zeta}{4 \eta_0 l} \tag{3.18}$$

The negative sign shows that when the wall (and ζ) are negative, the flow of positive counter-ions constitutes a positive current. This approach avoids surface conductance problems, and constitutes a very satisfactory method of finding ζ. It may readily be applied to non-aqueous systems, though here especially the conditions listed above must be shown to apply, and, in particular, it must be ascertained that the thickness of the electrical double layer (which may be considerable in solvents of low dielectric constant) does not exceed a. For example[13], tubes for which a $>$ 0.12 cm. should be used for benzene solutions of $(i\text{-}C_5H_{11})_4N$ – picrate over the concentration range 0 to 10^{-3} M. To make sure that the electrodes in the vessels at each end of the glass capillary pick up all the streaming current or potential, these vessels may each be wrapped in tin-foil, this wrapping being connected to the electrode inside the vessel. The streaming currents in these systems are of the order 10^{-11} to 10^{-13} amp., and may be measured with an electronic voltmeter or with an electrometer; the corresponding ζ potentials are as high as 280 mV for dilute solutions.

Another example is furnished by 10^{-4} M-Zn(di-isopropyl-salicylate) in benzene. The ionic concentration is very low because of incomplete dissociation, being about 10^{-11} N according to the measured conductance if the ions are assumed univalent. For this solution in a glass capillary (a = 0.116 cm.)

turbulence sets in when u = 1.11 cm.^{3}sec.$^{-1}$, and this is reflected in the shape of the plot of $I_{str.}$ against u (Fig. 3-7). At low flow-rates when the flow is wholly laminar, both $I_{str.}$ and u increase linearly with P, and from eq. (3.19) we calculate ζ to be about 330 mV. At high flow-rates the flow becomes

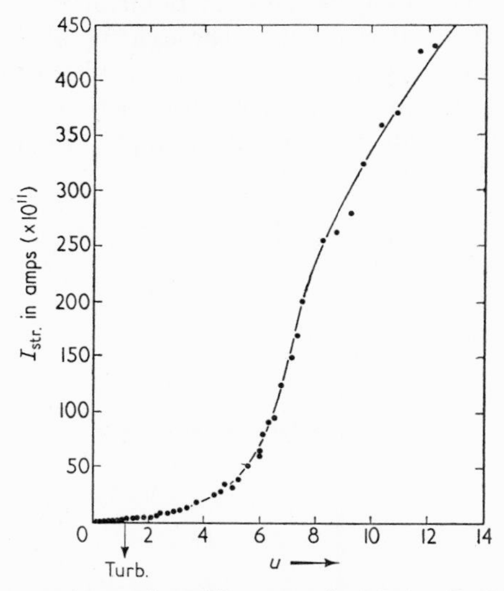

Fig. 3-7. Plot of $I_{str.}$ against volume flow-rate for 10^{-4} M-Zn(di-isopropyl-salicylate) flowing[13] through a glass tube of radius 0.116 cm. Here u is measured in cm.^{3}sec.$^{-1}$

turbulent through a core constituting most of the tube, and in this core the velocity is nearly constant. Only in a thin sublayer near the walls does the flow remain laminar: hydrodynamic theory (Rouse and Howe[16]) gives δx, the thickness of this laminar sublayer, as:

$$\delta x \approx 116 \text{ a } (\text{Re})^{-7/8} \tag{3.21}$$

where Re is the Reynolds number. When u = 8 cm.^{3} sec.$^{-1}$, this purely laminar sublayer will extend about 50 microns from the walls of the tube, a figure comparable with $1/\varkappa$ which is 15 microns. The velocity gradient in the laminar sublayer during turbulent flow is much higher than that in eq. (3.17), since most of the velocity gradient occurs in this region (Fig. 3-4c). If we rewrite eq. (3.17) with a velocity gradient $\bar{v}/\delta x$ (where $\bar{v} = u/\pi a^2$), we obtain the approximate relation:

$$I_{str.} \approx -\frac{\bar{v} D_0 a \zeta}{2\delta x} \tag{3.22}$$

which should apply when the flow is turbulent, provided that $\delta x > 1/\varkappa$. Thus, substituting the experimental values of u = 8 cm.^{3}sec.$^{-1}$, $D = 2$, a = 0.116 cm., $\delta x = 0.005$ cm., $\zeta = 330$ mV, we find $I_{str.} = -1.7 \times 10^{-9}$ amp., compared with the observed current[13] of -2.5×10^{-9} amp. for the dilute solution in benzene of Zn(di-isopropyl-salicylate). At still faster flow-rates, some of the ionic atmosphere will extend into the turbulent core, and the total streaming current will be given by eq. (3.22) only when an additional term is added to allow for this: a detailed analysis is due to Boumans[16].

When a hydrocarbon oil is pumped through metal pipes the streaming current may set up large potential differences, with consequent risk of a spark discharge and an explosion. Since these potentials may reach 70 kV, the streaming current can be used as a liquid high-voltage generator

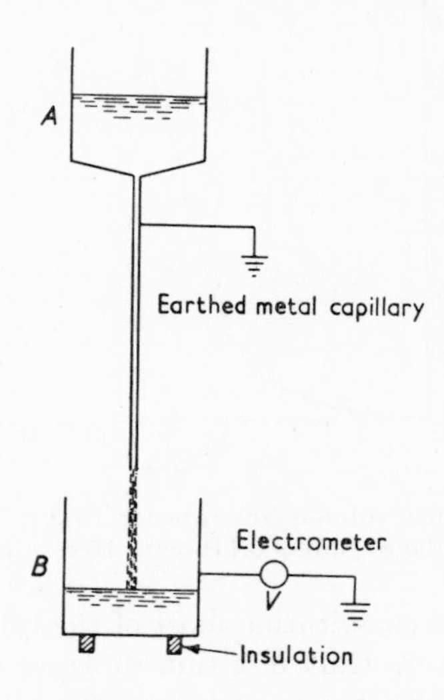

Fig. 3-8. Sketch of apparatus for the measurement of electric charging in metal capillaries.

(Boumans[16]). The effect of an additive in altering ζ at the oil-metal interface is therefore particularly interesting, and the following experimental technique[15] has been developed: it is generally applicable if the tube is of conducting material. The vessel A (Fig. 3-8) empties through a vertical capillary of mild-steel, 3 mm. in internal diameter and 50 cm. long. Vessel A and the capillary are earthed, and the liquid carrying the entrained charges of the diffuse double layer is collected in the insulated receiving vessel B. The total charge carried, after all the liquid has flowed out of A may be calculated from the potential and electrical capacity of vessel B. Electrical leakage back through the jet is usually unimportant if the jet is 10 cm. long and if the potential of vessel B is kept below 1 volt by condensers in parallel with it. If, however, electrical leakage through the jet is still important, the jet may be made to fall further, or even to break into drops before reaching

B. With gasoline flowing through a vertical tube of the above dimensions, the Reynolds number is between 4000 and 6000, well above the range of 2000 to 3000 at which the fluid flow becomes turbulent. Lower flow rates could be achieved by tilting the metal capillary away from the vertical.

The results of such measurements may be complicated by the current passing between earth and the liquid in the capillary tube and in A: this current must maintain the liquid at earth potential in spite of the loss of counter-ions but, in the process, products of electrolysis may be formed at the interface.

Typical results[15] of streaming currents are as follows:

10^{-6} M-Ca(di-isopropyl-salicylate) in benzene, $I_{str.} = -1 \times 10^{-9}$amp.
10^{-3} M-Ca(di-isopropyl-salicylate) in benzene, $I_{str.} = -5 \times 10^{-10}$amp.
10^{-6} M-Al(alkyl-salicylate) in gasoline, $I_{str.} = -1 \times 10^{-9}$amp.
10^{-6} M-Cr(alkyl-salicylate) in gasoline, $I_{str.} = -2 \times 10^{-11}$amp.

The precise values depend on the flow rates. That these streaming currents are negative shows that the mild-steel wall is positively charged, presumably by adsorption of a monolayer of ions such as Ca(di-isopropyl-salicylate)$^{+}$.

To interpret these figures quantitatively is not easy; a complete theory would have to allow for the separate effects in the turbulent core of liquid and in the laminar layer which persists near the solid wall[16].

Turbulence is not important in aqueous systems, since the electrical double layer is so thin that it lies almost completely within the laminar sublayer at all usual flow-rates: strict proportionality between $I_{str.}$ and P is thus expected, and is indeed always found.

FLOW THROUGH FINE PORES AND ζ

The electrical work done by the streaming potential may affect the rate of flow of liquid through a narrow capillary of non-conducting material. In the absence of any external electrical field, the work done per unit time is given by the product of the square of the streaming potential and the conductance, i.e.

$$\text{Work done by electrical forces per unit time} = \frac{P^2 a^2 D_0^2 \zeta^2}{16\pi l \eta_0^2 \left(K + \frac{2K_s}{a}\right)} \tag{3.23}$$

where l is the length of the tube. This electrical work will be added to the work of viscous flow of the liquid through a porous plug or capillary, appearing as an enhanced time of passage of a certain volume of liquid. As the ionic strength is reduced, ζ^2 will be increased considerably while K will be reduced, so that the liquid will flow more slowly through the plug. If, however, ζ can be reduced, the flow of liquid will be restored to that given by the Poiseuille equation below.

The exact relation between the flow time of a given volume of liquid and ζ is found as follows. In the absence of electrical effects, the volume of liquid flowing through the capillary tube per unit time is given by Poiseuille's equation:

$$u = \frac{\pi P a^4}{8\eta_0 l}$$

and the corresponding work is P times this volume. When the ζ potential retards the flow of liquid, however, the total work required per unit time becomes (using eq. 3.23)

$$\frac{\pi P^2 a^4}{8\eta_0 l} + \frac{P^2 a^2 D_0^2 \zeta^2}{16\pi l \eta_0^2 \left(K + \frac{2K_s}{a}\right)}$$

which is

$$\frac{\pi P^2 a^4}{8\eta_0^2 l}\left(\eta_0 + \frac{D_0^2\zeta^2}{2\pi^2 a^2\left(K + \frac{2K_s}{a}\right)}\right)$$

This total work per unit time is also related to the work in the absence of electrical drag by the ratio η_a/η_0, where η_a is the new, (apparent) viscosity. Hence

$$\eta_a = \eta_0 + \frac{D_0^2\zeta^2}{2\pi^2 a^2\left(K + \frac{2K_s}{a}\right)} \tag{3.24}$$

an equation derived, in a different manner, by Abramson and by Elton[18]. If ζ is high, and K, K_s, and a are small, η_a may thus be appreciably higher than η_0. For example, if for 10^{-4} N-KCl, $\eta_0 = 1.00$ c.p., $\zeta = -70$ mV, $D_0 = 80$, a = 0.15 micron, and $\left(K + \frac{2K_s}{a}\right) = 10^{-4}$ ohm^{-1} cm.$^{-1}$ (estimated with the data of Table 3-I), η_a is found by eq. (3.24) to be 1.01 c.p.; the time of passage of the liquid is therefore increased by 1% according to this calculation. Observed values[16,19] are 2.5% and 6%, the discrepancy possibly lying in the fact that the pores may have local narrow regions, which would increase the retardation. A layer of "ice" near the surface could also contribute to reducing a. The measured streaming potential may be very low in such a system, because of the still rather high total conductance term in the denominator of eq. (3.20).

This anomalous flow at low ionic strengths is most easily found experimentally in very fine-pored filters, such as the Chamberland 5.L.13 bacterial filter, of mean pore radius 0.15 micron. Typical results, of the order of magnitude calculated above, are shown in Fig. 3-9.

Various surface-active agents, adsorbing on the walls of the pores, may greatly decrease ζ and hence restore the flow of liquid to normal. Derjaguin and Krylov[20] claimed that the flow of water through a ceramic disc of very fine pore size was seven times faster after the pores had been coated with oleic acid and, although this has not been confirmed, it has been conclusively shown[16,19] that the flow of very dilute salt solutions through the L.13 filter can be altered by additives. Pretreatment of the ceramic pores with $C_{12}H_{25}NH_3^+$ ions, for example, increases the flow rate of distilled water by

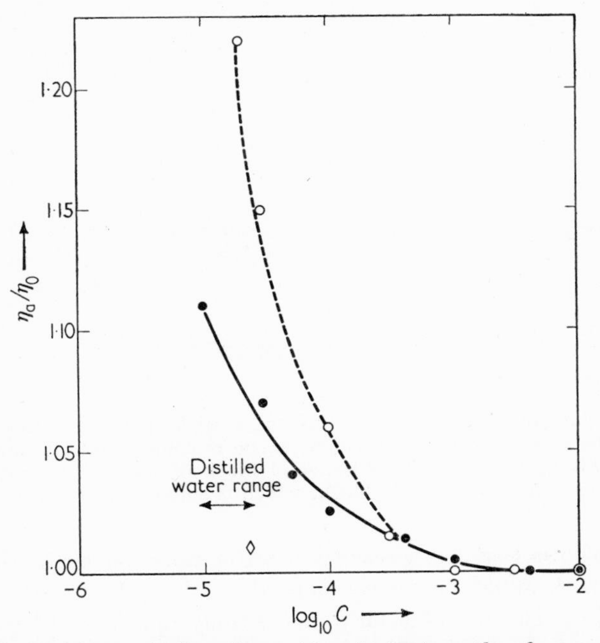

Fig. 3-9. Relative increase in flow times at low ionic strengths of aqueous KCl. Open circles are data of Henniker[16], with the diamond referring to pores treated with $C_{12}H_{25}NH_3^+$. Filled circles are from ref. 19.

about 20% (Fig. 3-9) to about the value for 10^{-3} N-KCl. Presumably the long-chain amine ions are strongly adsorbed on to the otherwise negative pore walls, thus neutralizing the ζ potential.

In the absence of precise numerical agreement of the results with eq. (3.24), an alternative explanation of the anomalously slow filtration is possible: the high potential gradient ($d\psi/dx$) near the charged surfaces of the pores may orientate and so solidify the adjacent water. In this way the pores would effectively become coated with a layer of "soft ice", which would reduce the pore radius and so retard the flow of liquid. If the retardation of flow of very dilute KCl solutions through a 5.L.13 filter is ascribed to this cause alone, the thickness of the "ice" layer would have to be 11 Å at 10^{-4} N and 40 Å at 10^{-5} N to account for the observed flow times. We can compare these figures with experimental data on the increased viscosity of certain organic liquids in electrical fields[20], the effect being proportional to the square of the dipole movement of the liquid and also to the square of the field strength.

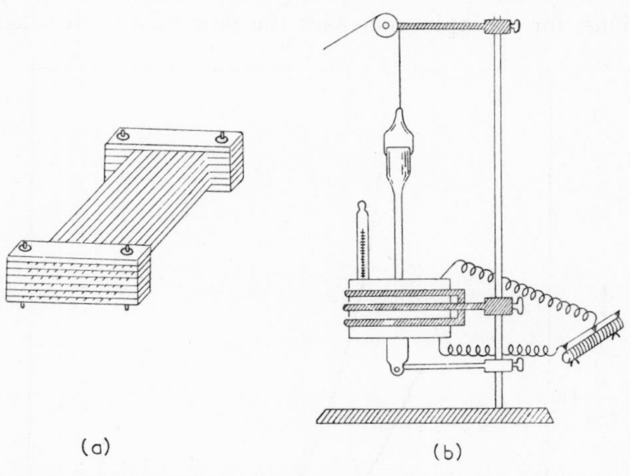

Fig. 3-10. Details of making filter of uniform, straight pores. (a) Regular array of metal wires between spacers, over which plastic is poured and allowed to solidify. (b) After removing the wires from the solid block of plastic, the pores are made narrower by stretching the plastic block, which is softened in a furnace (Zhukov[8]).

Calculation shows that the viscosity of water near the surface in a field of 10^5 volts cm.$^{-1}$ would be nearly normal (1.04 c.p.), using a reasonable estimate of ψ_0 and a value of $d\psi/dr$ from the Gouy equations. Thus calculated, the thickness of any "soft ice" layer must be much smaller than is required by the experimental results. We therefore believe that the retardation of flow through charged capillaries must be explained principally as an electrical drag, though more precise studies of ζ potentials and surface conductivity, and of the retardation of flow in single capillaries would be required to test this viewpoint rigorously. Alternatively, a suitable experimental technique would consist in using artificially prepared filters containing straight, uniform, cylindrical pores. These can be made from polystyrene[8]; this liquid is poured over a regular array of metal wires (Fig. 3-10a) which, when the polystyrene has set, are heated electrically and withdrawn from the polystyrene block. The latter is now left with an array of cylindrical holes[8], which can be narrowed to capillaries by stretching the polystyrene block in a furnace (Fig. 3-10b).

ELECTROPHORESIS

Electrophoresis is the converse of electro-osmosis. Whereas in electro-osmosis the liquid moves in relation to the solid, in electrophoresis the liquid as a whole is at rest while a particle moves through it under the action of an electric field. In both phenomena the forces acting in the electrical double layer control the relative movements of the liquid and solid, and hence eq. (3.8) will also apply to a first approximation to electrophoresis, v now being the velocity of electrophoretic movement:

$$v = \frac{E'_{app.} D_0 \zeta}{4\pi\eta_0} \qquad \left(a \gg \frac{1}{\varkappa}\right) \tag{3.25}$$

In this approximation the size and shape of the particle are not taken into account. To achieve this, one of the requisite corrections is a factor **R** to allow for the delayed relaxation of the ionic atmosphere in the wake of the moving particle. Another correction factor, most conveniently written as 2**F**/3, allows simultaneously for the interactions of the applied field with the electrical double layer (this effect being important if the particle is small), and with the surface of the particle (important if the surface conductance is appreciable).

For the electrophoresis of non-conducting solid spheres the complete equation is thus

$$v = \frac{\mathbf{R}.\mathbf{F}.E'_{app.} D_0 \zeta}{6\pi\eta_0} \tag{3.26}$$

where **R** and **F** are discussed in detail below. For rods exact equations are also available[21].

The relaxation effect is one consequence of the interaction of the applied electrical field with the ionic atmosphere of the particle. As the latter is pulled through the liquid, there are delays both in the formation of a new ionic atmosphere in front of it and in the relaxation of the ionic atmosphere behind it: the result is usually a retardation of the particle. Only if $\zeta < 25$ mV is this relaxation effect negligible[2,22]. If $\zeta = 50$ mV and $\varkappa a = 1$ for a spherical particle, the relaxation correction **R** will depress v by a few per cent at most, provided that the supporting electrolyte is uni-univalent[2,22]: the correction factor approaches unity if either $\varkappa a > 10^2$ or $\varkappa a < 10^{-1}$. If $\zeta = 50$ mV in solutions of polyvalent ions[2], the precise value of the correction factor **R** again depends on $\varkappa a$: it can be as small as 0.40, though if $\varkappa a > 10^3$ or if $\varkappa a < 10^{-2}$ it again approaches unity. When ζ is as high as 100 mV the relaxation correction is very important[22], the mobility of the particle being reduced by as much as 50% even in water, if $\varkappa a$ is of the order unity: the ζ potentials calculated by the approximate equation (3.25) are thus anomalously low. We may note that comparison of the effects of polyvalent ions on the mobilities of bacteria and on "model" surfaces (oil drops coated with various materials) is best made measuring the "reversal of charge" concentrations[23], which correspond to $\zeta = 0$: the relaxation effect is then necessarily also zero.

For glass spheres[9] of radius 0.8×10^{-4}cm. in 1.7×10^{-4} N aqueous KCl, $\varkappa a$ is about 34, and ζ is of the order -80 mV. In view of this value of $\varkappa a$ and the absence of polyvalent ions, taking $\mathbf{R} = 1$ introduces no great error. For bacteria with $a = 1 \times 10^{-4}$cm. in 0.14 N-NaCl, $\varkappa a$ is of the order 10^3, so $\mathbf{R} \approx 1$, even though ζ is of the order -60 mV. The same conclusion is true of emulsion drops if the salt concentration is not too low. Small particles, however, such as micelles (a $\approx$ 20 Å) and protein molecules have values of $\varkappa a$ of 0.07 or more if the ionic strength exceeds 10^{-4} N: since ζ is of the order 80 mV for micelles, the corrections for relaxation may well be quite important at ionic strengths greater than 10^{-4} N.

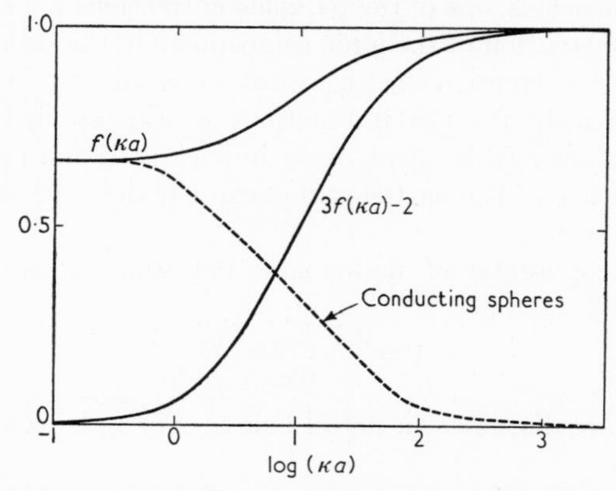

Fig. 3-11. The function **f** of eq. (3.27), as a function of $\log_{10}\varkappa$ a, for spherical particles which are non-conducting. The derived function, **3f** — 2 is also shown. If the spheres are electrically conducting, the value of **f** decreases towards zero (broken curve)[24].

For spheres of radius a, the factor **F** is given[24] by:

$$\mathbf{F} = \left\{1 + \frac{(K - 2K_s/a)(3\mathbf{f} - 2)}{2(K + K_s/a)}\right\} \tag{3.27}$$

where **f** is a function of a, given in Fig. 3-11 (upper curve). Certain limits of eq. (3.27) are of interest. If the particle is so small that $\varkappa a < 1$, we have from Fig. 3-11 that $\mathbf{f} = 2/3$, and hence $(3\mathbf{f} - 2) = 0$; consequently, whatever the surface conductance, $\mathbf{F} = 1$, and the 6π form of eq. (3.26) is therefore required. Only for large particles can the surface conductance correction be important: it makes **F** less than it would otherwise be, so that ζ potentials calculated without allowing for K_s may be too low. If the particle is so large that $\varkappa a > 300$, then $\mathbf{f} = 1$ and $(3\mathbf{f} - 2) = 1$ (Fig. 3-11): we now have

$$\mathbf{F} = \frac{3}{2}\left\{\frac{K}{K + K_s/a}\right\} \tag{3.28}$$

so that the 4π form of the equation (cf. eq. (3.25)) is again valid, with the correction factor $\left\{\frac{K}{K+K_s/a}\right\}$ for surface conductance.

As examples of the application of eq. (3.27) let us consider glass spheres[9] of radius 0.8×10^{-4}cm. and $K_s=0.5\times10^{-9}$ohm^{-1} suspended in 1.7×10^{-4} N aqueous KCl of conductance 2.6×10^{-5}ohm^{-1}cm.$^{-1}$ The ratio $K/(K+K_s/a)$ is now 0.81, so the exact expression (3.26) above, by allowing for surface conductance, gives a value of ζ 23% greater than would have been calculated as a first approximation by eq. (3.25). For the same spheres in 1.7×10^{-3} N-KCl, the correction factor increases ζ by only 2%. For bacteria[23], $a\approx1\times10^{-4}$cm. and $K_s=10^{-9}$ohm^{-1} are reasonable estimates, so that if their electrophoretic properties are studied in 0.14 N-NaCl ($K=2\times10^{-2}$) the surface conductance is relatively insignificant, and hence $\mathbf{F}=3/2$: the surface conductance effect becomes appreciable only if the ionic strength is below 10^{-3} N. Neither is the correction important for emulsion droplets[25] of radius 10^{-4}cm. undergoing electrophoresis in the presence of long-chain ions if the ionic concentration exceeds 10^{-2} N, even if by virtue of the adsorption of the charged groups, K_s is as high as 10^{-8}ohm^{-1}. Typical ζ potentials are listed in Table 3-III.

TABLE 3-III

	ζ(mV)	Reference
Drops of paraffinic oil in distilled water	—72	27
Drops of paraffinic oil with monolayers of cetyltrimethyl-ammonium bromide adsorbed from 0.05 M-acetate buffer with concentrations of long-chain ions of:		
5.8×10^{-5} M	+81	25
12.5×10^{-5} M	+93.5	25
Bact. coli in 10^{-2} N-KCl	—43	23
Nylon in 4.4×10^{-4} N-KCl	—43	12
Nylon in 3.6×10^{-4} N-KCl	— 9	12
Dacron in 4.7×10^{-4} N-KCl	—77	12
Dacron in 2.3×10^{-4} N-HCl	—31	12
Orlon in 4.9×10^{-4} N-KCl	—65	12
Porcelain in 3×10^{-3} N-KCl	—42	29
Porcelain in 10^{-2} N-KCl	—35	29
Porcelain in 10^{-1} N-KCl	—15	29
Porcelain in 10^{-2} N-NaCl	—40	29
Porcelain in 10^{-1} N-NaCl	—24	29
Porcelain in 10^{-3} N-KI	—41	29
Porcelain in 10^{-2} N-KI	—29	29
Porcelain in 10^{-1} N-KI	— 6	29
Kaolinite in aqueous HCl	—34	9

It has frequently been observed[9,23] that v passes through a maximum in diverse systems as the ionic strength is made very low, as in Fig. 3-12. This can be predicted by combining eqs. (3.26) and (3.28), and differentiating v with respect to $\varkappa$, finally setting $dv/d\varkappa = 0$. For the glass spheres mentioned above, this maximum is calculated to occur when the ionic strength is 0.4×10^{-3} N, the experimental figure being about 1×10^{-3} N. For bacteria calculation gives 0.23×10^{-3} N, compared with 0.5×10^{-3} observed, the agreement being satisfactory in view of the uncertainty concerning the surface conductance of bacteria. Possibly the position of the maximum could be used satisfactorily to calculate K_s.

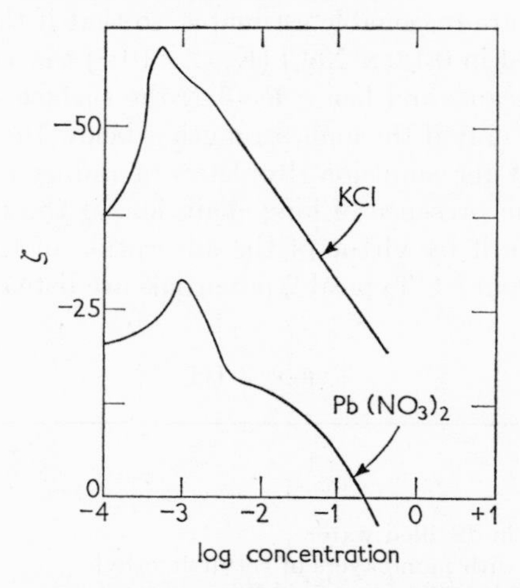

Fig. 3-12. Effect on ζ (in mV) of the normality (expressed logarithmically) of electrolyte in which B.coli are suspended[23]. The maximum is due to the effect of surface conductance in reducing **F** (eqs. (3.26) and (3.28)) when K is made small. The reversal of charge concentration for Pb^{++} is 0.2 N.

For micelles and protein molecules, where $\varkappa a$ is often of the order unity, the corrections are particularly important, especially as K_s may well be as high as 10^{-8}ohm^{-1} on an ionic micelle. Sodium lauryl sulphate micelles, for example, in electrolyte more dilute than 10^{-2} N, have a radius of about 20 Å and hence have $\varkappa a$ values of 0.7 or less. Under these circumstances, and at lower salt concentrations, **f** is close to ⅔, and hence **F** is close to unity. If, however, the micelles are in 10^{-1} N-NaCl, then $\varkappa a = 2$ and $\mathbf{f} = 0.71$; consequently, estimating K_s as 10^{-8}ohm^{-1}, we find $\mathbf{F} = 0.91$. This would increase ζ by about 10% above the values shown in Table 3-VI below.

Another correction to be included in eq. (3.26) arises from possible transfer of momentum across the surface of an oil drop or gas bubble undergoing electrophoretic movement. Booth[26] calculates that this will decrease v, and that for any gas bubble the internal viscosity will be so much less than that of the continuous aqueous phase that the electrophoretic velocity should always be negligibly small on account of almost complete transfer of the momentum to the gas phase. Jordan and Taylor[27], by contrast, deduce that the induced circulation in the drop or bubble should increase v, up to a calculated maximum (for a gas bubble) of 1.5 times that for a solid particle of the same ζ potential. Experimentally, gas bubbles usually migrate at speeds only a little lower than those of oil drops in distilled water[27,28], though when the purest conductivity water is used the gas bubbles may have values of $v/E'_{app.}$ as low as 0.6 micron cm.sec.$^{-1}$volt^{-1}. Further work, theoretical and experimental, is clearly required on such systems. When, however, there is a monolayer of adsorbed organic molecules or ions at the surface, the resistance to tangential momentum transfer (Chapter 7) is greatly increased both because of the enhanced interfacial viscosity (Chapter 5) and because of the surface pressure offering a resistance to local movements in the surface: under these conditions a drop or bubble can be assumed solid, and no further correction to eq. (3.26) is necessary.

The effective charge on the moving particle can be calculated if required. This charge represents the net charge on the surface, being the algebraic sum of the fixed charges on the surface and the strongly adherent counter-ions. Assuming the Gouy distribution of ions in the diffuse layer, one may use eq. (3.1) to calculate σ: these net charge densities are, in the absence of specific adsorption, rather lower than those of the potential-determining ions. At high potentials, the net σ thus calculated from ζ may be only 10% of the charge density of the fixed ions at the phase boundary, indicating that most of the counter-ions are tightly adsorbed within the plane of shear[30]. At lower potentials the figure approaches about 50%. If the counter-ions are specifically adsorbed, as are heavy-metal ions on quartz or glass, the zeta potential and the net σ may be decreased to zero or even reversed in sign.

Experimental Methods

There are two distinct approaches in measuring v, depending on whether or not the moving particles are visible in a microscope. If, as are protein molecules, they are invisible, special techniques depending on observing the movement of a refractive-index gradient are required[2]. These, including the familiar Tiselius method, lie outside the scope of this book. For particles about 1 micron or larger, however, direct observation with a travelling microscope is possible, the suspension being placed in a tube of rectangular or circular cross-section. A correction is necessary for the electro-osmotic movement of the liquid relative to the glass walls of the tube towards one of the electrodes, which necessitates a laminar return flow of the liquid if the tube is sealed at each end. The velocity profile resulting from this return flow is parabolic when steady state conditions have been achieved: the total transport of liquid is then zero, the flows along the walls and in the centre of the tube being in opposite directions. The velocity of electrophoretic movement is superimposed on this movement of the liquid as shown in Fig. 3-13, the curve being a parabola. To obtain the electrophoretic velocity of the particle it is usual to observe its movement at a region of the tube where the movement of the liquid support is zero. In a cell of circular cross-section this occurs at a distance from the wall of 0.147 times the diameter[2,31]. The appropriate points

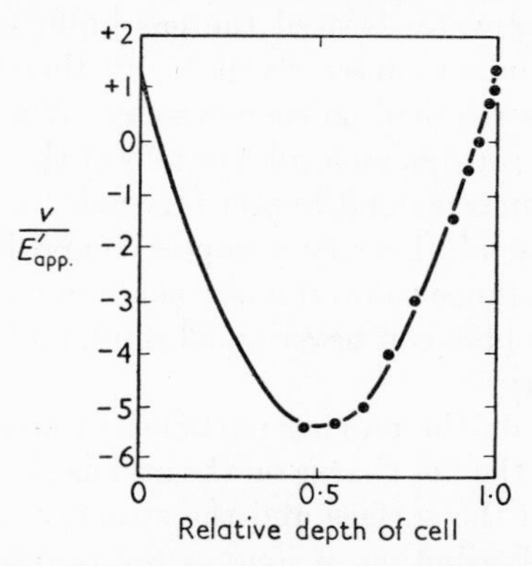

Fig. 3-13. Experimental variation of v/E′app., for particles moving along a tube of rectangular cross-section, at different depths across the tube[31]. Units of v are microns sec.$^{-1}$, and the potential gradient is measured in volts cm.$^{-1}$

of measurement in cells of rectangular cross-section have also been calculated[2]. A cylindrical cell can be made conveniently from a 10 cm. length of "Pyrex" tubing, 0.5 mm. internal diameter: electrodes may be of platinum if they are fairly large. It is important to correct for the lens effect in cylindrical cells, as the outer wall is conveniently ground flat in the region in which the electrophoresis is to be observed, and the correction is non-linear[32].

The particles are timed over (say) 100 microns, under a potential gradient of the order 10 volts cm.$^{-1}$, applied to the electrodes from high-tension batteries. These mobilites are converted to zeta potentials by eq. (3.26). If $\varkappa a > 300$ and K_s is negligible, $\mathbf{F}=3/2$, and if also $\mathbf{R}=1$, $\eta_0=0.01$ poise, and $D_0=80$, the resultant equation becomes, at 20°C,

$$\zeta = 14.2 \times \frac{v}{E'_{app.}}$$

while, at 25°C,

$$\zeta = 12.9 \frac{v}{E'_{app.}}$$

where v is the electrophoretic mobility, now in microns sec.$^{-1}$, $E'_{app.}$ is in volts cm.$^{-1}$, and ζ is in millivolts. The shape of the particles has apparently no effect on ζ, provided that the thickness of the double layer is small compared with the dimensions of the particles. If the ionic strength is very low, the above relations may require modification appropriate to a lower value of **F**.

In non-aqueous systems a special electrophoresis cell must be used[33], and the size of the particles (e.g. of carbon in oil) should be about 0.5 micron. Net charges on the particles may be very low, though, because of the very low ionic concentrations and the small dielectric constant of the oil, zeta potentials may be as high as or greater than those found in aqueous systems (cf. eq. (3.1)). The mobilities are also very low because of the low dielectric constant (eq. 3.26), and the experimental procedures call for extreme care, very high values of E′ (up to 1000 V.cm.$^{-1}$) being required. The velocity v may be found[33] by microscopic observation, by using radio-active tracers in conjunction with a Geiger counter, or by dye tracers[34]. A typical figure[15] for ζ is +50 mV for carbon-black particles dispersed in benzene containing 10^{-3}M Ca(di-isopropyl-salicylate).

When the mobilities of gas bubbles are required, these must be studied in a capillary tube continuously rotating about its long axis, to prevent the bubble rising and adhering to the glass wall of the tube[27,28].

There are two modifications of the procedure given above, applicable to very small and rather large particles. The former, e.g. protein molecules, may be allowed to adsorb on to larger particles such as those of quartz or carbon, or on to oil drops. The electrophoretic velocities of these coated particles are usually very close to, or identical with, those of the original small particles or molecules as measured by the Tiselius electrophoretic method. However, some proteins, such as insulin, show small differences, perhaps due to the exposure of different groups when adsorbed. Nylon shows large differences in potential between the stretched and the unstretched forms for the same reason. For materials such as this, one may also find ζ from the bending of a fine fibre under the influence of the applied electric field, one end of the fibre being fixed, while the deflection of the other end is observed with a microscope. The adsorption of long-chain ions or dyes may thus be rapidly and easily observed. A similar technique consists in attaching a sphere of the substance to be studied to the free end of an inert fibre[35].

Applications of Electrophoresis

(a) *Biological*

The identification of proteins by measuring ζ as a function of pH and ionic strength is well known[36]. More recently the outer layers of living cells have been studied in the same way, as well as by the more powerful method using the "reversal of charge" spectrum of concentrations of various heavy metal ions[23,37]. In this method the salt of the heavy metal, such as Pb^{++}, is added to the suspension of cells till ζ becomes zero instead of its usually negative value: this concentration of Pb^{++} (Fig. 3-12) is called the "reversal of charge" concentration, since higher concentrations make ζ positive, presumably by specific adsorption into a Stern layer at the surface. These specific adsorption potentials (Table 2-VII) are characteristic both of the heavy metal ion and of the chemical type of the cell surface, so that, by comparison with known surfaces, identification of the constituents on the surface of the cell is readily achieved.

Proteins are greatly influenced by small amounts of long-chain ions in solution, and electrophoresis studies have shown[38] that, while native ovalbumin adsorbs a maximum of three times its weight of sodium dodecylbenzene sulphonate at pH 6.5, the denatured protein adsorbs it in all proportions. Sodium dodecylsulphate will also adsorb on to horse serum albumin, changing its mobility in steps from -0.58 to -0.80, and to -0.94 micron sec.$^{-1}$volt^{-1}cm., corresponding to 0, 54, and 109 long-chain anions becoming adsorbed on to each protein molecule[39]. The latter has 110 positive groups on its surface at this pH, so, evidently, the long-chain anions are adsorbed on to these. The application of electrophoretic techniques to biological systems has recently been summarized[40]: the assay of bacterial variants is of particular interest.

(*b*) *Electrophoretic Deposition*

Metals have long been coated with rubber particles through electrophoretic movement followed by discharge on the electrode. Sulphur, accelerators, and fillers can be deposited simultaneously with the latex globules, to give a coating which may be subsequently vulcanized by heating. The electrophoretic deposition of rubber, lacquers, or synthetic resins has the advantage that, as parts of the surface become covered with the particles, the electrical resistance rises, the current thus being displaced to other parts of the surface which are still not well coated. This ultimately produces a pore-free coating[41].

(*c*) *The relation between ζ and emulsion stability*

Eilers and Korff[42] found that the stability of a dispersion decreased sharply below some critical surface potential ζ_c, such that $\zeta_c^2/\varkappa \approx 10^{-3}$ if ζ is in mV and $\varkappa$ (the Debye-Hückel distance) is in cm.$^{-1}$ In words, this implies that coagulation or coalescence sets in when added salt compresses the double layer and reduces ζ to a point where the repulsive forces can no longer prevent the surfaces of adjacent particles or drops coming into intimate contact. More generally,

$$\zeta_c \left(\frac{D_0}{\varkappa} \right)^{1/2} = \text{constant} \tag{3.29}$$

though, in the rigorous theoretical form[43], this equation is in ψ_0 rather than in ζ. The formulation in ζ is useful in practice, however, since ζ can readily be measured by electrophoresis of the dispersed phase[42,44]. Further, there is often a simple proportionality between ζ and ψ_0 as discussed below.

As a general approximate expression for the rate of coalescence of an emulsion we may write[45]

$$\text{Rate} = A_1 \exp\left(\frac{-B\psi^2}{RT}\right) \tag{3.30}$$

Here the potential to be used is ψ_δ if a Stern layer is formed, otherwise ψ_0. The application of this equation is described in Chapter 8.

Long-chain compounds are also added to lubricating oils to form a stabilizing monolayer on the particles of carbon formed from incomplete combustion of the fuel oil. The small quantities of oxidation products from the oil themselves stabilize somewhat these carbon particles by charging them negatively, with the result that if only small amounts of additive (e.g. 0.02% of Ca(di-isopropyl-salicylate)) are added, ζ may be reduced to nearly zero, with consequent deposition of unusually large amounts of sludge. With the normal concentrations of detergent additives (e.g. 15%), however, the carbon particles become strongly positively charged, and sludge deposition is almost eliminated[33].

(d) The Form of Macromolecules

Overbeek and Stigter[46] have used electrophoresis as a means of determining the configuration of poly-electrolyte molecules in solution: in solutions of low salt content the molecules behave as porous spheres, though at high ionic strengths they become cylindrical rods.

SEDIMENTATION POTENTIALS

Charged particles sedimenting in a liquid tend to leave behind them the diffuse part of the electrical double layer, and if electrodes (e.g. Ag/AgCl) are placed near the bottom of the vessel and near the top, the difference of potential gradient $E_{sed.}$ between these electrodes is the sedimentation potential. This is the converse of the streaming potential phenomena, from which (eq. (3.20)) we obtain $E_{sed.}$ by replacing the pressure P by the total driving force in the new system, i.e. by $(4/3)\pi a^3(\rho' - \rho'')n_0 g$; for $\kappa a \gg 1$, we have accordingly:

$$E_{sed.} = \frac{a^3(\rho' - \rho'')n_0 g\, D_0 \zeta}{3K\, \eta_0} \tag{3.31}$$

In this relation, first derived by Smoluchowski and extended by Booth[47], a is the radius of each falling sphere, ρ' and ρ'' are the densities of the spheres and of the liquid, n_0 is the number of spheres per unit volume, g is the gravitational constant, and K is the specific conductance of the liquid (K_s

being neglected). Further, since the volume fraction φ of dispersed phase is given by $(4/3)\pi a^3 n_0$, $E_{sed.}$ must be proportional to the amount of the dispersed phase, but independent of its degree of dispersion. This has been confirmed experimentally[48].

As with electrophoresis, the geometry of the particles must strictly be allowed for, and if $\varkappa a$ does not greatly exceed unity, we must write:

$$E_{sed.} = \frac{a^3(\rho' - \rho'')n_0 g \, \mathbf{f} \, D_0 \zeta}{3K \, \eta_0} \tag{3.32}$$

by exact analogy with electrophoresis[15], the function **f** varying with $\varkappa a$ as in Fig. 3-11.

For all particles such that $\varkappa a \ll 1$ we see that $\mathbf{f} = 2/3$, though, if $\varkappa a$ becomes large (i.e. $1/\varkappa$ becomes small), a relatively strongly conducting sphere will have **f** approaching zero[15]. Surface conductance should strictly also be allowed for[48].

The experimental results show that eq. (3.31) is obeyed closely for Pyrex glass powder (a ~50 microns) settling in dilute KCl solution, values of ζ calculated from it being -140 mV in 10^{-4} N-KCl and 10^{-5} N-KCl, and -190 mV in distilled water[49]; another value for the latter system[48] is -110 mV. The former figures agree well enough with those of Table 3-II. Also in agreement with theory is the finding that variation of particle size, over a limited range, does not affect $E_{sed.}$.

For particles of small size the sedimentation may only occur at a reasonable speed if the solution is centrifuged, in which case g in eq. (3.31) should be replaced by $\omega^2 r$, ω being the angular velocity of rotation and r the distance from the axis. For an aqueous colloid of AgI, for example, the calculated value of $E_{sed.}$ is 17.5 mV cm.$^{-1}$ at 100 r.p.m. with r = 10 cm., the observed value[50] being 15.6 mV cm.$^{-1}$ The proportionality of $E_{sed.}$ and ω^2 is also confirmed for colloidal As_2S_3[51]. In the presence of 10^{-2} N-KCl its ζ potential, by using eq. (3.31), is -50 mV[51], the best values by other methods being between -50 mV and -80 mV. Instead of centrifuging the solution, modified sedimentation potentials may be determined from the potentials generated when ultra-sonic waves are passing through a colloidal solution. For an AgI colloid, these measured ultrasonic vibration potentials are of the order 1 mV[50].

The settling of water drops in oil storage tanks may set up sedimentation potentials large enough to be dangerous, gradients between 10 and 100 volts cm.$^{-1}$ being observed in practice in gasoline tanks after filling, if 6% of added water in the gasoline has been broken into small drops during pumping[15].

These large potentials arise primarily because the conductivity K of the bulk phase is very low for petroleum oils (see eq. (3.31)). Also, the thickness of the ionic double layer in oils may be comparable with the particle radius;

as when gasoline ($1/\varkappa \approx 130$ micron) contains a suspension of water droplets of radius 50 microns. For such a relatively thick double layer, the 6π form of the electrophoresis equation is valid irrespective of the electrical conductivity of the particles, and hence the 2/9 form of eq. (3.32) is required to calculate the sedimentation potential. If, however, the ionic concentration of the oil is increased by additives, $E_{sed.}$ is reduced, both because K is greater and because the factor **f** in eq. (3.32) becomes smaller, tending in the limit to zero when $\varkappa a \gg 1$ (Fig. 3-11). Retardation effects also operate, but are important only for small particles, when a large $E_{sed.}$ may appreciably retard the sedimentation[15] (see below).

As an example, let us take $D_0=2$, $\zeta=25$ mV, $\varphi=0.06$, $g=1000$, $(\rho'-\rho'')=0.3$ g cm.$^{-3}$, $\eta=0.5$ c.p., $K=10^{-14}$ohm^{-1}cm.$^{-1}$, and $a=50$ microns; we then calculate that $E_{sed.}=1000$ volts cm.$^{-1}$ In a tank several metres high the potential can rise to dangerous levels even with a small value of ζ. If the droplets are much larger than 100 microns in diameter, they soon settle out completely and $E_{sed.}$ then becomes zero; if they are smaller the sedimentation potential gradient reduces the rate of settling to the point where a "haze" of water "floats" in the electric field it is itself producing; this effect reduces the field strength below that given by the above calculation. Numerical allowance can be made for this[15].

The addition of small amounts of Cr (alkyl-salicylate)$_3$ causes adsorption of positive ions (presumably Cr (alkyl-salicylate)$_2^+$): this increases ζ considerably and hence may increase $E_{sed.}$ several times. For example, adding 6×10^{-10} M additive to the oil is found to increase $E_{sed.}$ from 20 volts cm.$^{-1}$ to 55 volts cm.$^{-1}$ Further addition of the additive reduces $1/\varkappa$ and increases K, so that $E_{sed.}$ is greatly reduced[15], falling to only 3 volts cm.$^{-1}$ for an additive concentration of 10^{-7} M.

RETARDATION OF SETTLING VELOCITY AND ζ

Corresponding to the retardation of flow in narrow capillaries (p. 125) is the drag of the counter-ions on a sedimenting particle[47]. This is exactly equivalent to the reduction in sedimentation rate by the sedimentation potential gradient, as discussed above. If v_0 denotes the steady rate of fall when the surface of the particle is uncharged, and v the rate when the electrokinetic potential is ζ, then for a single sedimenting particle

$$v=v_0\left\{1-\frac{1}{\eta_0 K}\left(\frac{D_0\zeta}{4\pi a}\right)^2\right\} \tag{3.33}$$

Booth[47] has shown that this equation is strictly valid only if $\varkappa a \gg 1$; it neglects terms in ζ^3. Further, in suspensions, interactions between the particles may affect v.

According to eq. (3.33), if $\zeta=100$ mV and $a=10^{-1}$cm., the retardation of sedimentation velocity in water is only about 2%. For small particles, however, the effect is much larger, being 30% if $a=10^{-5}$cm., $\zeta=25$ mV, and $K=10^{-6}$ohm^{-1}cm.$^{-1}$; an ultracentrifuge may be required to study such small particles. Experimentally, Tiselius[52] has shown by the latter technique that for particles of $a=4\times10^{-7}$cm. agreement with theory is within an order of magnitude. If the particles are very small, it is clear from eq. (3.33) that the settling may become very slow; this is found for water "hazes" in oil, if the drop diameter is 15 microns or less[15].

INFLUENCE OF SURFACE ROUGHNESS ON ζ

A solid surface may ordinarily have "valleys" as deep as 10^{-7} or 10^{-6}cm.[41,53] During shear the moving liquid will not penetrate these depths, but will glide across them; consequently slight macroscopic roughness may increase the charge-bearing surface relative to the area of the hydrodynamic plane of shear, and the magnitude of ζ may also be increased. It is indeed found[54] that for calcite the potential is -18 mV over a normally smooth surface, but becomes -33 mV on a pitted surface, with a similar change in magnitude for $BaSO_4$. On the other hand, ordinarily smooth glass in contact with dilute aqueous KI has a potential of -162 mV, compared with only -143 mV after the surface has been roughened by chemical attack[8]: the valleys here may be so deep that the electrical double layer is partly confined therein, especially if the double-layer thickness $1/\varkappa$ is small. "Nestling" or penetration of counter-ions into gaps between the fixed charges will produce the same result.

THE RATIO ζ/ψ

This ratio may be calculated by evaluating the integral $\int_0^{\psi_0}\frac{D}{\eta}\mathrm{d}\psi$ of equation (3.10): by definition of ζ (see eq. (3.8)) this must be equal to $\frac{D_0\zeta}{\eta_0}$, where D_0 and η_0 are the dielectric constant and viscosity of the continuous phase in zero electrical field. This equality, i.e.

$$\zeta=\frac{\eta_0}{D_0}\int_0^{\psi_0}\frac{D\mathrm{d}\psi}{\eta} \tag{3.34}$$

is of interest also in showing what meaning can be attached to ζ. The ratio ζ/ψ_0 is a test of current theories of the variation of D and η with ψ' ($\psi'=\mathrm{d}\psi/\mathrm{d}x$).

Firstly, the movement of ions will be restricted if the viscosity of the liquid becomes high near the surface: Andrade and Dodd[20] have published data on the measured viscosity of organic liquids of various dipole moments as a function of electrical field strength. Using their relations, we find that the calculated viscosity of water rises steeply (Fig. 3-14) at a field strength of about 3300 e.s.u. cm.$^{-1}$ (i.e. 10^6 volts cm.$^{-1}$). Such a field is reached at 3 Å from the plane of the fixed charges if $\psi_0 = 200$ mV and $c_i = 10^{-2}$ N[55]. Closer to the surface than 3 Å there is effectively a monolayer of "soft ice"[16,20]. It is likely, however, that the many counter-ions in this region will disorientate somewhat this "soft ice", making it less viscous, though no correction is

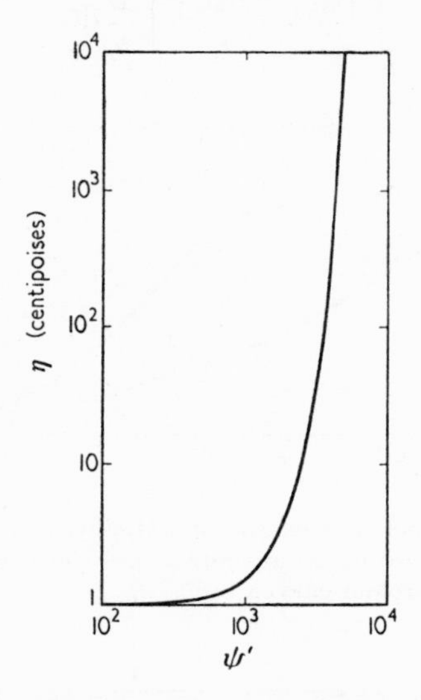

Fig. 3-14. Calculated viscosity of water as a function of the electrical field strength, expressed as e.s.u. cm.$^{-1}$ Both scales are logarithmic.

available for this factor; consequently the results can only be considered accurate to an order of magnitude. This concept of a viscous layer near the surface is supported by the observation that the zeta potential of spheres of poly-divinylbenzene, carrying chemically bound sulphonic acid groups on the surface, is greatly reduced by a little polyvinyl alcohol in the aqueous solution. Glycerol has a similar, though smaller, affect. The results are too large to be explained in terms of changes in surface conductance or dielectric constant, and are apparantly caused by these additives assisting in the formation of the highly viscous layer[56].

The dielectric constant also falls off as ψ' increases, several different estimates of this effect being available: these are shown in Fig. 3-15[55,57].

The net results are that D/η falls off sharply with ψ' (Fig. 3-16), and that, if D/η is plotted against the corresponding values of ψ (given any particular

ψ_0, here 200 mV, and given c_i), D/η has fallen to a negligibly small value when ψ is 100 mV (corresponding to the distance of 3 Å from the surface). At higher values of ψ, i.e. within this distance, there is virtually no contribution to electrokinetic phenomena.

The meaning of ζ is best seen as follows. In calculating ζ it is assumed that D and η are constant, with values D_0 and η_0 right up to a plane of shear, at which η suddenly becomes infinitely great; or, in symbols, that

$$\int_0^{\zeta} \frac{D_0}{\eta_0} d\psi = \frac{D_0 \zeta}{\eta_0} = \int_0^{\psi_0} \frac{D}{\eta} d\psi$$

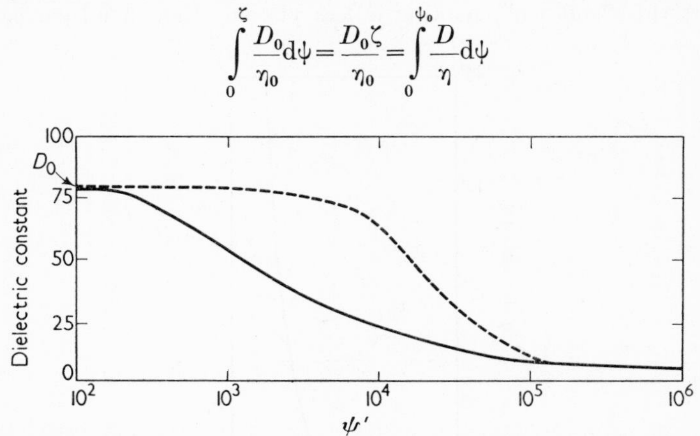

Fig. 3-15. Variation of dielectric constant of water with field strength (e.s.u. cm.$^{-1}$) plotted logarithmically according to authors of ref. 55 (full curve). The values of Booth[57] are shown by the broken curve.

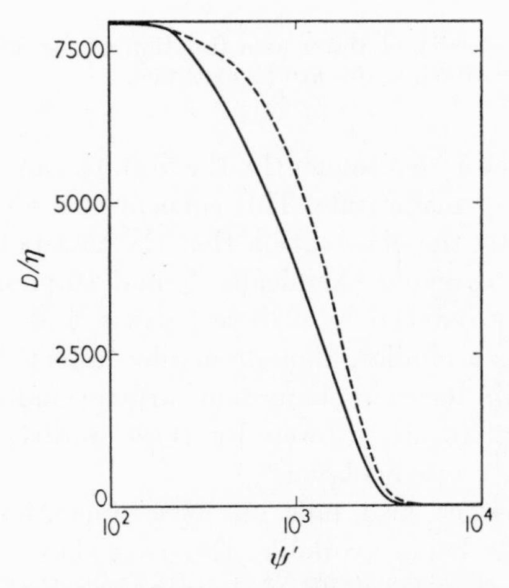

Fig. 3-16. Variation of D/η according to data of Figs. 3-14 and 3.15, with field strength (e.s.u. cm.$^{-1}$) plotted logarithmically. Here η is expressed in poises.

It is now apparent from Fig. 3-17 that the shaded area under the curve must be equated to the area of the rectangle shown by broken lines, from which ζ must be 55 mV (assuming that $\psi_0 = 200$ mV and $c_i = 10^{-2}$ N). At the same value of ψ_0 but with $c_i = 10^{-3}$ N, ζ increases to a calculated 99 mV.

Secondly, relation (3.34) permits a theoretical evaluation of the ratio ζ/ψ_0, as shown in Table 3-IV. The tendency is clearly for ζ/ψ_0 to increase as either the ionic strength or surface charge density is lowered.

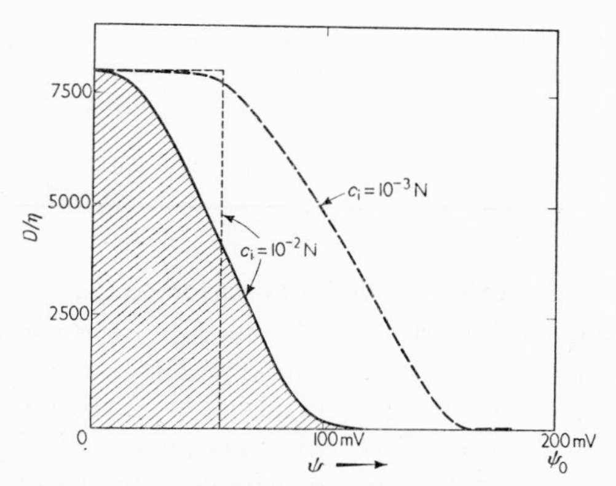

Fig. 3-17. Data of previous figure plotted as D/η against ψ, for the particular conditions of $\psi_0 = 200$ mV, and 10^{-2} N uni-univalent electrolyte, using a tabulated relation[55] between ψ and ψ'. At about 105 mV in this system, η is calculated to be 100 centipoises, and that part of the electrical double layer lying closer to the surface than this is therefore virtually immobile. The broken curve shows similar calculations, with ψ_0 again 200 mV, but with 10^{-3} N electrolyte: here more of the double layer is mobile.

All this assumes no specific interaction, which is probably justified as a first approximation for small inorganic counter-ions such as Cl′ or Na^+ against a liquid surface: any slight specific binding ($\sim$2kT) partly offsets the non-zero sizes of the counter-ions, as explained in Chapter 2, and the effect on the potentials is small. The ζ/ψ_0 ratio found experimentally in several such systems (see below) is usually about 0.5, confirming that specific effects are small at concentrations below about 10^{-1} N. Above this concentration, however, specific effects become more important, and "reversal of charge" occurs at concentrations of Na^+ between 1 N and 10 N.

Heavy metal ions, however, such as Th^{+4}, are strongly absorbed on a surface of negative ψ_0, forming a strongly bound Stern layer, as in Fig. 2-23 (position C). They may even cause ζ to be of the opposite sign from ψ_0 (Fig. 2-24d): for these systems it is safe to assume that D, η, and hence the ζ potential may be calculated (eq. (3.34)) from the plane of the adsorbed Th^+ ions, rather than from the plane of negative ions which determine ψ_0. The use of the reversal of charge concentrations of heavy metal ions in characterizing surfaces (specific interaction being a function of the surface as well as of the counter-ion) is discussed on pp. 135-6.

TABLE 3-IV

Calculated values of ζ/ψ_0 (*eq.* (3.34))

ψ_0	c_i	ζ/ψ_0
200 mV	10^{-2} N	0.28
200 mV	10^{-3} N	0.495
100 mV	10^{-2} N	0.61
100 mV	10^{-3} N	0.91

Experiments on ζ in relation to ψ_0.

The first experimental work[58] on ζ and ψ_0 was concerned with their difference, the changes in ψ_0 being measured with a glass electrode as changes in $\Delta\Phi$. For heavy ions such as thorium adsorbed on glass, the ψ_0 potential is little changed, even when the ζ potential becomes markedly positive. This suggests that the heavy metal ions are strongly bound to the solid surface, forming a solid Stern layer (the specific interaction λ_p being quite high); electrokinetic movement of liquid and ions must be possible only outside this layer. It is thus reasonable to suppose that ζ will be less than ψ_δ of the Stern theory (Figs. 2-23 and 2-24d).

While the term $(\psi_0 - \zeta)$ may be determined readily by the above method, the ratio ζ/ψ_0 has proved more difficult to evaluate experimentally[59]. It is known, however, that long-chain ions readily adsorb on to oil drops, of which the ζ potential may be found by electrophoresis. The oil drops, even in the absence of all long-chain ions, carry a slight negative charge produced by selective desorption of H_3O^+ ions from the water; consequently, to avoid this complication, the drops are best studied when most of the interface is well covered by the adsorbed monolayer. Typical results are shown in Table 3-V: they are selected at ionic strengths such that the correction factors **R** and **F** (eq. (3.26)) are known to be respectively unity and 1.5. The corresponding adsorptions may be used[59] to correlate these data with ψ_G, or ψ_0 may be measured across a plane interface at equilibrium (Chapter 2). For many long-chain ions the quotient ζ/ψ_0 is about 0.5, rather higher than according to calculation, which gives 0.3 to 0.4 by the method described above. The divergence in the ratio (by a factor of about 1.5) is probably caused by the non-zero sizes of the counter-ions, and by their disorientating effect, which may reduce the "ice"-forming tendency of the surface field.

It will be noted that the coated oil drops are treated as if they were solid, in that eq. (3.26) has been applied. This is equivalent to neglecting momentum transfer across the interface: there should thus be no circulation within the drops. The justification for this is that adsorbed monolayers usually possess sufficient interfacial viscosities and sufficiently high compressional moduli (p. 265) to prevent appreciable surface movements (cf. Fig. 7-5 and p. 335).

TABLE 3-V

	ζ(mV)	ψ_0(mV)	ζ/ψ_0 data of previous columns	ζ/ψ_0 calc. from eq. (3.34)
Drops of paraffinic oil with monolayers of cetyltrimethyl-ammonium bromide adsorbed from 0.05 M-acetate buffer with concentrations of long-chain ions of:				
5.8×10^{-5} M[59]	+81	151	0.54	~0.35
12.5×10^{-5} M[59]	+93.5	164	0.57	~0.35
Drops of paraffinic oil with adsorbed lauryl sulphate ions (no added salt) at concs. of:				
3×10^{-3} M[60]	—96	—214	0.45	~0.35
8×10^{-3} M[60]	—100	—211	0.47	~0.35
Ibid, in presence of 0.05 M-NaCl throughout, ($n = 2.0 \times 10^{14}$)[61]	—92	—160	0.57	~0.40
Ibid, in presence of 0.1 M-NaCl throughout,				
10^{-4} M[60]	—88	—133	0.66	~0.30
10^{-3} M[60]	—69	—147	0.47	~0.30
$n = 2 \times 10^{14}$)[61]	—76	—146	0.52	~0.30
Drops of paraffinic oil with adsorbed $C_{12}H_{25}N(CH_3)_3^+$ film, in 0.07 M-NaCl.	+60	+155	0.38	~0.35
Ibid, with 0.1 M-NaCl				
$n = 2.0 \times 10^{14}$)[61]	+50	+145	0.34	~0.35

Note: values at low ionic strengths are omitted because of the uncertainty of the correction for surface conductance, and values at low surface charge densities are omitted because of the uncertainty concerning the electrokinetic properties of the "bare" parts of the interface.

A modification of our method of finding the ratio ζ/ψ_0 was used by Stigter and Mysels[62] who studied the mobilities of micelles of sodium lauryl sulphate. The mobility values in different salt concentrations, extrapolated to zero concentration of micelles, are shown in Table 3-VI, together with the zeta potentials calculated using the equation of Booth[22] to obtain **R** (eq. (3.26)), which is here between 0.83 and 0.87. The fraction of sodium ions free to move independently is found from ζ to be 28.7%. The ψ_G values are calculated using the size of the micelle and the sizes of the component long-chain ions: these give a value of A for the charged groups in the micelle surface, from which, using eq. (2.27), ψ_G can be obtained directly. In fact, Stigter and Mysels modified the Gouy equations to allow for the curvature of the micelle surface, but the ψ_G values they obtained (Table 3-VI) are within a few millivolts of those calculated directly from eq. (2.27) with $A = 60 Å^2$, as assumed for the micelle surface. The resultant ratios of ζ/ψ_G are all close to 0.5, with NaCl concentrations from zero to 0.1 M. Again, at higher salt concentrations, the calculated ratio falls below the experimental figures.

TABLE 3-VI

Electrokinetic Potentials of Micelles of Sodium Lauryl Sulphate

Conc. of NaCl	v in micron sec.$^{-1}$ volt^{-1}cm.	ζ in mV (eq. (3.26)) (neglecting K_s)	ψ_G*(mV)	ζ/ψ_G from previous columns	ζ/ψ_G calc. from eq. (3.34)
0.00 M	—4.55	—101.2	—190	0.53	0.55
0.01 M	—4.28	— 92.3	—178	0.52	0.38
0.03 M	—3.84	— 80.9	—161	0.50	0.35
0.05 M	—3.63	— 75.0	—150	0.50	0.32
0.10 M	—3.42	— 68.3	—136	0.50	0.25

* A is assumed to be 60 Å² throughout.

One may also relate the Donnan potential, calculated for a sulphonate ion-exchange resin, and the zeta potential as calculated from the mobilities of particles of the resin[63]. The results, however, cannot be directly interpreted as ζ/ψ_0 values, because the Donnan potential is the mean potential throughout the particles of resin, and is not necessarily identical with the potential drop at the surface (Chapter 2).

Freedom of the Counter-ions

That the calculated viscosity of water increases steeply, if ψ' exceeds 3000 e.s.u. cm.$^{-1}$ (when $\eta \sim 25$ c.p.), implies that many, if not most, of the counter-ions will normally be incapable of independent motion. For example, if $\psi_0 = 200$ mV and $c_i = 10^{-2}$ N, this occurs about 3 Å from the surface, within which distance about 50% of the counter-ions may lie (their closeness to the surface being possible because of its penetrability, see Fig. 2-23). Even outside this region the viscosity of the water is still enhanced, and the counter-ions are still somewhat restricted. This restriction of many of the counter-ions, which is largely responsible for making ζ so much less than ψ_0, is well known in other branches of physical chemistry. Various studies[64,65], including those of the osmotic properties of micelles, show that only between 17% and 25% of the counter-ions can move independently in an electrical field, the remainder being carried with the micelle. Transport numbers, measured[66] for micelles of sodium lauryl sulphate, show directly that only 28% of the counter-ions are free to move independently, in good agreement with the figure of 28.7% (quoted above) calculated from ζ. This concordance also confirms directly the validity of the Booth correction factor **R** of about 0.8, used in calculating ζ for the micelles, since, without allowing for this correction (i.e. putting $\mathbf{R}=1$), the zeta potential method would give the fraction of Na^+ ions free to move independently as only 20%.

Origin of Charges on Surfaces

It is clear from the figures of Table 3-II that the effects on glass of hydroxyl and hydrogen ions are particularly important. Further, 10^{-5} N alkali has the unusual effect of making ζ larger numerically by 16 mV than for distilled water, whereas acid in the same concentration decreases ζ numerically by as much as 80 mV. For an electrolyte such as KCl, in the same concentration, the numerical decrease in ζ is only 20 mV. These results are explicable in terms of the glass surface containing silicic acid groups, which ionize further in alkali but less in acid, while the ψ_0 potential is decreased in the usual way (see eq. (2.32)) as the ionic strength is increased for any given surface charge density. From Table 3-II follows the relation

$$\left[\frac{\partial|\zeta|}{\partial \log_{10} c_i}\right]_{\mathrm{pH,T}} = -45\,\mathrm{mV}$$

(where KCl is taken as the "inert" electrolyte, and where c_i refers to the concentration of uni-univalent electrolyte). Comparison with the similar equation for ψ_0 (2.32) at these concentrations (about 10^{-4} N) suggests that for this system

$$\frac{\zeta}{\psi_0} \approx \frac{45}{60}, \text{ i.e. } \zeta/\psi_0 \approx 0.75.$$

For certain bacteria[23], whose surfaces are known to contain ionogenic carboxyl groups, the data of Fig. 3-12 show that

$$\left[\frac{\partial|\zeta|}{\partial \log_{10} c_i}\right]_{\mathrm{pH,T}} = -16\,\mathrm{mV}$$

This applies with uni-univalent electrolytes at about 10^{-2} N, suggesting that here $\zeta/\psi_0 \approx 0.28$. Crystals of $BaSO_4$ (for which ζ is positive) also show a decrease in ζ of about the same magnitude[54].

Among other aqueous systems whose charge arises from ionogenic groups are oil drops whose surfaces have adsorbed long-chain ions, proteins, or poly-carboxylic acids[67]. Beads of poly-divinylbenzene[56] of 5 microns radius and with chemically bound sulphonic acid groups on the surface provide a convenient system for fundamental studies, though surface conductance may be relatively important here. Micelles of long-chain ions[62] are also of considerable interest (Table 3-III); the differential is again -16 mV.

Crystals of $BaSO_4$ have a positive ζ potential arising apparently from preferential release of sulphate ions[54] from the surface of the crystal lattice: this is equivalent to postulating an excess of vacant lattice sites or interstitial ions[68], possibly due to defects and dislocations. The same explanation may be applied[69] to colloidal Fe_3O_4, TiO_2, ZrO_2 and ThO_2: ζ is positive with a charge density, calculated by the Gouy treatment (eq. (2.27)), of about 1 electronic charge per 1000 Å^2. The differential $(\partial|\zeta|/\partial \log c)$ is -16 mV for Fe_3O_4, consistent with the assumption of an ionogenic surface.

These ionogenic surfaces are characterized by high sensitivity to heavy-metal ions and to polyvalent ions of the opposite sign: as little as 10^{-6} M Th^{+4} reduces the zeta potential of glass to zero, presumably by forming a Stern layer in which these counter-ions are strongly-bound because of a high value of λ_p (Chapter 2). In a typical non-aqueous system, the adsorption of a monolayer of Ca(di-isopropyl-salicylate)$^+$ on to water drops in oil confers the positive charge[15].

By contrast, the origin of the negative charge on a drop (not covered with an adsorbed film) of a highly purified paraffinic oil[27], or on a gas bubble[28], must lie in the selective desorption of one of the ions in the supporting water from the non-polar surface. That desorption, rather than adsorption, is operative is suggested[70] by the fact that simple electrolytes raise the interfacial tension of water against air or oils; consequently, by the Gibbs equation (eq. (4.54)), ions must tend to keep away from the oil-water interface. Evidently H_3O^+ ions, being large and hydrated, keep further from the non-polar phase than do the OH' ions, since drops of pure hydrocarbon oil in pure water[27] carry a potential of about -72 mV (allowing for both **F** and **R**, since κa is in the range 2 to 150, and **R** of the order 0.7). At an ionic strength of 0.01 N (at pH 7), however, this is numerically reduced[71] to about -56 mV. Derived also from the experimental results[27,71] on pure oils is the relation

$$\left[\frac{\partial|\zeta|}{\partial \log_{10} c_i}\right]_{\substack{pH=7\\ T=25°C}} = -3\,\text{mV}$$

showing that this differential is very much smaller than for surfaces containing ionogenic groups. Further, these potentials, because of preferential desorption of certain ions, are characterized by their extreme sensitivity to traces of non-ionized organic additives: as little as 10^{-4} M-ethanol[27] will reduce the numerical value of ζ on an oil drop by about 20%, while 2% of a sugar[27,71] will reduce it by 65%! The explanation lies in the fact that the ions from the aqueous phase are tending to raise the interfacial tension, and adsorption of the organic material on to this interface will therefore proceed very readily. This makes the interfacial region more polarizable, so that strongly hydrated ions (e.g. H_3O^+) are not so strongly desorbed. It follows also that these desorption potentials can be assumed unimportant at polar, ionogenic interfaces.

For surfaces at which no monolayer is adsorbed, the mobility of the drop must be corrected for the momentum transfer across the interface. This correction, which is in dispute[26,27], becomes negligible when the liquid in the drop is very much more viscous than the continuous medium or if there is a monolayer adsorbed at the surface. For non-ionogenic surfaces there should be no excess surface conductance.

SPRAY ELECTRIFICATION

In the Austrian Alps in 1890, Elster and Geitel[72] noticed that there was intense electrification, accompanied by sparks, around certain waterfalls. Two years later Lenard studied this effect in more detail, finding that the fine mist carried upwards by the updrafts was predominantly negatively charged, while nearer to the surface of the water there was some positively charged spray[72].

The explanation of this is now known to be the negative charging (see above) of the surface of water in contact with air: disruption of this surface by shattering or bubbling causes the release of very fine drops, each consisting of some hundreds or thousands of water molecules, and each containing a slight excess of negative charge from the original surface[73]. Since, however, the depth of the surface required to form these droplets is comparable with $1/\varkappa$, the process is rather inefficient, and as little as 0.01% or 0.001% of the original charge on the surface may be found on the droplets. The current associated with the very fine negative spray, being about 10^{-9} or 10^{-10} coulombs per gram of water, is still sufficiently large to lead to sparking if no short-circuits are available. The same explanation may well apply to the charge generation in the turbulent, wet regions of thunderstorms.

FREEZING POTENTIALS

If a bent copper rod is dipped partly into liquid air and partly into water, ice will gradually form around the rod (Fig. 3-18). The electrical potential between the copper rod and a reference electrode in the water remains only small when either very pure water or 0.1 N aqueous electrolyte is studied, but potentials as great as 100 volts (the ice-coated electrode positive) arise if the water contains ammonium hydroxide of concentration about 10^{-4} N or

10^{-5} N. These freezing potentials increase with the rate of freezing of the water, and are caused by a few positive ions being frozen into the advancing ice layer: the high specific resistance of the ice phase (about 10^{11} ohm cm. under these conditions) permits the trapped cations to move back into the liquid water only slowly, so that there results a space charge of the order of 3×10^7 e.s.u. cm.$^{-3}$ in the ice, corresponding to 1 cation being occluded per 60,000 Å^2 of the advancing ice surface. A study of the entropy, probability and energy of this process would be very interesting[73;74]. Fluorides give up to 34 volts with the ice negative, and alkali chlorides and bromides give somewhat smaller negative potentials. Apparently it is the ions most readily incorporated into the lattice of the ice which are potential-determining.

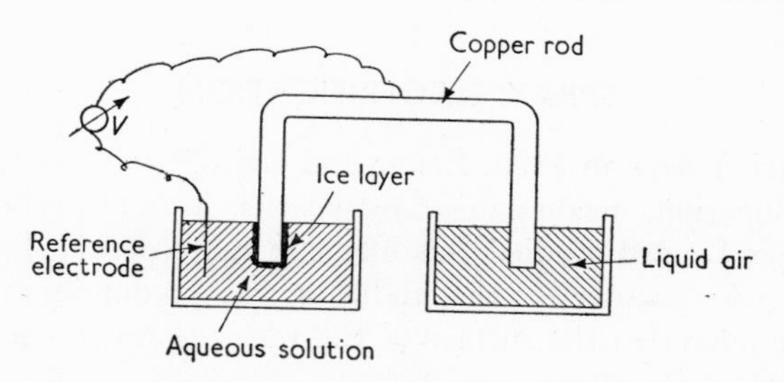

Fig. 3-18. Method of studying freezing potentials.

Salts (e.g. NaCl and $Ca(HCO_3)_2$) which exist as impurities in rain may well be responsible for atmospheric static electricity: falling hailstones, striking droplets of supercooled water, add to themselves further shells of ice, and, in doing so, become negatively charged, while the remaining liquid water may be broken into small droplets by the collisions, and then be carried away by updraughts[74].

The number of adsorbed ions per cm.^{3} of ice is, significantly, of the same order ($\sim 10^{15}$ per cm.^{3}) as the number of impurity centres in the semi-conducting crystals used as rectifiers, and ice samples prepared from dilute aqueous solutions do indeed show significant rectifying properties. Further, freezing the ice on to different substrates, of ice, Pt, or Ni, significantly affects the sign and amount of charge separated, suggesting that the crystal dislocations may be the absorption centres: this observation is supported by the finding that the polarity of a freezing substrate block of ice is transmitted unchanged to the freezing of a second solution on to the substrate, regardless of the electrolyte in the second solution.

REFERENCES

1. Reinold and Rücker, *Proc. roy. Soc.* **53**, 394 (1893);
Smoluchowski, *Phys. Z.* **6**, 529 (1905).
2. Overbeek, in "Colloid Science" I, Ch. 5, (ed. Kruyt) Elsevier, Amsterdam (1952); *J. Colloid Sci.* **8**, 420 (1953);
Fox, T.R.C.—private communication.
3. Urbain, White, and Stassner, *J. phys. Chem.* **39**, 311 (1935).
4. Bikerman, *Z. physik. Chem.* **163A**, 378 (1932); *Kolloidzschr.* **72**, 100 (1935); *Trans. Faraday Soc.* **30**, 154 (1940;
Müller, *Cold Spr. Harb. Symp. quant. Biol.* **1**, 1 (1933);
Cole, *Cold Spr. Harb. Symp. quant. Biol.* **1**, 23 (1933);
Rosenhead and Miller, *Proc. roy. Soc.* **A163**, 298 (1937);
Rutgers and Janssen, *Trans. Faraday Soc.* **51**, 830 (1955).
4a. van Olphen, *J. phys. Chem.* **61**, 1276 (1957).
5. Bikerman, A.C.S. Meeting **21** 4 (September 1953).
6. McBain and Peaker, *Proc. roy. Soc.* **A125**, 394 (1929);
McBain, Peaker, and King, A. M., *J. Amer. chem. Soc.* **51**, 3294 (1929);
McBain and Foster, *J. phys. Chem.* **39**, 331 (1935);
Krishnamurti and Srinivasarao, *Research, Lond.* **6**, 62S (1953).
7. Rutgers and de Smet, *Trans. Faraday Soc.* **43**, 102 (1947).
8. Zhukov, "Electrokinetics in Capillary Systems", pp. 62–3, Moscow (1956);
Watillon, *J. Chim. phys.* **54**, 130 (1957).
9. Fricke and Curtis, *J. phys. Chem.* **40**, 715 (1936);
Henry, *Trans. Faraday Soc.* **44**, 1021 (1948);
Street, *Aust. J. Chem.* **9**, 333 (1956).
10. Zhukov and Fridrikhsberg, *Colloid J. Voronesh.* **11**, 163 (1949);
Klinkenberg, *J. Petrol. Tech.*, April 1957.
11. Mysels and McBain, *J. Colloid Sci.* **3**, 45 (1948);
Bikerman, *J. phys. Chem.* **46**, 724 (1942).
12. Mossman and Mason, *Canad. J. chem.* **37**, 1153 (1959);
Biefer and Mason, *J. Colloid Sci.* **9** 20 (1954); *Trans. Faraday Soc.* **55**, 1239 (1959).
13. Rutgers and de Smet, *Trans. Faraday Soc.* **41**, 758 (1945);
Rutgers and de Smet, *Trans. Faraday Soc.* **43**, 102 (1947);
Rutgers, de Smet, and de Myer, *Trans. Faraday Soc.* **53**, 393 (1957).
14. Overbeek and Wijga, *Rec. Trav. chim. Pays-Bas* **65**, 556 (1946), also quoted in ref. 2.
15. Klinkenberg and van der Minne, "Electrostatics in the Petroleum Industry" Chap. VI, Elsevier, Amsterdam (1958).
16. Henniker, *J. Colloid Sci.* **7**, 443 (1952);
Rouse and Howe, "Basic Mechanics of Fluids" p. 137, Chapman and Hall, New York (1953);
Boumans, *Physica, 's Grav.* **23**, 1007, 1027, 1038, 1047 (1957).
17. Ueda, Tsuji, and Watanabe, *Proc. 2nd Internat. Congr. Surface Activity* **3**, 3, Butterworths, London (1957).
18. Abramson, *J. gen. Physiol.* **15**, 279 (1932);
Elton and Hirschler, *Proc. roy. Soc.* **A198**, 581 (1949).
19. Davies and Guldman, Research Project in Department of Chemical Engineering, Cambridge (1956).

20. Derjaguin and Krylov, *Akad. Nauk. S.S.S.R., Otdel. Tekh. Nauk. Rastvorov,* **2**, 52 (1944);
Andrade and Dodd, *Proc. roy. Soc.* **A204**, 449 (1951).
21. Henry, *Proc. roy. Soc.* **A133**, 106 (1931);
Alfrey, Berg, and Morawetz, *J. Polym. Sci.* **7**, 543 (1951).
22. Booth, *Proc. roy. Soc.* **A203**, 514 (1950);
Overbeek, *Kolloid beih.* **54**, 287 (1943).
23. Davies, Haydon, and Rideal, *Proc. roy. Soc.* **B145**, 375 (1956).
24. Henry, *Proc. roy. Soc.* **A133**, 106 (1931); *Trans. Faraday Soc.* **44**, 1021 (1948);
Booth, *Trans. Faraday Soc.* **44**, 955 (1948).
25. Alexander, A. E. and McMullen, in "Surface Chemistry", p. 309, Butterworths, London (1949).
26. Booth, *J. chem. Phys.* **19**, 1331 (1951).
27. Jordan, D. O. and Taylor, A. J., *Trans. Faraday Soc.* **48**, 346 (1952);
Taylor, A. J. and Wood, F. W., *Trans. Faraday Soc.* **53**, 523 (1957).
28. McTaggart, *Phil. Mag.* **28**, 367 (1914); **44**, 386 (1922);
Alty, *Proc. roy. Soc.* **A106**, 315 (1924);
Gilman and Bach, *Acta phys. chim. U.R.S.S.* **9**, 27 (1938).
29. Velisek and Vasicek, *Coll. Trav. chim. Tchécosl.* **4**, 428 (1932).
30. Philippoff, *Disc. Faraday Soc.* **11**, 96 (1951).
31 Ellis, *Z. physik. Chem.* **78**, 321 (1912);
Mattson, *J. phys. Chem.* **37**, 223 (1933).
32. Henry, *J. chem. Soc.* 997 (1938);
33. Garner et al., *J. Inst. Petrol.* **38**, 974, 986 (1952); *Nature, Lond.* **172**, 259 (1953);
Kaufman and Singleterry, *J. Colloid Sci.* **7**, 5 (1952);
Koelmans, *Philips Res. Rep.* **10**, 161 (1955);
Koelmans and Overbeek, *Disc. Faraday Soc.* **18**, 9, 52 (1954);
Gemant, *J. phys. Chem.* **56**, 238 (1952);
Van der Minne and Hermanie, *J. Colloid Sci.* **7**, 600 (1952).
34. Brady, *J. Amer. chem. Soc.* **70**, 911 (1948);
Mattoon and Mathews, *J. chem. Phys.* **17**, 496 (1949);
Hoyer, Mysels, and Stigter, *J. phys. Chem.* **58**, 385 (1954).
35. Sumner and Henry, *Proc. roy. Soc.* **A133**, 130 (1931);
Mossman and Rideal, *Canad. J. Chem.* **34**, 88 (1956).
36. Aronsson and Grönwall, *Scand. J. clin. Lab. Invest.* **9**, 338 (1957);
Alexander, A. E. and Johnson, P., "Colloid Science", Oxford University Press, Oxford (1949).
37. Bungenberg de Jong, *Arch. néerl. Physiol.* **25**, 431, 467 (1941);
Bangham, Pethica, and Seaman, *Biochem. J.* **69**, 12 (1958);
Douglas and Shaw, *Trans. Faraday Soc.* **53**, 512 (1957); **54**, 1748 (1958).
38. Lundgren, Elam, and O'Connell, *J. biol. Chem.* **149**, 183 (1943).
39. Putnam and Neurath, *J. biol. Chem.* **159**, 195 (1945).
40. Bier "Electrophoresis", Academic Press Inc., New York (1959);
Hartman et al. *Appl. Microbiol.* **1**, 178 (1953).
41. Bikerman, "Surface Chemistry", Academic Press Inc., New York (1958).
42. Eilers and Korff, *Trans. Faraday Soc.* **36**, 229 (1940).
43. Derjaguin, *Trans. Faraday Soc.* **36**, 209, 730 (1940);
Derjaguin and Landau, *Acta. phys.-chim. U.R.S.S.* **14**, 633 (1941);
Derjaguin, *Disc. Faraday Soc.* **18**, 85, 181 (1954);
Verwey and Overbeek, "Theory of the Stability of Lyophobic Colloids", Elsevier, Amsterdam (1948).

44. Graham and Benning, *J. phys. Chem.* **53**, 846 (1949).
45. Davies, *Proc. 2nd Internat. Congr. Surface Activity* **1**, 426, Butterworths, London (1957).
46. Overbeek and Stigter, *Rec. Trav. chim. Pays-Bas* **75**, 543 (1956).
47. Smoluchowski "Graetz Handbuch der Elektrizität und des Magnetismus II", p. 366, Leipzig (1914);
Booth, *J. chem. Phys.* **22**, 1956 (1954).
48. Quist and Washburn, *J. Amer. chem. Soc.* **62**, 3169 (1940).
49. Elton and Peace, *J. chem. Soc.* **53**, 22 (1956).
50. Hermans, *Phil. Mag.* **26**, 674 (1938);
Rutgers, *Nature, Lond.* **157**, 74 (1946);
Causse, *C. R. Acad. Sci., Paris* **230**, 826 (1950);
Rutgers and Vidts, *Nature, Lond.* **165**, 109 (1950);
Yeager et al. *Proc. phys. Soc., Lond.* **B64**, 83 (1951);
Rutgers and Rigoll, *Trans. Faraday Soc.* **54**, 139 (1958).
51. Jacobs, *Trans. Faraday Soc.* **48**, 355 (1952).
52. Tiselius, *Kolloidzschr.* **59**, 306 (1932).
53. Bikerman, *J. chem. Phys.* **9**, 880 (1941).
54. Douglas and Walker, *Trans. Faraday Soc.* **46**, 559 (1950);
Buchanan and Heymann, *J. Colloid Sci.* **4**, 137, 151 (1949).
55. Grahame, *J. chem. Phys.* **18**, 903 (1950);
Conway, Bockris, and Ammar, *Trans. Faraday Soc.* **47**, 756 (1951).
56. Kitchener and Schenkel, *Nature, Lond.* **183**, 78 (1959).
57. Booth, *J. chem. Phys.* **19**, 391, 1327, 1615 (1951).
58. Freundlich and Ettisch, *Z. Phys. Chem.* **116**, 401 (1925).
59. Davies and Rideal, *J. Colloid Sci. Suppl.* **1**, 1 (1954).
60. Anderson, P. J., *Trans. Faraday Soc.* **55**, 1421 (1959).
61. Haydon, *Proc. roy. Soc.* **A258**, 319 (1960).
62. Stigter and Mysels, *J. phys. Chem.* **59**, 45 (1955).
63. Kramer and Freise, *Z. Phys. Chem.* **7**, 40 (1956).
64. McBain "Colloid Science", D. C. Heath, Boston (1950).
65. Philippoff, *Disc. Faraday Soc.* **11**, 96 (1951).
66. Mysels and Dulin, *J. Colloid Sci.* **10**, 461 (1955).
67. Douglas and Shaw, *Trans. Faraday Soc.*, **54**, 1748 (1958)
68. Grimley and Mott, *Disc. Faraday Soc.* **1**, 3 (1947).
69. Anderson, P. J., *Proc. 2nd Internat. Congr. Surface Activity* **3**, 67, Butterworths, London (1957); *Trans. Faraday Soc.* **54**, 130, 562 (1958).
70. Haydon—private communication.
71. Douglas, *Trans. Faraday Soc.* **46**, 1082 (1950).
72. Elster and Geitel, *Wein. Ber.* **94** (1890); *Ann. Phys. Chem.* **47**, 496 (1892);
Lenard, *Ann. Phys. Lpz.* **46**, 584 (1892).
73. Loeb "Static Electrification", pp. 25–31 and 122–4, Springer-Verlag, Berlin (1958).
74. Workman and Reynolds, *Phys. Rev.* **78**, 254 (1950); N. Mex. Inst. Min. Technol: Thunderstorm report **9** (1955);
Gill, *Brit. J. appl. Phys. Suppl.* **2**, S16 (1953); *Nature, Lond.* **169**, 203 (1952);
Schaefer, V. J., *Phys. Rev.* **77**, 721 (1950).

Chapter 4

Adsorption at Liquid Interfaces

ADSORPTION PROCESSES

If a solute such as propanol comes in contact with a clean air-water surface, either approaching from the air or from solution in the water, the surface tension is lowered. The propanol molecules concentrate at the surface, being held there by energy barriers which oppose both the removal of the polar hydroxyl group from the water and the immersion of the hydrocarbon chain in the water (Fig. 4-1). The approach of the propanol molecules is thus

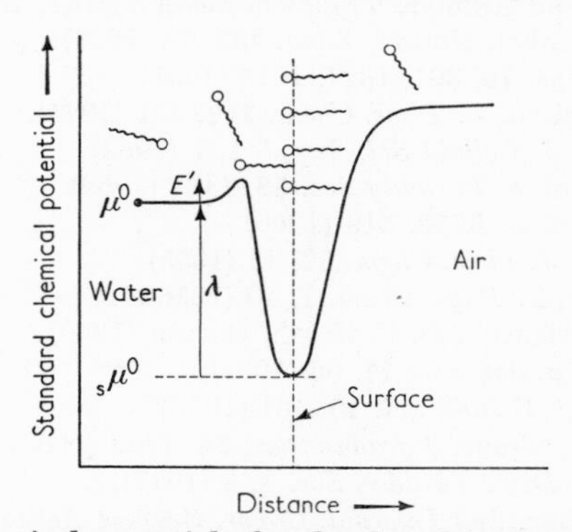

Fig. 4-1. Standard chemical potential of molecules of (say) propanol at an interface. The polar head is shown as a circle, and the hydrocarbon tail by ~~~~~.

governed by diffusion alone, while the desorption of the propanol monolayer is controlled by quite a high energy barrier. The amount of propanol in the surface thus increases, till the chemical potentials in the surface and bulk are equal; this corresponds to a greater concentration of solute in the surface than in the bulk, as expressed by the equations (with activity coefficients omitted)

$$\mu = \mu^0 + RT \ln c$$

$$_s\mu = {_s\mu^0} + RT \ln {_sc}$$

where μ is the chemical potential of the solute in solution, superscript zero denotes that the chemical potential refers to the standard state, and subscript s refers to material in the surface. The terms c and ${}_{s}c$ refer to bulk and surface concentrations. At equilibrium the surface concentration builds up to the level where

$${}_{s}\mu = \mu$$

and hence

$${}_{s}c/c = \exp\{(\mu^0 - {}_{s}\mu^0)/RT\} = \exp\{\lambda/RT\} \tag{4.1}$$

where λ, the standard free energy of desorption, is a convenient abbreviation for $\mu^0 - {}_{s}\mu^0$.

This isothermal relation between the equilibrium surface and bulk concentrations is often called an "adsorption isotherm". The quantity ${}_{s}c$ is, like c, expressed in gm.moles $l.^{-1}$, while the amount n of propanol adsorbed is usually expressed in units of molecules $cm.^{-2}$ To convert from one to the other, the thickness d of the monolayer is required, giving:

$$n = d \times {}_{s}c \times N/1000$$

where N is the Avogadro number.

We thus have (using eq. (4.1)) an "ideal adsorption isotherm":

$$n = d.c(N/1000)e^{\lambda/RT} \tag{4.2}$$

If now, for normal alcohols of different chain length, c is kept constant, and if we assume that d is also unchanged and that λ is of the additive form

$$\lambda = \lambda_p + W = \lambda_p + mw \tag{4.3}$$

(where subscript p refers to the (un-ionized) polar group, and W to the hydrocarbon chain, consisting of m $-CH_2-$ groups, of energy of desorption w each), then by combining (4.3) and (4.1), we find for the equal adsorption of two similar solutes, the first of m $-CH_2-$ groups and the second of (m+1), that

$$w = RT \ln\left(\frac{c_1}{c_2}\right) \tag{4.4}$$

Using two or three different alcohols, one can now obtain the appropriate values of c for some particular value of n, and hence find w. To a first approximation, one may study, following Traube[1], the concentrations required to produce the same lowering of the surface tension. He found that, for compounds of increasing chain-length, the requisite concentrations were about one-third of the previous value each time the chain was increased by one $-CH_2-$ group.

Langmuir[2] interpreted Traube's results quantitatively, and from the tabulated results[2] we have obtained w for each $-CH_2-$ group in the normal paraffin series as close to 600 cal. $mole^{-1}$, for films at great areas at the air-water surface.

When the areas are not limitingly great, special methods of determining n must be used: these are discussed below. Further, eq. (4.4) will apply accurately only to low concentrations and to very sparsely covered surfaces, since eq. (4.1) ignores activity coefficients.

The adsorbed molecules lower the surface tension further the greater the number present in the surface, and the relation between these quantities is expressed by a "surface equation of state" of the adsorbed molecules. The latter equation can be deduced directly from the properties of adsorbed films, as shown in Chapter 5.

Inter-chain Cohesion and Desorption Energies

If W_0 is the energy of desorption of the chain in the surface film of limitingly high area, and if W is the corresponding energy at any particular state of the film (Π_1, A_1), then

$$W = W_0 - W_e$$

where W_e is the increment in the desorption energy arising from cohesive forces within the monolayer. Written in this form W_e represents an energy of expulsion that will assist desorption: in practice cohesive forces will make W_e negative, as explained below, and so $W > W_0$.

The energy of expulsion, W_e, of a molecule will be the difference between the potential energy W_2 of a monolayer of unit area containing n molecules and the energy W_1 of a monolayer of unit area containing $n-1$ molecules[3]:

$$W_e = W_2 - W_1$$

Further, the cohesive potential energy per molecule in the monolayer is given[3] by $\int_{\mathbf{A}}^{\infty} \Pi_s \, d\mathbf{A}$; here $\mathbf{A}$ is conveniently expressed in absolute units (cm.^{2} per molecule) and Π_s is the contribution to the surface pressure due to cohesion (see p. 230 for details). For n molecules and an area $\mathbf{A}_2$ $(=1/n)$,

$$W_2 = n \int_{\mathbf{A}_2}^{\infty} \Pi_s \, d\mathbf{A}$$

Similarly,

$$W_1 = (n-1) \int_{\mathbf{A}_1}^{\infty} \Pi_s \, d\mathbf{A}$$

where $\mathbf{A}_1 = \dfrac{1}{n-1}$.

Hence

$$W_e = W_2 - W_1 = n\int_{A_2}^{\infty} \Pi_s\, d\mathbf{A} - (n-1)\int_{A_1}^{\infty} \Pi_s\, d\mathbf{A} =$$

$$= n\int_{A_2}^{A_1} \Pi_s\, d\mathbf{A} + n\int_{A_1}^{\infty} \Pi_s\, d\mathbf{A} - n\int_{A_1}^{\infty} \Pi_s\, d\mathbf{A} + \int_{A_1}^{\infty} \Pi_s\, d\mathbf{A}$$

or

$$W_e = \int_{A_1}^{\infty} \Pi_s\, d\mathbf{A} + n\int_{A_2}^{A_1} \Pi_s\, d\mathbf{A}$$

But $\mathbf{A}_1$ and $\mathbf{A}_2$ are very nearly identical, so that

$$W_e = \int_{A_1}^{\infty} \Pi_s\, d\mathbf{A} + n(\mathbf{A}_1 - \mathbf{A}_2)\Pi_{s(A_1)}$$

or, since $\mathbf{A}_1 - \mathbf{A}_2 = \dfrac{1}{n(n-1)} = \mathbf{A}_1/n$,

$$W_e = \int_{A_1}^{\infty} \Pi_s\, d\mathbf{A} + \mathbf{A}\Pi_{s(A_1)}$$

i.e.

$$W_e = \int_{0}^{\Pi_{s(A_1)}} \mathbf{A}\, d\Pi_s \tag{4.5}$$

We see also by comparison with the cohesive data of Chapter 5 (e.g. Figs. 5-9 and 5-13) that Π_s is negative, i.e. that Π is less than Π_k. Thus W_e is negative, and W at area $\mathbf{A}_1$ is therefore greater than W_0, as expressed by the relation

$$\frac{W}{RT} = \frac{600m}{RT} - \frac{1}{kT}\int_{0}^{\Pi_{s(A_1)}} A\, d\Pi_s \tag{4.6}$$

where W is here expressed per mole. We have divided by RT to simplify the change of units: R is 2 cal. in the first two terms, and $kT = 408$ at room temperature in the last, if A is here in Å^2 per chain and Π_s in dynes cm.$^{-1}$

As an example[4] of the use of (4.6) we may use the data of Fig. 4-2 to calculate w for each $-CH_2-$ group in a monolayer of octadecyltrimethylammonium ions at 90 Å^2 per long chain. By numerical integration

$$\int_{0(A \approx 500 Å^2)}^{\Pi_{s(A=90 Å^2)}} A\, d\Pi_s = -1303 \text{ c.g.s. units, i.e. } -1960 \text{ cal.mole}^{-1}.$$

For each of the 20 $-CH_2-$ groups which are effective the energy correction is therefore -98 cal.mole^{-1}, or $w = 600 + 98 = 698$ cal.mole^{-1}. By applying eq. (4.5) to adsorption of amines into monolayers at 40 Å^2 (essentially a "penetration" process, as explained in Chapter 6), the value of 800 cal.mole^{-1} has been found[5] for w at areas of 35-40 Å^2.

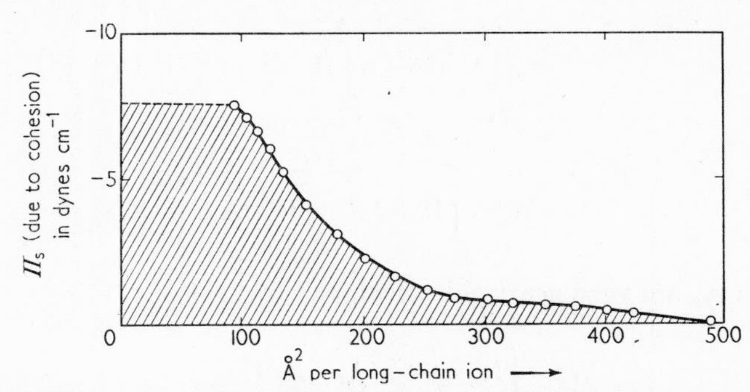

Fig. 4-2. Cohesive pressure in an air-water film of $C_{18}H_{37}N(CH_3)_3^+$: the shaded area is a measure[4] of the cohesive energy at 90 Å^2.

The Oil-Water Interface

Just as an alcohol can lower the surface tension of water, so it can lower the interfacial tension between an oil and water: equations (4.1) to (4.5) are again applicable.

To take a molar $-CH_2-$ group from the bulk of a paraffinic oil to the bulk of an aqueous phase requires a ΔG of $+860$ cal.[6] This is almost equal to the energy of desorption of films from the oil-water interface to oil (810 cal.), suggesting that the $-CH_2-$ groups in the adsorbed film are free to move laterally in the oil phase, the latter penetrating freely between the chains. The slight discrepancy represents the loss of entropy perpendicular to the surface in the adsorbed film. The same conclusion of free lateral movement is reached from the lack of cohesion of films at the oil-water interface (Chapter 5): the energy of desorption is thus independent of area.

If a polar oil is used there is a lower energy of desorption (Table 4-I), though the cohesion is not usually appreciable unless the oil is very polar[7,8].

TABLE 4-I

Non-aqueous phase	w
Paraffinic oil	810
Nitrobenzene	585
Air (sparsely covered surface)	600
Air (1 long-chain per 90 Å^2)	698

Energies of desorption (calories per mole $-CH_2-$ group) from different interfaces[2,4,8].

Polar Groups and Desorption Energies

The polar group, because of van der Waals bonds to the water, also contributes to the energy of desorption of the molecule; some values[2,8] of λ_p are shown in Table 4-II. These are for neutral molecules of 3 to 5 carbon atoms, such as alcohols or un-ionized acids, and are obtained by applying eqs. (4.3) and (4.4), taking d as 6 Å. If the polar group is ionized, as for example with long-chain sulphates, sulphonates, or quaternary amines, one must allow for the electrical repulsive potential of the interface, which tends to expel the adsorbed long-chain ions, and so leads to lower values of ${}_sc$ or n than with similar substances which are un-ionized. The electrical term in such charged films outweighs the purely dipolar desorption energy of the polar group, and, in practice, it is often possible to neglect the latter:

$$\lambda = \lambda_p + W - z_1 \varepsilon \psi_0 \approx W - z_1 \varepsilon \psi_0$$

and hence, from eq. (4.4),

$$\lambda \approx mw - z_1 \varepsilon \psi_0 \tag{4.7}$$

where z_1 is the valency of the long-chain surface-active ion, ψ_0 is the electrical potential of the surface, and ε is the electronic charge. Since z_1 and ψ_0 are always of the same sign, the electrostatic term is always negative in charged films, indicating reduced adsorption as compared with uncharged films. In Table 4-II values of $z_1 \varepsilon \psi_0$ for sodium lauryl sulphate are included.

TABLE 4-II

End-Group	λ_p (A/W)	λ_p (O/W)	$z_1 \varepsilon \psi_0$ (cal. per gm. mole)
—COOH	+ 437[2] +1035[8]	+1630	
—OH	− 50[2] + 575[8]	+ 800	
$—NH_2$	− 25		
—COO— (ester)	+ 470		
$—CO—NH_2$	− 510		
—CO— (ketone)	+ 295		
$—SO_4'$ (1mM-NaLS, $n = 0.74 \times 10^{14}$) (with no added salt)	+ 300*	+ 160* +1800[8]	—5000
$—SO_4'$ (0.01mM-NaLS, $n = 0.6 \times 10^{14}$) (with 0.145 M-NaCl added)	+ 300*	+ 160* +1800[8]	—1750
$—N(CH_3)_3^+$	—	+ 950	

Free energies (cal. per mole) for different polar end-groups desorbing into water at 20°C[2,8,9,10]. A negative value indicates a tendency to desorb: a positive value means that the polar group resists desorption. Values marked with an asterisk are obtained later in this chapter. Values of 1800[8] refer to 50°C.

Surface Concentrations

We can now estimate the surface concentrations of various substances using eq. (4.3), remembering, however, that its use depends on activity coefficients in both the surface and bulk phases being close to unity, a condition which is valid only at low surface coverages. For example, lauric acid is very strongly adsorbed, and we shall use therefore a value for w of 730 cal. mole^{-1} for each of the 11 $-CH_2-$ groups at the air-water surface, to which we add the 437 cal. for the carboxyl group. A typical value of c is 0.84×10^{-6} M for lauric acid in water. From eq. (4.1), ${}_sc/c$ is now found to be 1.8×10^6; the surface concentration is thus 1.6 M. If, further, the adsorbed surface film is an orientated monolayer, its thickness must be about 18 Å, and so with this value for d we find from eq. (4.2) that n is 1.7×10^{14} molecules cm^{-2}. Spread monolayers of lauric acid at the same surface pressure (2 dynes cm.$^{-1}$) have $n = 2.3 \times 10^{14}$, the agreement being close enough, in view of the neglect of activity coefficients, to confirm that the adsorbed film is indeed a monolayer, identical with a spread film.

There is other evidence besides this comparison with spread films which is in accord with the adsorbed film being a monolayer. This is discussed below, in connection with studies of surface potential, radioactive surface films, and surface viscosity.

Fluorinated Compounds

Perfluorocarbon derivatives with $-CF_2-$ groups instead of $-CH_2-$ groups form relatively insoluble monolayers; the strong adsorption of monolayers of $C_9F_{19}COOH$ suggests a very high energy of desorption. Indeed, the corresponding hydrocarbon acid, $C_9H_{19}COOH$, cannot be spread to form stable films. The energy of desorption of the $-CF_2-$ group[11], measured from rates of desorption as described below, is 1400 cal.mole^{-1} at 90 Å^2. This is considerably greater than the value of about 700 for the $-CH_2-$ group.

Aggregation has been found to occur in solutions of all perfluoroacids which have been prepared (even in solutions of perfluoroacetic acid). The decrease in the critical micelle concentration with added $-CF_2-$ groups in the fluoroacids is much larger than the corresponding decrease with added $-CH_2-$ in the paraffin chain salts. The contribution to the energy of disaggregation is therefore much higher per $-CF_2-$ than per $-CH_2-$, being about 1300 cal. per mole $-CF_2-$ as compared with 800 cal. per mole $-CH_2-$[12]. These latter figures are approximately of the same relative magnitudes as have been quoted above for the energies of desorption of a $-CF_2-$ group and of a $-CH_2-$ group.

Whether this strong adsorption and aggregation of the perfluorocompounds arises from a tendency to escape from the water or from attraction to themselves can be resolved as follows. Studies of the boiling points and entropies of vaporization of perfluorocarbons have shown that there is less attraction between perfluorocarbons than between paraffins. Further, surface tension studies[13,14,15] have shown that the monolayers of the perfluoroacids are in the "gaseous" state. These lines of evidence suggest that the attraction between the chains is low, and that it is the strong expulsion of the fluorocarbon chains by the water (which tends energetically to form hydrogen bonds with itself) which causes the high energies of desorption and of disaggregation.

Measurement of Adsorption

The surface tension lowering, Π can be readily measured for adsorbed films, either using the Wilhelmy plate method (if the contact angle is unaffected by the solute) or using the drop volume (drop weight) method. The Π-c relation so obtained may be converted to an n-c relation ("adsorption isotherm") if either the Π-n (i.e. Π-A) or the Π-n-c relation is known: the former may be obtained by comparison with insoluble films with an appropriate correction for the cohesion difference (eq. (5.11)), or by spreading the solute from a spreading solution as a monolayer and working quickly enough to obtain the Π-A plot before appreciable desorption has occurred. The latter (Π-n-c) relation is the Gibbs equation (4.54).

Other procedures use direct measurements of n (microtome method, foam method): these are explained in more detail below.

Finally, one may measure the interfacial potential, ΔV, from which, if μ_D is known from studies on insoluble films, n may be found by eq. (2.33).

THERMODYNAMICS OF ADSORPTION AND DESORPTION

For equilibrium between a dilute aqueous solution and a sparsely covered, uncharged surface, combination of eq. (4.2) with the appropriate equation of state (5.1) gives:

$$\Pi = \frac{\boldsymbol{k}\mathrm{TdNc}\, e^{\lambda/RT}}{1000}$$

This is strictly correct only in the limit when $A \to \infty$, i.e. as $c \to 0$, when it may be written:

$$\left(\frac{d\Pi}{dc}\right)_{c\to 0} = \frac{\boldsymbol{k}\mathrm{TdN}e^{\lambda/RT}}{1000} \qquad (4.8)$$

In this form it is useful for finding λ for any substance at any temperature. The quantity $\left(\frac{d\Pi}{dc}\right)_{c\to 0}$ is $-\left(\frac{d\gamma}{dc}\right)_{c\to 0}$, and this may conveniently be written as $\boldsymbol{\alpha}$.

From the temperature variation of λ (which, by eq. (4.1), must be a free energy) we can find the entropy of desorption as follows. In logarithmic form eq. (4.8) becomes:

$$\ln\alpha - \ln\left(\frac{kdN}{1000}\right) - \ln T = \frac{\lambda}{RT}$$

and this may be differentiated, assuming that d is independent of temperature, to:

$$\frac{d\ln\alpha}{dT} - \frac{1}{T} = \frac{-\lambda}{RT^2} + \frac{1}{RT} \cdot \frac{d\lambda}{dT}$$

which, since $\frac{d\lambda}{dT} = -\Delta S$, becomes

$$\Delta S = \frac{-\lambda}{T} + R - RT\frac{d\ln\alpha}{dT} \tag{4.9}$$

Aqueous Solutions and the Air-Water Surface

Table 4-III shows the figures that Ward and Tordai[16] calculated by applying eq. (4.9) to published data for the surface tension lowerings at different temperatures. As the hydrocarbon chain in the different molecules becomes longer, λ becomes greater, i.e. desorption is more difficult, as would be expected. The entropy figures show that the smaller molecules gain entropy on desorption into the water, i.e. the surface evidently restricts their movements much more than does the bulk system. But with larger molecules the entropy of desorption becomes negative: an important factor is the orientation and partial immobilization of the water molecules round the

TABLE 4-III

Thermodynamic quantities for Desorption from the Air-Water Surface into Water

Solute		λ (K. cal. mole^{-1})	ΔS (cal. (°C)$^{-1}$mole^{-1})	ΔH (K. cal. mole^{-1})
Propionic acid[16]	20°C	1.63	+14.4	5.9
n-Butyric acid[16]	20°C	2.45	+ 6.2	4.3
n-Valeric acid[16]	20°C	3.18	− 1.1	2.8
n-Caproic acid[16]	20°C	3.72	− 6.6	1.7
n-Heptanoic acid[16]	20°C	4.52	− 7.2	2.4
n-Octanoic acid[16]	20°C	5.40		
n-Nonanoic acid[16]	20°C	6.10		
n-Decanoic acid[16]	20°C	7.06		
n-Butanol[17]	25°C	2.66	+ 0.1	2.7
n-Pentanol[17]	25°C	3.36	+ 3.5	4.4
n-Hexanol[17]	25°C	4.05	− 0.6	3.9
n-Heptanol[17]	25°C	4.72	− 1.7	4.2
n-Octanol[17]	25°C	5.42	− 1.8	4.9
Phenol[16]	20°C	2.28		

hydrocarbon chains. By analogy with the entropy losses on solution of gases and vapours[16a], one may calculate that about 5 e.u. are lost by the water molecules per $-CH_2-$ group immersed[16b]. Each $-CH_2-$ group has two molecules of water round it, so that the loss of entropy per molecule of water is about -2.5 e.u., comparable with the value at the oil-water interface (p. 18), and about half the value for the formation of ordinary ice.

Besides those due to the restriction of the movements of the molecules held in the surface and the hydration effect, there is a third entropy change, a loss of about 1 e.u. per $-CH_2-$ group on desorption, due to the hydrocarbon chain becoming somewhat coiled-up (less random) when in the water[16b].

From λ (i.e. ΔG) and ΔS one can also calculate ΔH for the desorption process (Table 4-III). This remains positive at about 2 to 5 K.cal. mole^{-1}, both for the acids and the alcohols, and is related to the energy of making a hole in the water.

Aqueous Solutions and Oil-Water Interfaces

For each $-CH_2-$ group passing into aqueous solution, from the oil-water interface, ΔG is 810 cal.mole^{-1}, as described above. This figure appears to be independent of the surface concentration, at least if $n < 3 \times 10^{14}$, indicating no cohesion in the surface film. That this figure is lower than the ΔG for transfer from bulk oil to water (860 cal.mole^{-1}) suggests that the entropy of the two processes is slightly different. Calculation indicates that the $-CH_2-$ groups in the surface each have an entropy 0.17 e.u. less than those in the bulk of the oil, the smallness of this figure showing that the hydrocarbon chains of the molecules adsorbed at the oil-water interface are surprisingly free.

Oil Solutions and the Oil-Water Interface

Long-chain acids, desorbing from the hexane-water interface into oil, have to break the hydrogen bonds which the polar groups of the monolayer form with the water. The value of ΔG may be expected, therefore, to be about 5 K.cal.mole^{-1}, a figure supported by the experimental result of 5.5 K.cal. mole^{-1} found[18] for lauric, palmitic, and stearic acids. The length of the hydrocarbon chain is not important, as it remains immersed in the oil both in the monolayer and in the bulk of the oil.

Organic Vapours and the Air-Water Surface

The thermodynamic quantities for the desorption of organic vapours, from monolayers at the air-water interface, into the vapour, are shown in Table 4-IV. For hydrocarbons the value of ΔG per $-CH_2-$ group is about 420 cal. mole^{-1}, and the entropy of desorption of the molecules is fairly constant at about 9 e.u. per molecule: the latter figure corresponds to a gain of one degree of translational freedom, confirming that in the monolayer the adsorbed molecules have translational freedom in the two dimensions of the plane of the surface.

TABLE 4-IV

Thermodynamic quantities for Desorption from the air-water surface to the gas phase

Solute		λ (K.cal. mole^{-1})	ΔS(e.u.)
n-Pentane[19]	15°C	2.65	about 9
n-Hexane[19]	15°C	3.05	about 9
n-Heptane[19]	15°C	3.42	about 9
n-Octane[19]	15°C	3.91	about 9
n-Butanol[17]	25°C	7.32	38
n-Pentanol[17]	25°C	7.78	47
n-Hexanol[17]	25°C	8.36	49
n-Heptanol[17]	25°C	8.91	54
n-Octanol[17]	25°C	9.46	59
Anisole[20]	15°C	5.6	16.3
Chloroform[20]	15°C	3.8	11.0
Carbon tetrachloride[20]	15°C	3.3	8.3
Fluorobenzene[20]	15°C	4.0	9.0

For alcohols, as is to be expected, the energy of desorption is greater than for the hydrocarbons, because of the necessity of detaching the hydroxyl group from the surface: this requires an extra 5.2 K.cal. mole^{-1}. The $-OH$ group evidently restricts the rotational freedom of the alcohol molecules on the surface, since the entropy gain on desorption is now so high: part of this entropy increase may also arise from the change in the structure of the water at the surface when the alcohol leaves.

Other polar molecules are intermediate between the paraffins and the alcohols: the entropies are suggestive of the surface restricting only one degree of translational freedom, except for anisole. Here the $-OCH_3$ group of the adsorbed molecules evidently restricts the rotation as well as the translation.

The Vapour-Mercury Surface

In Table 4-V are summarized results for the desorption of vapours from the mercury surface. The rather high gain in entropy of benzene on desorption is accounted for quantitatively on the assumption that the molecules in the adsorbed film preserve two-dimensional mobility, and rotation is possible only in the plane of the ring, which is therefore adsorbed flat on the mercury surface. The entropy changes in these systems are particularly valuable in determining the conditions obtaining in the adsorbed film.

TABLE 4-V

Thermodynamic quantities for Desorption from the mercury surface to the vapour phase[21], *at 25°C*

Material	λ (K. cal. mole^{-1})	ΔS (e.u.)
Benzene	8.51	26.6
Toluene	9.63	39.6
n-Heptane	8.82	15.3
Methanol	7.88	27.2
Ethanol	8.06	11.2
n-Propanol	8.45	7.9
n-Butanol	9.82	27.8
n-Pentanol	10.51	29.0
n-Hexanol	11.23	33.4
Water	6.85	35.9

ADSORPTION KINETICS

At the moment that a fresh surface of a solution is exposed, there is no surplus solute adsorbed, and the surface tension is close to that for pure water. Since, however, solute molecules are continually diffusing towards the surface, from which they will only desorb with difficulty, solute accumulates at the surface till, at equilibrium, the rates of diffusion to the surface and of desorption from the surface are equal. Adsorption rates are determined by diffusion, and sometimes by eddy diffusion, from the solution to the surface, and theoretical interpretations of rates of adsorption require a knowledge of three quantities: the first is the relation between changes in γ and changes in n, the second is the diffusion coefficient of the solute in the bulk of the solution, and the third is the equilibrium condition, from which the relative importance of the adsorption and desorption effects can be calculated. As mentioned above, it is possible to obtain n from γ, using the Gibbs relation (eq. (4.54) below), though a surface equation of state (γ vs. n relation) can be used instead if such is known for the film under study.

The rate of arrival of molecules at the surface, in the absence of either an energy barrier or of stirring, is given[22] by

$$dn/dt = (D/\pi)^{1/2}\, ct^{-1/2} \times (N/1000) \qquad (4.10)$$

where c is the bulk concentration far from the surface in moles l^{-1}, and n represents the number of adsorbed molecules per cm.^{2} of surface after a time t from the beginning of adsorption. D is the bulk phase diffusion coefficient, and $\pi = 3.14$. Integrated, eq. (4.10) becomes

$$n = 2(D/\pi)^{1/2}\, ct^{1/2} \times (N/1000) \qquad (4.11)$$

which can readily be tested. One must note, however, that there is no allowance here for desorption, so that this simple relation between n and t can apply only to the initial stages of adsorption. For example, if D is 6×10^{-6} cm.^{2} sec.$^{-1}$ (which is the value for a compound with about 14 carbon atoms), an adsorption of $n = 2 \times 10^{14}$ from a 1.5 millimolar solution requires about 7 milliseconds according to eq. (4.11). Agreement with experiment is often satisfactory to within a factor of 3 or 4. If the surface-active material has a very high molecular weight, c on the molar scale may be very low in practice, with the result that adsorption is rather slow.

To improve further the accuracy of the theory, it is necessary to allow for desorption from the surface: due to the desorption of the film, n does not build up as quickly as is predicted by eq. (4.11). The full equation, due to Ward and Tordai[22], is

$$n = 2\left(\frac{D}{\pi}\right)^{1/2}\left\{ct^{1/2} - \int_0^{t^{1/2}} \mathbf{\Phi}(\mathbf{Z})\,d[(t-\mathbf{Z})^{1/2}]\right\}\left\{\frac{N}{1000}\right\} \tag{4.12}$$

where $\mathbf{\Phi}(t)$ represents the time-variable concentration of solute in the solution immediately underlying the film, t is the time elapsed since the adsorption began, and $\mathbf{Z}$ is a variable ranging from 0 to t. This relation shows that n must always be lower than according to eq. (4.11) and, unlike those above, it also shows that after an infinite time an equilibrium will be established, with n reaching a definite limit; the first term in the bracket (allowing for adsorption) and the second term (allowing for desorption) ultimately become equal. The latter term has to be evaluated graphically. As an illustration, the adsorption to the rather high figure of 2×10^{14} molecules cm.$^{-2}$ in the same system as described above, will require a calculated 28 milliseconds, i.e. four times as long as estimated before. Clearly this correction for desorption is always important if the extent of adsorption is high. The curves of Figs. 4-5 and 4-6 below show respectively the plots of t vs. c and, more conveniently, 1/t vs. c, both calculated from eq. (4.12).

Since, according to the simple relation (4.11), t will vary as $1/c^2$ for any given adsorption n, one may predict that, as the concentration of solute decreases, the time required for adsorption will increase markedly. That this is so in practice is shown by the results of Langmuir and Schaefer[23] (Table 4-VI). They studied a monolayer of stearic acid spread over very dilute unstirred solutions of aluminium ions, and it was assumed that every ion of aluminium arriving at the surface reacted to form aluminium stearate, so that back diffusion was negligible. Analysis was performed after skimming off the film. That agreement between theory and experiment is not better is due primarily to convection currents, arising both from slight inequalities in temperature and from the mechanical disturbance inherent in spreading the stearic acid film. The slow ageing of films of long-chain ions, sometimes observed when the water has not been quartz-distilled nor treated with ion-exchange resins to remove traces of polyvalent ion impurities, is now ascribed to a slow diffusion of the latter ions into the monolayer[24,25]: their concentration in the bulk may be perhaps 10^{-7} or 10^{-8} M, which would explain the periods of hours or even days during which some workers have found surface tension changes occurring. Adsorption of the monolayer of long-chain ions is normally complete within a few seconds.

TABLE 4-VI

Concentration of Al^{+3} in bulk, moles $l.^{-1}$	Time for stearic acid monolayer at air-water surface to be converted to aluminium salt	
	Calc. from eq. (4.11)	Observed[23]
10^{-4}	0.08 sec.	—
10^{-5}	8 sec.	—
10^{-6}	13 min.	very small.
10^{-7}	22 hr.	considerably less than 22 hr.
10^{-8}	89 days	—

When circulation currents, due either to thermal convection or to mechanical stirring, are such that the bulk solution is kept stirred to within a distance δx of the surface, diffusion to the surface may be expressed by a steady state equation. This will be possible when some depletion of the layer δx near the surface has already occurred, but when new solute is being brought to the lower limit of the layer δx by convection (Fig. 4-3): if back diffusion and depletion of the bulk solution can be neglected, the diffusion relation now becomes:

$$dn/dt = (D/\delta x)(cN/1000) \tag{4.13}$$

or, writing B_1 for $(DN/1000\delta x)$,

$$dn/dt = B_1 c \tag{4.14}$$

In this equation there is no term in t, in contrast to the relation (4.10), applicable to completely unstirred systems. Applying eq. (4.14) to the stearic acid monolayers spread over dilute solutions of aluminium ions, the latter solution now being stirred with a rod twenty times a minute, Langmuir and Schaefer[23] found $B_1 = 2.3 \times 10^{18}$, or, with $D = 5 \times 10^{-6}$ cm.^{2} sec^{-1}, $\delta x = 0.0013$ cm.

A correction term should be included in eq. (4.13) to allow for the fraction θ of the surface already covered with molecules. If a diffusing molecule strikes the surface where a molecule is already adsorbed, its chance of bouncing back without being adsorbed is high. If one assumes that on the liquid surface there is a definite number of adsorption "sites", as for solid surfaces, the fraction of the surface free for further adsorption is simply $(1-\theta)$, and eq. (4.14) becomes:

$$dn/dt = B_1c(1-\theta) \tag{4.15}$$

Although this assumption is not strictly true for the mobile films which adsorb at liquid surfaces, it is good approximation if $\theta<0.3$.

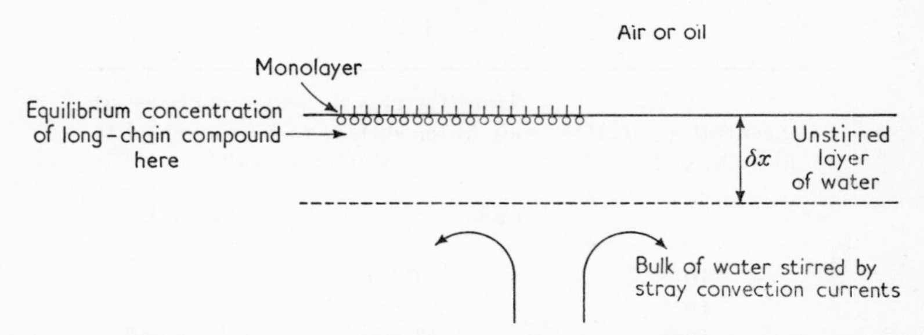

Fig. 4-3. Convection currents in the liquid maintain the supply of surface-active material to within a distance δx of the surface. The monolayer has been enlarged here compared with the rest of the system: the sublayer in equilibrium with it will be found at about 20 Å below the surface.

Experiments on Rates of Adsorption at the Air-Water Surface

Measurements of adsorption at times greater than 30 secs. may be undertaken with any of the usual techniques for measuring surface tension (Chapter 1), surface potential (Chapter 2), or extent of adsorption (with radioactive tracers, described below). In particular the latter method has shown[26] that 60% of the adsorption occurs in 9 minutes from a 0.9×10^{-6} M unstirred solution of sodium di-n-octylsulphosuccinate, although from a 9×10^{-6} M solution 60% of the final adsorption has occurred in less than 1 minute (in accord with the calculations from eq. (4.11), Table 4-VI). That the time enters this equation as the half power was also confirmed, though divergencies become apparent in the more dilute systems, presumably because of stray convection currents.

For times of adsorption less than 30 secs., one must use streams of liquid moving at different speeds, flowing as jets or as drops through a nozzle, or running along a channel. These techniques will now be described in detail.

Jet methods[27-30] are useful in measuring adsorption occurring within 1-50 milliseconds. The first of these is based on the observation that, if a stream of pure liquid be forced at constant speed through a small nozzle of elliptical cross-section, the issuing jet vibrates to form a series of equidistant nodes (Fig. 4-4). The reason for this lies in the tendency of the surface of the liquid (and therefore of the circumference of the jet) to contract as far as possible, forming a jet of circular cross-section. In attaining this end the momentum of the liquid perpendicular to the direction of the jet carries the system past the equilibrium position, and the momentarily circular jet again becomes elliptical, but the long axis perpendicular to that before. After a further time interval, the jet again becomes circular, then approaches the initial

Fig. 4-4. Form of a jet of liquid flowing from an elliptical orifice.

ellipticity, and so on, Fig. 4-4. The rate at which the jet oscillates in this way depends on the ratio of the driving force to the resisting force (approximately the square root of the surface tension and the square root of the density of the liquid with a term for the viscous resistance of the liquid, respectively); if we determine the last quantities, the surface tension can thus be found from the time of oscillation of the jet. The most convenient way to determine the latter is from the linear velocity of the liquid jet and from the distance between the nodes as determined photographically.

The nodes for a pure liquid are equidistant but, with a solution of surface-active material, the inter-nodal spacing increases down the jet, showing that the driving force of the oscillations is decreasing, i.e. that the surface tension is decreasing due to adsorption. In this way one finds the adsorption at various distances along the jet, and the results can be expressed as adsorption as a function of time. There are three principal difficulties in ascribing exact values to the rates of adsorption obtained in this way. Firstly, it must be assumed that the forward velocity is unaltered from the interior of the jet to the surface; secondly, before the liquid leaves the nozzle adsorption may already have occurred at the liquid-solid interface, and this adsorbed material may be carried out on the surface of the jet, so that the zero of time for the adsorption process is indefinite; thirdly, the rate of diffusion of the solute to the surface may be seriously and incalculably affected by the oscillations of the jet: it is necessary, for the system to be at all tractable mathematically, to assume that the flow of the liquid does not affect the diffusion to the surface[30].

Although the oscillating jet method has been widely used, the steady jet method[31] has the advantage of being free from the third difficulty. Here the jet issues from a circular nozzle, and the adsorption n at different distances along it is detected by a surface-potential apparatus. The radioactive source method has been used up to the present, but the greater steadiness of the vibrating plate apparatus when the air is moving or when the humidity and carbon dioxide content vary, renders the latter technique preferable. As with the vibrating jet method, the time for which adsorption has been occurring is assumed to be the ratio of the distance along the jet from the nozzle outlet to the velocity of the liquid in the jet. In all jet methods great care should be taken to prevent evaporation of the more volatile solutes[32].

Some of the results obtained by these two methods are summarized in Figs. 4-5 and 4-6. It is clear that adsorption rates depend primarily only on

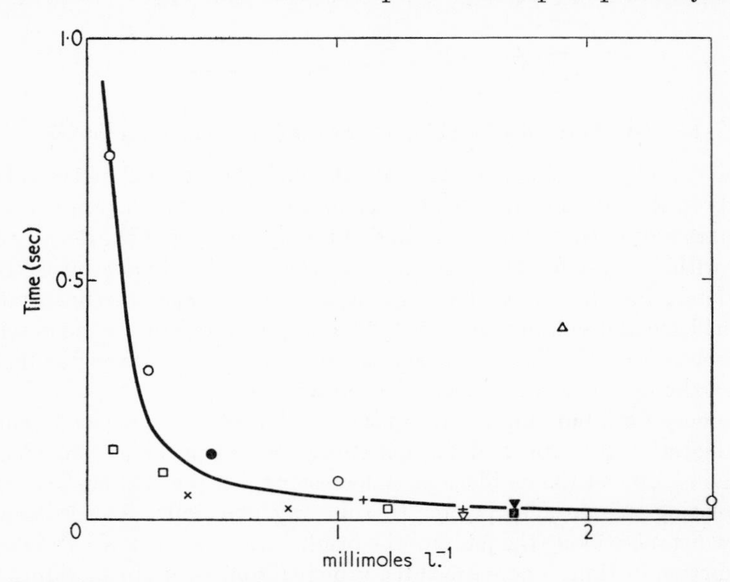

Fig. 4-5. Times for adsorption of various long-chain substances at room temperature. The times are those in which 60% of the final total change in surface-tension is attained, and the adsorption is from aqueous solution to the air-water surface unless otherwise stated. The full curve is derived from eq. (4.12) with D having a typical mean value of 6×10^{-6}cm.^{2}sec.$^{-1}$ Points have the following significance:
□ Cetyltrimethylammonium bromide adsorbing from 0.01 N-KCl[31] (steady jet).
△ Sodium lauryl sulphate from water[33] (drop method).
▽ sec-Octyl alcohol from water[31] (steady jet).
+ sec-Octyl alcohol from water[29] (oscillating jet).
× Aerosol OT from 10^{-3} N-HCl[31] (steady jet).
○ Cetyltrimethylammonium bromide from water to CCl_4 interface[24] (drop method).
▼ n-Heptanol from water[29] (oscillating jet).
■ n-Heptanol from water[30] (oscillating jet).
▲ n-Heptanol from water[36] (oscillating jet).
⁕ n-Octanol from water[31] (steady jet).
● Sodium lauryl sulphate from water[35] (oscillating jet).

the solute concentration, and that 60% of the surface tension lowering (a convenient figure) has often occurred between 10 and 40 milliseconds from the time a 1.5 millimolar solution leaves the nozzle. This agrees with the figure of 27 millisec. calculated (with assumptions) from the theory of Ward and Tordai. Adsorption equilibrium is thus attained rather rapidly; such rate processes must be very important if a foam is to be produced by bubbling a gas rapidly through a solution. Quantitatively, there is fair agreement with theory, at least within an order of magnitude; the tendency for experimental results to be lower than the theoretical in Fig. 4-5 may be due to convection and mechanical stirring (cf. Table 4-VI).

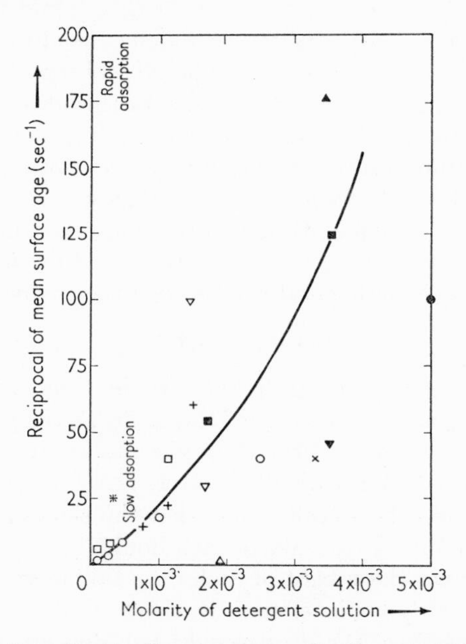

Fig. 4-6. Rates of adsorption to the air-water and oil-water interfaces (measured by the reciprocal of the mean surface age when 60% of the final total change of surface tension has occurred). Points and line as in Fig. 4-5.

The *channel method*[31] is very similar to the steady jet method. A stream of liquid flows from a rectangular orifice through a trough, the adsorption at different distances from the orifice being found from surface potentials. The rate of adsorption of solute follows by again assuming that there is no adsorption at the moment the liquid leaves the orifice, and that the flow-rate is uniform across the flowing liquid. In practice surface viscosity may well reduce the flow-rate in the surface, though this could be tested with talc particles. The flow in the channel method is slower than from a nozzle, and adsorption can be studied at times of 20 milliseconds to about 0.5 secs. The method is not easy to employ accurately, owing to spurious electrical fields from the sides of the trough: again, the vibrating plate apparatus might prove useful.

The *drop method* conveniently measures the amount of adsorption which has occurred at times between 0.1 sec. and half a minute. This method (McGee[33]) is a modification of the drop-weight method, which, as originally developed in 1919 by Harkins and Brown[37], requires the drop to be released infinitely slowly (Chapter 1). At appreciable rates of drop formation, however, the drop weight will depend not only on the surface tension at the time of release, but also on the extent to which further liquid is pushed into the "tail" of the drop as it leaves the tip on which it has formed. The faster the rate of formation of the drops, the more liquid will tend to enter the drop in this way, and, although no exact calculation of this effect has been possible, McGee[33] found an empirical correlation between drop weight and the time of formation of the drop. Since for pure liquids the surface tension will have reached its final value at times much shorter than we are concerned with here (only about 10^{-9} sec. is required), the correlation must therefore reflect only the effect of the liquid flow, i.e. the entry of additional liquid into the detaching drop. The experimental results obey the equation:

$$w = w_0 + \mathrm{b}(1/\mathrm{t})^{0.75} \tag{4.16}$$

where b is a constant for a given liquid, t is the time of formation of each drop, and w and w_0 are the weights of the drops at the given flow rate and zero flow rate. The slopes b, if the drop weights are in milligrams and the times in seconds, are 3.19 for distilled water, 1.22 for benzene, and 0.76 for dodecane. These and other results show that the slopes are proportional to the surface tension and the density of each liquid.

We can now correct the weight of each drop falling at a specified rate, to allow for this hydrodynamic effect. If there is also slow adsorption of material at the interface, the drop weight will depend on the lowering of surface tension that occurs during the time of formation of the drop, as well as on the extra liquid being pushed into the "tail". The appropriate slope b can give the corrected weight w_0 from which, by the Harkins and Brown formula[37] (Chapter 1), the surface tension at that time is readily found. Ageing is slower than at a plane interface because diffusion is occurring outwards on to a nearly spherical surface[33].

To carry out this calculation a method of successive approximations is necessary. At first, a slope b for the pure solvent is assumed, and the corresponding drop weight is calculated for $1/\mathrm{t}^{0.75}=0$. A surface tension corresponding to this drop weight is now found from the corrections of Harkins and Brown, from which one can obtain a more accurate slope b and, hence, a more accurate drop weight. Since not all the surface has been present during the total time of formation of the drop, the "mean surface age" will be rather less than t, the time of formation of the drops; in fact it is only 40% of t[24].

The rate of surface ageing is of interest in connection with foam formation (Chapter 8), and Brady and Brown[34] have shown that when $t = 1$ sec. (mean surface age 0.4 sec.), 60% of the total equilibrium lowering of surface tension has occurred from a 1.9 mM aqueous solution of sodium lauryl sulphate. At higher concentrations considerably more than 60% of the surface-tension lowering occurs in this time. The adsorption of lauryl alcohol, present as impurity in the detergent, is very much slower than that of the detergent, as expected from the very low concentration of the alcohol. At concentrations of lauryl alcohol in the solution of 0.25 millimolar, there is no appreciable adsorption of this material up to times of about 1 second, whereas the sodium lauryl sulphate (5 millimolar) is by then strongly adsorbed. This is consistent with theory; the times calculated for the alcohol and sulphate to adsorb (eq. (4.12)) are approximately 1 second and 3 milliseconds.

Rates of adsorption are, to within the usual limits of concordance of such results, the same for long-chain ions and for molecules (Figs. 4-5 and 4-6). Consequently, even when the surface is charged by some of the long-chains already adsorbed there, the long-chain ions still approaching are not appreciably retarded. In fact we should have predicted this, since the electrical potential at (say) 20 Å from the surface will generally be very considerably less than ψ_0 (Chapter 2): at distances less than about 20 Å the hydrocarbon chain can come into contact with the surface and pull the ion up the remaining potential gradient. As shown in Chapter 7, the energy of activation E' cannot be detected experimentally if less than 3000 cal.: this would correspond to an activation potential of 130 mV, which could occur as far as 20 Å from the surface only in the final stages of the adsorption of long-chain ions and at very low ionic strengths. The earlier published figures of very high energy barriers in the adsorption of long-chain ions are now explicable in terms of experimental errors and of the diffusion of polyvalent, specifically adsorbed ions to the film. The latter lower the surface tension, and will diffuse very slowly from low concentrations, as explained above.

Experiments on Rates of Adsorption at the Oil-Water Interface

The earlier experimental work on this subject was carried out with the ring or drop methods, with which adsorption can be studied at times greater than about 15 seconds. Though time effects extending up to several hours have occasionally been reported for certain systems, the interface usually ages so quickly that no ageing effects can be detected by these techniques. Recent interest in liquid-liquid extraction, however, has focused attention on the properties of freshly formed liquid-liquid interfaces[24], since as much as 60% extraction may occur from a small drop within 1 second of its formation.

Although the drop method[24] is simplest, the jet method can be adapted empirically to the oil-water interface[38]; the latter technique is laborious, however, and we prefer the former. With this drop method one can study liquid-liquid interfaces at times as low as 0.1 second after their formation, and, provided that the solute is in the external phase, results are comparable with those calculated for plane interfaces. Basically this method is similar to that of McGee for studying rapid adsorption at the air-water surface: as before, the drop weight method is modified in that the drops are formed at various rates, instead of infinitely slowly as required in calculating tensions directly with the formula of Harkins and Brown. At appreciable dropping rates, however, the drop weight is no longer a function only of interfacial tension. The two new factors affecting the drop weight are:

(i) that the drop, just as it begins to detach itself from the tip on which it was formed, will have its "tail" blown up with more liquid as it flows from the tip, thus enlarging the drop, and

(ii) that circulation currents, caused by the stream of falling drops, will detach the drops before they would otherwise fall, thereby causing them to be smaller than when formed very slowly.

For pure CCl_4 drops being formed at various rates in distilled water the second factor is predominant, and smaller drops are detached at high flow rates. At the air-water surface the second factor is negligible, and the drops are slightly heavier at faster dropping rates (see p. 172). The form of the calibration curve is also sensitive to the shape and size of the receiving vessel[24 39]. From the "calibration curve" for CCl_4 drops in distilled water, one knows the extent of the effect of hydrodynamic factors (tabulated above) on the drop weight, since here the interfacial tension is independent of time. If now a surface-active agent is added to the water, one can find the interfacial tension as a function of surface age by varying the time of drop formation. The "calibration curve" thus allows for the hydrodynamic factors: the variation of interfacial tension and hence of adsorption with time can therefore be simply found.

Experimentally, the steel tip is carefully cleaned in carbon tetrachloride and then coated with silicone to give it the correct wetting properties to make the drops reproducible. The whole apparatus must be maintained to within 0.5°C, and the depth of immersion of the tip below the water surface must be carefully adjusted, 0.7 cm. being a suitable distance.

For a clean interface the empirical equation is:

$$w = w_0 + b_i(1/t)^{1.5}$$

where w_0 is the drop weight at zero flow rate, and t is the time of formation of each drop. That the drops are lighter at high dropping rates is reflected in negative values of b_i: for the CCl_4 drops entering water b_i is -0.7 when w is in mg. and t is in seconds.

Fig. 4-7 shows the results[24] when cetyltrimethylammonium bromide

(CTAB) solutions are used instead of distilled water. These curves reflect the variation of interfacial tension, γ, with time as well as the circulation effect expressed by the above relation. The data are analysed as follows. The value of b_i should be proportional to γ as for the air-liquid system, but, since γ is not yet known for the CTAB solutions, the curves of Fig. 4-7 must be interpreted by successive approximations, as for results at the air-liquid interface. With b_i as for the CCl_4-water system, the drop weight corresponding to $(1/t)^{1.5}=0$ is first calculated, and from this and the Harkins and

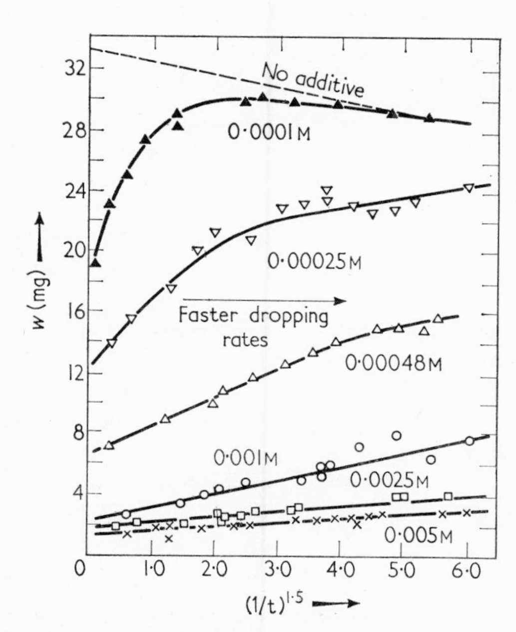

Fig. 4-7. Weights of CCl_4 drops formed at different rates in water containing various concentrations of $C_{16}H_{33}N(CH_3)_3^+$. The broken line is the calibration curve for zero amount of surface-active additive.

Brown formula[37] a provisional interfacial tension is obtained. A new value of b_i is now derived, and hence a better value for the drop weight, and so on.

Fig. 4-8 shows the final results for the ageing of the CCl_4-water interface as the CTAB adsorbs from aqueous solutions of different concentrations. The mean age of the surface is only 40% of the total time required to form a drop, and with this factor the data of Fig. 4-8 can be converted to mean surface ages, plotted in Figs. 4-5 and 4-6. In these figures the mean interfacial ages required for 60% of the equilibrium interfacial tension lowering to have occurred are summarized. The turbulence, associated with forming the drops quickly, makes comparison with theory difficult, but the calculated ageing effects (due to smooth diffusion to and from the interface) are of the order of magnitude of the experimental figures. We believe that any hindrance other than diffusion is therefore unimportant, and that no allowance need be made for possible orientational or electrical barriers to adsorption.

The adsorption equation ($n=f(c,t)$) for long-chain ions at the oil-water interface can be found from data on the rate of adsorption of cetyltrimethylammonium ions from water to the carbon tetrachloride interface. The

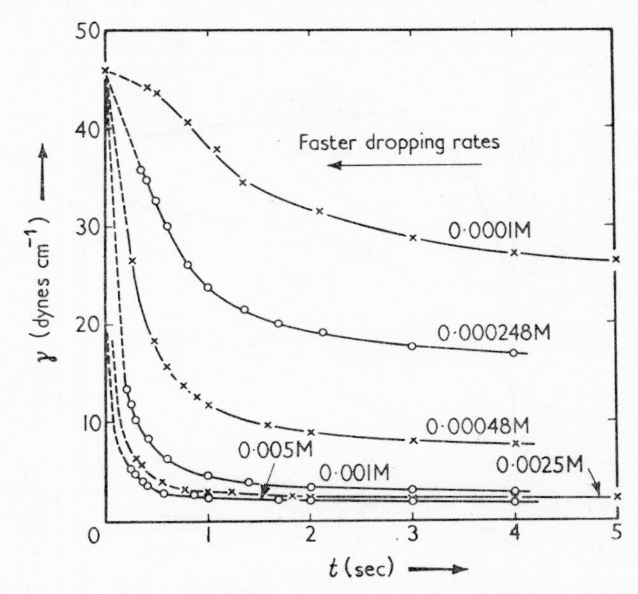

Fig. 4-8. Interfacial tensions at different times of drop formation for $C_{16}H_{33}N(CH_3)_3^+$ adsorbing from aqueous solutions of various concentrations to the interface with CCl_4[24].

experimental quantity is the interfacial tension as a function of time, measured in the flowing system. From the slopes (Fig. 4-8) in the early stages of adsorption, values of $d\gamma/dt$ are found, and these lead to dn/dt figures as follows.

Firstly, A is evaluated from eqs. (5.10a) and (2.51) by successive approximations, and with $A_0=33$ Å^2. Next, from differentiation and substitution of the appropriate value of A, a value for $d\Pi/dA$ is found, which, since $A=10^{16}/n$, leads at once to $d\Pi/dn$ (i.e. $-d\gamma/dn$). This, combined with the experimental value of $d\gamma/dt$, gives dn/dt. The results in Fig. 4-9 show that dn/dt is proportional to c in the early stages of adsorption, before θ and the desorption rate become appreciable[39]. We can therefore assume that eq. (4.15) applies to this flowing system, even at small times. From the slope of the line in

Fig. 4-9, B_1 is 1.8×10^{18} if dn/dt is in molecules cm.$^{-2}$ sec.$^{-1}$ and c is in moles l^{-1}.

The kinetics of the adsorption of a protein, bovine plasma albumin, on to drops of CCl_4 from dilute aqueous solutions[40] are of double interest: such materials may be present as chance impurities in many industrial processes involving liquid interfaces, while the rate of unfolding of the globular protein molecule at the interface may possibly alter the measured rate of change of

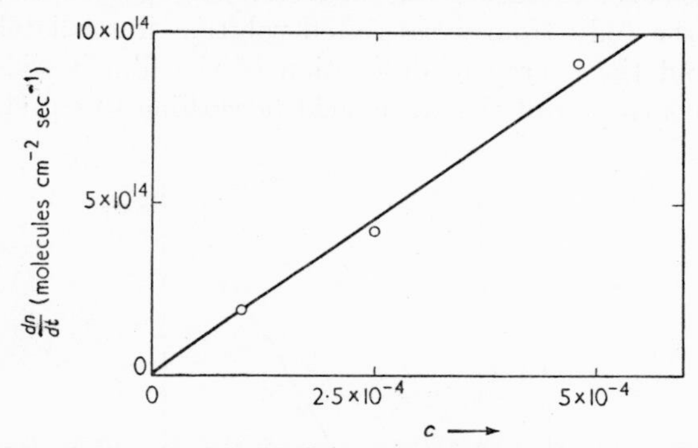

Fig. 4-9. Adsorption rates, in the early stages of adsorption, for $C_{16}H_{33}N(CH_3)_3^+$ from aqueous solution to the interface with CCl_4[39]. The proportionality to the concentration of this substance is consistent with eq. (4.15) when θ is small.

interfacial tension. Experiment shows[40] that, for aqueous 3×10^{-6} M protein, B_1 of eq. (4.15) is 1.9×10^{18} at the surface of the falling drops, a value about the same as that quoted above for cetyltrimethylammonium ions. For molecules as large as those of this protein, however, the value would be expected to be only one-quarter of this figure: the difference is possibly due to the use of a receiving vessel of different shape in the experiments with protein, with a consequent change in the hydrodynamic flow round the drops as they are formed. For 60% of the lowering of interfacial tension to have occurred, about 3 secs. are required for a 10^{-6} M solution, and about 0.2 sec. for a solution of concentration 2×10^{-5} M (which, because of the high molecular weight, is about 1 g.$l.^{-1}$). For substances of high molecular weight it is generally true that surface ageing effects are important, even with quite appreciable weight concentrations.

The very slow adsorption of lauric acid and palmitic acid from hexane to the interface with water has not yet received a satisfactory explanation[41].

DESORPTION KINETICS

A film of lauric acid, spread on a Langmuir trough and maintained at a constant pressure of 4 dynes cm.$^{-1}$, desorbs into the underlying water at a measurable speed[42]. For the first few minutes the desorption process obeys the equation[43]

$$-\frac{dn}{dt}=\frac{cN}{1000}\left(\frac{D}{\pi t}\right)^{1/2} \tag{4.17}$$

where c is the molar concentration in the lauric acid immediately below the film. The lauric acid diffuses into the bulk phase, which is effectively unstirred near the surface. Kinetically, therefore, the process is controlled by diffusion into the bulk from a saturated solution immediately below the surface film, and the energy of desorption of the film is reflected in this saturation concentration of the lauric acid (according to eq. (4.34) below).

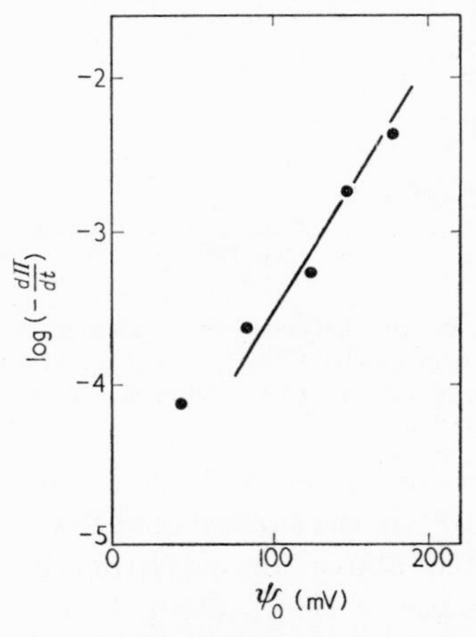

Fig. 4-10. Test of eq. (4.20) for the desorption of films of $C_{18}H_{37}N(CH_3)_3^+$, at a constant mean area of 85 Å^2 per long-chain, from the air-water surface into NaCl solutions of different concentrations c_i. The slope of the line is that calculated by eq. (4.20)[39,44].

After the first few minutes the diffusion kinetics change, and then follow the relation[43]

$$-\frac{dn}{dt}=\text{constant} \tag{4.18}$$

showing that a stagnant layer has been established below the film, and that diffusion is occurring from the saturated region immediately subjacent to the film through this layer, at the extremity of which it is removed by convection currents in the bulk of the solution (in which the concentration may be taken as zero). This implies that a steady state has now been reached, dependent on the convection currents. From eq. (4.18) one can calculate that δx, the thickness of this stagnant layer, is about 1 mm., though this may be decreased by increasing the convection currents by slight heating of the solution with a torch-bulb. The situation for steady-state desorption is as in Fig. 4-3.

For steady-state desorption in general, the desorption rate is given[4] by

$$-\left(\frac{dn}{dt}\right)_{desorption} = B_2 n \exp\left\{\frac{z_1 \varepsilon \psi_0 - W}{\boldsymbol{k}T}\right\} \tag{4.19}$$

where B_2, like B_1, is a constant depending on the hydrodynamics of the diffusion process, W is the energy of desorption of the hydrocarbon chain, and z_1 is the valency of the long-chain ion: this may be written, for small intervals of time t (when n changes but slightly), as

$$\log\left(-\frac{d\Pi}{dt}\right) = \frac{z_1 \varepsilon \psi_0}{2.3\boldsymbol{k}T} + \text{constant} \tag{4.20}$$

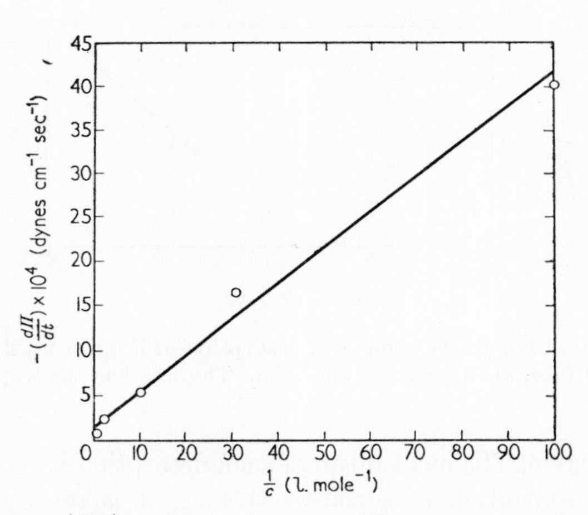

Fig. 4-11. Plot of $-\left(\frac{d\Pi}{dt}\right)$ vs. $\frac{1}{c_i}$ for films of $C_{18}H_{37}N(CH_3)_3^+$ at 85 Å² per long-chain ion on solutions of NaCl[44].

The constant includes W, and this will in fact be constant for films at the air-water interface only when A is sensibly constant. At lower areas W increases, due to inter-chain cohesion. Fig. 4-10 shows the plot of eq. (4.20), when the mean A is constant at 85 Å² for air-water films. Here the different values of ψ_0 are obtained by varying the concentration of salt in the substrate. By equating ψ_0 and ψ_G and using the approximation of eq. (2.31) at the appropriate temperature, one finds that, if ψ_0 is large, that $-(d\Pi/dt)$ should vary linearly with $1/c_i$. The experimental plot (Fig. 4-11) fully confirms this[44].

In general, eq. (4.19) may be written:

$$-(1/n)(dn/dt)_{desorption} = B_2 \exp((z_1 \varepsilon \psi_0 - W)/\boldsymbol{k}T) \tag{4.21}$$

The desorption of films of cetyltrimethylammonium ions from the oil-water interface into 0.01 N HCl (with no stirring) gives data suitable for testing this equation[4,39]. Results are shown in Fig. 4-12, in which the line drawn through the experimental points has the slope required by eq. (4.21). The value of the constant is -6.5, and from this B_2 may be calculated since, by eq. (4.21), this constant is

$$\left(\log B_2 - \frac{W}{2.3kT}\right),$$

where W is the (constant) total energy of adsorption of the hydrocarbon chain at the oil-water interface. This is 810 cal.mole^{-1} per $-CH_2-$ group, and in the cetyltrimethylammonium structure there are 18 $-CH_2-$ groups adsorbed. Thus $W/2.3kT$ is 10.8, and hence B_2 is 2×10^4 sec.$^{-1}$ for desorption

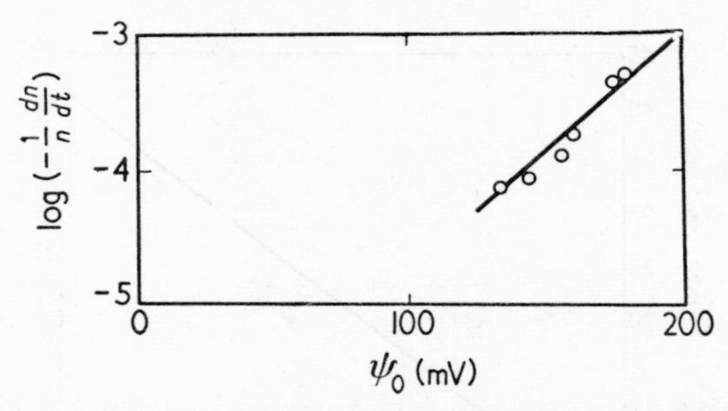

Fig. 4-12. Test[4,39] of eq. (4.21) applied to desorption of films of $C_{16}H_{33}N(CH_3)_3^+$ from the oil-water interface, at different values of n. The line has the slope calculated by eq. (4.21).

in a "static" system. The mechanism of such desorption is, as usual, that an equilibrium concentration of surface-active agent is set up in the water immediately subjacent to the interface, further desorption being limited by the rate of diffusion of the long-chain ions from the equilibrium concentration through a thin unstirred layer of solution. At greater distances from the surface, unless special precautions are taken, convection currents rapidly remove the long-chain ions, distributing them through the bulk of the liquid. It is to such a "static" system that the value of B_2 of 2×10^4 applies.

Hexadecanol monolayers, of interest in retarding evaporation, desorb at a rate[43] of about 0.5×10^{10} molecules cm.$^{-2}$ sec.$^{-1}$, in excellent accord with eq. (4.21): here $\psi_0 = 0$, $W = 16 \times 800$ cal. mole^{-1}, and B_2 is unchanged at 2×10^4 sec.$^{-1}$ At very high film pressures, the molecules desorb about twice as fast as this, being expelled by the high pressure. This effect is not allowed for in eq. (4.21). Evaporation of the monolayer also becomes important at high film pressures[43].

From equilibrium concentrations (p. 189) we shall see that, for any system, $B_1/B_2 = 1.2 \times 10^{14}$, so that, in the "static" system, B_1 must be 2.4×10^{18}, when the units of dn/dt and c are respectively molecules cm.$^{-2}$ sec.$^{-1}$ and moles l^{-1}. This value of B_1 for a "static" system is close to the value ($B_1 = 1.8 \times 10^{18}$) measured directly from the flow system.

The solution immediately below the film attains equilibrium with the film very rapidly, and, after the first few minutes of desorption, a steady state will be set up in which the loss from this equilibrium region is equal to the diffusion to the bulk of the water through the stagnant layer δx (Fig. 4-3). The former loss can be measured directly from the change in surface tension with time, while the latter is directly proportional to c, the equilibrium concentration of solute immediately below the film. This is true whatever may be the hydrodynamics of the diffusion process. Consequently, for the steady-state desorption,

$$-(dn/dt)_{desorption} = B_3 c \tag{4.22}$$

where B_3 is a constant, the value depending on convection currents in the bulk of the water, on the diffusion coefficient, and on the temperature[4]. By comparison of eqs. (4.19) and (4.22), and that relating equilibrium concentrations with n when θ is small (eq. (4.34) below), we see that B_3 must be identical with B_1, i.e. 2.4×10^{18} if n is in molecules cm.$^{-2}$ and c in moles l.$^{-1}$ In absolute units the permeability constant becomes 4×10^{-3} cm. sec^{-1}, which must be related to the thickness δx of the unstirred layer of liquid through which diffusion is occurring. The upper boundary of this layer is in equilibrium with the film, and contains solute of concentration c. The lower boundary of the layer contains zero concentration of solute, a condition which remains true even after considerable diffusion because of convection currents. The thickness of this liquid layer is calculated from D, the diffusion coefficient of the solute (about 6×10^{-6} cm.^{2} sec.$^{-1}$) and the permeability: it is $(6 \times 10^{-6}/4 \times 10^{-3})$ cm., i.e. 1.5×10^{-3} cm.; this is rather lower than the 1 mm. quoted by Saraga[43].

The fractional rate of loss of molecules from the air-water film of $C_9F_{19}COO^-$ at 90 Å^2 is 5.2×10^{-4} sec.$^{-1}$, similar to that from a hydrocarbon compound such as $C_{16}H_{33}COO^-$; the energy of adsorption of a $-CF_2-$ group must therefore be nearly twice as high as that of a $-CH_2-$ group. More exactly, comparison may be made[8,11] with the fractional rate of desorption of the quaternary amine, $C_{18}H_{37}N(CH_3)_3^+$ which at 90 Å^2 and under similar conditions of ionic strength (1.0N) is 1.6×10^{-5} sec.$^{-1}$ On the basis of slow diffusion away from the interface this rate is given by eq. (4.21) in which W = mw, where m is the number of carbon atoms adsorbed at the interface and w is the energy of adsorption of each group, here either $-CH_2-$ or $-CF_2-$.

Now the figures for the rates and numbers of adsorbed carbon atoms (20 $-CH_2-$ groups in the quaternary compound and 9 $-CF_2-$ groups in the perfluoroacid) may be inserted into the appropriate equations of the type (4.21) for fully ionized groups on a substrate of unit ionic strength. The potentials ψ_0 are thus the same for the two films, and if the value of the energy of adsorption of a $-CH_2-$ group be taken as 700 cal. mole^{-1} (as above), a value of about 1400 cal. mole^{-1} is found for the energy of desorption at an air-water interface for a $-CF_2-$ group in a film at 90 Å^2.

NET RATES OF ADSORPTION OR DESORPTION

For a completely stagnant system the quantity $(dn/dt)_{net}$, the net rate of adsorption, is found by differentiating eq. (4.12). If only a thin stagnant layer separates the surface film from the stirred bulk, however, one subtracts eqs. (4.15) and (4.19) to obtain[16b]

$$\left(\frac{dn}{dt}\right)_{net} = B_1c(1-\theta) - B_2n \exp.\left\{\frac{z_1\varepsilon\psi_0 - W}{\boldsymbol{k}T}\right\} \tag{4.23}$$

When equilibrium is established, $(dn/dt)_{net} = 0$, and

$$B_1c(1-\theta_e) = B_2n_e\exp\left\{\frac{z_1\varepsilon\psi_{0(e)} - W_e}{\boldsymbol{k}T}\right\} \tag{4.24}$$

where subscript e refers to equilibrium conditions, and W will vary somewhat with n for films at the air-water surface, but will be constant at the oil-water interface.

If θ, the surface covered, is fairly small, and W is sensibly constant (for small changes in n close to n_e), combination of eqs. (4.23) and (4.24) gives:

$$\left(\frac{dn}{dt}\right)_{net} = \left\{B_2 \exp\left\{\frac{-W}{\boldsymbol{k}T}\right\}\right\}\left\{n_e\exp\left\{\frac{z_1\varepsilon\psi_{0(e)}}{\boldsymbol{k}T}\right\} - n\exp\left\{\frac{z_1\varepsilon\psi_0}{\boldsymbol{k}T}\right\}\right\} \tag{4.25}$$

If the film is of electrically neutral molecules this reduces to:

$$(dn/dt)_{net} = \{B_2 \exp\{-W/\boldsymbol{k}T\}\}\{n_e - n\} \tag{4.26}$$

though strictly W should be replaced by λ to allow for the adsorption energy of the polar group. For example, for lauryl alcohol $m=12$ and, if $w=700$ cal. mole^{-1} for $-CH_2-$ group, then with $B_2 = 2\times10^4$ sec.$^{-1}$ the rate of adsorption or desorption is:

$$(dn/dt)_{net} = 1.7\times10^{-2}(n_e - n).$$

Hence, for a 10% change in n from (say) 10^{14} molecules cm.$^{-2}$ at equilibrium, the net rate of adsorption or desorption is 1.7×10^{11} molecules sec.$^{-1}$, i.e. 60% of the change back to equilibrium occurs in 60 secs. This time is independent of the extent of the change from equilibrium. For derivatives with longer chains, the rates become correspondingly smaller, and the times correspondingly longer. The figures calculated in this way will refer to dilute systems only, as the term θ has been neglected in comparison with unity, and eq. (4.23) should strictly be used. A positive sign for $(dn/dt)_{net}$ denotes adsorption; a negative sign desorption.

For adsorption on to an initially clean surface (i.e. n=0 when t=0), integration of eq. (4.26) gives for the change in n with t:

$$\ln\left(\frac{n_e}{n_e - n}\right) = tB_2 \exp\{-W/\boldsymbol{k}T\}$$

which should apply to uncharged molecules if θ is small (when also W will be sensibly constant during the adsorption). More exactly, eq. (4.23) must be integrated.

For adsorption of long-chain ions eq. (4.25) can be simplified if W is constant (θ being negligible again), if ψ_0 is greater than 100 mV. Then the logarithmic approximation (of the form of eq. (2.31)) may be used, to give (for 20°C),

$$\left(\frac{dn}{dt}\right)_{net} = \frac{B_2 \times 134^2 \times 4 \times 10^{-32}}{c_i}\left\{\exp\left(\frac{-W}{\boldsymbol{k}T}\right)\right\}(n_e^3 - n^3)$$

For lauryl sulphate ions (m=12) at the air-water surface, with w=700 cal. $mole^{-1}$ per $-CH_2-$ group, $B = 2 \times 10^4 sec.^{-1}$, $c_i = 10^{-3}$ M (no added inorganic salt), and $n_e = 0.7 \times 10^{14}$, a sudden 10% change in n away from n_e gives a calculated net rate of adsorption or desorption of 1.1×10^{15} long-chain ions $sec.^{-1}$, showing that equilibrium will be 60% restored in 6.4 milliseconds. Again, this time depends only slightly on the extent of the initial displacement of equilibrium, provided this does not exceed 10%: in the limit of very small changes the time tends to 5.6 milliseconds. When the concentration of lauryl sulphate is increased to 2×10^{-3} M, n_e is 1.25×10^{14}, and the calculated time for 60% recovery from a 10% change of area is about 4 milliseconds, which should be constant at higher concentrations[16b].

If, however, 0.145 M-NaCl is present with the lauryl sulphate, the times for the above two values of n are increased respectively to 1 sec. and 0.3 sec.

The time taken to regain equilibrium is important in wave-damping and in mass-transfer in turbulent systems.

ADSORPTION EQUATIONS FOR NON-ELECTROLYTES

The amount of adsorption of any solute from solution is related to its bulk concentration, c, by the "adsorption isotherm". For very dilute solutions of a un-ionized solute, the number of molecules n adsorbed per $cm.^2$ of surface is given[45] by the simple relation

$$n = \alpha c \tag{4.27}$$

where α is an arbitrary constant and c is in moles $l.^{-1}$ This corresponds to the "ideal" isotherm (eq. (4.3) above).

At high solute concentrations this simple equation is no longer valid, and the complete equation of Langmuir[2] is required for un-ionized solutes. This equation is

$$n = \frac{\beta(c/a)}{(1 + c/a)} \tag{4.28}$$

where β and a are again arbitrary constants. When c and n are small eq. (4.28) reduces to eq. (4.27), with $\alpha = \beta/a$. It may be derived from the equilibrium relation (4.24) with $\psi_0 = 0$, and with $\theta = n/n_0$, where equilibrium conditions are now understood without subscripts, and where n_0 is the adsorption density of surface-active molecules in a close-packed monolayer. Comparison of the constants shows that $\beta = n_0$, and $a = n_0(B_2/B_1)\exp(-W/\boldsymbol{k}T)$.

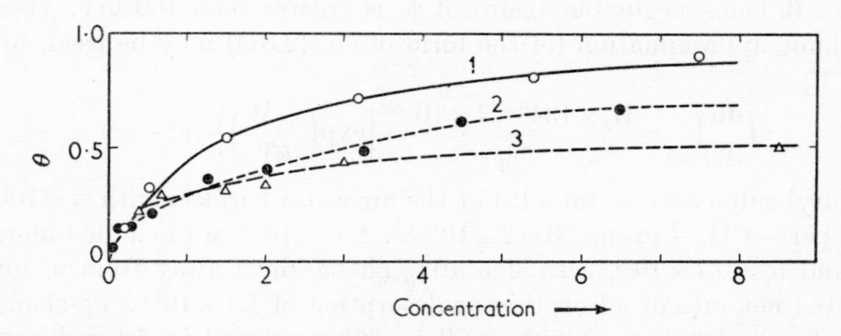

Fig. 4-13. Curve 1 represents the Langmuir adsorption equation (4.28) fitted[50] to the experimental data (o) of Harkins & King on the adsorption of un-ionized butyric acid molecules to the benzene-water interface. The constants a and β of eq. (4.28) have the values 0.79 and 3.04×10^{14} respectively. One unit on the concentration scale represents 67 millimoles $l.^{-1}$, and units of the ordinate are n/β, (corresponding to θ, the fraction of the interface covered by the adsorbed molecules). Curve 2 refers empirically to the adsorption of lauryl sulphate ions at the oil-water interface in the absence of added salt; data are calculated[50] from the experimental results of Kling & Lange[51]. With unity on the concentration scale representing 1 millimole $l.^{-1}$, the adsorption data fit the Langmuir isotherm at low concentrations, though falling considerably below the latter at higher surface coverages, when the electrical repulsion becomes significant. Curve 3 refers to the adsorption of lauryl sulphate ions in the presence of excess (8×10^{-3} M) NaCl, again calculated[50] from Kling & Lange's figures. With unity on the concentration scale representing 0.1 millimole $l.^{-1}$ the data again fit the Langmuir isotherm at low concentrations, though the adsorption is considerably less than according to the Langmuir equation at higher concentrations.

The Langmuir equation, by its derivation, assumes that the film is immobile (cf. eq. (4.15)), but the error introduced by this assumption can be shown (p. 230) to be small if $n < 1.2 \times 10^{14}$. It holds reasonably well for the adsorption of un-ionized butyric acid at the benzene-water interface (Fig. 4-13).

Another adsorption equation, for moderately high concentrations of solute, is that suggested by Küster[46] in 1894:

$$n = Kc^{1/\nu} \tag{4.29}$$

where K and ν are constants characteristic of the system. Usually ν lies between 2 and 10. This relation is sometimes called the Freundlich adsorption isotherm, though Freundlich rejected it at first[47]. While it is often in close accord both with the Langmuir equation (4.28) and with the experimental data over moderate ranges of c, it cannot be valid at very low concentrations (since it does not reduce to eq. (4.27)), nor at high concentrations (since n must reach some definite limit when the surface is fully covered by a monolayer). It is particularly useful, however, for representing the adsorption of monolayers at liquid interfaces in that both theoretically and experimentally it represents the adsorption of long-chain ions at the oil-water interface under certain conditions (see below).

In 1908 Szyszkowski[48] put forward his empirical equation for the adsorption of films at the air-water surface; in this relation the adsorption is measured by Π, the surface tension lowering. If, as before, c is the bulk concentration, Szyszkowski's equation is (for 20°C),

$$\Pi = 30 \log_{10}\left(1 + \frac{c}{a}\right) \tag{4.30}$$

Here Π is in dynes cm.$^{-1}$, and a is a constant for any given chain-length, and is the same as a in eq. (4.28): this conclusion follows by combination of the Szyszkowski equation with the approximate Gibbs differential adsorption relation (derived below, eq. (4.54)):

$$\frac{d\Pi}{d\ln c} = \boldsymbol{k}Tn$$

These equations give for adsorption at 20°C:

$$n = \frac{3.18 \times 10^{14}(c/a)}{(1 + c/a)}$$

which is identical with the Langmuir expression of eq. (4.28) and shows that β is 3.18×10^{14}.

For the adsorption of un-ionized acids at the air-water surface, Szyszkowski's values of a are empirically related to the number m of $-CH_2-$ groups in the molecule by

$$a = 1.9e^{-710m/RT} \tag{4.30a}$$

and hence, if c is small, eq. (4.28) becomes

$$n = \frac{\beta c}{a} = 1.67 \times 10^{14} c \exp\{710m/RT\} \tag{4.31}$$

which is similar in form to eqs. (4.27) and (4.3). Numerical comparison with the latter, allowing for λ_p for the carboxyl group (Table 4-II), gives a calculated value of d of 14 Å, a slightly high figure for butyric acid. This may be because the film is mobile (pp. 228-30).

ADSORPTION EQUATIONS FOR LONG-CHAIN IONS

For the adsorption of ions (such as lauryl sulphate, $C_{12}H_{25}SO_4'$) the Langmuir equation no longer holds (Fig. 4-13): such adsorption equations are necessarily invalid because they contain no term for the energy of the electrical double layer which is constituted by the charged surface and the underlying counter-ions. This electrical energy makes the adsorption considerably less than predicted at higher concentrations. Further, the figure of 710 calories for the adsorption of a molar $-CH_2-$ group (eq. (4.31)) is for any solute only approximate: the exact value is known to depend on n. If, for example, n is about 0.1×10^{14}, the adsorption energy is 600 cal. mole^{-1}, but when n is 1.11×10^{14} hydrocarbon chains adsorbed per cm.^{2}, the adsorption energy has risen to 698 cal. mole^{-1}, as shown on p. 157.

The isotherm of Temkin[49] allows for varying interaction of solute molecules with the surface. It may be written

$$n = b \ln c + \text{constant} \tag{4.32}$$

where b is a constant. Clearly such an equation cannot be valid at very small or very high values of c, though, as explained below, it does allow for electrical factors over a limited range of concentrations of adsorbing long-chain ions.

To obtain the general adsorption equation, however, one must consider the equilibrium between the bulk and the surface; then from eq. (4.24), with $\theta = n/n_0$, may be written[50]

$$n = \frac{(B_1/B_2)\, c \exp\left\{\dfrac{W - z_1 e\psi_0}{\boldsymbol{k}T}\right\}}{1 + c(B_1/B_2 n_0) \exp\left\{\dfrac{W - z_1 e\psi_0}{\boldsymbol{k}T}\right\}} \tag{4.33}$$

This is of the form of the Langmuir equation (eq. (4.28)) although, unless ψ_0 is zero or negligibly small, the quantity a in the Langmuir equation is no longer constant (because a includes ψ_0, and this is a function of σ and hence of n). If c and n are small eq. (4.33) approximates to

$$n = \left(\frac{B_1}{B_2}\right) c \exp\left\{\frac{W - z_1 \varepsilon\psi_0}{\boldsymbol{k}T}\right\} \tag{4.34}$$

which will be useful in practice if $n < 10^{14}$.

In logarithmic form this equation becomes

$$\log\left(\frac{n}{c}\right) = \log\left(\frac{B_1}{B_2}\right) + \frac{W}{2.3\boldsymbol{k}T} - \frac{z_1 \varepsilon \psi_0}{2.3\boldsymbol{k}T} \tag{4.35}$$

where the known influence of added salt (e.g. NaCl) on the adsorption isotherm (e.g. in Fig. 4-13) is allowed for in the ψ_0 term: this is decreased by added salt and so the adsorption (n) is increased.

In the limit when n is relatively very small, either with or without added salt, ψ_0 tends to zero, and eq. (4.35) becomes

$$\log\left(\frac{n}{c}\right)=\log\left(\frac{B_1}{B_2}\right)+\frac{W}{2.3\boldsymbol{k}T} \tag{4.36}$$

and hence, for any given chain-length, n is proportional to c, as in eq. (4.27), i.e. as in the isotherm for un-ionized substances. This limit of very small n values has never been reached experimentally with adsorbed films of long-chain ions.

For films at the *air-water surface*, W increases with n because the inter-chain cohesion increases. The relation between W and n (or A) is:

$$\frac{W}{\boldsymbol{k}T}=\frac{600m}{RT}-\frac{1}{\boldsymbol{k}T}\int A d\Pi_s \tag{4.6}$$

where W is now expressed per molecule and where 600 cal. is the adsorption energy per $-CH_2-$ group for hydrocarbon chains at high areas; the second term allows for cohesion between the hydrocarbon chains. Here Π_s is the "cohesive pressure" of the film at the air-water interface for any given value of A. The last term of eq. (4.6) can be evaluated using the empirical expression for Π_s (Chapter 5):

$$\Pi_s=\frac{-400m}{A^{3/2}} \tag{5.8}$$

where A is in Å^2 per long-chain, and Π_s is in dynes cm.$^{-1}$ By insertion of Π_s from eq. (5.8) into eq. (4.6) and integrating between the limits A and 500 Å^2 (at which area a value of 600 cal. may be taken to apply), one obtains W at any value of A:

$$\frac{W}{\boldsymbol{k}T}=\frac{600m}{RT}+\frac{1200m}{A^{1/2}\boldsymbol{k}T}-\frac{53.5m}{\boldsymbol{k}T} \tag{4.37}$$

Insertion of this value of W into eq. (4.36), with $A=10^{16}/n$, gives the complete isotherm for adsorption at the air-water surface:

$$\log\left(\frac{n}{c}\right)=\log\left(\frac{B_1}{B_2}\right)+\frac{521m}{2.3RT}+\frac{1200\times10^{-8}mn^{1/2}}{2.3\boldsymbol{k}T}-\frac{z_1\varepsilon\psi_0}{2.3\boldsymbol{k}T} \tag{4.38}$$

This is tested in Fig. 4-14. The value of (B_1/B_2) found from the plot is 2×10^{14}.

We may note that adsorption at the air-water interface is characterized by a relatively constant ratio n/c: as c and n increase, the term in ψ_0 tends to reduce n/c, while the term in $n^{1/2}$ tends to increase it. The net result is that n/c varies only slightly with ψ_0 or c.

At the *oil-water interface* monolayers of long-chain ions should obey eq. (4.35) with W constant at 810; consequently, for any given long-chain electrolyte the plot of log (n/c) against ψ_0 should be rectilinear[50]. The data of Kling and Lange[51] are exactly those required to test this, and Fig. 4-15 shows that the plots of ln(n/c) against ψ_G (eq. (2.30)) are indeed linear. In this figure the straight lines have been drawn with theoretical slope of $-\varepsilon/2.3\boldsymbol{k}T$ lying as far as possible through the experimental points, for the adsorption of various long-chain sulphates at the interface between water and n-heptane at 50°C, both without and with added NaCl. The data for adsorption from water alone (using the Gibbs equation with a factor of 2 as explained below) and from salt solutions (with a factor of 1 in the Gibbs equation) agree satisfactorily with eq. (4.35).

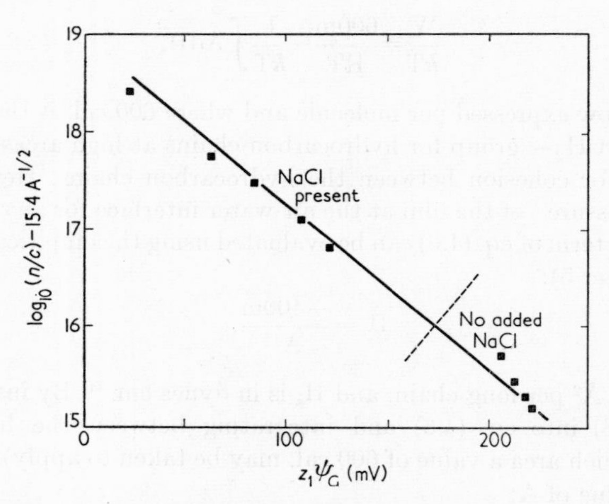

Fig. 4-14. Test of eq. (4.35) for the adsorption of lauryl sulphate ions at 20°C at the air-water surface, both with and without added salt[10,50]. The line has the theoretical slope, and ψ_G is calculated by eq. (2.28). A is in Å^2 per long-chain ion.

Further, the family of lines for different long-chain ions whose chains differ by 2 carbon atoms are separated, at constant ψ_0, by mean intervals of 1.13 on the logarithmic scale. The theoretical value from eq. (4.35) is

$$\frac{2 \times 810}{2.3 \times 2 \times 323}, \text{ i.e. } 1.09$$

If the lines of Fig. 4-15 are extrapolated to zero ψ_G, and each intercept is then reduced by $810m/2.3\boldsymbol{k}T$, the constant $\log(B_1/B_2)$ is found[50] to have a mean value of 1.6×10^{14}, compared with the calculated figure of 1.2×10^{14} (see below). The discrepancy here, and for films at the air-water interface, could result from a λ_p of the order $+0.4RT$ for the polar head group.

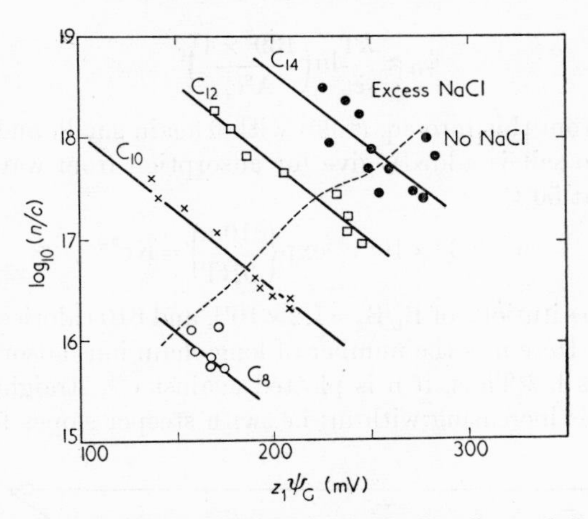

Fig. 4-15. Test of eq. (4.35) for straight-chain (C_8, C_{10}, C_{12}, and C_{14}) sulphate ions at the oil-water interface at 50°C, both with and without added salt[50,51]. The lines have the theoretical slope, ψ_G having been calculated by eq. (2.29). Units of n,c, and ψ_G are respectively molecules cm.$^{-2}$, moles l.$^{-1}$ and millivolts, and the NaCl concentrations are as in Table 4-VIII.

Calculation of B_1/B_2

The value of the constant B_1/B_2 in the above equations can be estimated theoretically as follows. If the energy of desorption W, and ψ_0 were both zero, then the surface and bulk concentrations would be equal. Hence, if c were expressed in molecules per cm.^{3}, n/c would be equal to the thickness of the surface film, i.e. about 20 Å. Usually, however, c is expressed in moles per litre, as in the data used in Figs. 4-14 and 4-15, so that n/c (and hence B_1/B_2 by eq. (4.35)) is

$$\frac{20 \times 10^{-8} \times 6.02 \times 10^{23}}{1000}, \text{ i.e. } 1.2 \times 10^{14}$$

This ratio is a general result, independent of the hydrodynamics of the adsorption and desorption processes: the hydrodynamics will alter B_1 and B_2 separately, but their ratio must remain constant.

EXPLICIT ISOTHERMS FOR ADSORPTION AT THE OIL-WATER INTERFACE

Adsorption in the Absence of Salts

Although in general eqs. (4.33) and (4.35) must be solved by successive approximations, under certain conditions an explicit function can be derived for the adsorption of long-chain ions at the oil-water interface. Thus, if ψ_G exceeds 100 mV,

$$\psi_G \approx \frac{\boldsymbol{k}T}{\varepsilon}\ln\left(\frac{139^2 \times 4}{A^2 c_i}\right) \tag{2.31}$$

Insertion of ψ_G from this into eq. (4.35) with θ again small, and substitution of $c_i = c$ (since no salt is added), give for adsorption from water to the oil-water interface at 50°C

$$n = 5.35 \times 10^{13} c^{2/3} \exp\left\{\frac{810m}{3RT}\right\} = Kc^{2/3} \tag{4.39}$$

including the substitutions of $B_1/B_2 = 1.2 \times 10^{14}$, and 810 calories per $-CH_2-$ group adsorbed. Here n is the number of long-chain ions adsorbed per cm.^{2}, and c is in moles l.$^{-1}$ Thus, if n is plotted against $c^{2/3}$, straight lines should result, with slopes increasing with m: i.e. with steeper slopes for ions of in-

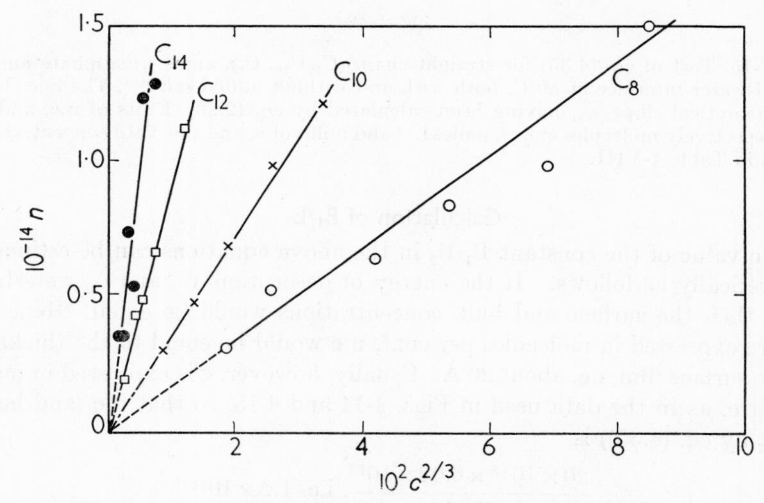

Fig. 4-16. Plot of n against $c^{2/3}$ for adsorption of long-chain sulphate ions at the oil-water interface[50,51] in the absence of added salts, and with $\theta < 0.3$ and $\psi_0 > 100$ mV. Cf. eq. (4.39) and Table 4-VII.

creasing chain-length. This relation, identical with the Kuster isotherm (4.29) with $\nu = 1.5$, is tested in Fig. 4-16: the slopes of the straight lines drawn through the points are compared in Table 4-VII with those calculated from eq. (4.39).

Since eq. (4.39) involves neglect of the term in θ in eq. (4.24) as well as the approximation of eq. (2.31), it should apply only when n is in the range $10^{14} > n > 10^{11}$. The latter figure is calculated for $m = 10$ from eq. (4.35): at lower values of n the potential ψ_0 decreases below about 100 mV and the approximation (eq. (2.31)) is therefore no longer valid. In practice, however, adsorptions as low as 10^{11} long-chain ions per cm.^{2} cannot be detected, so that the lower theoretical limit to n is not significant.

TABLE 4-VII

System	$\left(\frac{\partial n}{\partial(c^{2/3})}\right)_T$ measured from Fig. 4-16	$\left(\frac{\partial n}{\partial(c^{2/3})}\right)_T$ calculated from eq. (4.39)
$C_8H_{17}SO_4'$ (m = 8)	1.7×10^{15}	1.6×10^{15}
$C_{10}H_{21}SO_4'$ (m = 10)	3.6×10^{15}	3.5×10^{15}
$C_{12}H_{25}SO_4'$ (m = 12)	10×10^{15}	8.1×10^{15}
$C_{14}H_{26}SO_4'$ (m = 14)	22×10^{15}	20×10^{15}

Slopes of adsorption isotherms for long-chain sulphates at the oil-water interface, with no added inorganic salt[50,51]. Units of n and c are molecules cm.$^{-2}$ and moles litre^{-1} respectively.

Adsorption from Salt Solutions

If an excess of inorganic uni-univalent salt is present in the water ($c_{salt} \approx c_i$ and $c_i \gg c$), an approximate isotherm for adsorption at the oil-water interface at 50°C follows from eqs. (4.35) and (2.31) with $B_1/B_2 = 1.2 \times 10^{14}$:

$$n = 5.35 \times 10^{13} c^{1/3} c_i^{1/3} \exp\left\{\frac{810m}{3RT}\right\} \qquad (4.40)$$

This approximate isotherm, which is again true only when $\psi_0 > 100$ mV, is

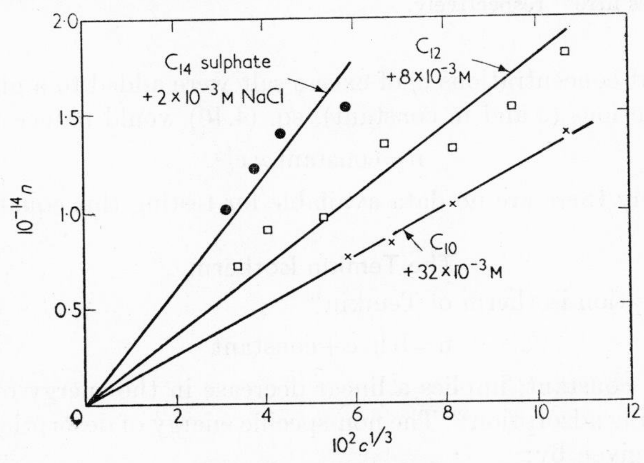

Fig. 4-17. Plot of n against $c^{1/3}$ for the adsorption of long-chain (C_{10}, C_{12}, C_{14}) sulphate ions at the oil-water interface[50,51] in the presence of excess NaCl, with $\theta < 0.3$ and $\psi_0 > 100$ mV. Straight lines are predicted by eq. (4.40); cf. Table 4-VIII.

the Küster isotherm with $\nu = 3$. Here c is the concentration of long-chain ions, and c_i the total electrolyte concentration.

Thus, if excess inorganic electrolyte is present, this approximate explicit oil-water isotherm predicts rectilinear plots of n against $c^{1/3}$ at fixed values of c_i and m: this is shown in Fig. 4-17. The experimental points lie closely about straight lines, and the measured slopes compare well with those required by theory, as shown in Table 4-VIII. The adsorption equation (4.40) may also be tested by plotting n against $c_i^{1/3} \exp\{810m/3RT\}$ at constant c. Such a plot for $c = 0.20 \times 10^{-3}$ M has a measured slope of the line of 3.4×10^{12}, compared with 3.1×10^{12} calculated from eq. (4.40).

TABLE 4-VIII

System	$\left(\frac{\partial n}{\partial(c^{1/3})}\right)_T$ measured from Fig. 4-17	calculated from eq. (4.40)
$C_{10}H_{21}SO_4'$ (m = 10) ($c_i = 32 \times 10^{-3}$ M)	1.3×10^{15}	1.2×10^{15}
$C_{12}H_{25}SO_4'$ (m = 12) ($c_i = 8 \times 10^{-3}$ M)	1.7×10^{15}	1.7×10^{15}
$C_{14}H_{29}SO_4'$ (m = 14) ($c_i = 2 \times 10^{-3}$ M)	2.9×10^{15}	2.5×10^{15}

Slopes of adsorption isotherms for long-chain sulphates at the oil-water interface, with excess NaCl present[50,51]. Units of n and c are molecules $cm.^{-2}$ and moles $litre^{-1}$ respectively.

If different concentrations c_i of excess salt were added to a given solution of long-chain ions (c and m constant), eq. (4.40) would reduce to

$$n = \text{constant} \times c_i^{1/3} \tag{4.41}$$

Unfortunately there are no data available for testing this equation.

The Temkin Isotherm

The adsorption isotherm of Temkin[49]

$$n = b \ln c + \text{constant} \tag{4.32}$$

where b is a constant, implies a linear decrease in the energy of desorption with increasing adsorption[52]. The non-specific energy of desorption of a long-chain ion is given by:

$$\lambda \approx mw - z_1 \varepsilon \psi_0 \tag{4.7}$$

where mw is the constant energy of taking the hydrocarbon chain from water to oil. Only when ψ_0 is smaller than 25 mV ($c_{salt} \approx c_i$, and c_{salt} is large), so that eq. (2.28) reduces, at 20°C, to $\psi_G = 268\boldsymbol{k}T/Ac_i^{1/2}\varepsilon$ will the energy vary linearly with θ: the latter is given by A_0/A, and eq. (4.7) now becomes:

$$\lambda = mw - \frac{268\boldsymbol{k}T\,\theta}{A_0 c_i^{1/2}}$$

whence the Temkin equation (4.32) follows with the constant b given[50] by

$$b = \frac{A_0 c_i^{1/2}}{268}$$

As before, c_i is the total uni-univalent electrolyte concentration. Thus in the non-specific adsorption of long-chain ions from concentrated salt solution to the oil-water interface, the Temkin isotherm should be obeyed. This has not been tested, but for the specific adsorption of ions at the mercury-water interface the Temkin isotherm is valid[53].

SURFACE EQUATIONS OF STATE FROM ADSORPTION ISOTHERMS

The Linear Isotherm

One may eliminate c between the isotherm for *neutral films*, $n = \alpha c$ (eq. (4.27)) and the Gibbs equation (4.54) below, obtaining the equation of state

$$\Pi A = \boldsymbol{k}T \qquad (5.1)$$

For *films of long-chain ions* at exceedingly low values of n the linear limiting relation (4.36) (i.e. $\psi_0 \to 0$) is approached only very gradually as c and n are decreased. Eq. (4.35) can be solved by successive approximations to show that ψ_0 must still be about 110 mV when c is as low as 10^{-9} M ($n = 10^{10}$) for sodium lauryl sulphate in the absence of salt, and only when one substitutes $c = 1.3 \times 10^{-16}$ M ($n = 10^5$) does ψ_0 reduce to 6 mV, making the linear isotherm a valid approximation. If c is now eliminated between the linear isotherm and the Gibbs equation, one must again obtain (5.1): the Gibbs equation (4.54) is used with a factor 1 (not 2, see below) because the counter-ions tend (as $c \to 0$ and $\psi_0 \to 0$) to be infinitely far from the surface, i.e. they are in the bulk phase.

The Langmuir Isotherm (un-ionized films)

We have seen that the Langmuir, Gibbs, and Szyszkowski equations are consistent with each other. Since there are three variables (n, c and Π), these three equations must also be consistent with a surface equation of state for the adsorbed monolayer, relating Π with n (or A). From the Langmuir adsorption equation (4.28) and the Gibbs equation (4.54) one obtains

$$\Pi = (\boldsymbol{k}T/A_0)\ln\left(\frac{A}{A - A_0}\right), \qquad (5.5)$$

where A_0 is the limiting area of the film at high pressures, and is equal to $10^{16}/\beta$. At large areas, $\Pi \rightarrow \boldsymbol{k}T/A$.

This equation, derived by Volmer and by Frumkin (see Chapter 5), implies that adsorption occurs on to *fixed sites* on the surface. Although different in form from the more usual surface equation of state for *mobile monolayers*

$$\Pi(A-A_0)=\boldsymbol{k}T, \tag{5.4}$$

eq. (5.5) gives results for films above about 80 Å^2 in close numerical agreement with eq. (5.4), so that the latter may therefore often be used regardless of whether the film is completely mobile or not. To modify eq. (4.28) for mobile films, one should strictly write[8] in eq. (4.28) that $a=e^{\theta/(1-\theta)}e^{-\lambda/RT}$ (cf. eq. (4.30a)).

The Küster Isotherm

By combining the Küster isotherm (4.29) with the Gibbs equation, one may eliminate c and obtain surface equations of state.

In the absence of salt, a factor 2 is required in the Gibbs equation (as shown below):

$$-c\left(\frac{\partial \gamma}{\partial c}\right)=c\left(\frac{\partial \Pi}{\partial c}\right)=2\boldsymbol{k}Tn \tag{4.57}$$

This gives for the equation of state for charged films at the oil-water interface in the absence of inorganic salt

$$\Pi A=2\nu\boldsymbol{k}T,$$

where $A=10^{16}/n$. By eq. (4.39) $\nu=1.5$, so the required relation becomes

$$\Pi A=3\boldsymbol{k}T.$$

This equation of state for charged monolayers is valid, as is eq. (4.39), only when $n<10^{14}$ and $\psi_0>100$ mV. With these restrictions it has been checked experimentally. It may also be derived (Chapter 5) from properties of the charged monolayer, being made up of a kinetic contribution from the long-chain ions (approximately $\boldsymbol{k}T/A$) and an electrical term for the counter-ion system (approximately $2\boldsymbol{k}T/A$).

If excess inorganic salt is present, the Gibbs equation has the same form (eq. (4.54)) as for uncharged films:

$$-c\left(\frac{\partial \gamma}{\partial c}\right)=c\left(\frac{\partial \Pi}{\partial c}\right)=\boldsymbol{k}Tn.$$

From this, by elimination of c using (4.29),

$$\Pi A=\nu\boldsymbol{k}T.$$

For this system $\nu=3$ (by eq. (4.40)) so that, although the adsorption isotherm is different when the excess salt is added, the resulting equation of state is the same as without salt, i.e.

$$\Pi A=3\boldsymbol{k}T.$$

Again there are the restrictions that $n < 10^{14}$, $\psi_0 > 100$ mV, the latter being true if the total ionic strength is fairly small.

The derivation of the relation $\Pi A = 3\boldsymbol{k}T$, which is the limiting equation of state of a charged monolayer (Chapter 5), shows that the adsorbed film at the oil-water interface, like that at the air-water interface, is a monolayer.

RELATIONS BETWEEN SURFACE PRESSURE AND CONCENTRATION

It is of interest to revise Szyszkowski's $\Pi - c$ relation (4.30) to apply to charged films at the *oil-water* interface, both without and with added salt. *If no salt is added* the Küster isotherm for films at the oil-water interface (eq. (4.39)) and the Gibbs equation yield, by elimination of n, the required $\Pi - c$ relation

$$\log_{10}\Pi = \tfrac{2}{3}\log_{10}c + \log_{10}(3\boldsymbol{k}TK) \qquad (4.42)$$

This is tested in Fig. 4-18.

If now an *excess of electrolyte is present*, and the total uni-univalent ion concentration is c_i, the $\Pi - c$ relation for films at the oil-water interface can be obtained from a comparison of eq. (4.40) and the Gibbs equation, giving

$$\log_{10}\Pi = \tfrac{1}{3}\log_{10}c + \log_{10}(3\boldsymbol{k}TKc_i^{1/3}) \qquad (4.43)$$

This is also tested in Fig. 4-18. It is important to note that these equations

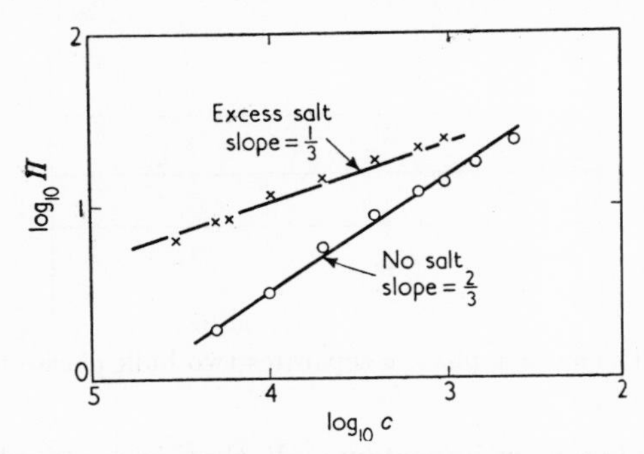

Fig. 4-18. Test of eqs. (4.42) and (4.43). The salt used is 8×10^{-3} M-NaCl. The lines are drawn to have the theoretical slopes[50,51].

are applicable only under the same conditions as are the parent equations (4.39) and (4.40). In Fig. 4-18 straight lines of the theoretically required slopes are drawn through the experimental points for sodium lauryl sulphate. The constants in eqs. (4.42) and (4.43) are calculated to be 3.04 and 2.34 respectively, while the experimental figures from the lines in Fig. 4-18 are 3.14 and 2.33 respectively.

Adding electrolyte must always increase the surface pressure of an ideal ionized film: this follows from eqs. (4.42) and (4.43) and is verified by comparing the lines of Fig. 4-18. The relative increase due to the salt is $(c_i/c)^{1/3}$, which, if $c_i = 8.4 \times 10^{-3}$ M and $c = 0.4 \times 10^{-3}$ M, is 2.7. The experimental figure for lauryl sulphate ions at the oil-water interface is 18.2/8.6, i.e. 2.1, and since $c_i \gg c$, the effect of NaCl must always be to increase the film pressure. Use of eqs. (4.42) and (4.43) implies, however, that the potential ψ_0 exceeds 100 mV: with large excess of NaCl, ψ_0 will fall below this figure.

THE GIBBS EQUATION

In 1878 J. W. Gibbs[54] derived thermodynamically a differential equation relating n, c, and $(\partial\gamma/\partial c)_T$; strictly, activities should be used. The derivation is as follows.

Consider the vapour-water interface to be situated along the plane CC′ in Fig. 4-19. In practice, it will have a thickness of 10 Å or more, over which distance the physical properties will vary continuously. The only precise statement that can be made, therefore, is that the surface of the water is situated somewhere between the planes AA′ and BB′, defined such that at and near AA′ the properties of the system are identical with those of the water phase, while at and near BB′ the properties are identical with those of the upper bulk phase. The distance BA may be perhaps as much as 100 Å,

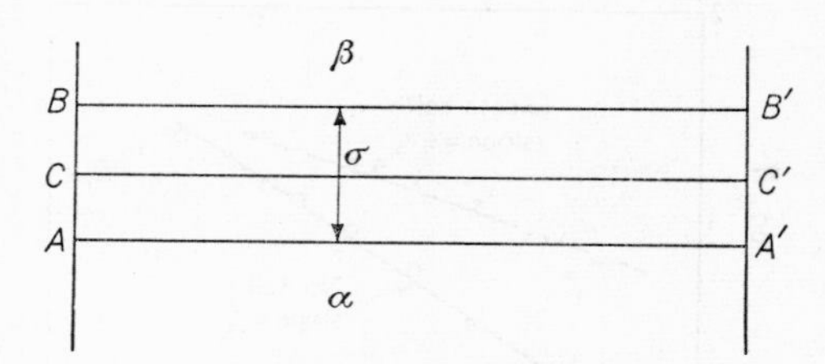

Fig. 4-19. The surface phase σ separates two bulk phases 'α' and 'β'.

but the exact value is unimportant: all that is required is that all the variation across the interface takes place in the surface phase σ between AA′ and BB′. To this surface phase we can apply the general equation of the additivity of energies

$$U^\sigma = TS^\sigma - PV^\sigma + \gamma A + \Sigma\mu n^\sigma \tag{4.44}$$

where U represents the total internal energy, S entropy, P pressure, V volume, A the total area, μ the chemical potential, and n the number of molecules of any component, including the water and any solutes added. Now one may subtract from this equation the corresponding quantities which would have contributed to U had the bulk phases persisted unchanged up to the mathematical plane CC′. The resultant equation, after subtraction of the two similar equations for the bulk phases, is:

$$U_s = TS_s + \gamma A + \Sigma \mu n_s \qquad (4.45)$$

where the subscript s denotes the excess quantity associated with the real surface. We may note that the intensity factors, temperature, pressure, and chemical potential, are necessarily constant throughout the system, that the term in A does not enter the equations for the corresponding amounts of bulk phases, and that the volumes between BC and CA together amount to V^σ.

Differentiation of eq. (4.45) leads to

$$dU_s = TdS_s + S_s dT + \gamma dA + A d\gamma + \Sigma \mu dn_s + \Sigma n_s d\mu \qquad (4.46)$$

The Gibbs equation now follows by eliminating dU_s, using the first and second Laws of Thermodynamics; namely that if a small reversible change occurs in the surface phase σ, very close to equilibrium, the change of internal energy must be given in general by:

$$dU^\sigma = TdS^\sigma - PdV^\sigma + \gamma dA + \Sigma \mu dn^\sigma \qquad (4.47)$$

From this equation, as before, may be subtracted the two equations for the corresponding amounts of bulk phases between BC and CA, leaving the relation for the surface excess quantities

$$dU_s = TdS_s + \gamma dA + \Sigma \mu dn_s \qquad (4.48)$$

Elimination of dU_s between eqs. (4.46) and (4.48) gives:

$$S_s dT + A d\gamma = -\Sigma n_s d\mu \qquad (4.49)$$

or, in the usual nomenclature, with n the number of adsorbed molecules per unit area written for n_s/A, and with the temperature constant,

$$\partial\gamma)_T = -\Sigma n \partial\mu)_T \qquad (4.50)$$

This is the strict *Gibbs equation* for adsorption, and, unless there are factors other than those allowed for in the above derivation, it is necessarily as correct as are the Laws of Thermodynamics. We may note that any electrical terms (as for the energy in the condenser system set up by an electrical double layer) may influence γ, although, since the surface phase as a whole is electrically neutral by definition of the planes AA′ and BB′, no specific mention of this is necessary. The summation on the right hand side of eq. (4.50) must, however, include all adsorbed species, including the counter-ions if the solute consists of long-chain ions.

Derived Equations for the Air-Water Surface

For a two-component system, such as a solute in a solvent (including solvent vapour), eq. (4.50) becomes

$$\partial\gamma)_T = -n\partial\mu)_T - n_{solv.}\partial\mu_{solv.})_T \qquad (4.51)$$

where n and μ refer to solute (μ necessarily being the same in the bulk and in the surface at equilibrium, as assumed in the derivation), and $n_{solv.}$ and $\mu_{solv.}$ refer to solvent. If we now make the extra-thermodynamic assumption that the plane CC′ can be chosen in such a position that the excess or deficit of solvent, usually water, is zero, then:

$$\partial\gamma)_T = -n\partial\mu)_T \tag{4.52}$$

Into this equation we can substitute for μ using the thermodynamic relations:

$$\mu^\sigma = \mu = \mu^0 + \boldsymbol{k}T\ln c + \boldsymbol{k}T\ln f \tag{4.53}$$

where μ^0 is the standard chemical potential and f is the activity coefficient of the solute in the bulk solution. In dilute solutions f is close to unity, and its change with concentration will be zero to a first approximation, so that we may write

$$\partial\gamma)_T = -\boldsymbol{k}Tn\ \partial\ln c)_T$$

or

$$c\left(\frac{\partial\Pi}{\partial c}\right)_T = -c\left(\frac{\partial\gamma}{\partial c}\right)_T = \boldsymbol{k}Tn \tag{4.54}$$

In the adsorption of long-chain salts into the surface phase, equal numbers of cations and anions must enter the surface. Several examples of this type of adsorption are of interest.

Adsorption of long-chain salts in the absence of inorganic electrolyte is treated as follows, subscripts R′ and Na^+ being used to illustrate the adsorption of long-chain anions (e.g. lauryl sulphate ions) and of counter-ions. Eq. (4.50) may now be expanded to

$$\partial\gamma)_T = -(n_{R'}\partial\mu_{R'} + n_{Na^+}\partial\mu_{Na^+})_T \tag{4.55}$$

which, with eq. (4.53) applied to each type of ion, becomes

$$\partial\gamma)_T = -\boldsymbol{k}Tn_{R'}\left(\frac{\partial c_{R'}}{c_{R'}} + \frac{\partial c_{Na^+}}{c_{Na^+}}\right)_T \tag{4.56}$$

in which activity coefficients have been neglected and n_{Na^+} has been put equal to $n_{R'}$, since the surface phase is defined as being electrically neutral. Further, since no inorganic electrolyte is present, $c_{R'} = c_{Na^+}$, $\partial c_{R'} = \partial c_{Na^+}$, and so finally, omitting subscripts,

$$c\left(\frac{\partial\Pi}{\partial c}\right)_T = -c\left(\frac{\partial\gamma}{\partial c}\right)_T = 2\boldsymbol{k}Tn \tag{4.57}$$

Adsorption of long-chain salts in the presence of excess electrolyte (e.g. NaCl) is treated similarly, including now a term for the chloride ions:

$$\partial\gamma)_T = -(n_{R'}\partial\mu_{R'} + n_{Na^+}\partial\mu_{Na^+} + n_{Cl'}\partial\mu_{Cl'})$$

Now as the concentration of long-chain salt in the solution is varied, that of chloride ion (in excess) is not appreciably changed: hence $\partial\mu_{Cl'}$ is zero. We also know, from extra-thermodynamic knowledge of charged films, that $n_{Cl'}$ is, in practice, very small compared with $n_{R'}$, so that to a first approximation

$$\partial\gamma)_T = -\boldsymbol{k}Tn_{R'}\left(\frac{\partial c_{R'}}{c_{R'}}\right)_T - \boldsymbol{k}Tn_{Na^+}\left(\frac{\partial c_{Na^+}}{c_{Na^+}}\right)_T \tag{4.58}$$

Addition of further long-chain salt to the system increases equally the concentrations of R′ and of Na^+, i.e. $\partial c_{R'} = \partial c_{Na^+}$, while, since the salt is postulated to be present in excess, $c_{Na^+} \gg c_{R'}$. The second term of eq. (4.58) is thus negligible compared with the first, and so

$$\partial\gamma)_T = -\boldsymbol{k}Tn\frac{\partial c}{c}\bigg)_T$$

or

$$c\left(\frac{\partial \Pi}{\partial c}\right)_T = -c\left(\frac{\partial \gamma}{\partial c}\right)_T = \boldsymbol{k}Tn \tag{4.54}$$

where the quantities n and c without subscripts refer to the long-chain ions and where the effective constancy of the activity of the counter-ions has reduced the factor 2 of eq. (4.57) to a factor of 1.

Surface-active weak electrolytes, such as, for example, stearic acid, may be adsorbed both as the ion and as the uncharged molecule. Equation (4.50) above, if excess salt is present, then becomes

$$\partial\gamma = -n_{HSt}\,\boldsymbol{k}T\,d\ln c_{HSt} - n_{St'}\,\boldsymbol{k}T\,d\ln c_{St'} - n_{H^+}\,\boldsymbol{k}T\,d\ln c_{H^+} \tag{4.59}$$

For the bulk solution, to which the terms c now apply, we have the usual equilibrium relation

$$\frac{c_{St'} \cdot c_{H^+}}{c_{HSt}} = K$$

from which, if c_{H^+} is kept constant,

$$\partial \ln c_{St'} = \partial \ln c_{HSt}$$

Hence, under these conditions,

$$\partial\gamma = -\boldsymbol{k}T(n_{HSt} + n_{St'})\partial \ln c_{HSt} = -\boldsymbol{k}T(n_{HSt} + n_{St'})\partial \ln c \tag{4.60}$$

where $c = c_{HSt} + c_{St'}$ and the pH is still held constant.

The experimental curves of surface-tension vs. concentration for air-water systems fall into three broad classes, as McBain[55] pointed out. In Fig. 4-20, the "Type I" curve is typical of such materials as propanol, butyric acid, and un-ionized organic compounds generally. "Type II", in which there is an increase in surface tension, shows (according to eq. (4.54)) that n must be negative: this negative adsorption implies that the solute is strongly hydrated, and tends to keep away from the vapour or air phase (with its lower dielectric constant): the surface phase is thus less concentrated than is the bulk phase. This behaviour is shown by inorganic salts, as well as by sucrose and glycerol. "Type III" curves, with or without the minimum, are often found with long-chain ions: the surface tension reaches a constant value at higher concentrations. According to eq. (4.57) a zero or positive value of $(\partial\gamma/\partial c)$ should imply zero or negative adsorption, although, since the surface tension is very low in this region, clearly this cannot be correct. Indeed, radioactive tracer studies (described below) show conclusively that there is strong positive adsorption over the whole region of the curve.

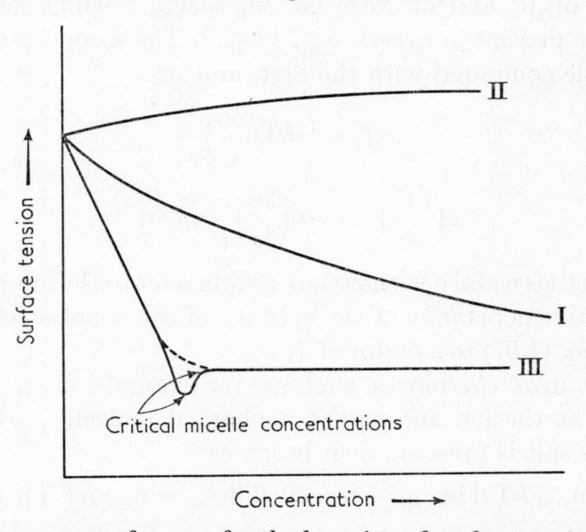

Fig. 4-20. The three types of curves for the lowering of surface tension. The minimum in the Type III curve is due to traces of impurity, and disappears after rigorous purification (broken line).

The explanation of the minimum in a "Type III" curve lies in the fact that it is very difficult to prepare either highly pure water or highly pure long-chain salts. When one remembers that 1 m.g. of impurity can cover 1 m.^{2} of surface, the standards of purity required are seen to be exacting. Recent work with water that has been passed through ion-exchange columns or has been distilled several times in all-quartz apparatus, and with long-chain salts purified to better than 99.99% by foam fractionation (see Chapter 8), shows that the minimum is progressively eliminated as the purity of the system increases. With the knowledge that adsorption of the long-chain ions is always strongly positive, and that the minimum is due to impurity, we can interpret "Type III" curves in terms of the general Gibbs equation (4.50).

Suppose, firstly, that the water is pure and that the long-chain salt contains no trace of long-chain alcohol—a common impurity either remaining from its manufacture or else formed in small but significant amounts by hydrolysis. Application of eq. (4.50) to the system gives

$$-\partial\gamma = n_1\partial\mu_1 + n_2\partial\mu_2 + n_3\partial\mu_3 + n_4\partial\mu_4 \tag{4.61}$$

where subscripts 1, 2, 3, and 4 refer respectively to water, long-chain ions (e.g. lauryl sulphate), counter ions (e.g. Na^+), and micelles. The latter are agglomerates, of perhaps 100 long-chain ions surrounded by counter-ions, that are formed when the bulk concentration of single ions reaches a "critical micelle concentration". Further addition of long-chain salt above this concentration results in no further increase in the concentration of the single ions: all the additional material goes into the micelles. This explains why there is a flat portion of the curve above the critical micelle concentration, since now μ_2 and μ_3 are constant, n_1 is zero by the convention discussed above, and n_4 is negligible (the micelles are not adsorbed as they are strongly charged and hydrated, with all the hydrocarbon chains lying inside them). Hence $\partial\gamma = 0$, i.e. γ is constant as in the broken line of the "Type III" curve of Fig. 4-20. Surface tension studies thus form an effective way of finding the critical micelle concentration for pure substances.

If now a trace of long-chain alcohol is present as impurity, it will also be adsorbed, and must be allowed for in the Gibbs equation, and is here denoted by subscript 5:

$$-\partial\gamma = n_1\partial\mu_1 + n_2\partial\mu_2 + n_3\partial\mu_3 + n_4\partial\mu_4 + n_5\partial\mu_5 \tag{4.62}$$

Just above the critical micelle concentration, as before, the first four terms on the right are zero, while the fifth is negative since the micelles are known to "solubilize" within themselves impurities: the addition of long-chain ions therefore reduces the concentration (and hence the chemical potential) of the long-chain alcohol impurity, and $(\partial\gamma/\partial c)$ is positive, passing through a minimum as observed. Further addition of long-chain salt (increase of c) soon removes *all* the alcohol from the surface by solubilization; n_5 is thus reduced to zero, and the surface tension becomes constant. Exactly the same argument applies if the impurity is a trace of heavy metal ion from the water, with subscript 5 now referring to this instead of to the long-chain alcohol impurity. The interpretation of the "Type III" curves illustrates very clearly the need for extra-thermodynamic knowledge in applying the Gibbs equation.

If we study the adsorption of a vapour (e.g. $CHCl_3$) on to a water surface, the Gibbs equation (4.52) may be combined[20] with the relation $\mu = \mu^* + \boldsymbol{k}T\ln p$, giving

$$p\left(\frac{\partial\gamma}{\partial p}\right)_T = -\boldsymbol{k}Tn \tag{4.63}$$

where p is the vapour pressure of the $CHCl_3$, and n is the number of adsorbed molecules per cm.^{2}

Checking the Derived Equations for the Air-Water Surface

While eq. (4.50) is necessarily correct, the derivation of eq. (4.54) or eq. (4.57) from it is less certain, as extra-thermodynamic postulates are required.

Experimental tests of the validity of these equations are therefore interesting; the left-hand term is obtained directly from the γ vs. c or ln c plots, though considerable accuracy is necessary if the measured slope is to be reliable. Estimation of the right-hand term, n, however, requires either special apparatus or special techniques. Early workers in this field passed a stream of bubbles through a column of liquid and measured the concentration changes in the liquid after a certain total interface had been exposed to the solute. Results were not reliable, however, because the system was not static and hence equilibrium was not fully established at the bubble surfaces, and also because bubbles rising through a liquid pass through many shapes and configurations, of uncertain surface area.

McBain's microtome method[56,57] avoids these difficulties: a microtome blade skims at 35 ft. sec.$^{-1}$ across the surface of a large trough. The blade cuts off 0.05 mm. of the surface, which is then analysed. There are two advantages of this method: the very rapid movement of the blade ensures that the equilibrium is not appreciably disturbed by ripples or diffusion of surface-active material during the slicing process, and also the thinness of the layer

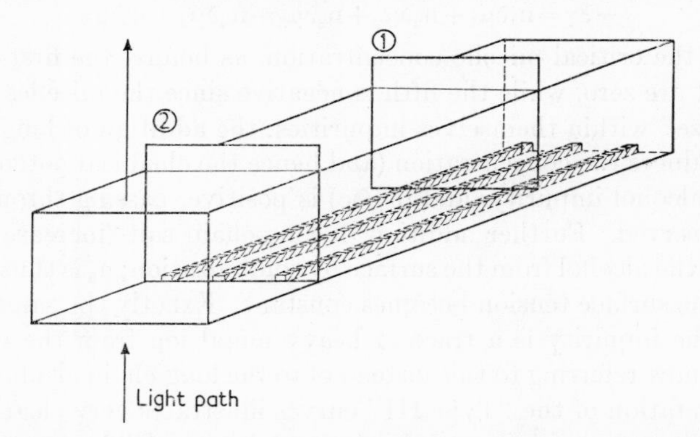

Fig. 4-21. Special trough[58] for studying the amount of adsorption. The barrier is moved along the skids from position 1 to position 2, the water moving back between the skids at the bottom of the trough. The excess material associated with the surface which has been compressed is estimated optically.

of surface removed greatly facilitates analysis for the excess material adsorbed. Interferometric analysis of the surface layer skimmed off gives good agreement with eq. (4.54) for both positive and negative adsorption, a typical result[56] being that for the adsorption of 0.05 N phenol: n calculated by eq. (4.54) is 1.35×10^{14} molecules cm.$^{-2}$, while the value of n found directly is 1.42×10^{14} molecules cm.$^{-2}$ The microtome method, however, involves very considerable mechanical difficulties before use, and several months' work in levelling the rails carrying the microtome blade are required, to ensure that only 0.05 mm. of surface shall be sliced off. McBain and his collaborators had constantly to re-align the rails, and for this reason McBain finally abandoned this spectacular "railroad apparatus", preferring the greater difficulties of analysing deeper layers of liquid for the surface excess to the mechanical difficulties of keeping the rails in alignment.

McBain's surface compression method[58] is ingenious but simple: the surface of a liquid is fairly rapidly compressed in a special trough (Fig. 4-21), followed by desorption of the surplus material from the surface. The consequent small increase in bulk concentration due to this desorption can be estimated interferometrically, whence n is deduced. For lauryl sulphonic acid at a concentration of 0.5 g.l.$^{-1}$, n calculated by the Gibbs equation (4.54) is 1.44×10^{14} molecules cm.$^{-2}$, the observed figures in different experiments being 1.30, 1.78, 1.46, 1.60, 1.62, 1.60, and 1.60×10^{14} molecules cm.$^{-2}$ This is satisfactory agreement.

Direct measurement with a foam of the amount of solute adsorbed is possible using a column of equilibrated foam of a single bubble-size: the decrease in concentration of the bulk liquid gives the amount of adsorption

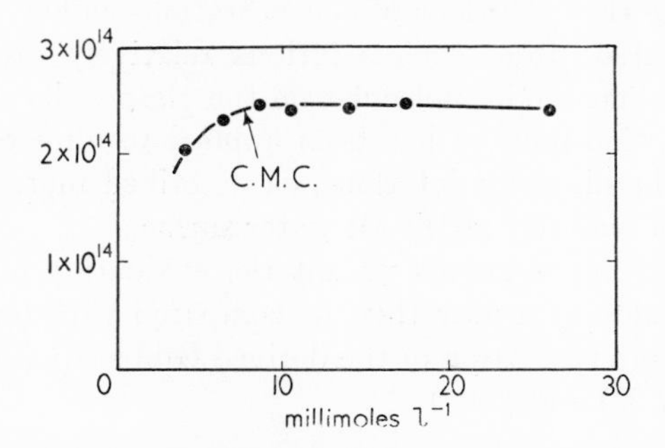

Fig. 4-22. Test of eq. (4.57), by the foam method[59], for sodium lauryl sulphate in the absence of salt, at 25°C. The adsorption calculated at the critical micelle concentration is 1.9×10^{14} molecules cm.$^{-2}$, compared with the observed value of 2.4×10^{14} from this figure.

on a known surface area of foam. The latter may conveniently be as great as 5×10^4 cm.^{2}, when the bulk solution may decrease in concentration by perhaps 3×10^{-4} M. Results on sodium lauryl sulphate[59] (Fig. 4-22) show that adsorption, n_2, increases to 2.4×10^{14} long-chain ions per cm.^{2}, but does not increase above the micelle concentration (8×10^{-3} M). This result is in accord (Table 4-X) with both eqs. (4.57) and (4.61), confirming the factor 2 in the former, and agrees also with results by the tritium method discussed below.

Force-Area curves of sparingly soluble films can be used to evaluate directly n on the right-hand side of eq. (4.54): a further assumption is required, however, in that all adsorption must be confined to a monolayer. Frumkin[60] calculated from the slope of the surface-tension vs. concentration curve for solutions of un-ionized lauric acid that n should be 3.4×10^{14} molecules cm.$^{-2}$, in good agreement with the figure of 3.1×10^{14} molecules cm.$^{-2}$ calculated from rapid measurements of the area (32 Å^2) of spread monolayers of the same substance at the same value of the surface tension. The agreement implied in these figures has been verified by Mme. Saraga[61], who found similarly values of 3.4×10^{14} and 3.6×10^{14} molecules cm.$^{-2}$ respectively at 23 dynes cm.$^{-1}$, and 2.0×10^{14} and 2.3×10^{14} molecules cm.$^{-2}$ respectively at 2 dynes cm.$^{-1}$ The validity of the interpretation (4.54) of the Gibbs equation as applied to this system is thus clearly demonstrated; the concentrations of lauric acid are of the order 10^{-6} N, so the neglect of activity coefficients is justifiable. The agreement also confirms that all adsorption is in the monolayer.

Desorption kinetics of sparingly soluble films similarly give a check on eq. (4.54). Here the rates of desorption of a monolayer are used to give values of the equilibrium concentrations, while n is obtained directly from rapid measurements of the force-area curve of a spread monolayer. This method has the advantage that solutions of concentrations as low as 10^{-6} N do not have to be prepared: from such solutions relatively great depletion by adsorption of the long-chain molecules on the glass walls of the vessel may occur. The desorption method has been applied to films of $C_{16}H_{33}N(CH_3)_3^+$ desorbing from the oil-water interface, as described more fully below: the method is of equal validity at the air-water surface.

Surface potential measurements permit the evaluation of n in eqs. (4.54) and (4.57): the value of n may then be compared with $(\partial\gamma/\partial c)$ to obtain a direct test of the interpretation of the derived Gibbs equations. As shown in Chapter 2, eqs. (2.23) and (2.28)

$$\Delta V = 4\pi n \mu_D + (2\boldsymbol{k}T/\varepsilon)\sinh^{-1}(134/Ac_i^{1/2})$$

from which, knowing μ_D for spread films of materials of various chain-lengths, one may calculate n from the measured ΔV and the ionic concentration c_i. The second term on the right also depends on n (since $A = 10^{16}/n$), so that successive approximations are necessary to obtain n from the experimental data: this is relatively easy, however. This method is applicable to any substances for which μ_D can be shown, from studies on insoluble monolayers, to be independent of n and the chain-lengths. For example, for normal alcohols μ_D is constant at 220 millidebyes, and, as for other un-ionized films, the second term on the right in the above relation is not required. For 0.16 M-butyl alcohol n is 2.05×10^{14} from eq. (4.54), and 2.1×10^{14} from ΔV measurements[62]. Slightly modified when applied to adsorbed films of sodium lauryl sulphate, this method shows good agreement with eq. (4.54) in the presence of added salt[63].

Radioactive tracers provide a means of checking the derived Gibbs relations (equations (4.54) and (4.57)) that is free from the possible uncertainties in the use of sparingly soluble monolayers, in the assumptions of the adsorption equations, or in the use of the Gouy equation. Like the microtome and interferometer methods, however, the radioactive tracer method requires exceptionally accurate experimental techniques and the most stringent precautions both in preventing evaporation of water and in counting techniques. Consequently, such experiments cannot lightly be undertaken. The principle of using radioactive isotopes is that, if the emitted particles are of short range in water, a counter placed in the air just above the liquid will register only those particles coming from the adsorbed film and from the immediately underlying solution (Fig. 4-23). The radiation from the latter is measured with isotopes in non-surface-active form, and is subtracted from the total radiation found over the solution of surface-active agent. The

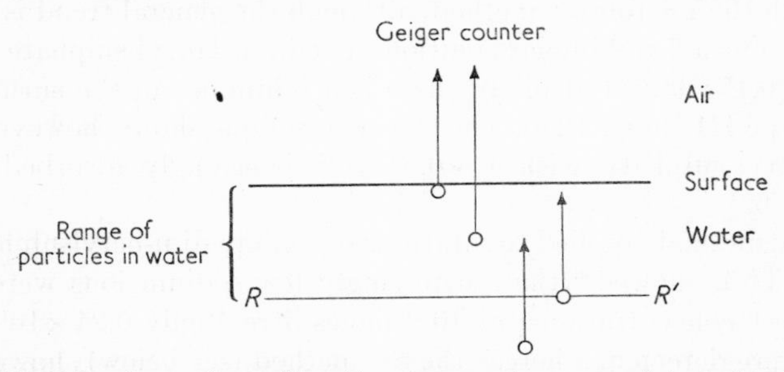

Fig. 4-23. Particles from the radioactive tracer atoms are detected only from the monolayer and the solution just below it. Particles emitted from atoms below the line RR′ are absorbed into the water without being registered by the Geiger counter.

effect of adsorption is relatively important: the activity from the adsorbed layers is of the same order as that from the bulk solution, the particles emitted from the material in the bulk of the solution far from the surface being absorbed in the liquid without escaping from the surface. The shorter the range of the emitted particles in water, therefore, the more accurate is the measurement of n, since the thickness of the underlying solution contributing to the radioactivity becomes less. Table 4-IX shows some of the possible systems for such studies, the last columns allowing for the fact that not all the particles within the "range depth" will emerge: those at small angles to the horizontal will be absorbed in the liquid. The most accurate results will clearly be obtainable when the radiation is of extremely short range, when the correction to be subtracted for the isotopes in the solution underlying the adsorbed film is insignificant.

Curiously, however, among the first published measurements using radioactivity to check eq. (4.57) was one based on the use of Na^{22}. The ready availability of this isotope and the fact that it could very easily be incorporated as the counter-ion of a long-chain sulphate led Hutchinson[65] to adopt an ingenious special procedure for overcoming the otherwise overwhelming effect of radiation from the bulk of the solution. He constructed a platinum loop from wire 0.5 mm. in diameter, which he carefully withdrew from the surface of a solution of "tagged" sodium lauryl sulphate so that carried out with it was a liquid film, spread across the ring. The weight of this liquid film was found by transferring both the loop and its film to a bottle and weighing: the radioactivity was then corrected for an equivalent amount of bulk solution, from which the adsorption on the two faces of the film followed. By withdrawing a thin film of liquid from the solution before measuring the radioactivity it is thus possible to reduce the otherwise prohibitively high count from the solution underlying the adsorbed film. Results are shown in Table 4-X. The accuracy does not seem to be quite as high as with the microtome method, although the general trend is to confirm eq. (4.57). For a 7 millimolar solution of sodium lauryl sulphate (the latter containing 0.1% lauryl alcohol), there is a minimum in the surface tension curve ("Type III", Fig. 4-20). The tracer technique shows, however, that the sodium lauryl sulphate, with $n_3 = 0.5 \times 10^{14}$, is strongly adsorbed, in accord with eq. (4.62).

The same method, applied to solutions of sodium di-n-octylsulphosuccinate (Aerosol OTN), showed[66] that surprisingly few sodium ions were adsorbed at the lowest concentrations; at 10^{-4} moles litre^{-1} only 0.24×10^{14} adsorbed Na^+ ions were detected, whereas the S^{35} method (see below) showed that the adsorption of the long-chain sulphosuccinate ions was 1.2×10^{14}. The interpretation of this difference is that hydrogen ions, present in the water, are preferentially adsorbed, suggesting that λ_p of the Stern equation is appreciable here. As the bulk concentration of H^+ is virtually unaltered, the interpretation of the Gibbs equation is that only one term should be allowed for, and the form of eq. (4.54), rather than that of eq. (4.57), is correct.

The first measurement of adsorption at an undisturbed surface using tracers counted di-n-octylsulphosuccinate ions, tagged with S^{35}, adsorbed at the air-water interface[66,67]. At low concentrations, about 0.1 millimolar, agreement with the Gibbs equation (with a factor 1 for the reason discussed above) was satisfactory, experiment giving n as 1.7×10^{14} and the Gibbs relation (4.54) giving 1.44×10^{14} sulphosuccinate ions cm.$^{-2}$ At higher concentrations the adsorption as detected by radioactive counts apparently increases much further—to two or three monolayers. Not only is this in conflict with the Gibbs equation, however, but also it is not born out either by surface viscosity studies or by the necessarily more accurate measurements with tritiated material in the systems discussed below. Part of the difficulty lies in the small relative contribution of the surface material at higher bulk concentrations (Table 4-IX): if the solution is 0.01 millimolar the total count and bulk phase correction may be, for example, 2800 and 512 particles per minute[66], whence the count from the adsorbed film must be 2288. At the higher concentration of 1 millimolar a lower specific activity has been used,

TABLE 4-IX*

Isotope	Radiation	Range in water	Ratio of activity from adsorbed film ($n = 0.6 \times 10^{14}$) and bulk solution (within range) of concentration: 10^{-3} M	10^{-2} M
Bi^{212}	Tl^{208} (recoil)	0.1μ	20	2
H^3	β	6μ	1.8	0.18
Po^{210}	α	32μ	6×10^{-2}	6×10^{-3}
C^{14}	β	300μ	3.6×10^{-2}	3.6×10^{-3}
S^{35}	β	340μ	3.6×10^{-2}	3.6×10^{-3}
Ca^{45}	β	650μ	1.8×10^{-2}	1.8×10^{-3}
P^{32}	β	about 1 cm.	$\sim 10^{-3}$	$\sim 10^{-4}$
Na^{22}	γ	very large	very small	very small

* Most of the data in this table are those cited by Aniansson[64,71].

and these figures are (for example) 600 and 530, the surface contribution now being given by the relatively small figure of 70 counts per minute. This is clearly rather sensitive to any evaporation from the surface of the solution, which is known[68] to increase by as much as 6% the total count from a solution of long-chain sulphate. In the absence of surface-active agent, the contribution from the inorganic sulphate ions is unaltered[68], since, even if the evaporation is occurring, convection currents rapidly equalize concentrations in the bulk of the liquid and in the uppermost layer. But when the solution is surface-active, the surface compressional modulus and the viscosity of adsorbed film will greatly retard convection currents, and any evaporation will, therefore, preferentially concentrate the upper region of the solution. As an example, if, before measurements were made on the more concentrated solution cited above, the small amount of 0.002 cm. of liquid had evaporated, then the count of 600 from the surface-active agent would be in error: without evaporation it would have been 565 per minute, a difference of 6% resulting from evaporation. For the inorganic sulphate the count is unaltered[68], remaining in this example at 530 per minute, giving a contribution from the adsorbed material of 35 counts, instead of 70. Since a count of 70 corresponds to 2 monolayers, a correction for this small amount of evaporation could have given agreement both with the Gibbs equation and with the idea that such adsorbed films are always monolayers. While this effect has not been verified experimentally for the di-n-octylsulphosuccinate solutions, it is supported by the constancy of the surface-viscosity results even at higher concentrations. Because of the difficulty of maintaining a high accuracy at higher concentrations with the S^{35} method, the more accurate measurements with tritiated derivatives are to be preferred as a test of the Gibbs equation: the accuracy is much higher because the correction from the bulk phase immediately underlying the adsorbed film is much less important.

The isotope S^{35} was also used in an independent investigation of eq. (4.57) by Anianson and Lamm[69]. Tagged $C_{18}H_{37}SO_4Na$ adsorbed from a 1.6×10^{-5} N solution at 18°C to give 3.64×10^{14} long-chain ions cm.$^{-2}$ whereas the value of n from $(\partial\gamma/\partial c)$ and eq. (4.57) is 3.5×10^{14}.

The most accurate work in checking the interpretation of the Gibbs equation with radioactive tracers uses tritium, H^3. As seen in Table 4-IX, this gives a very favourable ratio of activity from the adsorbed film relative to that from the underlying solution, because of the very low range of the β-particles emitted. Nilsson[70], studying the adsorption of tritiated sodium lauryl sulphate from water at 25°C, has shown that the adsorption (Table 4-X and Fig. 4-24) is virtually constant at higher concentrations, reaching

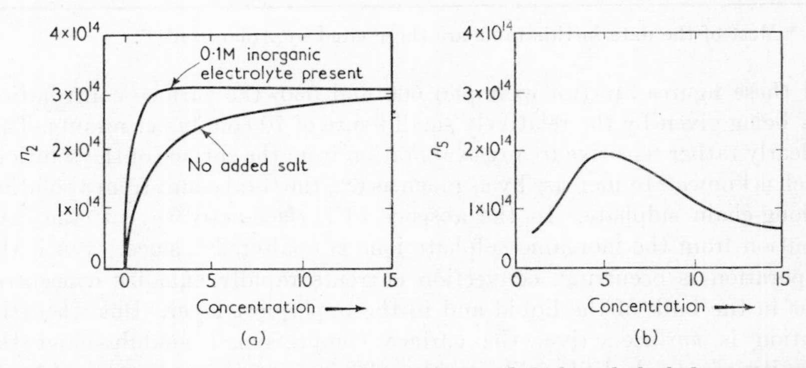

Fig. 4-24. Adsorption of (a) lauryl sulphate ions and (b) lauryl alcohol present as 1% (mole) impurity in the NaLS. Both are plotted as molecules cm.$^{-1}$, as measured at the air-water surface at 25°C by the tritium method[70], against the concentration of sodium lauryl sulphate in millimoles l.$^{-1}$ The critical micelle concentration is known to be about 8 millimoles l.$^{-1}$ in pure water.

about 3×10^{14} long-chain ions cm.$^{-2}$: the factor 2 must be used in the Gibbs equation. Further, the addition of 1% (mole) of lauryl alcohol to the tritiated sodium lauryl sulphate below the critical micelle concentration scarcely affects the adsorption of the latter substance (i.e. n_2 and n_3 of eqs. (4.61) and (4.62) are effectively unaltered). The lauryl alcohol must thus affect the surface tension by penetrating into the film of long-chain sulphate ions, a conclusion elegantly born out by studying the adsorption of tritiated lauryl alcohol from solutions made up from sodium lauryl sulphate, the latter containing as "impurity" 1% (mole) of the tritiated lauryl alcohol (Fig. 4-24b). From a 5 millimolar solution of the sodium lauryl sulphate the adsorption of the lauryl alcohol (n_5 in eq. (4.62)) is 1.5×10^{14} molecules cm.$^{-2}$, close to the figure for the adsorption of the long-chain ions. The adsorption of the tritiated lauryl alcohol increases up to about the micelle point, showing that $\partial\mu_5$ is increasing in this region, and that the negative slope of the surface tension curve must therefore be steeper than for pure sodium lauryl

TABLE 4-X

Adsorption of Sodium Lauryl Sulphate

Concentration of sodium lauryl sulphate, millimoles litre^{-1}	n_2 from analysis[59]	From radioactive counts, ions cm.$^{-2}$ n_3 (Hutchinson[65])	From radioactive counts, ions cm.$^{-2}$ n_2 (Nilsson[70])	$n_2(=n_3)$ from $(\partial\gamma/\partial c)$ using eq. (4.57) (refs. 34, 63)
1.0	—	0.66×10^{14}	1.5×10^{14}	1.0×10^{14}
2.4	—	3.42×10^{14}	1.9×10^{14}	1.5×10^{14}
4.9	2.1×10^{14}	4.5×10^{14}	2.5×10^{14}	1.8×10^{14}
6.0	2.3×10^{14}	—	2.7×10^{14}	1.9×10^{14}
12	2.4×10^{14}	—	3.0×10^{14}	1.8×10^{14}
Higher concentrations	2.4×10^{14}	—	3.0×10^{14}	1.8×10^{14}

Data refer to the air-water surface at about 25°C in the absence of added salt. The terms n_2 and n_3 are defined as in eq. (4.61); the critical micelle concentration of pure sodium lauryl sulphate is 8 millimolar.

sulphate in this region. Around or just below the micelle point (about 6 millimolar in this system) n_5 reaches a maximum of about 2×10^{14}, and $\partial\mu_5 = 0$. Above the micelle point n_5 decreases due to "solubilization" of the alcohol in the ionic micelles; though the amount of the adsorption of the lauryl alcohol n_5 is still appreciable, $\partial\mu_5$ is negative and so the explanation of the rise in the "Type III" surface tension plot (Fig. 4-20 and eq. (4.62)) receives direct experimental confirmation.

Our conclusion must be that, as radioactive techniques become more refined with the introduction of tritium and with the use of closed systems and proportional flow counters, the adsorption data are more nearly in accord with the requirements of eqs. (4.54) and (4.57). The recoil atom technique[71], though very accurate because of the short range of the particles, involves the presence of heavy metal ions in the solution, thereby complicating the physical chemistry of the system. For this reason the recoil technique has not been used to check the interpretation of the Gibbs equation.

Derived Equations for the Oil-Water Interface

The Gibbs equation (4.50) is applied to films adsorbed at the oil-water interface by substituting into it terms appropriate to solute (no subscript), oil, and water:

$$\partial\gamma)_T = -n\gamma\mu)_T - n_{oil}\partial\mu_{oil})_T - n_{H_2O}\partial\mu_{H_2O})_T \quad (4.64)$$

We must now invoke the extra-thermodynamic assumption that the plane CC′ (Fig. 4-19) can be so chosen that the excess or deficit of both water and oil is zero; i.e. that both oil and water have uniform concentrations up to a certain plane, where these concentrations each fall abruptly to zero[7,72].

Consequently, if this assumption is physically reasonable, i.e. if adsorption of a solute is confined to a monolayer,

$$\partial\gamma)_T = -n\partial\mu)_T \quad (4.52)$$

This we shall call the derived Gibbs equation, and from it eqs. (4.54) and (4.57) follow as before, applicable now to adsorption at the oil-water interface.

Checking the Derived Equations for the Oil-Water Interface

While eqs. (4.50) and (4.64) are necessarily correct, the derived forms (4.54) and (4.57) depend on the assumption that we can choose n_{oil} and n_{H_2O} as simultaneously zero in eq. (4.64).

The emulsion method[73] of checking (4.57) corresponds to the foam method for air-water adsorption. It is designed to overcome the difficulties of analysing the excess on small areas of surfaces. Purely analytical, the method depends on preparing an emulsion of ascertainable total interfacial area from an oil such as n-decane, dispersed in (say) 10 mM-sodium lauryl sulphate. To one sample of the emulsion excess sodium chloride is added, while the other is left without excess electrolyte. Conditions are chosen so that coalescence is negligible in both emulsions for several hours. A small amount of creaming agent, such as alginate, facilitates analysis of the emulsions to find the respective amounts of sodium lauryl sulphate taken up by each sample of the emulsion: the clear aqueous solution under the creamed (but not coalesced) droplets of oil is analysed by titration, whence the amount adsorbed on the oil drops is found by difference.

For the sample of the emulsion with excess electrolyte the total interfacial area may now be found from the equilibrium concentration of sodium lauryl sulphate in the clear aqueous solution, combined with interfacial tension measurements over a range of concentrations: the latter results can be interpreted using the derived Gibbs equation (4.54) with a factor unity to give areas per long-chain ion adsorbed. Hence, from the amount of long-chain ion adsorbed, the area of the oil-water interface is computed. The value may be about 4×10^4 cm.^{2} per cm.^{3} of oil.

This interfacial area is the same in the emulsion without added salt, since no coalescence has occurred: from the analytical figure for the sodium lauryl sulphate adsorbed on this area in the absence of added salt, the areas per long-chain ion are known. We can now compare these areas with those calculated by the Gibbs equation with the factor 2 (eq. 4.57)) applied to data on the lowering of interfacial tension at the appropriate equilibrium concentrations in the absence of added salt. Cockbain's data[73] are shown in Table 4-XI, agreement being satisfactory in establishing the use of the derived Gibbs equation (4.57) for this system.

TABLE 4-XI

Equilibrium concentration of sodium lauryl sulphate in water, in moles $l.^{-1}$ (20°C)	Area in $Å^2$ per long-chain ion from emulsion analysis[73]	Area in $Å^2$ per long-chain ion from eq. (4.57) applied to interfacial tension data[73]
0.0079	48.8	44.2
0.0069	49.1	47.6
0.0052	50.8	54.1
0.0035	56.8	64.9
0.0017	69.4	78.2

Equilibrium concentration of dodecyltrimethyl-ammonium bromide in water, in moles $l.^{-1}$ (20°C)	Area in $Å^2$ per long-chain from surface potentials[74]	Area in $Å^2$ per long-chain ion from eq. (4.57) applied to interfacial tension data[74]
0.0100	60	59.5
0.0063	63.7	63.8
0.0040	67.8	70.0
0.0025	76.0	80.0
0.0016	85.0	96.0
0.0010	103	112
0.0006	132	131

Desorption kinetics[4] can also be used to check the interpretation of the Gibbs equation, assuming that adsorption is confined to a monolayer. This method applies equally well to measurements at the air-water or oil-water interfaces, but has only been applied to films at the latter. The method depends on relation (4.22) and its consequences, i.e.

$$-dn/dt = B_3c = B_1c = 2.1 \times 10^{18}c$$

which shows that the net rate of desorption can be used as a means of measuring c in the solution immediately below the film, with the practical advantage that very sparingly soluble films can be used. We can thus check the derived Gibbs equation, first integrating (4.54) to give

$$\ln c + \text{constant} = \int (A/\boldsymbol{k}T)\,d\Pi \qquad (4.65)$$

and then substituting for the left-hand side $\ln(-dn/dt)$ at different values of Π, according to eq. (4.22). The right-hand side is obtained from the force-area plot, working quickly enough so that desorption is unimportant during the measurements. The exact value of the constant is not required in checking the differential form of the Gibbs equation, though the slope of the plot of $\ln c$ against $\int (A/\boldsymbol{k}T)\,d\Pi$ should be unity, $\boldsymbol{k}T$ here being equal to 408 at 20°C.

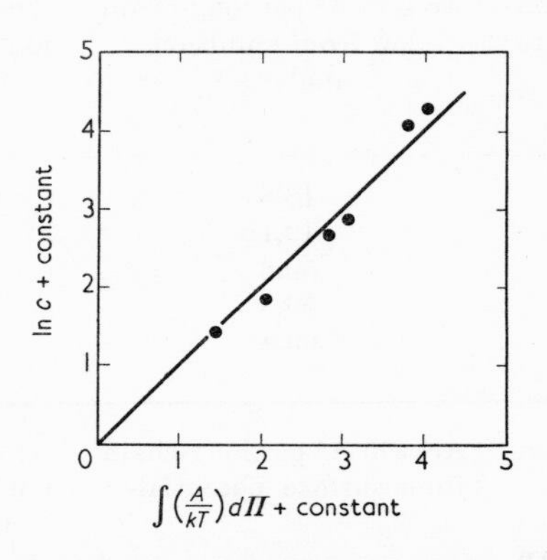

Fig. 4-25. Test of eq. (4.65) by desorption kinetics at the oil-water interface. The film is $C_{16}H_{33}N(CH_3)_3^+$ desorbing into aqueous 0.01 M-HCl, and the constants are chosen to make the straight line drawn through the points pass through the origin. The measured slope of unity supports eq. (4.65), and hence eq. (4.54).

Fig. 4-25 shows that this is so for monolayers of $C_{16}H_{33}N(CH_3)_3^+$ desorbing into aqueous 0.01 N-HCl. The latter constitutes an excess of inorganic electrolyte, so theory and experiment agree that the derived Gibbs equation should be used with a factor unity (eq. (4.54)). The method is not applicable to the evaluation of the factor in the Gibbs equation when no excess electrolyte is present: eq. (4.22) above would still apply, but since dn/dt must be in the range 10^9 to 10^{11} molecules cm.$^{-2}$ sec.$^{-1}$ for desorption measurements to be practical, c would always be less than 10^{-7} M in this type of experiment. There would thus always be an excess of inorganic ions in the water, so the factor unity is still to be expected.

Surface Potential Measurements permit evaluation of n in eqs. (4.54) and (4.57) using eqs. (2.23) and (2.28) as explained on p. 204. Results[74] for dodecyltrimethylammonium bromide adsorbing from highly purified water to the interface with petroleum ether, show agreement with n, calculated by the derived Gibbs equation (with a factor of 2), to within 10%, though for most points the agreement is better than this (Table 4-XI). It is important, when working with a small interface (order of 100 cm.^{2}), to eliminate all traces of counter-ions other than those added, otherwise the factor 2 in the derived Gibbs equation will not apply. Double distillation of the water in a silica still is recommended[44,74].

ADSORPTION IN AN ELECTRIC FIELD

When an electrical potential ($E_{pot.}+E_{ref.}$) is applied as an external variable (as in Fig. 2-27), eq. (4.50) becomes, for the adsorption of a salt,

$$-d\gamma = \sigma dE_{pot.} + \sigma dE_{ref.} + n_+ d\mu_+ + n_- d\mu_- \qquad (4.66)$$

where the temperature is understood to be constant and n_{H_2O} is, as usual, taken as zero. This equation applies to the adsorption of ions at the mercury-water electrode[75,76]: it is also an extension of eq. (2.56), and some results obtained from its use have been mentioned in Chapter 2. To avoid variable liquid junction potentials, let the liquid in the reference electrode now be the same as in the aqueous solution (Fig. 2-27), and let this reference electrode be reversible with respect to the anion (electrolytes being assumed uni-univalent). Then for this electrode, if the salt concentration is changed,

$$-\boldsymbol{F} dE_{ref.} = d\mu_- \qquad (4.67)$$

where $\boldsymbol{F}$ is the Faraday unit.

Further, the mercury surface and the ionic double layer must be, as a whole, electrically neutral, so that

$$\sigma + n_+ \boldsymbol{F} - n_- \boldsymbol{F} = 0 \qquad (4.68)$$

while, by definition of μ for the salt,

$$d\mu = d\mu_+ + d\mu_- \qquad (4.69)$$

Substitution for μ_+, μ_-, and n_- from eqs. (4.67), (4.68), and (4.69) into eq. (4.66) leads to

$$-d\gamma = \sigma dE_{pot.} + n_+ d\mu \qquad (4.70)$$

which is an extension of eq. (2.57), and leads to the Lippmann equation (2.58) if μ is constant. Alternatively, if $E_{pot.}$ is kept constant,

$$\left(\frac{\partial\gamma}{\partial\mu}\right)_{E_{pot.}} = -n_+$$

implying that, at any given point on the electrocapillary curve, n_+ can be obtained. Similarly, using a reference electrode reversible to the cations, the value of n_- can be found. Typical values (Fig. 2-30) for NaCl on mercury show that at positive applied potentials the specific adsorption of Cl′ is so strong that the potential must locally be negative, as witnessed by the positive adsorption of Na^+ in this region[76].

Though more conducting, the interface between water and nitrobenzene can also support an imposed electrical potential applied between the phases. If a long-chain electrolyte such as cetyltrimethylammonium bromide is dissolved in the oil, the interfacial tension decreases when the electrode in the oil is made positive relative to that in the water (Fig. 4-26). This is because long-chain cations are pushed through the oil towards the water

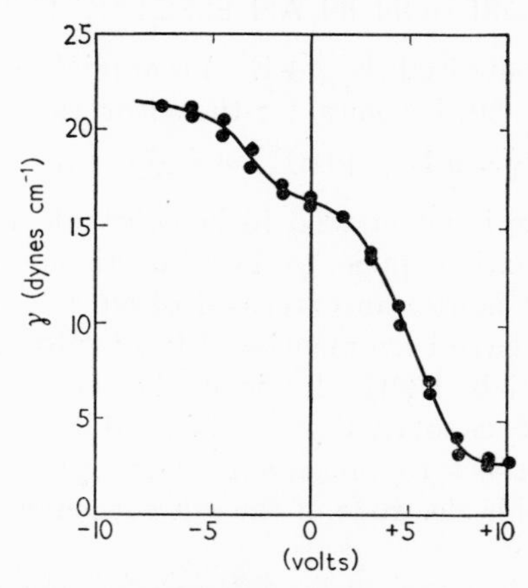

Fig. 4-26. Effect of applied potential (to the electrode in the nitrobenzene) on the interfacial tension against water. The solute is cetyltrimethylammonium bromide[77].

phase until they reach the interface. Here they desorb only slowly since they must overcome the rather high energy barrier opposing the removal of the hydrocarbon chains from the oil. The result is that there is an increased density of long-chain ions in the interface at any time, and γ is correspondingly lowered. Just as spontaneous emulsion may form at the mercury-water interface (p. 363) if the potential pulls into the surface so many ions that Π exceeds the interfacial tension of the clean interface, so oil-water emulsions may perhaps be prepared purely electrically.

REFERENCES

1. Traube, *Liebigs Ann* **265**, 27 (1891).
2. Langmuir, *J. Amer. chem. Soc.* **39**, 1848 (1917): see particularly Table III, p. 1892.
3. Guastalla, *C. R. Acad. Sci., Paris* **228**, 820 (1949).
4. Davies, *Trans. Faraday Soc.* **48**, 1052 (1952).
5. Crisp, in "Surface Chemistry" p. 65, Butterworths, London (1949).
6. Schüler, Private communication.
7. Hutchinson, E., in "Monomolecular Layers" p. 161, A.A.A.S., (1954).

8. Haydon and Taylor, F. H., *Phil. Trans.* **252**, 225 (1960).
9. Davies, *Proc. roy. Soc.* **A245**, 417 (1958).
10. Pethica, *Trans. Faraday Soc.* **50**, 413 (1954).
11. Klevens and Davies, *Proc. 2nd Internat. Congr. Surface Activity* **1**, 31, Butterworths, London (1957).
12. Phillips, *Trans. Faraday Soc.* **51**, 561 (1955).
13. Scholberg, Guenthner, and Coon, *J. phys. Chem.* **57**, 923 (1953).
14. Klevens and Raison, *J. Chim. phys.* **51**, 1 (1954).
15. Arrington and Patterson, *J. phys. Chem.* **57**, 241 (1953).
16. Ward and Tordai, *Nature, Lond.* **158**, 416 (1946); *Trans. Faraday Soc.* **42**, 408, 413 (1946).
16a. Hammerschmidt, *Industr. Engng. Chem.* **26**, 851 (1934);
Butler, J. A. V., *Trans. Faraday Soc.* **33**, 229 (1937);
Frank and Evans, M. W., *J. Chem. Phys.* **13**, 507 (1945).
16b. Davies, unpublished calculations.
17. Posner, Anderson, J. R. and Alexander, A. E., *J. Colloid Sci.* **7**, 623 (1952).
18. Ward and Tordai, *Rec. Trav. chim. Pays-Bas* **71**, 482 (1952).
19. Cutting and Jones, D. C., *J. chem. Soc.* 4067 (1955);
Jones, D. C. and Ottewill, *J. chem. Soc.* 4076 (1955).
20. Jones, D. C., Ottewill and Chater, *Proc. 2nd Internat. Congr. Surface Activity* **1**, 188.
21. Kemball and Rideal, *Proc. roy. Soc.* **A187**, 53 (1946);
Kemball, *Proc. roy. Soc.* **A187**, 73 (1946); *Proc. roy. Soc.* **A190**, 117 (1947).
22. Ward and Tordai, *J. chem. Phys.* **14**, 453 (1946).
23. Langmuir and Schaefer, V. J., *J. Amer. chem. Soc.* **60**, 1351 (1938).
24. Davies, Collis-Smith, and Humphreys, *Proc. 2nd Internat. Congr. Surface Activity* **1**, 281, Butterworths, London (1957).
25. Haydon and Phillips, *Nature, Lond.* **178**, 183 (1956); *Trans. Faraday Soc.* **54**, 698 (1958);
Sutherland, *Aust. J. Chem.* **12**, 1 (1959).
26. Salley, Weith, Argyle, and Dixon, *Proc. roy. Soc.* **A203**, 42 (1950).
27. Rayleigh, *Proc. roy. Soc.* **29**, 71 (1879).
28. Bohr, *Phil. Trans.* **A209**, 281 (1909).
29. Addison, *J. chem. Soc.* 98 (1945); 3090 (1950).
30. Rideal and Sutherland, *Trans. Faraday Soc.* **48**, 1109 (1952).
31. Posner and Alexander, A. E., *Trans. Faraday Soc.* **45**, 651 (1949); *J. Colloid Sci.* **8**, 575, 585 (1953).
32. Hommelen, *Bull. Soc. chim. Belg.* **66**, 476 (1957); *J. Colloid Sci.* **14**, 385 (1959).
33. McGee, unpublished results (1949) quoted in ref. 34;
Harrold, *J. Colloid Sci.* (1960).
34. Brady and Brown, A. G., "Monomolecular Layers" p. 33, A.A.A.S. (1954).
35. Burcik, *J. Colloid Sci.* **5**, 421 (1950).
36. Defay and Hommelen, *J. Colloid Sci.* **13**, 553 (1958); **14**, 411 (1959).
37. Harkins and Brown, F. E., *J. Amer. chem. Soc.* **39**, 499 (1919).
38. Garner et al., *Trans. Faraday Soc.* **55**, 1607, 1616, 1627 (1959).
39. Davies, in "Surface Phenomena in Chemistry and Biology" p. 55, Pergamon Press, London and New York (1958).
40. Davies, Bouette, and Gerrard, Research Project in Department of Chemical Engineering, Cambridge (1958).
41. Ward, in "Surface Chemistry" p. 55, Butterworths, London (1949);
Ward and Tordai, *Rec. Trav. chim. Pays-Bas*, **71**, 396, 572 (1952).

42. Sebba and Briscoe, *J. chem. Soc.* 114 (1940);
Addison and Hutchinson, S. K., *J. chem. Soc.* 3395 (1949).
43. Guastalla and Saraga, in "Surface Chemistry" p. 103, Butterworths, London (1949);
Saraga, *J. Chim. phys.* **52**, 181 (1955);
Roylance and Jones, T. G., *Proc. 3rd Internat. Congr. Surface Activity*, Cologne (1960);
Brooks J. H. and Alexander, A. E., *Proc. 3rd Internat. Congr. Surface Activity*, Cologne (1960).
44. Davies, *Proc. roy. Soc.* **A208**, 224 (1951).
45. Adam "The Physics and Chemistry of Surfaces", Clarendon Press, Oxford (1941).
46. Küster, *Liebigs Ann.* **283**, 360 (1894).
47. McBain, "The Sorption of Gases and Vapours by Solids", Routledge, London (1932).
48. Szyszkowski, *Z. phys. Chem.* **64**, 385 (1908).
49. Temkin, *J. phys. Chem., Moscou.* **15**, 296 (1941); *Chem. Abstr.* **36**, 6392 (1942).
50. Davies, *Proc. roy. Soc.* **A245**, 417, 429 (1958).
51. Kling and Lange, *Proc. 2nd Internat. Congr. Surface Activity* **1**, 295, Butterworths, London (1957).
52. Trapnell, "Chemisorption" London (1955).
53. Anderson, W., and Parsons, *Proc. 2nd Internat. Congr. Surface Activity* **2**, 45, Butterworths, London (1957).
54. Gibbs, "Collected Works" **1**, p. 219, Yale University Press (1948).
55. McBain, "Colloid Science" D.C. Heath, Boston (1950).
56. McBain and Swain, *Proc. roy. Soc.* **A154**, 608 (1936).
57. McBain and Wood, L. A., *Proc. roy. Soc.* **A174**, 286 (1940).
58. McBain, Mills, G. F., and Ford, *Trans. Faraday Soc.* **36**, 931 (1940).
59. Wilson, A., Epstein, and Ross, *J. Colloid Sci.* **12**, 345 (1957).
60. Frumkin, *Z. phys. Chem.* **115**, 499 (1925).
61. Saraga, *Mémor. Serv. chim. 'Etat.* **37**, 27 (1952).
62. Posner, Anderson, J. R., and Alexander, A. E., *J. Colloid Sci.* **7**, 623 (1952).
63. Pethica and Few, *Disc. Faraday Soc.* **18**, 258 (1954).
64. Aniansson, *J. phys. Chem.* **55**, 1286 (1951).
65. Hutchinson, E., *J. Colloid Sci.* **4**, 600 (1949).
66. Dixon, Judson, and Salley, in "Monomolecular Layers" p. 63, A.A.A.S. (1954).
67. Dixon, Weith, Argyle, and Salley, *Nature, Lond.* **163**, 845 (1949).
68. Nilsson and Lamm, *Acta chem. scand.* **6**, 1175 (1952).
69. Aniansson and Lamm, *Nature, Lond.* **165**, 357 (1950).
70. Nilsson, *J. phys. Chem.* **61**, 1135 (1957).
71. Steiger and Aniansson, *J. phys. Chem.* **58**, 228 (1954).
72. Guggenheim, *Trans. Faraday Soc.* **36**, 397 (1940).
73. Cockbain, *Trans. Faraday Soc.* **50**, 874 (1954).
74. Haydon and Phillips, *Trans. Faraday Soc.* **54**, 698 (1957).
75. Frumkin, *Proc. 2nd Internat. Congr. Surface Activity* **3**, 58, Butterworths, London (1957).
76. Grahame, *Chem. Rev.* **41**, 441 (1947).
77. Guastalla, *Proc. 2nd Internat. Congr. Surface Activity* **3**, 112, Butterworths,
Brooks J. H. and Alexander, A. E., *Proc. 3rd Internat. Congr. Surface Activity*, Cologne (1960).

Chapter 5

Properties of Monolayers

Benjamin Franklin's experiments on the spreading of oils on the pond at Clapham Common were carried out in 1765: he found[1] that the oil would not spread thinner than 25 Å. More than a century was to pass, however, before Frl. Pockels in 1891 published her paper on further quantitative aspects of spread films. She developed the technique of confining insoluble films between barriers extending the whole width of a trough of water filled to the brim, and consequently was able to measure the relation between the surface tension and the area quickly and reproducibly. With this technique Lord Rayleigh showed in 1899 that the surface tension fell steeply only when the surface was covered with a close-packed monomolecular film. Devaux confirmed the results of Pockels and Rayleigh, and showed that the movements of the film could be made visible if it were lightly sprinkled with a fine powder. In this way he showed that even mono-molecular films could become solid when compressed.

Sir William Hardy[1] pointed out in 1913 that monolayers are formed from molecules consisting of a hydrophobic and a hydrophilic part, and hence that at the air-water surface the molecules must be orientated with the polar, hydrophilic parts buried in the water, while the remainder of the molecules will tend to leave the water. It remained for Langmuir to provide conclusive support for this hypothesis of orientation: he showed in 1917, using the now well-known Langmuir trough, that monolayers of fatty acids of various chain-lengths compress to the same limiting area, an indication that the different acids must all form films in which the molecules are orientated identically with respect to the surface[1].

The study of monolayers has been pursued vigorously in the intervening years: Adam has made many refinements in technique and published comprehensive data[1]; Rideal and his school have studied potential changes (Chapter 2) which again confirm that the molecules in the film are orientated, and in recent years attention has been focused on films of polymers at surfaces, the study of monolayers at the oil-water interface, electrically charged films, and the flow properties of spread and adsorbed monolayers.

SURFACE PRESSURE

The surface pressure of a monolayer is the lowering of surface tension due to the monolayer. The molecules constrained in the monolayer may be regarded as exerting a two-dimensional osmotic pressure; there is a repulsion in the plane of the surface, which is measured on a floating barrier acting as a semi-permeable membrane permeable to water only (Fig. 5-1). It is this

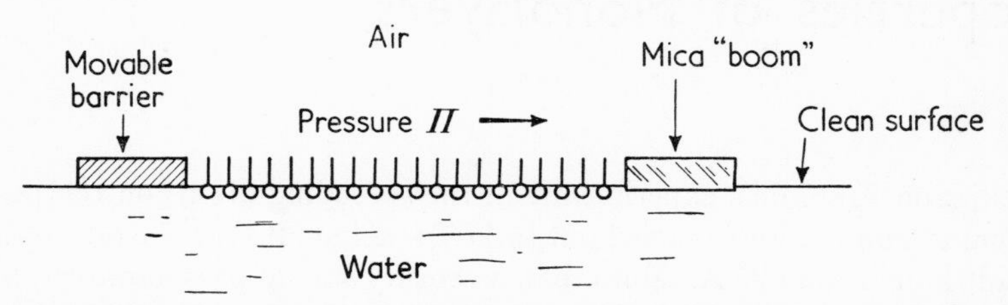

Fig. 5-1. Principle of the Langmuir Trough. The lowering of surface tension is measured directly, as a surface pressure Π exerted on the mica "boom" and balanced by applying an equal pressure from a torsion wire attached to the "boom". The barrier on the left is used to compress the film as required.

pressure, opposing the contractile tension of the clean interface, that is called the surface pressure. In symbols, as shown above,

$$\Pi = \gamma_0 - \gamma$$

where γ_0 is the surface tension of the clean interface. The variation of Π with the area available to the surface-active material is represented by a Π-A curve (sometimes called a force-area curve). It is usual to express Π in dynes $cm.^{-1}$ and A either in $Å^2$ for long-chain (for simple molecules) or in $m.^2 mg.^{-1}$ for more complex molecules.

Experimental Methods of Spreading Films at the Air-Water Surface

A light mica "boom", waxed to prevent wetting and attached to a torsion wire, separates two regions of a shallow vessel (Langmuir trough), which is filled to the brim with water (Fig. 5-1). In the original work a glass or silica vessel with lightly waxed edges was used, though in modern work Teflon troughs have proved eminently suitable[2]. If to the surface on the left of the "boom" a little stearic acid (or other insoluble surface-active material) is spread from solution (about 1 mg. per cm^3, i.e. about 0.1%) in petrol-ether, the latter quickly evaporates leaving a monomolecular film of stearic acid, this film exerting a pressure Π on the "boom". Benzene is not recommended as a spreading solution if the film is to be studied at low pressures, as it is slightly soluble in water, and some may therefore remain in the surface for long periods of time. Other spreading solutions are: gliadin in ethanol containing 40% water, poly-dl-alanine in 50% ethanol-water, poly-glutamic-acid-lysine or cetyltrimethylammonium bromide in 50% isopropanol-water, and ceryltrimethylammonium iodide in a mixture of chloroform and isopropanol with a little ethanol and water. Proteins generally spread well from a 0.1% solution in a 60% (v.v) solution of propanol (or isopropanol) in water, 0.5 M with respect to sodium acetate[3], or from water containing a trace (about 0.1%) of amyl alcohol to assist spreading[4]. Many poly-amino-acids must be first dissolved in dichloracetic acid, to which 10-20% propanol can then be added to assist spreading. Some long-chain compounds (e.g. ω-bromohexadecanoic acid and sodium docosyl sulphate) spread well from solution in isopropanol containing small quantities of ethanol and petrol-ether.

It is essential that scrupulous care be exercised in preparing the trough and the solutions, since as little as 0.1 mg. of impurity per square metre can constitute a serious source of error. This cleanliness is achieved on all-glass apparatus by soaking in warm chromic acid till the surface remains completely wetted on withdrawal. It should then be washed with syrupy phosphoric acid to remove adsorbed chromium, followed by extensive washing in tap water and finally redistilled water. Teflon also may be cleaned in hot chromic acid.

All organic liquids must be distilled in all-glass apparatus to remove any surface-active impurities, and the water must also be redistilled from alkaline permanganate in all-glass apparatus; if charged monolayers are being studied, the water must be finally distilled in all-quartz apparatus (some particles of resin may remain in water from ion-exchange columns). It is often best to use the purest solid reagents commercially available without further purification, as laboratory recrystallization may introduce more surface-active contamination, both from the air and from the fingers, than is present during the preparation of the material in large batches. In general all glass-ware and other parts coming into contact with the liquid in the trough must be handled with forceps cleaned in chromic acid, as grease from the fingers can prove a serious source of contamination.

To find precisely how much material has been spread on the surface it is convenient to employ a micrometer syringe, e.g. the Agla model. This is essentially a 1 cc. hypodermic syringe, the movement of the piston being controlled by a micrometer, so that known quantities of liquid as low as 2×10^{-4} cc. can be ejected. In practice it is convenient to use spreading solutions of such a strength that about 10^{-2} cc. of spreading solution is required, this being ejected in small amounts at various points in the interface through a stainless-steel hypodermic needle, initially touched into a separate, clean water surface to remove any excess solution at the tip of the needle.

Care must be taken in purifying the wax for the edges of the trough and the "boom", so that no oxidized material, which might spread as impurity over the surface, is present. A satisfactory method of purification is to warm the wax with petrol-ether and extract the resultant solution with aqueous NaOH.

When the water surface has been formed, any contamination remaining must be removed before spreading is begun. The surface may be conveniently swept two or three times with the glass slides, these having been carefully cleaned and then waxed to prevent wetting. Unwaxed Teflon slides may also be used. If the "constant total area" technique is used (it is more commonly employed at the oil-water interface), or if low values of Π are to be measured using one of the special techniques described below, slides cannot be employed to remove contamination. Instead, the surface is lightly sprinkled with talc (previously ignited for 4 hours at red heat to remove any greasy material) and the contamination, its movement thus made visible, is "pushed" into one region of the dish by a stream of air (preferably filtered) from a capillary[5]. It may then be sucked off by means of another capillary joined to a water pump. This process may be repeated two or three times, when virtually no contamination will remain.

By means of the slides of waxed glass or Teflon the film may be compressed to any required area. It is important to move the slides slowly and smoothly when compressing the film, and for this purpose various mechanical devices have been developed[6,7,8].

Experimental Methods of Measuring Π at the Air-Water Surface

Langmuir's method[1] of having a "boom" of waxed mica in the surface, the film pressure then being measured directly, has already been mentioned: Adam's modifications are very satisfactory, Usually one mounts a small

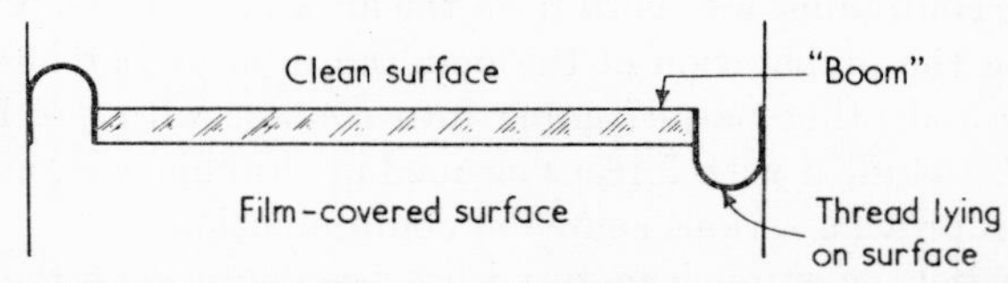

Fig. 5-2. Plan of Langmuir Trough, showing how the film is prevented from leaking past the ends of the "boom".

mirror on the "boom" or its attachments, and uses an optical beam to determine exactly the displacement of the "boom". A calibrated torsion wire is used to return the "boom" to its original position, following the spreading of the film: the torque on the torsion wire is then converted into dynes. To prevent leakage around the gaps ($\sim$0.5 cm.) at each end of the "boom", one may use very thin platinum ribbons, but Guastalla's modification[9] of using silk threads, covered with petroleum jelly and lying on the surface, has been widely adopted in recent years (Fig. 5-2). Although this apparatus is fairly simple and quite accurate, the possibility of changes of water level and contact angle affecting the results is always present; the Wilhelmy plate method (below) is both easier to use and slightly more accurate[10]. The ring method (used for pure liquids, Chapter 1) is not satisfactory for studying surface films: the interface is extended during measurement, and the film may also cause the ring to become incompletely wetted.

Use of *the hanging plate* (Wilhelmy[10,11]), perfectly wetted by the liquid, is rapid, simple and accurate. Long, thin glass cover-slips may be used, or platinum or mica sheets. The former should be cleaned in warm chromic acid, while the latter two should be slightly roughened with very fine grade emery paper to make them more easily wetted by water[12] (cf. Chapter 1). The plate

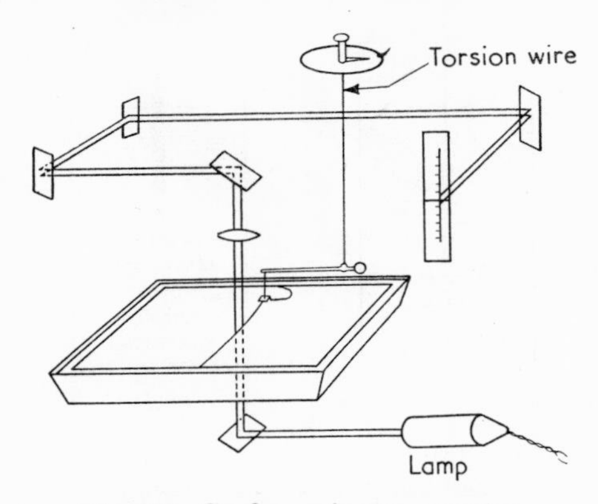

Fig. 5-3. Surface micromanometer.

may be used either attached to one arm of a balance[13] or hung on an arm attached to a torsion wire: correction may be made for buoyancy effects, or they may be eliminated completely by returning the slide always to its original immersion in the liquid after each compression of the film. If the latter course is adopted the contact angle must be carefully watched. Optical levers, from mirrors mounted on the beam of the balance or on the torsion wire, define accurately the position of the slide, and generally the results are accurate to 0.01 dyne cm.$^{-1}$ Sometimes this technique is combined with spreading at total constant area, as is more frequently done for the oil-water interface.

For *very low pressures*, such as are required for the determination of the molecular weights of large molecules in surface films, special techniques are required. Guastalla[5,14] has devised a very sensitive "surface micromanometer" shown in Fig. 5-3. With this instrument pressures as low as 0.001 dyne cm.$^{-1}$ can be recorded, by measuring the displacement of a fine silk thread, lightly covered with petroleum jelly, which divides the film-covered surface from the clean surface in the trough. Such an apparatus is useful in studying charged monolayers at very large areas[15]. An apparatus based on the same principle has been described by Kalousek, with a more sensitive modification by Tvaroha[16].

Experimental Methods of Spreading Films at the Oil-Water Interface

The surface pressure of a monolayer at the oil-water interface, Π, is $\gamma - \gamma_0$, where γ_0 is the interfacial tension of the clean interface. The experimental techniques of measuring Π for spread interfacial films require special mention, as they are less easy in practice than those for studying films at the air-water interface.

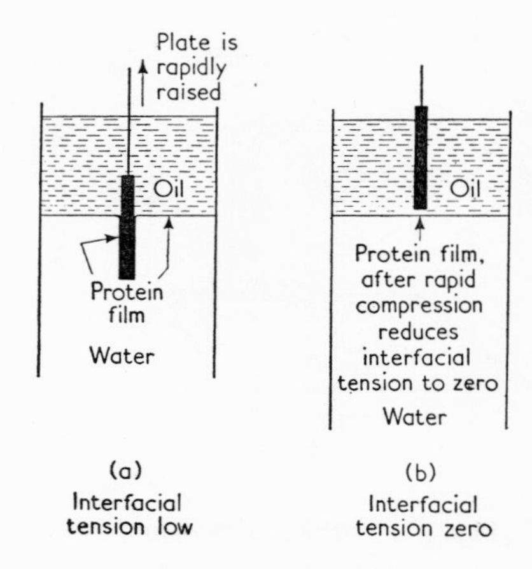

Fig. 5-4. Langmuir's experiment for obtaining zero interfacial tension.

The Langmuir trough, so useful for air-water surfaces, has been modified by Askew and Danielli[17] for films at the oil-water interface; the trough containing one liquid phase is completely immersed in a second liquid phase. The difficulties of maintaining appropriate contact angles and of preventing leakage past the "boom" are considerable, and the method is too time-consuming to be considered really satisfactory.

Langmuir in 1938[18] developed the technique illustrated in Fig. 5-4: an *oleophilic plate* is lowered through an oil-water interface, increasing the area of the interface because the plate carries with it on immersion a layer of oil; the total area is therefore decreased when the plate is raised. In this method Π must be measured separately: the plate in Fig. 5-4 merely alters the area of the interface. It is of some interest to note in passing that Langmuir in one experiment allowed a protein film to adsorb at the interface when the plate was in position (a): on rapidly raising the plate the film was so compressed that Π became equal to γ_0, i.e. γ was zero. This did not lead to spontaneous emulsification, presumably because of the very high interfacial viscosity of the monolayer, but the interfacial tension rose slowly as the protein film crumpled up[18].

The usual method, and the simplest, of studying films at the oil-water interface, is that of *Alexander and Teorell*[19]. This consists in measuring the film pressures set up by the injection into the clean interface of small amounts of a spreading solution of the film-forming material. The success of such a method depends on obtaining a sufficiently "activated" spreading solution, since spreading of the material must, in general, occur against the pressure of that material which has already been spread in the interface. The area available to each molecule in the interface is reduced by successive spreading, and the Π-A curve can thus be obtained as long as more material can be spread.

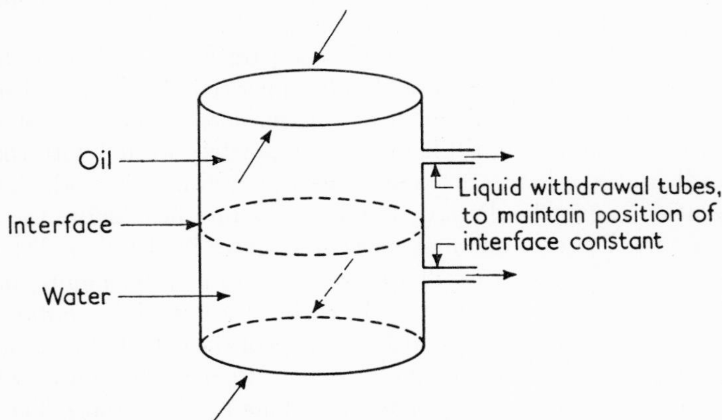

Fig. 5-5. Suggested method of studying insoluble films at the oil-water interface: a plastic vessel is squeezed in the direction of the arrows to compress the film. The position of the interface may be seen by a long-focus microscope above the vessel, and the interfacial tension may be measured by a hanging plate: neither of these devices is shown here.

To overcome the difficulty of spreading additional material against an appreciable surface pressure, one could perhaps use a flexible vessel of inert plastic (without plasticiser, which would spread at the interface), as illustrated in Fig. 5-5. Teflon of 1 mm. thickness might be a suitable material. As the vessel is squeezed in one direction, the area of the interface therein must decrease, and the change of Π could be measured at various surface areas, following a single spreading of the film at the maximum area. To prevent deposition of the film on the walls, the level of the interface should not rise: a constant-level device should therefore be included.

Assuming, however, that we are spreading a film at constant total interfacial area, and decreasing A by successive injections (Alexander and Teorell's method), we must use an "activated" solvent for spreading. Propyl alcohol (either normal or iso-) is often included in the spreading solutions, on account of its high spreading coefficient (Chapter 1), combined with the virtual disappearance after a few minutes of this material (in the amounts used) from the interface by solution in the adjoining bulk phases. Thus proteins spread readily from 0.1% solution in 60% (v.v) propanol in water, 0.5 M with respect to sodium acetate[3]. Other spreading solutions containing propanol or ethanol are mentioned above and in reference 20. As far as possible the spreading solution should not be readily miscible with the bulk phase from which it is injected into the interface, and it should always move towards the interface under the action of gravity[20]. For example, if one is spreading a protein at the benzene-water interface, using the above spreading solution, the tip of the needle of the Agla micrometer syringe should be in the benzene, as close as possible to the interface[13].

All solvents must be distilled in all-glass apparatus, as explained above, to remove possible traces of high-molecular material. The glass dish, conveniently about 12 cm. in diameter and 3 cm. deep[13], must be soaked in a bath of warm chromic acid till it remains completely wetted with the latter on removal from the bath. Since no slides are used in this method, surface contamination cannot be directly swept off the interface before the experiment is begun. Instead, one first places the water in the glass dish, cleans this using an air-jet and a capillary connected to a water-pump, and then pours on a few drops of benzene. If there is appreciable contamination remaining, this will be forced towards the sides of the dish, the benzene drops acting as pistons. The contaminated area is then sucked clean with the capillary, and the benzene drops now spread across the surface. The rest of the benzene (to about 1 cm. depth) is now added, followed by a final sucking with the capillary, held in the interface so that a coarse emulsion of oil and water is being removed. If an oil heavier than water is used, one must rely on more prolonged sucking off of the coarse emulsion.

Experimental Methods of Measuring Π at the Oil-Water Interface

The ring method (Chapter 1), though sometimes used, suffers from the disadvantages already mentioned. More usual and more satisfactory is the use of the Wilhelmy hanging plate: this may hang from one arm of a balance, a mirror being attached to the centre of the beam.

If, for example, one is studying spread films at the benzene-water interface, a glass cover-slip (e.g. $4 \times 3 \times 0.015$ cm.), cleaned in chromic acid, phosphoric acid, and distilled water, and attached to threads by a suitable adhesive, may be suspended through the interface. With this technique[13] and an optical lever of 150 cm., the sensitivity in measuring Π is about 0.01 dyne cm.$^{-1}$ Instead of hanging from the arm of a balance, the plate may be hung from a 10 cm. arm joined to a fine torsion wire at its other end: a calibrated scale attached to one end of the torsion wire gives the force required to restore the plate to its initial condition. In general, however, if the contact angle is zero measured through the water, it is advisable to measure the rise of the plate as Π is increased, converting the deflections of the light spot into dynes cm.$^{-1}$ In this way the contact angle is always a receding one, which facilitates its remaining zero, as does also the use of alkaline solutions[21].

Sometimes, however, the monolayer is such that it alters the contact angle on a glass plate, and it is then necessary to use a hydrophobic and oleophilic plate. For this purpose glass[22] or very thin mica[21,23] may be coated with carbon-black deposited from a flame of burning paraffin wax. This carbon-black adheres strongly, especially if the mica is first roughened with a fine emery paper, and the contact angle of zero (measured now in the oil phase) is less labile in the presence of monolayers. Close to pH 7 either a hydrophobic plate or a hydrophilic plate may be used, with identical results. It is curious that no better material than carbon-black has yet been found: silane derivatives are less satisfactory[21], possibly because they do not impart the granular structure of carbon-black to the surface of the plate: this roughness may assist the oil to wet the plate completely (cf. eq. (1.36)).

Types of Force-Area Curve

Figure 5-6 shows three typical Π-A curves: for stearic acid the curve characterizes a "condensed film", the force remaining very small till the area available to each molecule (i.e. A) becomes very close to 20 Å^2, the so-called "limiting area". When, however, the area is reduced yet a little further, the pressure rises steeply and the film becomes quite solid, the low compressibility indicating that there is now strong repulsion between the chains. In fact 20 Å^2 is close to the cross-sectional area of hydrocarbon chains in the crystalline state, deduced from X-ray studies. At areas appreciably greater than 20 Å^2 per chain, the film consists of rather large "islands" up to several millimetres in diameter, held together by the van der Waals forces of cohesion between the hydrocarbon chains, which are still nearly vertical (Fig. 5-7).

If now we introduce a double bond into the chain, the cohesion is greatly reduced. Thus the force-area curve of oleic acid (Fig. 5-6) shows that this substance gives "expanded", rather than condensed films. In such expanded films the molecules are orientated at high pressures as in Fig. 5-8a, and at low pressures as in Fig. 5-8b.

The presence of electrical charges on the molecules forming the film greatly augments the film pressure. The third curve in Fig. 5-6 shows that the

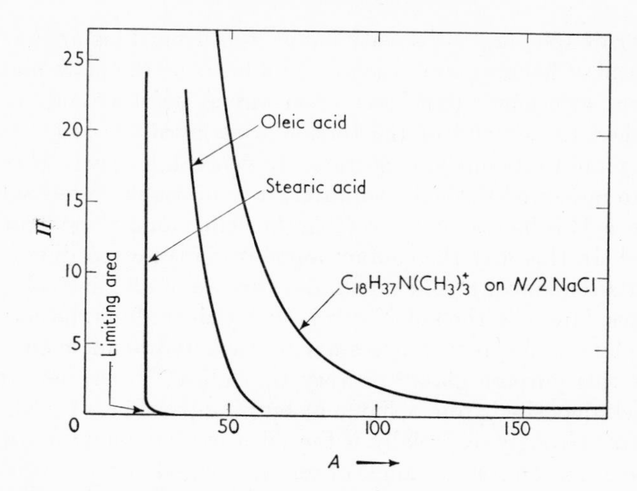

Fig. 5-6. Three typical air-water curves of Π against A, where A is the mean area (in square Ångstroms) available to each molecule on the surface. Stearic acid molecules cohere strongly in the surface, and Π rises only when this "condensed" film is compressed till A is reduced to approximately the cross-sectional area of each hydrocarbon chain. Oleic acid gives an "expanded" film—there is but little cohesion and so Π is greater than for stearic acid at any value of A. The "gaseous" film of $C_{18}H_{37}N(CH_3)^+$ is electrically charged, however, and these long-chain ions all tend to repel each other, so that Π is relatively great at all points in the diagram.

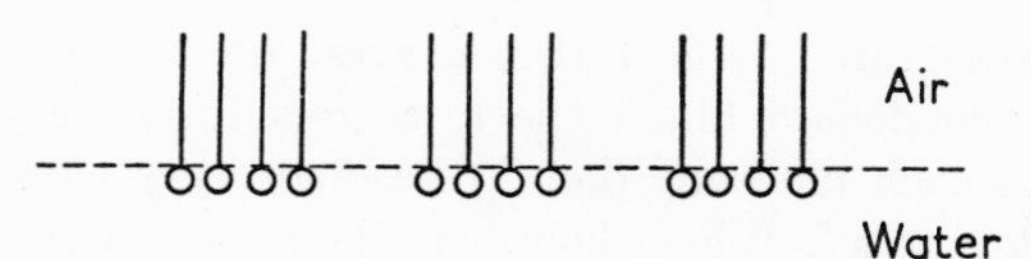

Fig. 5-7. "Islands" of stearic acid molecules, often several millimeters in diameter, float on the water surface. The strong cohesion between the hydrocarbon chains is responsible for this.

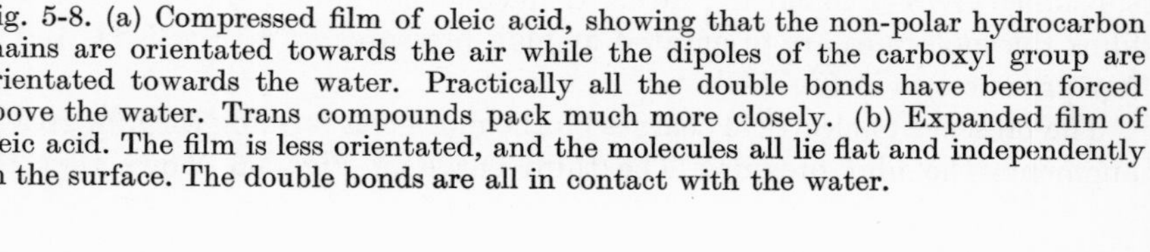

Air
Water
(a)

Air
Water
(b)

Fig. 5-8. (a) Compressed film of oleic acid, showing that the non-polar hydrocarbon chains are orientated towards the air while the dipoles of the carboxyl group are orientated towards the water. Practically all the double bonds have been forced above the water. Trans compounds pack much more closely. (b) Expanded film of oleic acid. The film is less orientated, and the molecules all lie flat and independently on the surface. The double bonds are all in contact with the water.

pressure exerted by an ionized film is very high compared with that of an un-ionized monolayer, due, as shown in Chapter 2, to the internal repulsion within the surface film. Such films are classified as "gaseous".

Films are generally more expanded at the oil-water than at the air-water interface, the area being greater for a given film pressure (Fig. 5-9). As at the air-water surface, an electrical charge on the molecules forming the film greatly augments the film pressure. Adsorbed monolayers may be studied quantitatively by assuming the Gibbs equation.

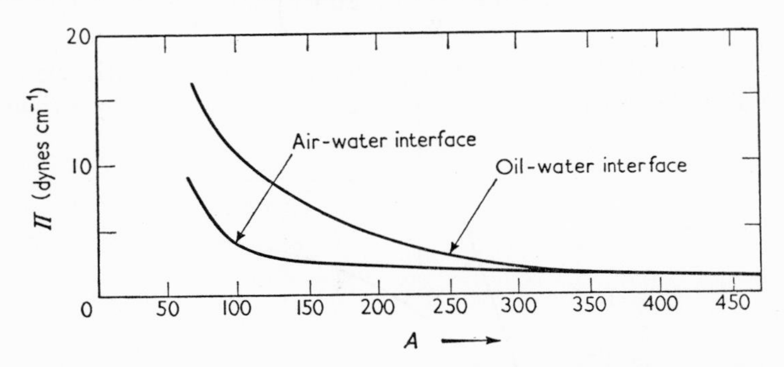

Fig. 5-9. The film pressure Π is usually greater, for any value of A, at the oil-water interface than at the air-water surface. The reason is that the molecules of oil penetrate between the hydrocarbon chains and remove all inter-chain attraction. The film here is $C_{16}H_{33}N(CH_3)_3^+$, and A is in square Ångstroms per long chain.

"Gaseous" Films

The equation of the Π-A curve is sometimes termed the surface equation of state, by analogy with the P-V curve for a gas. It also resembles the curve of osmotic pressure against concentration of a solution. The simplest molecular equation of state is:

$$\Pi A = \boldsymbol{k}T \tag{5.1}$$

where Π is the calculated film pressure allowing only for kinetic movement in the surface film, which is restricted to take place in two dimensions only with kinetic energy $\frac{1}{2}\,\boldsymbol{k}T$ each. This relation assumes that there is neither cohesion nor repulsion between the molecules in the film, and it is in practice, therefore, a limiting law, true as $\Pi \rightarrow 0$ (or as $n \rightarrow 0$, i.e. as $A \rightarrow \infty$). It is illustrated in Fig. 5-10. In equation (5.1) if A is in Å^2 per molecule, and Π is in dynes cm.$^{-1}$, $\boldsymbol{k}T$ has the value of 405 at 20°C, and 412 at 25°C. If the molecular weight is not known, we may deduce it by writing instead of equation (5.1) the form applicable to N moles.

$$\Pi A = NRT \tag{5.2}$$

where, at 25°C, the term NRT becomes 2478×10^7 N ergs. If, experimentally, we find the limit y to which the product ΠA tends as Π tends to zero (Fig. 5-22), then $y = 2478 \times 10^7$ N ergs. The molecular weight M may now be simply calculated since, if we work with 1 mg. of material, $M = 10^{-3}/N$. If, further, A is expressed in square metres of surface occupied by the mg. of material, we have

$$\left.\begin{aligned} M &= \frac{2478}{y} \text{ (at 25°C)} \\ \text{or } M &= \frac{2438}{y} \text{ (at 20°C)} \end{aligned}\right\} \tag{5.3}$$

We may note that according to equation (5.1), at sufficiently high values of Π, A should approach to zero. However, for a straight hydrocarbon chain

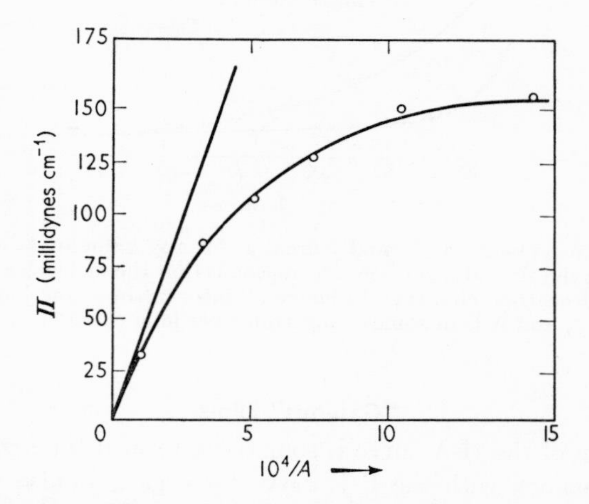

Fig. 5-10. Plot of Π against $10^4/A$ for myristic acid on 10^{-2} N-HCl at 17°C. The straight line represents the equation $\Pi A = \boldsymbol{k}T$.

the "limiting area" (Fig. 5-6) is about 19.5-20 Å, and allowing for this, still assuming that the monolayer is completely mobile, Langmuir obtained[25] for the kinetic energy of the molecules in the "gaseous" film:

$$\Pi(A - A_0) = \Pi_k(A - A_0) = \boldsymbol{k}T \tag{5.4}$$

where Π_k is the calculated pressure due to the kinetic energy of a molecule in the film, and where A_0 is the area actually occupied by the molecules in the surface. When Π is small, $\Pi A \rightarrow \boldsymbol{k}T$ as before, but as Π becomes very great, $A \rightarrow A_0$. Equation (5.4) is often obeyed by electrically neutral films at the oil-water interface, where there is no inter-chain cohesion[23] (Figs. 5-11, 5-12).

Instead of assuming that the monolayer is completely mobile, one may

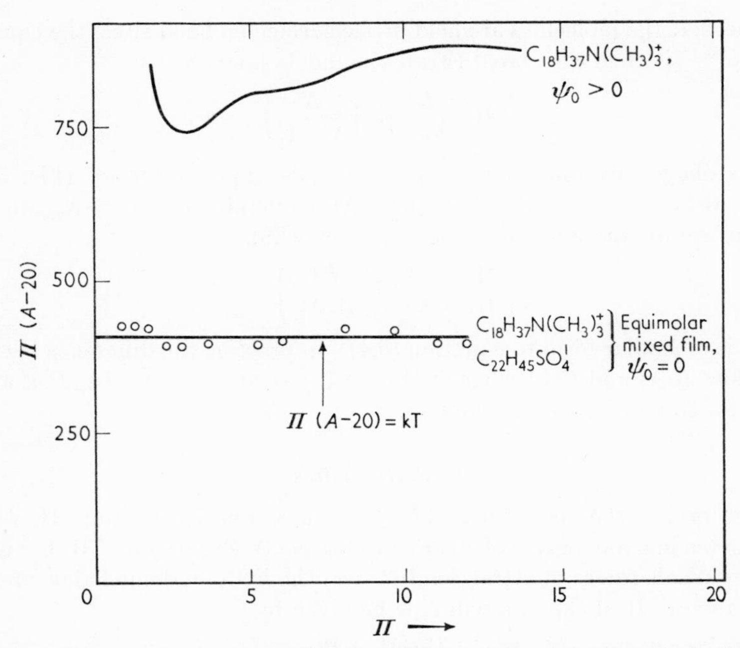

Fig. 5-11. Characteristics of films spread at the interface between petrol-ether and 0.01 N-HCl at 20°C. The mixed film, which has no net charge, obeys eq. (5.4)[23].

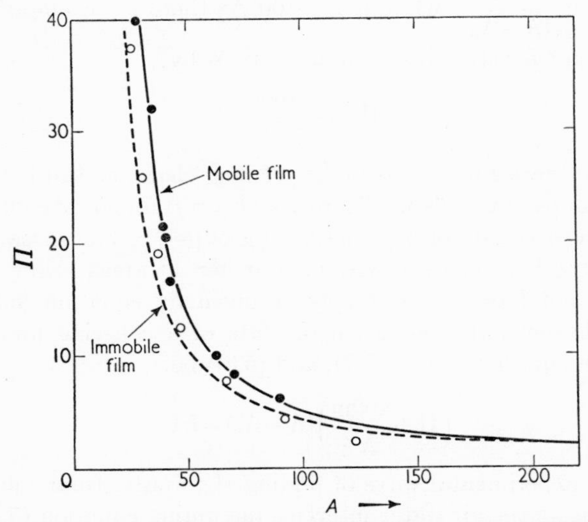

Fig. 5-12. Plots of eqs. (5.4) and (5.5). Filled circles are experimental points for films of butyric acid adsorbed at the oil-water interface[28], and open circles for this material adsorbed at the interface between paraffin wax and water[6].

assume that the molecules are held in the surface on fixed sites; the equation of state[26,27] is then calculated (Frumkin and Volmer) to be:

$$\Pi=\left(\frac{\boldsymbol{k}T}{A_0}\right)\ln\left(\frac{A}{A-A_0}\right) \tag{5.5}$$

This is obeyed by films adsorbed at the solid-liquid interface (Fig. 5-12). Again, at high A (or low Π), $\Pi A \rightarrow \boldsymbol{k}T$. At moderate areas ($A \gg A_0$) we may use the approximate expansion of equation (5.5).

$$\left.\begin{array}{r}\Pi(A-A_0/2)=\boldsymbol{k}T\\ \text{or } \Pi A=\boldsymbol{k}T+\tfrac{1}{2}\Pi A_0\end{array}\right\} \tag{5.6}$$

which is of similar form to equation (5.4). In practice the difference between equations (5.4) and (5.5) is small (Fig. 5-12) at areas above $4A_0$ (i.e. above 80 Å^2 for single-chain compounds)[28].

Cohering Films

In general at the air-water surface there is a cohesive pressure Π_s within the film (as in a monolayer of stearic or oleic acid). The origin of Π_s lies in the van der Waals forces of attraction between the hydrocarbon chains, and the total pressure Π of the film will now be given by

$$\Pi=\Pi_k+\Pi_s \tag{5.7}$$

where we may expect Π_s to be negative, as in Fig. 5-13. It is clear from eq. (5.4) that Π_k is $\dfrac{\boldsymbol{k}T}{(A-A_0)}$, while if $A>100$ Å^2 there is an empirical finding that Π_s is given[29] for straight-chain derivatives by:

$$\Pi_s=\frac{-400m}{A^{3/2}} \tag{5.8}$$

where Π_s is in dynes cm.$^{-1}$, A is in Å^2 per long-chain, and m is the number of $-CH_2-$ groups in the chain. There are also certain specific effects due to cohesion between the dipoles of the head-groups[29-32], but these are not as important as the inter-chain cohesion. Further, at areas below 100 Å^2 the cohesion does not increase as steeply as given by equation (5.8). Within these limits, however, the equation of state of a cohering film must (by combination of equations (5.4), (5.7), and (5.8)) be:

$$\left(\Pi+\frac{400m}{A^{3/2}}\right)(A-A_0)=\boldsymbol{k}T \tag{5.9}$$

There are two experimental ways of finding Π_s: firstly, from equation (5.7) applied to films at the air-water interface (assuming equation (5.4) for Π_k). Secondly, Π_s may be determined using equation (5.7), with Π_k following from a comparison of data at the air-water and oil-water interfaces (at the latter Π_s is always zero). In the latter method, either neutral or charged films may be used since the repulsion Π_r in these is the same at the air-water and oil-water interfaces. For the air-water surface equation (5.7) becomes:

$$\Pi_{A/W} = \Pi_k + \Pi_s + \Pi_r \tag{5.10}$$

while at the oil-water interface the equivalent films would obey:

$$\Pi_{O/W} = \Pi_k + \Pi_r \tag{5.10a}$$

so that from the difference between these equations Π_s may be obtained. The result for octadecyltrimethylammonium ions (Fig. 5-13) is in accord with eq. (5.8) for neutral films.

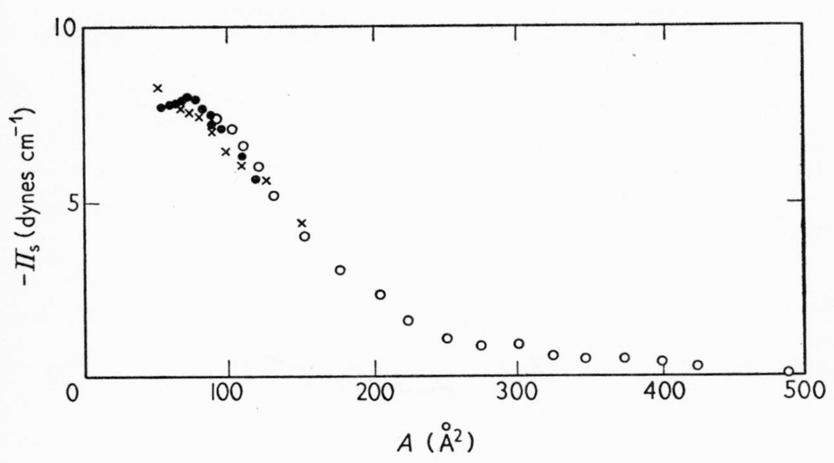

Fig. 5-13. Cohesive pressure ($-\Pi_S$) for $C_{18}H_{37}N(CH_3)_3^+$ ions[29].

Charged Films

If the film bears a net electrical charge, this will increase Π at any given area A per long-chain ion (Fig. 5-6). Quantitatively this repulsive energy is given by the Davies equation

$$\Pi_r = 6.1\, c_i^{1/2} \left\{ \cosh \sinh^{-1} \left(\frac{134}{A c_i^{1/2}} \right) - 1 \right\} \tag{2.51}$$

at 20°C, or as a simple approximation if $Ac_i^{1/2} < 38$, by

$$\Pi_r = \frac{2kT}{A} - 6.1\, c_i^{1/2} \tag{2.52}$$

This repulsive pressure must be substituted in equation (5.10) to obtain the total pressure in a charged film at the air-water interface. Multiplication of both sides by $(A-A_0)$ and substitution and approximation for Π_r and Π_s from eqs. (2.52) and (5.8) gives:

$$\Pi(A-A_0)=3kT-\frac{400m(A-A_0)}{A^{3/2}}-6.1\,c_i^{1/2}(A-A_0)-\frac{2kTA_0}{A} \tag{5.11}$$

This is the complete (but approximate) surface equation of state for charged monolayers at the air-water surface. At the oil-water surface the cohesive term in $A^{-3/2}$ is to be omitted. For a more precise value of Π_s we may use equation (2.51) for Π_r. The equations of state for both spread and adsorbed films are tested in Figs. 5-14 and 5-15. Data for films on salt of different

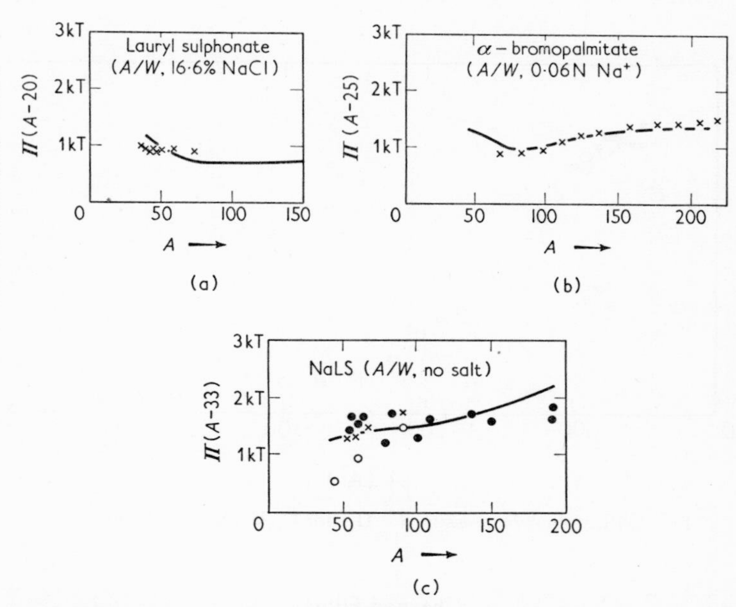

Fig. 5-14. Test of theory (full lines in (a) and (b)) for spread films[33,34]. In (c) are shown data for lauryl sulphate adsorbing at the air-water interface in the absence of inorganic salt[32,35-7].

concentrations are also available[29]. For the charged adsorbed films there is always uncertainty about the value of A_0 to be used[28,29,32]: the persistence of electrical repulsion even at low pressures makes graphical estimation of A_0 virtually impossible. It is widely accepted, however, that A_0 is increased in highly soluble films (m or c_i small), being about 30 Å^2 for adsorbed films of sodium lauryl sulphate. It is therefore of interest to make measurements at high areas where A_0 can be neglected in comparison with A. In Fig. 5-16 such results are compared with theory as c_i is varied.

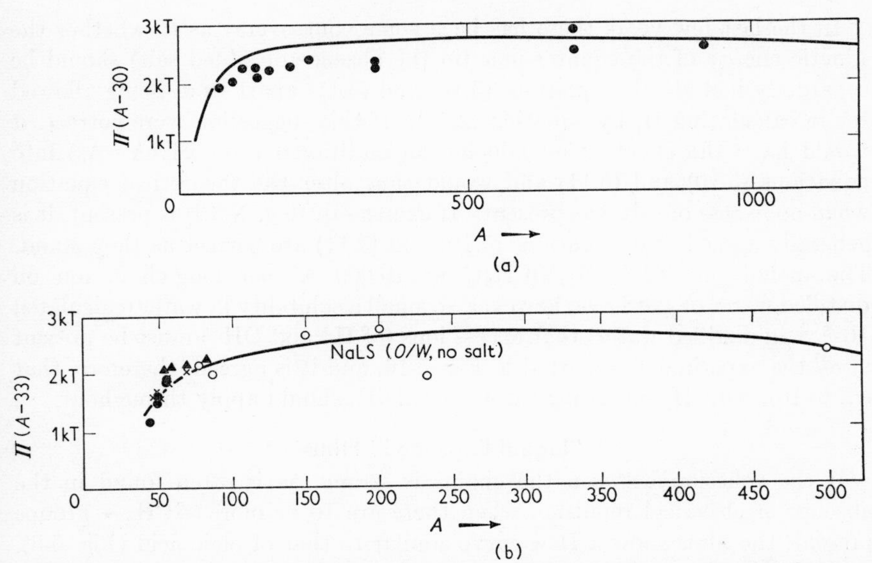

Fig. 5-15. Equation of state for films at the oil-water interface. In (a) the behaviour of films of $C_{26}H_{53}N(CH_3)_3^+$ spread on 10^{-3} N-NaCl is shown, the full line being calculated[38] from eq. (5.11) omitting the cohesive term. In (b) data for adsorbed films of sodium lauryl sulphate are compared with eq. (5.11) (full line), again omitting the cohesive term[29,32,39-41].

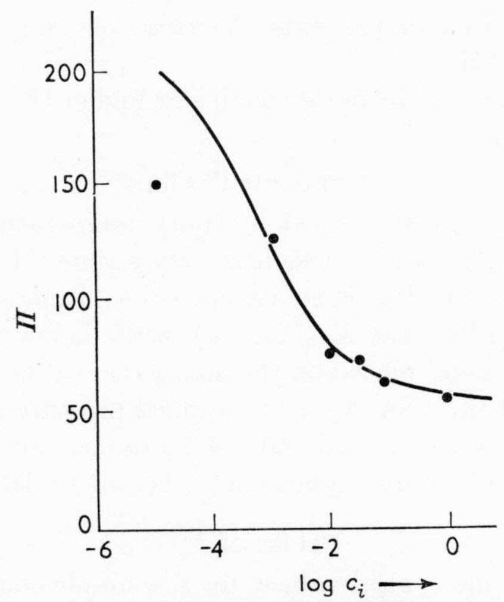

Fig. 5-16. Plot of Π in millidynes cm.$^{-1}$ against log c_i for spread monolayers of $C_{18}H_{37}N(CH_3)_3^+$ at the air-water surface. Points are taken from experimental curves[42], while the line is theoretical, calculated from eqs. (5.8), (5.10), and (2.51). The concentration c_i refers to added KCl, and A is constant at 5,000 Å^2 per long-chain ion.

In the last few years there has been some controversy as to whether the kinetic energy of the counter-ions (in the absence of added salt) should be separately included in equations (5.10) and (5.11) apart from being allowed for in calculating Π_r by equation (2.51). If this suggestion were correct, it would have the effect of introducing an additional term $\boldsymbol{k}T/(A-A_0)$ into equations (5.10) and (5.11) and would thus alter the theoretical equation when no excess of salt was present. If excess salt (e.g. NaCl) is present, it is generally agreed that equations (5.10) and (5.11) are correct as they stand. The monolayers of $C_{18}H_{37}N(CH_3)_3^+$ at 10,000 $Å^2$ per long-chain ion on distilled water or salt have, however, so small a solubility in water (calculated 10^{-8} N in distilled water) that excess ions (of H^+ and OH^-) must be present in *all* the experiments reported in Fig. 5-16, and it is agreed, therefore, that eq. (5.10), with Π_r substituted from eq. (2.51), should apply throughout.

"Liquid Expanded" Films

If the cohesion between the chains is strong, as is often found in the absence of electrical repulsion when there are 10 or more $-CH_2-$ groups present, the films show a Π-A curve similar to that of oleic acid (Fig. 5-6). In such films the cohesion is approximately constant over a considerable range of areas[25,43], and the equation of state is therefore

$$(\Pi - \Pi_s)(A - A_0) = \boldsymbol{k}T \tag{5.12}$$

this following from eqs. (5.4) and (5.7). For the C_{10} hydrocarbon chain Π_s is about -10 dynes cm.$^{-1}$, while for the C_{22} chain it is -22 dynes cm.$^{-1}$ Unsaturation of the chain prevents the close packing of the chains, and reduces the value of Π_s.

At greater areas and with shorter chains or higher temperatures, the films become "gaseous"[43].

"Condensed" Films

Materials such as stearic acid, at ordinary temperatures, give curves as shown in Fig. 5-6. Here the cohesion is very strong, Π_s being of the order -50 or -100 dynes cm.$^{-1}$ This cohesion causes "islands" to form at areas greater than the limiting area A_0 (Fig. 5-6), while below this area the film is so solid that talc particles placed on the monolayer cannot readily be moved. On the steep part of the Π-A curve the applied pressure (measured in dynes cm.$^{-2}$) is very high: a surface pressure of 20 dynes cm.$^{-1}$ applied to a film 20 Å thick corresponds to 10^8 dynes cm.$^{-2}$, i.e. about 100 atm.

Values of A_0

Table 5-I shows some typical values: for the simple compounds first listed the values correspond to the cross-sectional areas of the orientated molecules, though the di-esters quoted later in the table are anchored flat on the surface by the polar groups.

TABLE 5-I

Compounds	A_0 (in Å^2)
Straight-chain acids on water[43]	20.5
Straight-chain acids on dilute HCl[43]	25.1
Esters of saturated acids[43]	22.0
Primary n-alcohols[43]	21.6
Long-chain p-phenols[43]	24.0
Cholesterol[43]	40.8
Lecithins[43,44]	about 50
Dilauryl oxalate[45]	90
Didecyl adipate[45]	120
Dioctyl sebacate[45]	150
Dibutyl thapsate[45]	190
Various proteins[46]	about 1 m.^{2}mg.$^{-1}$

Other Films

Bile acids as well as cholesterol can be spread at the air-water surface, generally giving rather condensed films[47], as do also long-chain phosphates[48]. The alkyl esters of sucrose are strongly adsorbed[49], and sucrose monolaurate gives monolayers of very low viscosity (see below), indicating rather non-coherent films.

Perfluoro-derivatives at the air-water surface give rather "gaseous" films, even at high surface pressure, suggesting that the cohesion in such films is low[50,51,52]. On a Teflon trough the fluorocarbon derivatives may be studied as spread monolayers on hexadecane, white oil, or tricresylphosphate[53]. On various oils siloxanes also spread well to give insoluble films[53-55], a property related to their ability to prevent foaming of oils by displacing a foam-stabilizing film from the oil-air interface. Siloxanes will also spread on water[56], the spreading pressure being 10-17 dynes cm.$^{-1}$

On solutions of different salts[57] stearic acid films show systematic changes of molecular packing with changes of pH. Even at pH 4.5 the presence of small amounts of Cu^{++} reduces A_0 from about 26.5 Å^2 to 24 Å^2 (for 0.5mM. Cu^{++}), and to 22.5 Å^2 on 5mM. solution. This is related to the tendency to form soaps of the long-chain acid.

Molecular Complexes

At the air-water surface, sodium cetyl sulphate and sodium cetyl sulphonate both form 1:1 complexes with cetyl alcohol, the free energy of mixing to form these being -300 cal. mole^{-1} (Fig. 5-17a)[58]. A similar complex is formed from $C_{14}H_{29}NH_3Cl$ and tetradecanol (Fig. 5-17b). This energy of mixing of the monolayers of ionic and non-ionic molecules to form these complexes may be evaluated using insoluble films, spread on concentrated ammonium sulphate solutions. The equation, similar to (4.5), gives the energy of mixing at constant surface pressure Π:

$$\Delta F_{\text{mixing}} = \int_0^{\Pi} A_{12}\, d\Pi - x_1 \int_0^{\Pi} A_1\, d\Pi - x_2 \int_0^{\Pi} A_2\, d\Pi$$

where x_1 and x_2 are the mole fractions of the two components constituting the film, whose molecules before mixing occupy areas A_1 and A_2 respectively, and area A_{12} in the mixed monolayer. The energy of mixing is small compared with the total cohesive energy of a monolayer (300 cal. mole^{-1} compared with

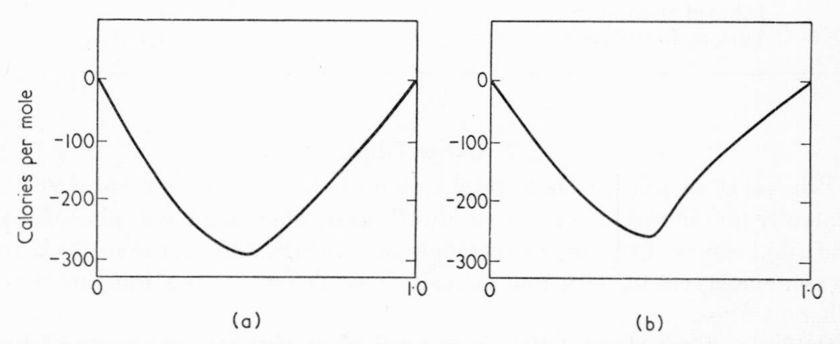

Fig. 5-17. Energy of complex formation of monolayers of (a) Sodium cetyl sulphate against mole fraction of cetyl alcohol, and of (b) $C_{14}H_{29}NH_3Cl$ against mole fraction of tetradecanol. Both curves refer to monolayers at the air-water surface at 32 dynes cm.$^{-1}$, at 25°C, spread on concentrated ammonium sulphate[58].

about 1600 cal. mole^{-1}): the complexing is equivalent to increasing the chain length of the components by about 3 $-CH_3-$ groups. The origin of the energy of complex formation may lie either in hydrogen-bonding between the hydroxyl groups of the long-chain alcohol and the charged groups, or in the screening of the latter from the repulsion of their neighbours. The lack of specificity of the ionic head-groups and the negligibly small entropy of complex formation (from the temperature coefficient of the energy) suggest that no new hydrogen bonds are formed in the complex: screening of the head-groups is therefore apparently responsible, though Dervichian[59] prefers the explanation of hydrocarbon-chain interactions. Certainly it is difficult to detect complexing with fewer than 8 carbon atoms in the chains. Possibly experiments on complexing at the oil-water interface would be able to distinguish these views: here there are no inter-chain forces, and complexing could readily be detected from changes in interfacial viscosity[60]. Present evidence supports the view that complexing does occur at the oil-water interface, but further experiments are required before the issue can be regarded as decided[60,61].

Partial Ionization of Surface Films

The process of ionization of a monolayer can be traced[62-64] by the change of ΔV, the contact potential (Chapter 2). It must be remembered, however, that once the film bears a net charge, the pH at the surface is no longer equal to that in the bulk (cf. eq. (2.53)). If one corrects for this, one generally finds that half-ionization occurs at a pH close to that expected for the

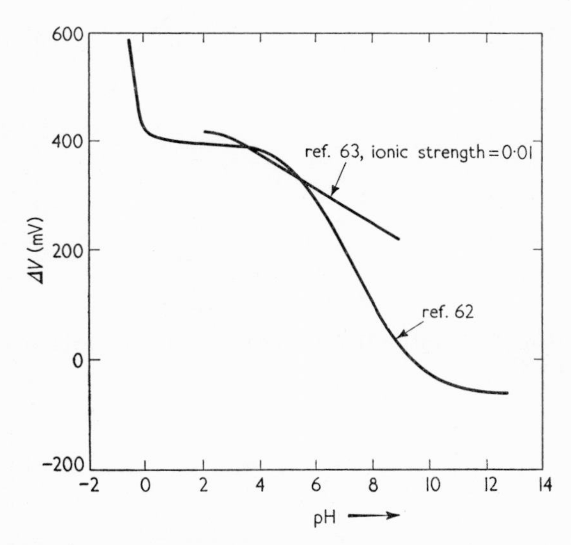

Fig. 5-18. Variation of ΔV with pH for carboxylic acid monolayers, at room temperature, A being 20 Å².

ionogenic group concerned. Figs. 5-18 and 5-19 show typical curves for carboxylic acids and for amines: for the former the bulk pH at half-ionization is about 7.5 units, the corresponding surface value being about 5.5 units. The forms of the curves corresponding to the partly ionized films in Figs. 5-18 and 5-19 vary markedly with ionic strength and the presence of polyvalent ions, even in very small quantities.

As a film becomes ionized, it also becomes more soluble (eq. (4.33)): this may be guarded against by using very long hydrocarbon chains. Films of $C_{21}H_{43}COOH$ would be insoluble under all conditions, and spread monolayers of $C_{22}H_{45}SO_4'$ and of $C_{26}H_{53}N(CH_3)_3^+$ are suitable for studying completely ionized films[23,38]

Monolayers of long-chain alcohols and ethers show slight charge effects at very high and very low pH values (Fig. 5-20). These changes reflect specific adsorption of OH′ (or CO_3'') and H^+ respectively, and could be interpreted by eq. (2.44).

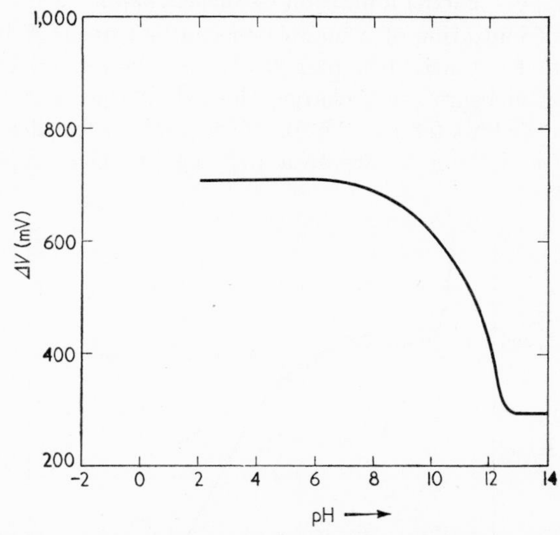

Fig. 5-19. Variation of ΔV with pH for monolayers of long-chain amine[63], at room temperature and with A = 25 Å².

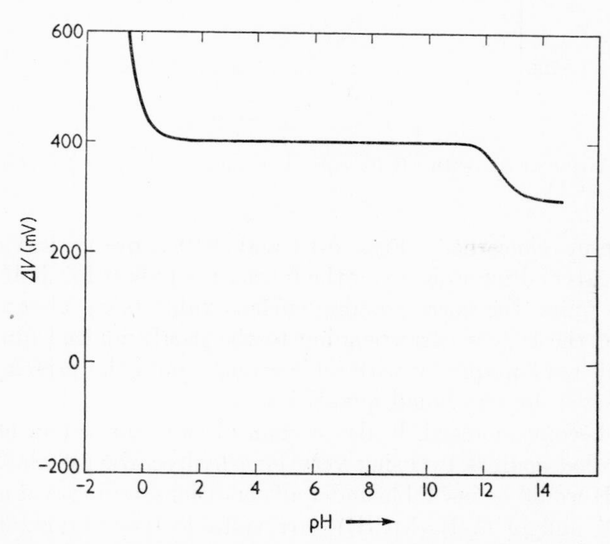

Fig. 5-20. Variation of ΔV with pH for monolayer of tetradecanol[62], with A = 20 Å².

Equilibrium in Films

We have noted that insoluble films at the air-water surface may be so coherent that, on expansion of the available area, discrete islands are left on the water surface. These islands may be either solid or liquid and are in equilibrium with a two-dimensional "gaseous" film. On raising the temperature the coherent films can undergo expansion, a phenomenon first noted by Labrouste[65]. A solid, coherent, condensed film can, on raising the temperature, undergo two-dimensional "melting" to a liquid expanded film.

There is now little doubt that, in condensed films of the aliphatic materials, the hydrocarbon chains are arranged parallel to one another as in the solid fatty acids, while, in the liquid expanded form, they are coiled and interlocked with one another to form a liquid hydrocarbon layer, below which the polar heads are immersed in the aqueous phase. It was this conception that led Langmuir[25] to term such liquid expanded films "duplex": the equation of state for these has already been given (5.12).

Both condensed and expanded films, being coherent, are unable to cover the complete surface at sufficiently large expansions, but exist as islands, which are in dynamic equilibrium with a two-dimensional "vapour" film. These surface equilibrium vapour pressures are very low, as seen in the following data of Adam and Jessop (quoted in ref. 43).

Acid	Surface vapour pressure at 14.5°C (dynes $cm.^{-1}$)
Tridecylic	0.30
Myristic	0.19
Pentadecylic	0.11
Palmitic	0.04

Though these pressures seem very low, when expressed in dynes $cm.^{-2}$ they are of the order of several atmospheres pressure in a bulk phase, not much smaller than the critical pressures of bulk liquids. The transition of monolayers from one state to another can be effected by alteration of one of the three independent variables: temperature, the two-dimensional pressure, and the extent of interaction of the polar head with the substrate, e.g. by changing the pH under a monolayer of a fatty acid.

The experimental force-area curves suggest that the separate states of matter in monolayers are to be regarded as separate phases which can come to equilibrium with one another and to which the phase rule is applicable. If the surface film is a one-component system, the surface-phase rule deduced by Crisp[66] shows that there must be an invariant point when two surface phases are present. We note that the transitions vapour-liquid or vapour-solid are such phase changes. They are called transitions of the first kind. Many Π-A curves, however, show discontinuities which indicate transitions of the second kind, i.e. extending over a range of temperatures, and transitions of even higher orders are found. These transitions of higher order have been detected by discontinuities in the compressibilities, surface potentials and surface viscosities[67-69]. In general, a progressive increase in the degree of freedom permitted in the hydrocarbon chains, on lowering the pressure on a condensed monolayer, may be said to be responsible for these higher order transitions.

The concomitant thermodynamic quantities in these phase changes can be calculated with the aid of an equation of the type of the Clapeyron-Clausius relation. For the first three acids listed above one obtains heats of "surface evaporation" of 2000, 3200 and 9500 cal. mole^{-1} respectively.

Similarly one may calculate the latent heat of spreading of solid palmitic acid from the temperature coefficient of the spreading pressure[70,71]: it is 8 K.cal.mole^{-1}. For spreading from a liquid lens the figure is -4.9 K.cal. mole^{-1}: the difference represents the heat of fusion of palmitic acid. The entropy changes are respectively $+28.9$ e.u. and -9.5 e.u., showing that, whereas the monolayer is disorientated relative to the solid, it is more orientated than in the bulk liquid palmitic acid. Results for cetyl alcohol are, respectively, -0.29 K.cal.mole^{-1} and -8.88 K.cal.mole^{-1}, and for the entropy changes $+3.6$ e.u. and -23 e.u.

It is of considerable interest that the entropy changes for the spreading of the acids suggest that the films formed behave as if immobile: the entropy of the film can be accounted for by rotation of the molecules about their long axes, but not by two-dimensional translation[71].

Films of Polymers

Many polymers, natural and synthetic, spread readily at surfaces[72]. The films usually show cohesion, either intermolecular or intramolecular, or sometimes both. A typical force-area curve for a protein is shown in Fig. 5-21, the value of Π at any given area usually being greater at the oil-water interface than at the air-water surface. The value of Π depends on the molecular weight, the intramolecular forces, and the intermolecular forces. Usually the films of polymers of high molecular weight are very resistant to desorption: the many hydrocarbon residues orientated away from the water make the total energy of desorption W in eq. (4.33) very high. Only when the total electrical force also becomes very high (as in polyacrylic acid on alkaline substrates) are the polymer monolayers soluble.

The **molecular weight** can be obtained from the intercept (by extrapolation) on the ΠA axis of the ΠA vs. Π plot, if intermolecular forces are not important in the range of areas under investigation. If this condition is satisfied, one should be able to apply equation (5.4) in the form

$$\Pi A = \boldsymbol{k}T + \Pi A_0 \tag{5.13}$$

This predicts a linear plot (as in Fig. 5-22) if A_0 is constant, from the intercept of which M is found by equation (5.3). Bull[73] has obtained the molecular weights of many proteins, *spread at the air-water surface*, in this way; very low pressure is often unnecessary, since the forces are largely intramolecular,

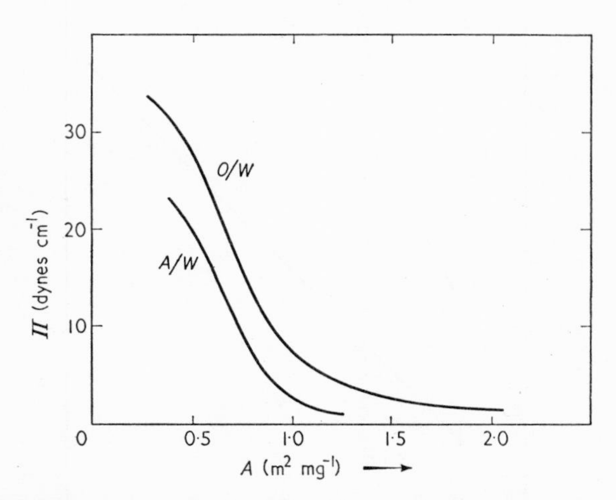

Fig. 5-21. Force-area curves for haemoglobin monolayers spread on 0.1 N-aqueous HCl, at the interfaces with air and with cyclohexane[22].

and the intermolecular effects are small. In particular, the non-polar side-chains of a protein molecule are folded inwards towards the centre of each spread macromolecule: these are therefore not "sticky" in the surface. The intermolecular repulsion corresponds only to the "edge" effect of an electrical condenser (constituted by the polymer molecule in the surface and the counter-ions below).

Guastalla[74] has studied various proteins with his most sensitive instrument, with results shown in Fig. 5-22. As seen from comparison with the results for poly-1:1:2-lysine-glutamic-acid-leucine[21] from Fig. 5-23, the measurements must be made at very low surface pressures when the film bears a large net charge, as when spread on acid; electrical repulsion causes deviations from linearity at higher pressures.

Poly-alanine and poly-leucine, having no intermolecular repulsion, are very coherent in monolayers[76]: sufficiently low pressures cannot be reached to study the molecular weights.

For proteins the molecular weights are usually the same in monolayers and in bulk solutions[72-75]: this is true of gliadin, ovalbumin, serum albumin, β-lactoglobulin, and pepsin. A detailed table is given by Cheesman and Davies[72]. For haemoglobin, the molecular weight is found to be near the normal value of 66,000 at pH 6.8, except on strong salt solutions, when the surface film gives a molecular weight of about half this figure[74,75,77,78]; this dissociation is rather similar to the phenomenon of molecular dissociation

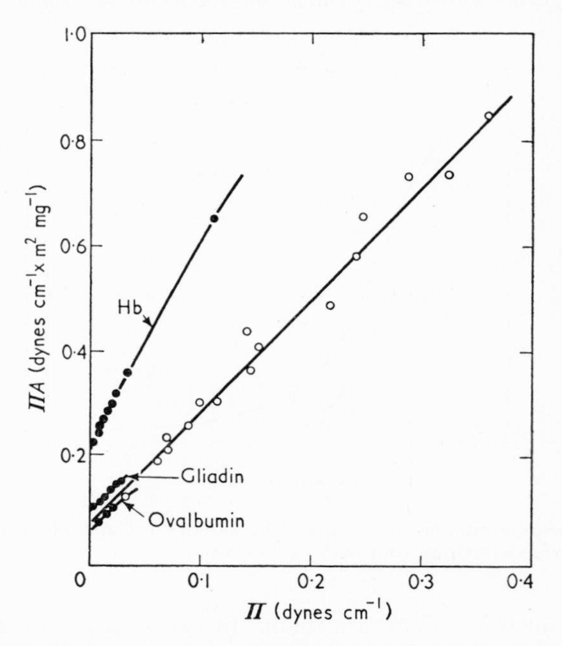

Fig. 5-22. Plot of ΠA against Π for synthetic copolymer of lysine, glutamic-acid, and leucine (open circles)[21], and data of Guastalla[74] for haemoglobin, gliadin, and ovalbumin. All films are spread at the air-water surface, on 0.01 N-HCl.

into two sub-units in bulk solutions of urea. On 0.01 N acid the monolayer gives a molecular weight of only about 11,000, suggesting more extensive dissociation[74].

At the oil-water interface the molecular weight can be determined in the same way provided that sufficiently low pressures can be attained to permit extrapolation. As shown in Fig. 5-24, there is marked curvature (due to high molecular flexibility) in the plot of ΠA vs. Π for the amino-acid copolymer[21], even at pressures below 0.1 dyne cm.$^{-1}$, and extrapolation is therefore difficult. However, the intercept y is approximately 0.07, whence, using the relation $M = 2438/y$, the molecular weight is calculated to be the accepted value of about 35,000. For proteins, where the molecular weights may be higher than this, and the intercepts correspondingly smaller, the molecular weight cannot be found from interfacial film studies. Fig. 5-25 shows the data of Cheesman[13] for films of human methaemoglobin spread at the oil-water interface (filled circles): the curvature below 0.1 dyne cm.$^{-1}$ is expected theoretically (see below) if the molecular weight is normal at 66,000. Clearly

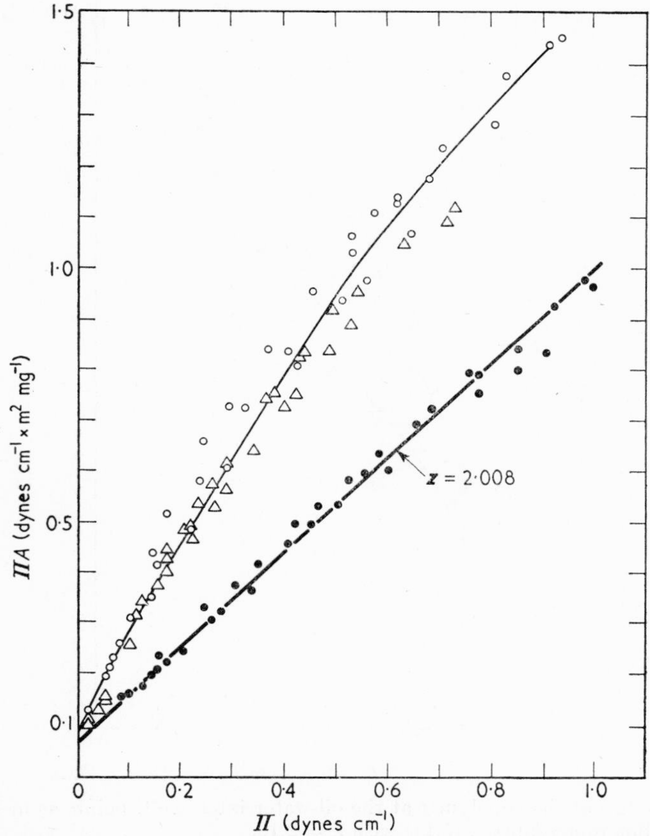

Fig. 5-23. Plots of ΠA against Π for poly 1:1:2-lysine-glutamic acid-leucine (isoelectric point about 7) at the air-water surface. ●, 10^{-2} N-NaCl (pH 7); ○, 10^{-2} N-HCl; Δ, 10^{-2} N-NaOH. The heavy line represents eq. (5.14) with $\mathbf{z} = 2.008$[21].

it would be necessary to work at very much lower surface pressures to obtain accurate molecular weights.

Intramolecular forces determine the shape of the polymer molecule in the surface film. If there is strong cohesion the polymer is coiled up in the plane of the surface, the molecules lying like pennies on the surface, as in the first diagram of Fig. 5-26. The pressure at any given area is then considerably less than if the molecules lie like random chains, as in the last diagram of Fig. 5-26. The treatment of Singer[79] allows for these effects quantitatively by assuming that adsorption of each monomer residue occurs on a fixed site on

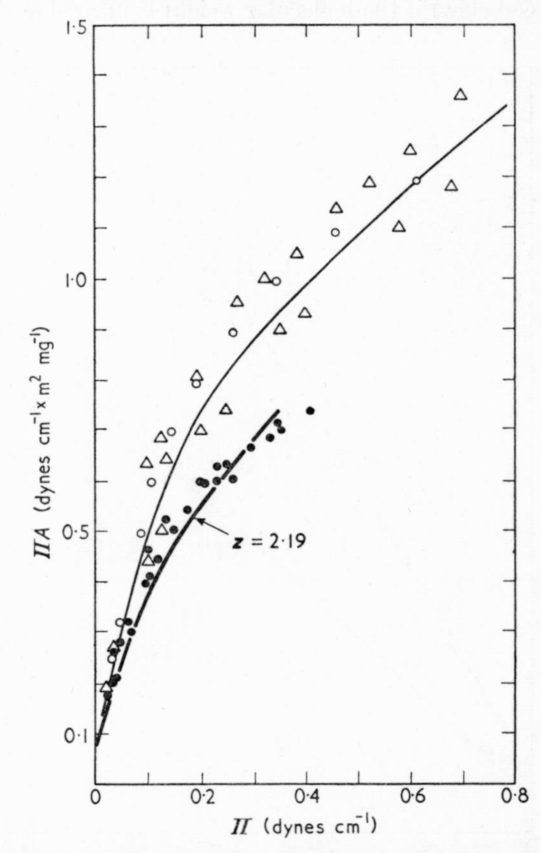

Fig. 5-24. Films of the copolymer at the oil-water interface[21]: points as in Fig. 5.23. The heavy line represents eq. (5.14) with z = 2.19.

the surface: the degree of unfolding is then reflected in **z**, the co-ordination number of the residues in the two-dimensional quasi-lattice in the surface. For a rigid, coiled molecule (with strong cohesion) $\mathbf{z}=2$.

Using the high-polymer theory of Huggins[80], Singer derived the equation

$$\Pi=\frac{\mathbf{x}k\mathrm{T}}{\mathrm{A}_0}\left[\ln\left(\frac{\mathrm{A}}{\mathrm{A}-\mathrm{A}_0}\right)+\left(\frac{\mathbf{x}-1}{\mathbf{x}}\right)\frac{\mathbf{z}}{2}\ln\left(1-\frac{2\mathrm{A}_0}{\mathbf{z}\mathrm{A}}\right)\right] \tag{5.14}$$

where A_0 is the actual molecular area ("limiting area") and $\mathbf{x}$ is the degree of polymerization. If either $\mathbf{x}=1$ or, for any $\mathbf{x}$, $\mathbf{z}=2$ (i.e. the molecule is rigid and coiled), equation (5.14) reduces to the relation (5.5) for a monolayer adsorbed on to sites on the surface.

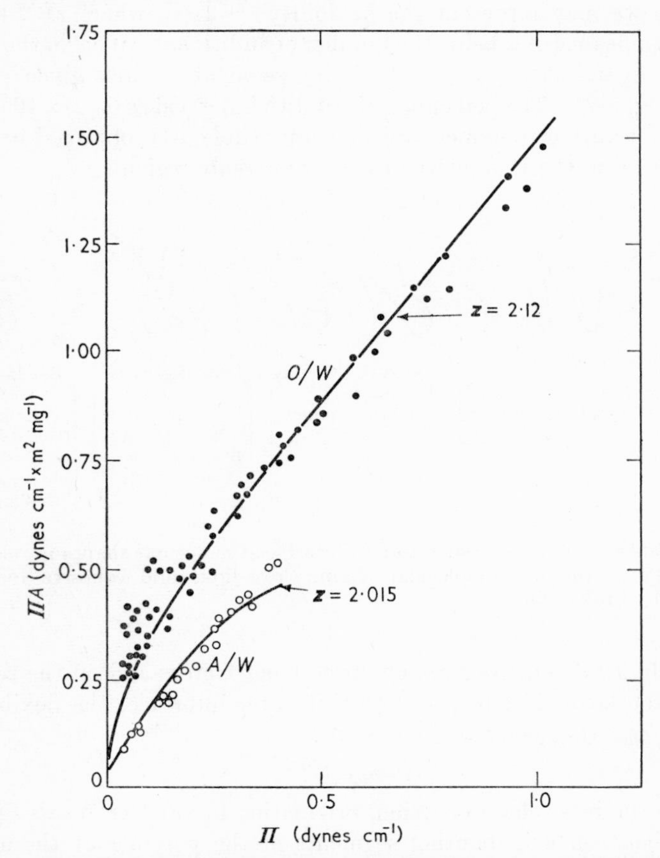

Fig. 5-25. Plots of ΠA against Π for haemoglobin films spread at the air-water and oil-water interfaces, with the lines obtained by fitting eq. (5.14).

If $\mathbf{z}$, though not equal to 2, is yet close to 2, equation (5.14) approximates to:

$$\Pi A = \boldsymbol{k}T + \tfrac{1}{2}\,\Pi A_0 \tag{5.15}$$

which suggests that under these conditions the plot of ΠA vs. Π should be approximately linear; this is often found for polymers at the air-water surface, as shown in Figs. 5-22 and 5-23. If $\mathbf{z}$ is not close to 2, equation (5.14) predicts a plot of ΠA vs. Π which is convex upwards, as in Fig. 5-24.

Whatever the value of $\mathbf{z}$, $\Pi A \to kT$ as $\Pi \to 0$, the only restriction being the experimental one of measuring sufficiently low values of Π.

In a completely flexible film[79] $\mathbf{z}=4$, while in a film in which the molecules are tightly coiled up into two-dimensional discs, with no internal flexibility, $\mathbf{z}=2$. Thus we may represent the flexibility[79,81] as ω, where $\mathbf{z}=2+\omega$, the physical significance of ω being the number of additional "sites" in the surface on which can be placed the neighbouring segments of any given polymer segment (Fig. 5-27). The percentage flexibility is defined as $(\omega/2)\times 100\%$, and the values for various polymers are shown in Table 5-II, obtained by fitting equation (5.14) to the Π-A curve in the low pressure region.

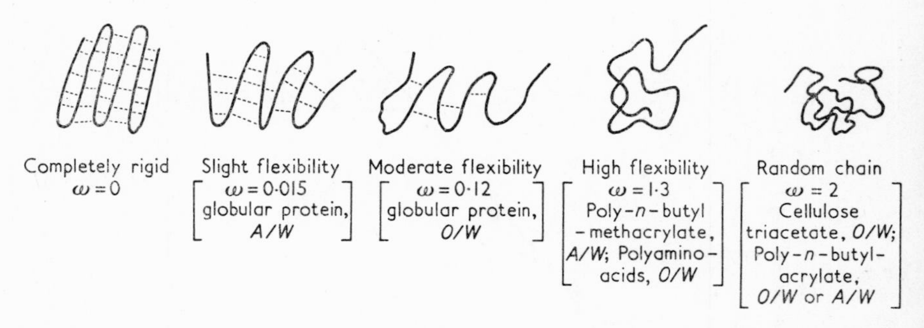

Fig. 5-26. Diagrammatic representation of interfacial molecular shapes in relation to the flexibility of the macromolecular chain. Cross-links and cohesive forces are represented by broken lines.

The flexibility ω must depend on steric bond factors and on the cohesive energy: if the latter is zero (as at the oil-water interface) the flexibility is denoted by ω_0. In general we may write

$$\omega = \omega_0 e^{-W/kT} \tag{5.16}$$

where W is the inter-chain cohesion, originating in van der Waals forces of attraction between neighbouring segments of the polymer at the surface. In poly-1:1:2-lysine-glutamic-acid-leucine at the air-water surface the cohesion W at pH 7 is 3.2 kT, or about 2.9 ergs cm.$^{-2}$ per $-CH_2-$ group[21,32], closely similar to the value of about 3.0 ergs cm.$^{-2}$ in close-packed films of simple long-chain molecules. If the film is spread on 1N-HCl, W is reduced from 3.2 kT to 1.6 kT, while the stronger intramolecular repulsion associated with a higher ψ_0 value (Chapter 2) reduces W to 0.62 kT in the film on 0.01 N-NaOH or 0.01 N-HCl[21,32], and increases the pressure as shown in Fig. 5-23. Figure 5-28 shows the same effect for a haemoglobin monolayer.

The entropy of surface denaturation of each amino-acid residue in a protein is called S_D, and is given[76] at large areas by:

$$S_D = (1 - 2/\mathbf{x}) R \ln(\mathbf{z} - 1)$$

For a molecule that remains rigid on unfolding at the surface (i.e. $\mathbf{z} = 2$), we have $S_D = 0$. Ovalbumin at the air-water interface has $S_D = +0.03$ e.u. When, however, the ovalbumin monolayer is compressed to 1m^2.mg.$^{-1}$, the entropy of surface denaturation is reduced to -0.01 e.u.[76], showing that compression reduces the ability of the protein chains to flex.

Table 5-II shows some typical values of the flexibilities of films at the oil-water interface: where these are higher than in films at the air-water surface, van der Waals cohesive forces within the molecule have been eliminated by the oil.

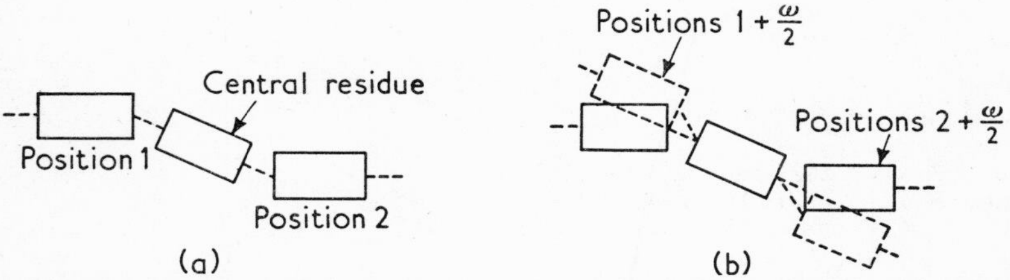

Fig. 5-27. Diagrammatic representation of the flexibility of a polymer chain in terms of the number of "sites" available for neighbouring units, i.e. the co-ordination number of each unit in the surface.

TABLE 5-II

Film	Flexibility (%) O/W	Flexibility (%) A/W
Poly-1:1:2-lysine-glutamic-acid-leucine	10	0.4
Cellulose triacetate	100	10(0)
Polyvinyl alcohol	2.5 to 5	25(7.5)
Poly-dl-alanine	65	—
Poly-dl-leucine	65	6
Poly-n-butylacrylate	100(70)	9
Haemoglobin (neutral)	6	1.5
Haemoglobin (on acid)	38	20
Insulin (neutral)	25	1.5
Gliadin	65	3
Ovalbumin	22	0.8
Bovine serum albumin	4.5	3
Pepsin	19	1.5

These are selected values from Cheesman and Davies[72,81], Dieu[75], and Crisp[82]. The last author lists results for many acrylates and vinyl derivatives.

That polymer molecules, particularly proteins, may be only partly unfolded at the surface has long been suspected; though it is still not known whether parts of each molecule retain the globular, "native" configuration, or whether certain molecules unfold completely while others are adsorbed in the original globular form.

Clear evidence of incomplete unfolding follows from experiments on pepsin[83]. Films spread from aqueous solution on to 0.01 N-HCl in a waxed Buchner funnel are finally deposited on a filter paper at the bottom of the funnel, by lowering the water level. The pepsin is then tested for enzymic

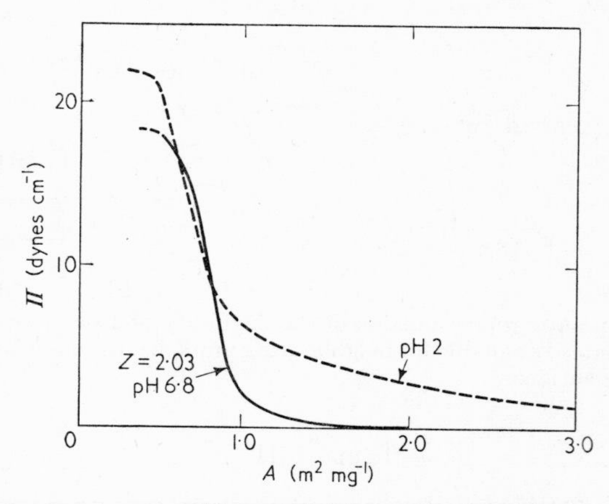

Fig. 5-28. Force-area curves of haemoglobin at air-water interface at pH 6.8 and pH 2, showing the effect of the net positive charge on the molecule spread on acid.

activity. As seen in Fig. 5-29, the activity falls off rapidly as the surface pressure increases, and tends to zero where the surface concentration of protein is very low. By comparison with completely spread films from an "activated" solution containing propyl alcohol, it is clear that, whereas a fully unfolded pepsin molecule irreversibly loses its enzymic activity, unspread material, loosely anchored below the monolayer, retains its activity. On tricresyl-phosphate substrates, zein may be spread from 90% isopropanol to a monolayer, and studied using a Langmuir trough made of Teflon[53]. From the Π-A curve (Fig. 5-30) we may calculate that the thickness of the film is 40-50 Å: it presumably consists entirely of "native" globular protein.

Polymeric acids show an interesting difference between monolayers spread and those adsorbed from solution: while spread films consist of molecules extended flat in the surface, molecules are adsorbed from solution as three-dimensional coils, which are apparently very slow to unfold in the surface[84]; presumably there is a high internal potential barrier. Copolymers of methacrylic acid and 2-vinyl-pyridine are adsorbed as coils, with some lateral compression tending to decrease their dimensions in the plane of the surface[85]; they thus form a two-dimensional gel. Many polymers may be studied at interfaces[72,82,86,87]: the interest lies in the study of the polymer films at different interfaces. The polar groups will tend to remain always in water, but some non-polar side-chains, though preferring the non-aqueous phase, will sometimes be pulled below the water (and undergo van der Waals attraction between themselves) if the "backbone" of the polymer cannot fold suitably to permit them all to enter the non-aqueous phase[76].

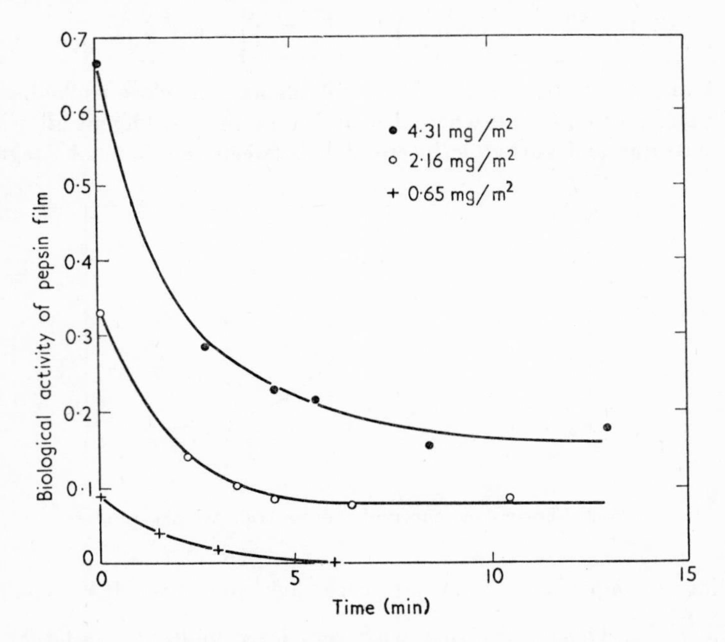

Fig. 5-29. Enzymic activity of pepsin films spread from aqueous solution on to the air-water surface, as a function of surface concentration and the time elapsed after spreading.

Because the polar groups remain in the water, we may expect that the effects of hydrogen-bonding, salt-bridging and hydration will be much the same as in bulk aqueous solution. At the air-water interface those non-polar groups which are in the air will interact by van der Waals forces of attraction at least as strongly as if the molecules were dissolved in the bulk aqueous phase. Oil will eliminate this cohesion, and thus, by working with different non-aqueous phases and with various aqueous solutions, we may find easily the various forces responsible for maintaining molecular form.

Some macromolecules, instead of lying flat in the interface, may "loop" when compressed. This affects the apparent values of A_0, and Frisch and Simha[88] have allowed for this as follows. A coefficient p is defined by

$$p = \frac{2\alpha_i}{(\pi f_i \mathbf{x})^{1/2}}$$

in which f_i characterizes the internal flexibility of the molecule, $\mathbf{x}$ is the degree of polymerization, and α_i is the intramolecular segment "condensation coefficient". The surface pressure is then given, according to the theory[88], by:

$$\Pi = \frac{\mathbf{x}\boldsymbol{k}\mathrm{T}}{\mathrm{A}_0}\left[\ln\left(\frac{\mathrm{A}}{\mathrm{A}-p\mathrm{A}_0}\right) - \left(\frac{p^2\mathrm{A}_0}{\mathrm{A}-p\mathrm{A}_0}\right)\right] - \mathrm{J}p^2\frac{\boldsymbol{k}\mathrm{T}\,\mathrm{A}_0^2}{\mathrm{A}^2} \tag{5.17}$$

where A and A_0 are respectively the available and the close packed areas of the polymer molecule, and where J is an interaction coefficient allowing for lateral entropic and enthalpic interaction between the "anchor" segments.

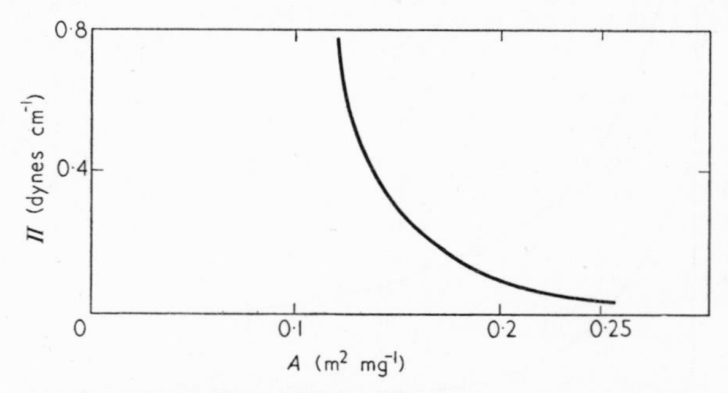

Fig. 5-30. Force-area curve of zein on tricresyl phosphate[53].

According to equation (5.17), at very low pressures $\Pi\mathrm{A} \to \mathbf{x}p\boldsymbol{k}\mathrm{T}$, i.e. $\Pi\mathrm{A} \to \frac{2\alpha_i \boldsymbol{k}\mathrm{T}\mathbf{x}^{1/2}}{(\pi f_i)^{1/2}}$. This implies that with very large molecules ($\mathbf{x}$ large), more parts of the molecule could make large "loops" and return to the interface to act as separate kinetic units. This has not been verified experimentally: equation (5.14) and its limits are valid when sufficiently low pressures are measured (Figs. 5-22, 5-23, 5-24): the pressures should be below 50 millidynes cm.$^{-1}$ if the plot is to be extrapolated to the limiting value of $\Pi\mathrm{A}$.

The effect of temperature on a polymer film will generally be to increase $\mathbf{z}$, according to the relation $\mathbf{z} - 2 = \omega_0 e^{-W/\boldsymbol{k}\mathrm{T}}$. This represents the effect of temperature on the intramolecular flexibility and hence on Π, although no intermolecular forces are allowed for here. Dimethylcellulose monolayers[89] at constant area show only slight increases of pressure at high temperatures, but polyvinylacetate films, while showing an increasing Π with increase T at

4 $m.^2mg.^{-1}$, have a minimum Π at 27°C when spread as more compressed films. Spread films of serum albumin, and γ-globulin also show minima, though polyvinyl alcohol exhibits maxima. Since incomplete unfolding and solubility effects during spreading may be important in determining the pressures of these films, and since the latter effect may be temperature-dependent, it would be interesting to repeat these experiments using a single spread film, the temperature of the substrate being increased and then, after measurements have been made, decreased again to check the reproducibility.

Intermolecular forces may, especially for un-ionized films at the air-water surface, be very high. Poly-leucine and poly-alanine give films that are so coherent that even at the lowest pressures attained the film is far from ideal. Quantitatively, one usually assumes that intermolecular attraction varies as A^{-2}, the pressure in the monolayer being reduced by an amount $QA_0^2\boldsymbol{k}T/A^2$, where Q is a constant. Films of proteins usually have low intermolecular cohesion: the molecules are unfolded into a configuration in which the hydrocarbon chains cohere only intramolecularly. Electrical repulsion between molecules in the surface is less important than between single ions: for poly-1:1:2-lysine-glutamic-acid-leucine on 0.01 N-HCl the intermolecular repulsion is equivalent to that of only 16 single charges, whereas the true valency of the polymer molecule is 72 charges. The effective charge of 16 corresponds closely with the number of charges calculated to lie on the periphery of each polymer molecule in the surface[21].

For films at the oil-water interface, van der Waals forces are absent, so there is no intermolecular cohesion on this account: films of poly-alanine or poly-leucine can be readily studied[23]. Electrical repulsion is slightly greater than between molecules at the air-water surface[21].

SURFACE VISCOSITY

A monolayer is resistant to shear stress in the plane of the surface just as, in bulk, a liquid is retarded in its flow by viscous forces. The viscosity of the monolayer may, indeed, be measured in two dimensions by flow through a canal in a surface or by its drag on a ring in the surface, corresponding to the Ostwald and Couette instruments for the study of bulk viscosities. The surface viscosity η_s is defined by the relation

$$\text{Tangential force per cm. of surface} = \eta_s \times (\text{rate of strain})$$

and is thus expressed in units of $[mt^{-1}]$ (called surface poises or s.p.), whereas bulk viscosities (η) are in units of poises, $[ml^{-1}t^{-1}]$. The relationship between the two is

$$\eta = \eta_s/d \tag{5.18}$$

where d is the thickness of the "surface phase", (about 10^{-7}cm. for many monolayers). Since η_s is often of the order of $10^{-3}-1$ surface poise, this implies that over the thickness of the monolayer, assuming this is uniform, the viscosity is about $10^4 - 10^7$ poises. The film is thus rather solid, like butter, confirming the orientation of the hydrocarbon chains and the strong interaction between the orientated molecules.

Surface and interfacial viscosities are sometimes measured with a needle oscillating in the surface. Not only does this disturb the surface film, however, but the surface pressure gradients set up by the movement of the needle must severely affect its behaviour. We shall discuss below the most accurate techniques available for measuring surface viscosity.

Drag of a Monolayer on and by the Underlying Water

If a monolayer is flowing along the surface under the influence of a gradient of surface pressure, as in Fig. 5-31 at A, it carries some of the underlying water (B) with it. This is a consequence of the lack of slippage between the monolayer and the bulk liquid adjacent to it. For a monolayer of oleic acid

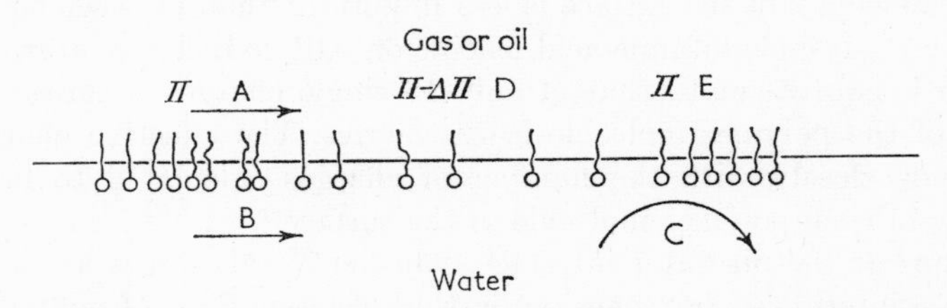

Fig. 5-31. Drag by a moving monolayer on the underlying liquid, and the inverse phenomenon.

moving at between 1 and 5 cm.sec.$^{-1}$, direct measurement[90] gives the thickness of the entrained layer of water as 3×10^{-3}cm., independent of film velocity, when the surface film flows over a barrier. If the viscosity of the underlying solution is increased, the thickness of the aqueous layer carried with the monolayer increases in direct proportion to the viscosity.

For a freely flowing monolayer, not passing over a barrier, direct experiment is more difficult. However, the system is closely analogous to a boundary at a solid surface, with a superimposed uniform velocity equal to that of the moving monolayer. On this assumption, the thickness of the entrained water layer is $4.8\ (\eta l/p_l \mathrm{v})^{1/2}$, where l is the length and v the velocity of the flowing sheet of film, η and p_l are the viscosity and density respectively of the bulk liquid[90].

Conversely, if liquid is moving, as at C (Fig. 5-31) below an otherwise stationary, uniformly spread monolayer, it carries the molecules in the surface along with it, compressing them at E until the stress due to such viscous traction is balanced by the back-spreading pressure of the monolayer, which tends to move back from E to D.

The first phenomenon, besides being basic to the study of the viscosity of the monolayer itself (without the underlying water), is important in foam-breaking (Chapter 8) and in the amplification of eddies (during unstable mass-transfer) (Chapter 7). The second, the basis of the "viscous traction" method below, is found to affect the approach of eddies to the surface and to reduce or prevent circulation in moving liquid drops (Chapter 7).

Insoluble Monolayers (A/W): the Canal Method

To measure the viscosity of a monolayer spread at the air-water surface, the surface of a Langmuir trough is divided into two compartments, connected by a narrow canal: a film on the surface of the trough will flow through the canal at a rate depending on (i) the surface pressure gradient along the canal, (ii) the width and length of the canal, (iii) the viscosity of the monolayer, and (iv) the drag on the underlying water. Experimentally the arrangement shown in Fig. 5-32 is convenient—the floating frame of plastic

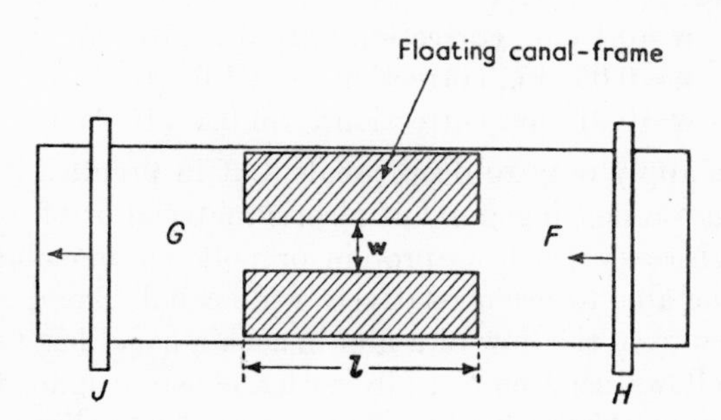

Fig. 5-32. Canal method of measuring surface viscosity of insoluble films at the air-water surface.

or mica ensures that the level of liquid is always constant, and the monolayer flows from the region of higher surface pressure (F) to that of lower pressure (G). The difference of pressure $\Delta\Pi$ across the ends of the canal is kept constant by moving the two barriers, H and J, during the experiment. If Q is the flow rate of the film through the canal (measured in cm.^{2}sec.$^{-1}$) and η_s is the surface viscosity, these would be related by the simple formula:

$$Q = \frac{(\Delta\Pi)\mathbf{w}^3}{12\eta_s \mathbf{l}} \tag{5.19}$$

provided there was slippage between the film and the immediately underlying water. Here $\mathbf{l}$ is the length of the canal, and $\mathbf{w}$ its width. In practice, however, the water beneath the surface canal will be dragged along with the flowing monolayer, and a correction must be made for this. This correction is quite considerable: Harkins[6] allowed for the lack of slippage as follows. The underlying water, with a viscosity η_0, increases the resistance to movement of the film in the ratio $\left(1+\frac{\mathbf{w}\eta_0}{3.14\eta_s}\right)$ and hence

$$Q=\frac{(\Delta\Pi)\mathbf{w}^3}{12\eta_s\mathbf{l}}\left(\frac{1}{1+\frac{\mathbf{w}\eta_0}{3.14\eta_s}}\right) \tag{5.20}$$

This formula holds strictly only if the depth and length of the canal are great compared with its width. As we should expect, the flow, Q of the film in the canal becomes very small if either η_s or the ratio η_0/η_s is large.

The above relation may be rewritten in the form

$$\eta_s=\frac{(\Delta\Pi)\mathbf{w}^3}{12Q\mathbf{l}}-\frac{\mathbf{w}\eta_0}{3.14} \tag{5.21}$$

where the correction term $\mathbf{w}\eta_0/3.14$ assumes the following values[6], expressed in surface poises:

$\mathbf{w}=0.1$ cm., correction term 3.2×10^{-4}s.p.
$\mathbf{w}=0.05$ cm., correction term 1.6×10^{-4}s.p.
$\mathbf{w}=0.01$ cm., correction term 0.3×10^{-4}s.p.

These formulae apply only to deep canals, but in practice it is much more convenient to use a floating canal of plastic material or of waxed mica; for films of high viscosity (such as protein or polymer monolayers) it is also convenient to be able to use a relatively wide canal. The calculation of the surface viscosity from the flow of a film in such a canal has been carried out by Joly[91] who allows, semi-empirically with a constant $\mathbf{c}$, for the drag of the underlying water on the flowing film. This constant $\mathbf{c}$ has dimensions of cm.$^{-1}$, and may be thought of intuitively as the reciprocal of the thickness of the water layer influenced by the moving film, though wall effects also influence $\mathbf{c}$. The final formula Joly obtains is

$$Q=(\Delta\Pi/\mathbf{l}\mathbf{c}\eta_0)\left[\mathbf{w}-2(\eta_s/\mathbf{c}\eta_0)^{1/2}\tanh\{(\mathbf{c}\eta_0/\eta_s)^{1/2}.\mathbf{w}/2\}\right] \tag{5.22}$$

where $\mathbf{c}\eta_0$ has the values shown in Table 5-III, being a function of $\mathbf{w}$. As before, η_0 is the bulk viscosity of the aqueous substrate.

For water the value of $\mathbf{c}$ is of the order 10 cm.$^{-1}$, corresponding to a layer of water 0.1 cm. deep below the flowing monolayer. Because of the complicated flow patterns of the water under the floating canal, it is not possible

TABLE 5-III

$\mathbf{w}$(cm.)	0.066	0.117	0.265	0.383	∞
$\mathbf{c}\eta_0$(g.cm.$^{-2}$sec.$^{-1}$)	0.225	0.191	0.109	0.083	0.050

to determine **c** except empirically: Joly carried out flow measurements with oleic acid monolayers in canals of various widths and, knowing that η_s must be constant, substituted his results into equation (5.22), using successive approximations to obtain $\mathbf{c}\eta_0$.

When the flow through the canal is very slow (with either **w** very small or η_s very high), the correction term in equation (5.22), like that in equation (5.20) above, becomes small and

$$Q \approx \frac{(\Delta\Pi)\mathbf{w}^3}{12\eta_s \mathbf{l}}\left(1 - \frac{\mathbf{w}^2\mathbf{c}\eta_0}{6.7\eta_s}\right) \tag{5.23}$$

This may be seen by expanding the hyperbolic tangent term in (5.22). In the limit that the correction term involving $\mathbf{w}^2/\eta_s$ becomes extremely small, eq. (5.23) reduces to eq. (5.19).

On solving eq. (5.23) for η_s, we obtain, by a further approximation,

$$\eta_s = \frac{(\Delta\Pi)\mathbf{w}^3}{12Q\mathbf{l}} - \frac{\mathbf{w}^2\mathbf{c}\eta_0}{6.7} \tag{5.24}$$

where the correction term ($\mathbf{w}^2\mathbf{c}\eta_0/6.7$) has the following values, expressed in surface poises:

$\mathbf{w} = 0.1$ cm., correction term $= 3 \times 10^{-4}$ s.p.
$\mathbf{w} = 0.05$ cm., correction term $= 1 \times 10^{-4}$ s.p.
$\mathbf{w} = 0.01$ cm., correction term $= 0.1 \times 10^{-4}$ s.p.

These values are only approximate, and they are listed to show that the corrections in the Harkins and the Joly formulae are of a similar order of magnitude. The full formula (5.22) should always be used in the interpretation of experimental data, together with the interpolated value of $\mathbf{c}\eta_0$ from the figures listed above.

The canal method is the most accurate available, particularly for films of very low surface viscosity, i.e. in the range 10^{-5} to 10^{-3} surface poises. The shear gradient is not constant across the canal, however, and the method is not satisfactory for adsorbed films, for which the solubility of the monolayer increases very considerably as the film pressure is raised. With such films, the higher surface pressure at one end of the canal would cause desorption of the film there, to re-adsorb at the end of the canal where the surface pressure was lower. In this way a spurious figure for surface viscosity would be obtained in terms of the area swept out per second by the barriers in Fig. 5-32, even if there were no flow at all through the canal.

Insoluble or Soluble Monolayers (A/W): Rotational Torsional Methods

If a circular dish containing water is rotated, the surface will tend to rotate at the same speed. However, if a stationary disc (usually hollowed out to reduce liquid drag, as in Fig. 5-33) is placed in the surface, the liquid surface between this and the wall of the dish is subjected to shear, since the disc is stationary while the side of the dish rotates. The strain applied to the disc to maintain the shear stress can be readily found by attaching the disc to a fine torsion wire, and measuring the deflection relative to that when the dish is stationary. Before the film is spread, however, the dish must be filled with clean water, and the deflection of the disc at various shear rates noted: these deflections represent the drag of the underlying water on the disc, and

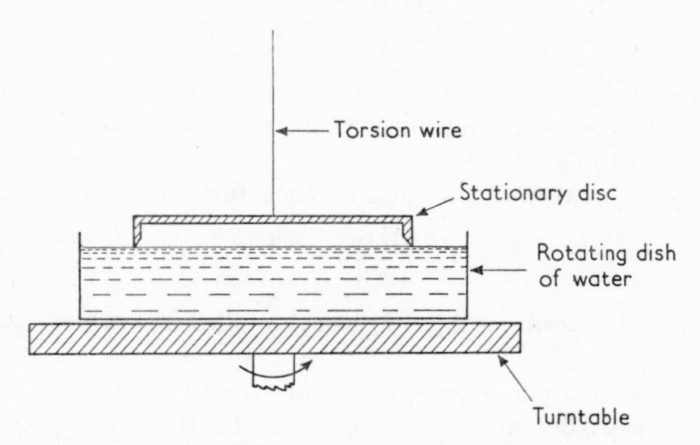

Fig. 5-33. The rotational torsional method for determining the surface viscosity of spread or adsorbed films at the air-water surface.

must be allowed for when a film-covered surface is studied. Thus, by subtracting the deflection for a clean surface from that for a film one, the surface viscosity of the monolayer should be calculable at different viscosity gradients[92,93]. The formula used is

$$\eta_s = \frac{(\Delta\boldsymbol{\theta})K_T}{4\pi\boldsymbol{\omega}_r}\left(\frac{a_2^2 - a_1^2}{a_1^2 a_2^2}\right) \tag{5.25}$$

where K_T is the torsion constant of the suspension, $\boldsymbol{\omega}_r$ is the angular velocity of rotation of the dish, and a_1 and a_2 are the radii of the disc and the dish respectively. The term $\Delta\boldsymbol{\theta}$ represents the difference in the deflections of the disc at the film-covered and clean surfaces. It is clear that with this apparatus the surface viscosity of a film at different angular velocities may be obtained.

Though this type of surface viscometer has been widely used, there are three limitations to its sensitivity and accuracy. Firstly, the correction for the torque of the clean water surface (and the immediately underlying water) on the disc is rather high, and is equivalent to a surface viscosity of about 10×10^{-4} surface poises. Secondly, addition of surface-active agent can alter the water level slightly, by altering the form of the menisci[92]. This slightly increases the depth of immersion of the disc, and may hence lead to increased drag from the underlying water. This level effect has been responsible for spurious readings when the film viscosity is very low. Thirdly, the formula (5.25) used to calculate the viscosity is based on the assumption that the movement of the underlying water is the same whether a film is present or not, i.e. that the film slips freely over the water, its movement being a function only of its viscosity, the radii of the disc and the dish, and the rate of rotation of the wall of the vessel. Now there is no authenticated example of such "slippage"; the velocity of the water immediately below the film must be the same as the velocity of the film. This implies that any small area of the film is being pulled round by the edge of the dish and also by the subjacent water, if this is also rotating. The measured torque is therefore higher than would have been found for the monolayer only, and the derived surface viscosities are thus too high.

Bernard[94], in a new version of this type of instrument, employs two concentric floating rings, the inner one being rotated while the torque is measured on the outer one. The underlying liquid is not rotated, and is prevented from taking up appreciable movement from the rotating ring by a sheet of mica placed about 1 mm. below the surface to act as a "brake" on liquid movements; this device reduces the torque to be subtracted for non-slippage of the substrate. This instrument is equally useful for spread or adsorbed films, and has reputedly a high sensitivity (about 1×10^{-4} s.p.).

A disc translation method[95], in which the disc is made to oscillate on its torsion wire while touching the surface of the water in a normal Langmuir trough, may also be used. This method is sensitive to about 10^{-2} s.p.

Insoluble or Soluble Monolayers (A/W): The "Viscous Traction" Method

Another method of measuring the viscosity of adsorbed or spread films employs a moving liquid substrate to cause the film to flow through a stationary canal in the surface (cf. right-hand side of Fig. 5-31). The movement of the film (shown with talc) in the canal depends partly on the viscous traction of the subjacent layers of bulk liquid; this traction tends to move the film in the canal at the speed of the liquid. This, however, is opposed by the drag of the stationary walls of the canal, acting on the viscous monolayer, and from the movements of the talc particles the surface viscosity may be calculated. This method has several advantages over the torsional instruments, as explained in detail below, following a description of the two principal types of "viscous-traction" surface viscometer.

The first of these employs a *linear canal*. With this instrument one may calculate the retardation of movement of talc particles on the surface in terms of the flow rate of the underlying liquid, the width of the stationary canal, and the surface viscosity of the film[96]. The simplest measurements are those on spread, insoluble monolayers.

With soluble films the procedure is complicated by the build-up of surface pressure at the down-stream end of the canal, due to the film having been carried down the canal. This higher surface pressure leads to increased desorption of the film, and re-adsorption may then occur at the up-stream end of the canal. To correct accurately for these effects is very difficult[96]. To deal more simply with such soluble, adsorbed films, a *circular "viscous-traction" instrument* has been devised[97]. This consists of a circular brass canal, 3.5 mm. wide and 12.5 cm. outside diameter, the edges being ground to knife-edges (Fig. 5-34). After the latter have been coated with purified wax, the circular canal is lowered just on to the surface of the solution. This is contained in a petri dish, which can be rotated (say) once every 38 secs.

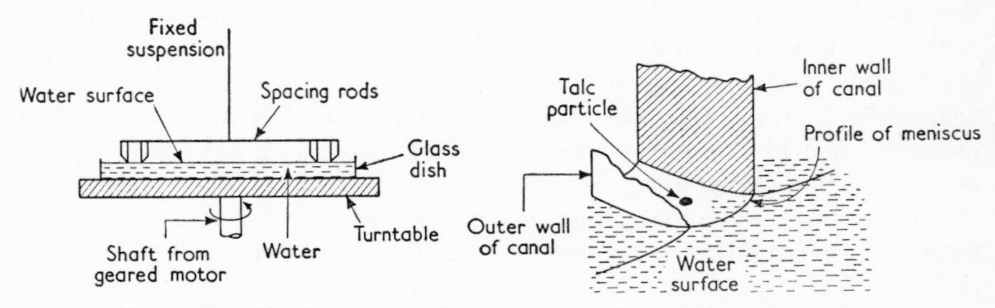

Fig. 5-34. Principle of circular viscous traction surface viscometer.

As before, the drag of the underlying liquid rotating at this speed tends to rotate the surface at the same speed, while the stationary circular canal tends to prevent this. The advantage for soluble films is that there is no surface pressure gradient set up: desorption and adsorption need not, therefore, be considered. In practice, even a clean water surface is retarded by the presence of the knife-edge canal in the surface, to one revolution in 78 secs. in the original instrument. This is measured with talc particles, as before. When an adsorbed or spread film is present, the talc particles rotate still more slowly, the additional retardation being a measure of the surface viscosity of the film (cf. Fig. 5-35).

As with the linear form of the instrument, the mathematical treatment of the retardation of the surface movement relative to the bulk of the liquid is rather complicated; strictly, the velocity profile is required at different distances across the canal, though in practice this is not easily obtained. Instead, the canal knife-edges are adjusted in the surface so that the meniscus in the canal is very slightly concave upwards, and the talc particles lie in the centre of the canal (Fig. 5-34). A calibration curve relating the retardation to the surface viscosity can now be obtained, using viscosity data on insoluble films studied by the canal method[98] (described above) in which corrections have been made for the lack of slippage between the monolayer and the water. Fig. 5-35 shows such a calibration curve, the sensitivity of the apparatus being about 1×10^{-4} surface poises: with a narrower canal the sensitivity would be increased.

Typical viscosities measured with this apparatus are zero for 0.1% sodium laurate at pH 10 alone, but 3×10^{-4} surface poises when 0.01% lauryl alcohol is present in the solution. These results are of interest in connection with the criteria of foam stability, as explained in Chapter 8.

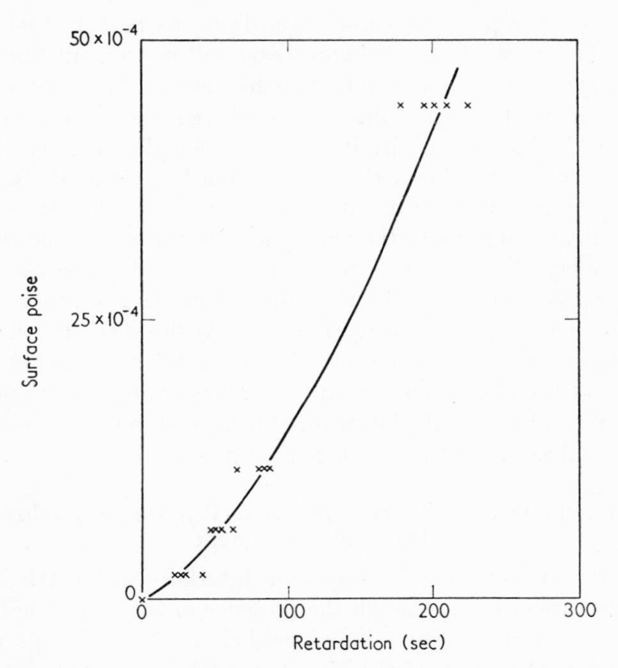

Fig. 5-35. Retardation of talc particles in centre of canal in instrument of previous figure, relative to a clean surface in the same canal. Dimensions and speed as in text.

Unlike torsional surface viscometers, the "viscous-traction" instrument has the advantage that the circular canal can readily be made as narrow as required, so that the sensitivity can be high even when the surface viscosities are very low. It also allows improved sensitivity; in the torsional instrument the correction for the drag of the underlying water on the torsion system is rather high, and this limits the sensitivity to about 10×10^{-4} surface poises, whereas for "viscous-traction" instruments the sensitivity is ten times better than this.

Apart from sensitivity, however, there seems to be a real difference in the magnitudes of surface viscosities obtained with the torsional and "viscous-traction" instruments. Results from the former are often about ten times higher than from the latter: for example, for sodium lauryl sulphate, the figures are 20×10^{-4} and 2×10^{-4} surface poises respectively. When lauryl alcohol is added to anionic films, these figures increase to 400×10^{-4} and 30×10^{-4} surface poises. Further, the torsional instrument gives a surface viscosity of 15×10^{-4} surface poises for 0.1% laurate at pH 10, while no viscosity is detected with the "viscous-traction" instrument.

This tenfold discrepancy is found, regardless of whether absolute calibration is used (as for the linear "viscous-traction" instrument) or whether an empirical calibration curve is required (as for the circular instrument). There are two reasons for the high values obtained with the torsional instrument. Firstly, since the limit of sensitivity of the torsional instrument is 10×10^{-4} surface poises, viscosities lower than this cannot be measured. Secondly, the theory of the torsional instrument requires that the film be sliding freely over the substrate; in practice the film adheres to the water, and the viscosity of the surface layers of the latter are measured as well as the viscosity of the film alone. Surface viscosities by the torsional method will therefore always be too high unless suitable corrections can be devised. In the "viscous-traction" apparatus, however, not only is the sensitivity considerably higher, but the drag of the film on the substrate is allowed for, both in the absolute calculation of results from the linear instrument and in the surface viscosities used in the calibration of the circular apparatus[97].

Insoluble or Soluble Monolayers (A/W or O/W): the Generalized "Viscous Traction" Instrument

Early work on films at the oil-water interface was carried out with oscillating needles or discs, though the presence of the second bulk phase so reduces the sensitivity that viscosities below about 1 s.p. cannot be measured[99]. A modification of the viscous-traction method, however, makes it suitable for either the oil-water or the air-water interfaces, for either spread or adsorbed films[60]. The sensitivity is still about 10^{-4} sp. The knife-edge system described above is replaced by rings of stainless steel wire (0.064 cm. in diameter), forming concentric circles of diameters 12.5 cm. and 11.6 cm. When the vessel containing the two liquid phases is rotated, the canal formed by the space between these rings which are situated in the interface (Fig. 5-36), retards the motion of the interface; the retardation is greater the higher the viscosity of the monolayer. The ring system is fixed in the interface in such a position that the meniscus within the canal is slightly concave upwards, so that the rotation of talc particles moving in the centre of the canal can be timed readily.

It is necessary to calibrate the apparatus, since absolute calculation of the

Fig. 5-36. Generalized "viscous traction" surface viscometer, suitable for use with soluble or adsorbed films, either at the air-water surface or the oil-water interface (as shown here)[60].

[*To face p. 260*

retardation time is not possible. By a fortunate coincidence, when benzene is used as the oil, the retardation of a talc particle in the clean interface is almost the same as in the same canal at the air-water surface: rotation times of 110 sec. and 100 sec. respectively are recorded when the vessel is rotated once in 38 secs. For the water-ethyl acetate interface the time is 105 sec. One may assume, therefore, as a first approximation, that the relations between film viscosity and retardation time are the same at the air-water and benzene-water interfaces: knowing the former relation (again by working with spread films of known surface viscosity) we therefore have the latter (Fig. 5-37).

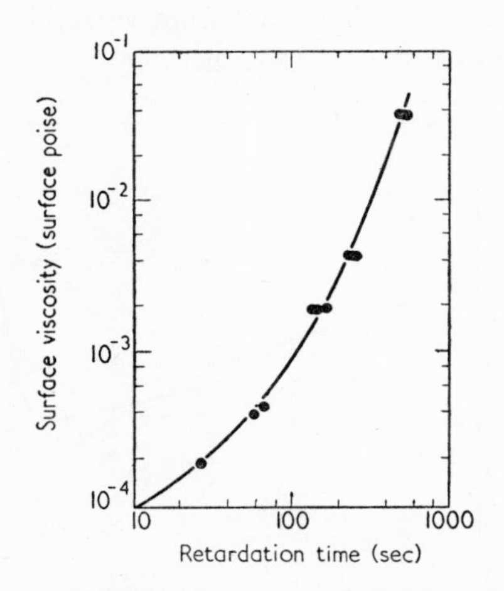

Fig. 5-37. Calibration curve (A/W) of instrument shown in Fig. 5-36. Retardation times are additional to the rotation times at the clean surface (100 sec.), the turntable speed being 1 revolution in 38 sec. The dimensions of the canal are given in the text. The same calibration curve is used at the oil-water interface[60].

Experimentally, it is convenient to immerse the ring system, previously cleaned in hot chromic acid, into the water before spreading the film, subsequently raising it carefully into the interface when the film has been spread. The apparatus is equally able to measure the viscosity of adsorbed monolayers. The width of the canal (here 0.9 cm.) and the speed of rotation may both be altered, to obtain results for films of widely differing viscosities, and at various shear rates.

Viscosities of Insoluble Films

Fatty acid esters, which give "*gaseous*" *monolayers* at great areas, are characterized by Newtonian flow and low viscosities (10^{-5} to 10^{-4} s.p.). The rather *more condensed monolayers* of the higher fatty esters and of oleic acid also show Newtonian viscosities of this order, increasing with the available surface area; the molecules are less orientated, and presumably cannot slide past each other so easily[69,91,98]. The *strongly coherent* and highly condensed films of stearic acid or myristic acid, spread on dilute HCl, have Newtonian viscosities all about 10^{-4} s.p., increasing to about 10^{-3} s.p., at high compressions[94,95,98] when the inter-chain cohesion becomes dominant. The viscosity of a monolayer of cholesterol[99a] is extremely low.

Long-chain alcohols and amides give non-Newtonian surface viscosities: cetyl alcohol on water at 20°C has $\eta_s = 7 \times 10^{-3}$ s.p. at a shear rate of 0.5 sec.$^{-1}$ but $\eta_s = 0.8 \times 10^{-3}$ s.p. at 2.6 sec.$^{-1}$ Changes of state of insoluble films of esters and acids are also readily found from variations of surface viscosity with area and temperature[98].

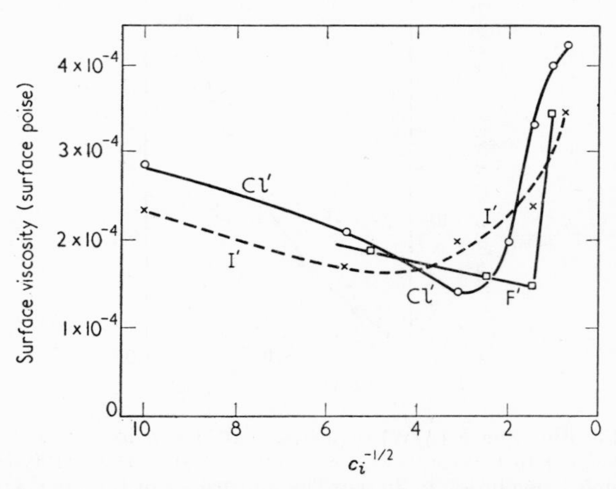

Fig. 5-38. Surface viscosities at the air-water surface for films of $C_{18}H_{37}N(CH_3)_3^+$ at 85 Å^2, spread on solutions of NaF, NaCl, and NaI of concentration c_i. The plots (against $1/c_i^{1/2}$) show minima, corresponding to higher concentrations for the more highly hydrated ions[100].

In general, the viscosity of *electrically charged monolayers* is very low, about 10^{-4} surface poises: the best method of measurement, therefore, is that using the floating canal. The variation of the viscosity of spread films of $C_{18}H_{37}N(CH_3)_3^+$ with added salt is of special interest, as it throws some light on the position of the counter-ions relative to the surface. The potential measurements discussed in Chapter 2 suggest that, at the higher salt concentrations when the thickness ($1/\varkappa$) of the ionic atmosphere below the film is small, some of the counter-ions penetrate into the film, between the charged headgroups of the quaternary salt. This is expressed by penetration into the position C shown in Fig. 2-23. If this occurs, and if the packing in the surface is sufficiently close, a rise in surface viscosity may be expected because a two-dimensional salt lattice, made up of ions of opposite sign attracting each other strongly, is formed in the surface[100].

The results of experiment (Figs. 5-38 and 5-39) confirm this view: increases in the viscosities of films of the quaternary ions at 85 Å^2 per long chain are strongly marked when the salt concentration is increased. Fluoride ions are evidently less effective in penetrating and linking up the charged heads than are the smaller, less hydrated ions of chlorine and bromine. At 180 Å^2 the viscosity plot does not show any rise, but decreases continuously at higher salt concentrations; at these great areas penetration of the counter-ions, if it occurs, is no longer able to cause the close linking required to raise the surface viscosity[100].

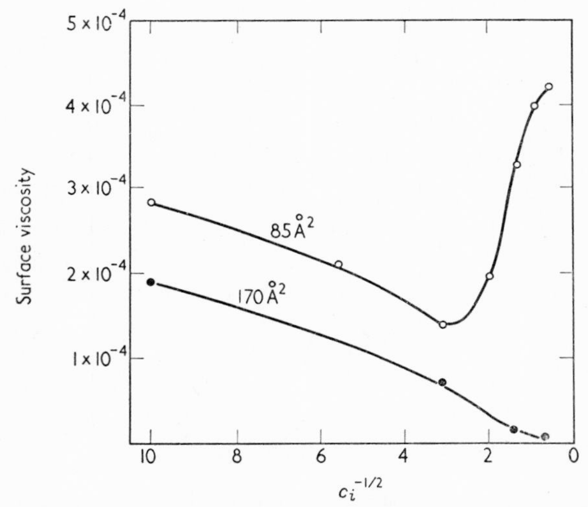

Fig. 5-39. Surface viscosity of spread films of $C_{18}H_{37}N(CH_3)_3^+$ spread on NaCl solutions at two different areas[100].

Spread films of proteins show viscosities which increase sharply if the area available is reduced below a certain limit. Fig. 5-40 shows that these limits are about 0.9 m.^{2} mg.$^{-1}$ for plasma albumin between water and air, and about 0.8 m.^{2} mg.$^{-1}$ when the air is replaced by benzene. Evidently the chains are more unfolded at the benzene interface, as is consistent with the flexibility data (p. 247). At the ethyl acetate-water interface there is evidence that unfolding is incomplete, as the limiting area is then rather low[60].

Viscosities of Adsorbed Films

Foam stability may depend largely on surface viscosity (Chapter 8). The large increases in surface viscosity (Figs. 8-43 and 8-44) for adsorbed films of sodium lauryl sulphate or sodium laurate at the air-water surface on the addition of very small amounts of lauryl alcohol or lauric isopropanolamide are roughly parallel to the efficacy of the latter materials as foam stabilizers[97].

Cetyl alcohol (0.125 M in benzene and adsorbing on to the interface with redistilled water above 22°C) gives films of interfacial viscosity between 1×10^{-4} and 2×10^{-4} s.p., though below this temperature the viscosity increases markedly[60].

For an adsorbed film from 4 millimolar sodium lauryl sulphate to the interface with benzene, the interfacial viscosity is only 1.7×10^{-4} s.p. at 22°C (cf. about 2×10^{-4} s.p. against air)[96]. This viscosity, like that of cetyl alcohol,

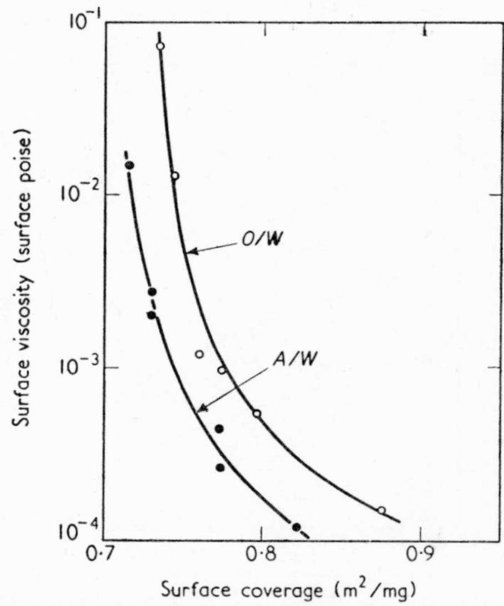

Fig. 5-40. Interfacial and surface viscosities of spread films of bovine plasma albumin spread at the oil-water (—O—) and air-water (—●—) interfaces, the aqueous phase being 0.01 N-HCl.

implies that the head-groups are embedded in a surface film of water of hydration, i.e. in a layer (not necessarily monomolecular) of "soft ice", since the hydrocarbon chains do not cohere in the oil phase and so should not contribute to the surface viscosity. If this interpretation is correct, such a layer, if 10 Å thick, would have an equivalent bulk viscosity of about 2×10^3 poises. The question of such layers of orientated water has already been raised in connection with the dipole contributions to ΔV (Chapter 2), and in explaining the relation of ζ to ψ_0 (Chapter 3). It is also invoked in mass transfer phenomena (Chapter 7, p. 313), and in stability of colloids (Chapter 8). A full discussion is given on p. 369.

The low interfacial viscosity of sodium lauryl sulphate may in part explain the ineffectiveness of this material alone as an emulsifying agent. With M/8 cetyl alcohol in benzene against water, the viscosity is again low, about 1×10^{-4} s.p., but when both cetyl alcohol is adsorbed (from the oil) and sodium lauryl sulphate is adsorbed from the water[60], the interface is very viscous, though less so at temperatures around 30°C. This is of interest in connection with the stabilization of emulsions by mixed monolayers—the interfacial complexes should be viscous enough to prevent the monolayers becoming easily pushed aside as droplets collide (Chapter 8). Experiment shows that emulsions are only markedly stabilized when both the anionic derivative and the cetyl alcohol are both adsorbed, giving the high interfacial viscosity.

COMPRESSIONAL MODULI OF MONOLAYERS

The two-dimensional compressibility C_s of a monolayer at any area A is defined by:

$$C_s = -\frac{1}{A}\left(\frac{\partial A}{\partial \Pi}\right)_T \tag{5.26}$$

It may thus be calculated directly from the slope of the Π-A plot. It is more convenient, however, to use the reciprocal of C_s in discussing the properties of surface films: this is called the surface compressional modulus, C_s^{-1}, and has the dimensions of dynes cm.$^{-1}$; for films in which $\Pi A = $ constant, eq. (5.26) shows that C_s^{-1} tends to Π. It follows that for a clean surface the modulus C_s^{-1} is zero, increasing with the amount of surface-active material present. In general C_s^{-1} will depend also on the state of the film, being greater for more condensed films.

TABLE 5-IV

Monolayer	C_s^{-1} (dynes cm.$^{-1}$)
Clean surface	0
Ideal	Π
Protein ($A \approx 1$ m.^{2} mg.$^{-1}$)	1 to 20
Liquid expanded	12.5 to 50
Liquid condensed	100 to 250
Solid condensed	1000 to 2000

The compressional moduli of protein films at the oil-water interface are very close to Π, confirming their ideal behaviour. The compressional modulus, which is a measure of the compressional elasticity of the film[101,102] is of the highest importance in kinetic properties of monolayers, including foam stability (p. 411), retardation of circulation in falling drops (p. 335), and the damping of waves and ripples.

The Elimination of Waves and Ripples

The calming of the waves at sea during a storm was noted by Pliny, who recorded the practice of the seamen of his time of pouring oil on to the turbulent sea. In 1762 Benjamin Franklin[1] was told by an old sea captain that the Bermudans, before they dived to spear fish below the surface of the sea, would first calm the ripples with a little oil to eliminate the otherwise distracting patterns of sun-light produced by the waves. The same captain told him how the fishermen of Lisbon used oil to reduce the surf at the bar of the river.

Later, at Clapham (London), Franklin used oil to calm the ripples on the pond, and in 1773 he astonished his friends by calming the waves on Derwent Water, a lake of moderate size in England.

The three phenomena, the formation of waves, their damping once formed, and the tendency of large waves to break, are all affected by monolayers on the water. The immediate effect noticed at sea seems to be largely the damping of the waves of short wave-length, which disturb the surface of the larger waves with consequent dangerous breaking of the latter. The wind drag on the large waves is also reduced.

In 1891 Fraulein Pockels[103] showed experimentally that a monolayer spread on the surface of water will damp ripples on the surface. In the laboratory waves may be conveniently produced in a Langmuir trough by inserting at one end a reed attached to a loud-speaker coil, the latter being actuated with an electronic oscillator[104,105].

The velocity v of a wave of length λ on the surface of a deep liquid is given[106] by

$$v^2 = \frac{g\lambda}{2\pi} + \frac{2\pi\gamma}{\rho_l \lambda} \tag{5.27}$$

where g is the gravitational constant and ρ_l is the density of the liquid. For example, on clean water, waves of length 0.5 cm. move at 31.5 cm. sec.$^{-1}$, the corresponding frequency being 63 sec.$^{-1}$ (Table 5-V).

If the liquid is not very deep, a corrective term must be inserted[107]. The first term is unimportant relative to the second if λ is very small: the waves are then known as capillary waves. If λ is large, however, the first term is dominant, and the waves are called gravity waves. It is clear from eq. (5.27) that for capillary waves ($\lambda < 0.5$ cm.) on the surface of water, a reduction in γ reduces the speed of the waves. The surface film also acts by introducing a surface compressional modulus, which is a measure of the work done in resisting the extensions and contractions of the surface during the passage of a wave. The associated components of liquid flow normal to the surface during this passage of the wave are illustrated by the observation[104] that induced waves increase the rate of adsorption of surface-active impurities

(a)

Fig. 5-41. Flow of clean water down the outside of a vertical length of Pyrex tube, 3.94 cm. diameter. The dark vertical streak is a permanganate trace, injected just below the surface through a capillary 0.01 cm. in diameter[111]; there is no turbulence just below the surface in any of the photographs, which show two consecutive frames of the ciné film moving at 80 frames sec.$^{-1}$ In (a), Re = 4; in (b), Re = 21.5; and in (c), Re = 112. Addition of a trace of surface-active agent eliminates the ripples, so that (b) and (c) then appear smooth, as in (a).

[*To face p. 266*

Fig. 5-41 (b)

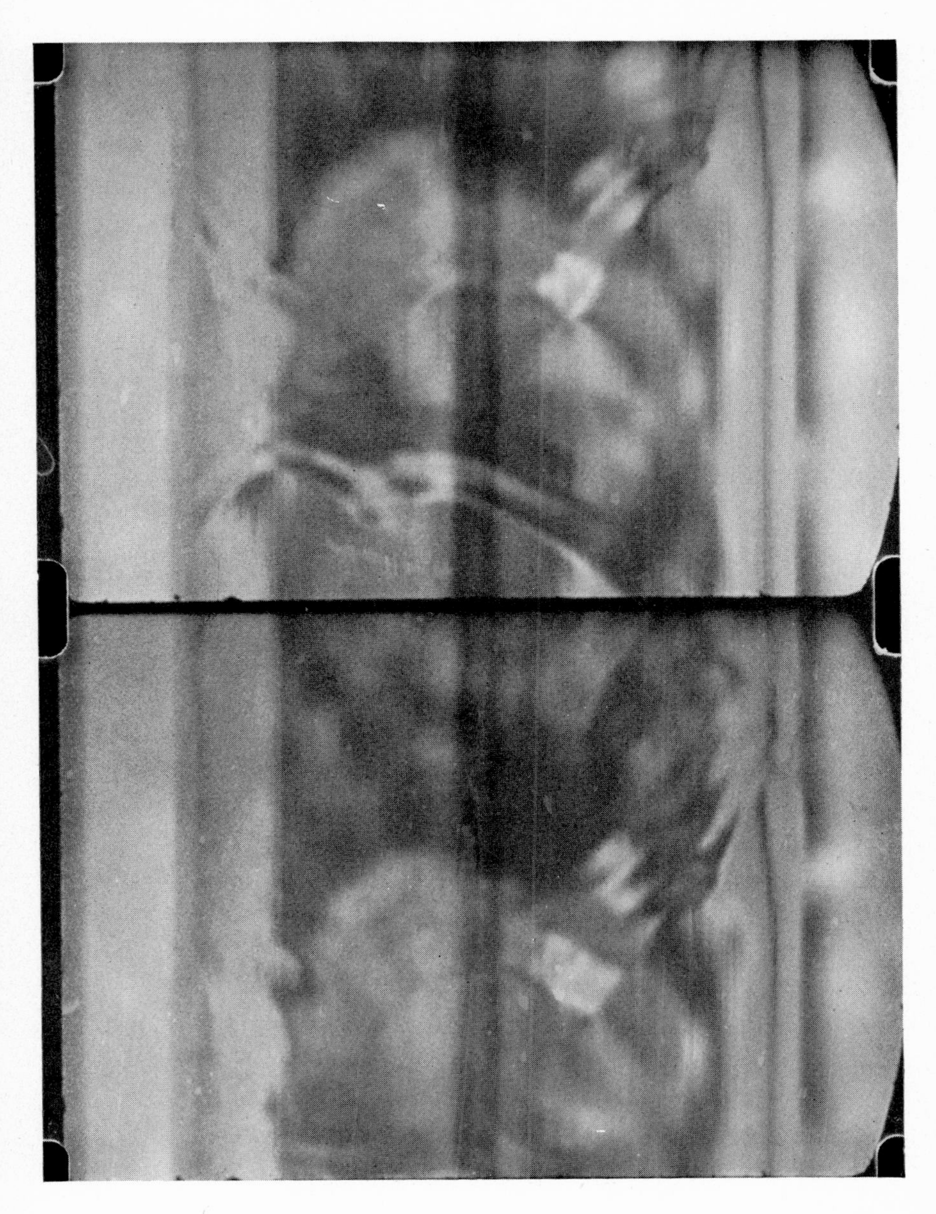

Fig. 5-41 (c)

(from very low concentrations in the bulk) by up to seven times. Further, the rate of adsorption of gases into liquids is greatly accelerated if the surface is disturbed by waves[108] or by ripples (Chapter 7).

The **formation of waves**[109] may be studied in a wind-tunnel, while the spontaneous formation of ripples[110] may be seen clearly on a falling film column at quite low values of Re (here defined as $4u/\nu l$, u being the volume flow-rate of the liquid, l the perimeter of the column, and ν the kinematic

TABLE 5-V

λ (cm)	v (cm.sec^{-1})	f (sec^{-1})	Percentage of v^2 due to surface tension term
0	∞	∞	100
0.1	67.8	678	99.97
0.3	39.4	131	97.0
0.5	31.5	63.0	92.2
0.59	29.5	50	89.5
1.0	24.8	24.8	74.5
1.71	23.1	13.5	50.0
2.0	23.2	11.6	42.4
5.0	29.5	5.9	10.5
10.0	40.0	4.0	2.86
100	125	1.25	0.03
1000	395	0.4	—

Velocity of waves on a clean surface of deep water.

viscosity). At Re=18 the ripples begin to be obvious, and are very marked at Re values of several hundred, well below the Re value (about 1200) at which turbulent flow begins. Typical ripples are shown in Fig. 5-41, with the hydrodynamic conditions of typical ripples shown in Fig. 5-42. Whether the surface-active agents, which are so effective in eliminating the ripples, prevent them from forming is not yet completely clear. The favoured explanation at present is that the ripples always form in this type of experiment, but are rapidly damped out if a surface-active agent is present. Experiments with a Langmuir trough containing at one end a vertical plate[104,105] which can be oscillated backwards and forwards in a horizontal direction might be used to decide this point: it is most important that the plate should always have the same wetting characteristics, and roughened platinum, cleaned by flaming, should prove suitable.

The shear stress exerted by a gas stream (or natural wind) on a water surface is expressed for wind velocities below about 5 metres sec.$^{-1}$ by:

$$\tau = c_h \rho_g v_h^2 \qquad (5.28)$$

where τ is the stress (dynes cm.$^{-2}$), ρ_g is the density of the gas, $\mathbf{c}_h$ is the drag coefficient, v_h being the velocity of the gas stream (or wind) at a height h metres above the surface.

The relative stress increases with wind velocity for clean water surfaces, owing to the development of rougher surfaces, but this is not observed if a monolayer is present[112]. Values of $\mathbf{c}_h$ (0.9×10^{-3} when h is several metres) increase slightly at lower h values, becoming about 2.5×10^{-3} when h is 0.05 m.; these figures apply to aerodynamically smooth surfaces whether

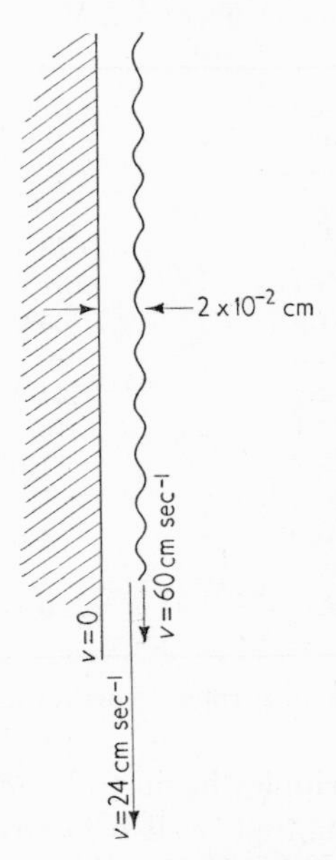

Fig. 5-42. Water flowing down a solid surface, with surface rippling, at Re = 112. From consecutive ciné-film frames[111], the total ripple velocity is about 60 cm. sec.$^{-1}$, whereas dye-trace experiments[110] show that the velocity of mass flow of the water just below the surface is about 24 cm. sec.$^{-1}$, while the mean liquid velocity is 12 cm.sec.$^{-1}$. The wavelength of the ripples[111] is of the order 0.35 cm., corresponding (Table 5-V) to a wave velocity (relative to the water) of about 37 cm.sec.$^{-1}$, if the liquid layer were deep. The total measured wave velocity should hence be about $(24 + 37)$ cm.sec.$^{-1}$, as observed.

covered with a film or not. The stress of the wind on the surface can be opposed by a monolayer, which is compressed downstream till the surface pressure gradient, $d\Pi/dl$, becomes equal to τ. For example, if τ is 0.1 dyne cm.$^{-2}$, a monolayer 400 cm. long will have clean surface upwind, and be compressed by the air flow to 40 dynes cm.$^{-1}$ at the downwind end of the containing vessel. This back-stress of the monolayer greatly reduces the net stress on the liquid surface, which is therefore moved less by the wind, and so does not break into waves so readily. The stationary region of the surface is readily seen by sprinkling a little ignited talc on the water.

Van Dorn's experiments on a 260 metre yacht pond during a steady wind have shown[112] that a commercial detergent, added at the upwind end, eliminates the ripples, so reducing the drag to that for a smooth surface, and consequently reducing the formation of the larger waves.

Wave-damping at a clean interface is related to the viscosity of the water. For a low-amplitude ripple propagated from a line source on deep water, hydrodynamic theory[112] gives the equations:

$$\boldsymbol{a}=\boldsymbol{a}_0 e^{-\Delta_c l} \tag{5.29}$$

$$\Delta_c = 8\pi^2\eta/\rho_l v_g \lambda^2 = 16\pi^2\eta/3\rho_l \boldsymbol{f}\lambda^3 = 8\pi\eta \boldsymbol{f}/3\gamma \tag{5.30}$$

where $\boldsymbol{a}$ is the amplitude of the wave, l is the distance from the line source, and Δ_c is the damping coefficient for a clean (i.e. uncontaminated) surface. In eq. (5.30), v_g is the group velocity of the wave-train, $\boldsymbol{f}$ is the frequency and η is the viscosity. The last two expressions are valid only for water, and for short ($\lambda < 0.5$ cm.) waves. The last expression in eq. (5.30) shows that Δ_c should increase linearly with frequency.

Experimentally, one observes the amplitude of successive waves along the trough by using stroboscopic illumination and measuring the focal length of the mirrors constituted by the troughs and crests of the waves.

With water alone, the only difficulty lies in obtaining a surface free from contamination: even very slight traces of impurities, having an almost negligible effect on the surface tension, may cause appreciable damping of the

TABLE 5-VI

Damping at Clean Surfaces.

System	Frequency (cycles sec.$^{-1}$)	Observed Δ (cm.$^{-1}$)	Calculated Δ from eq. (5.30)
Water[113,113a]	50	0.055	0.055
	98	0.110	0.109
	250	0.314	0.289
	500	0.560	0.565
	920	1.08	1.05
Ethanol[113b]	100	0.50	0.52

ripples. Only when talc is spread on the surface, and then sucked off (see p. 220) into a fine capillary attached to a pump, can the cleaning be considered as possibly satisfactory. This technique has been used in the experiments quoted in Table 5–VI; only in such runs is the damping coefficient as low as calculated by eq. (5.30). Theory and experiment are then in agreement.

Wave-damping by surface-active agents is very marked, even when very low concentrations (10^{-5}M to 10^{-3}M) are employed. The damping coefficient Δ_i can be easily calculated for the limit at which an insoluble monolayer completely immobilizes the interface[112]:

$$\Delta_i = (\pi/v_g\lambda)(\eta\boldsymbol{\sigma}/2\rho_1)^{\frac{1}{2}} \tag{5.31}$$

where
$$\boldsymbol{\sigma} = (2\pi g/\lambda + 8\pi^3\gamma/\lambda^3\rho_1)^{\frac{1}{2}}$$

By comparison with eq. (5.30), one sees that immobilizing the surface increases the damping coefficient in the ratio $(\lambda/8\pi\sqrt{2})(\rho_1\boldsymbol{\sigma}/\eta)^{\frac{1}{2}}$: for waves of $\lambda \sim 0.5$ cm., this ratio is about 2 or 3. The physical mechanism of immobilization is that, due to surface viscosity and surface incompressibility (measured by C_s^{-1}), liquid flow in the plane of the interface during the passage of a wave is prevented. Equation (5.31) is derived by equating the surface velocity to zero at all points in the surface. In Fig. 5-44 the damping by a monolayer of hexadecanol is seen to be very effective for the shorter waves. In Table 5-VII the first four results quoted show good agreement between experiment and Δ_i calculated by eq. (5.31).

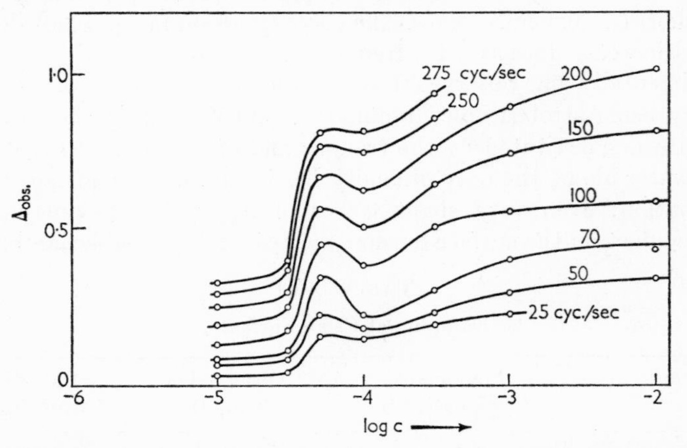

Fig. 5-43. Observed damping coefficient ($\Delta_{obs.}$ in cm.$^{-1}$) plotted against log concentration (moles l.$^{-1}$). Results, which are for sodium lauryl sulphate in water, are shown at various frequencies[113].

The adsorption of *small* amounts of surface-active material, however, may not render the interface completely immobile; the surface can remain mobile, though the alternate compression and expansion of the adsorbed monolayer as the waves pass will produce surface pressure gradients, the consequent movements dissipating some of the energy of the waves.

Two pertinent observations in this connexion are firstly that the damping coefficient may pass through a maximum as the concentration of surface-active agent is increased[105,113,113c] (Fig. 5-43), and secondly that the damping induced by very small amounts of surface-active agent may exceed that

(a)

(b)

Fig. 5-44. Spreading of monolayer of hexadecanol, with consequent wave-damping. In (a) a few particles of hexadecanol have been applied (middle right) to the surface of a reservoir. In (b), taken about half a minute later, the hexadecanol has spread to a monolayer in the lower part of the photograph, damping the shorter waves. Both photographs are reproduced by courtesy of Price's (Bromborough) Limited.

calculated in the limit when the surface becomes completely immobile. [112, 113, 113b]. The question may also be raised as to whether it is the surface compressional modulus or the surface viscosity which is responsible for the damping by a viscous monolayer of protein.

That the wave-damping reaches a maximum for a certain concentration of surface-active agent has been repeatedly found[105,113c,113d]. For example, for sodium lauryl sulphate the optimum concentration for wave-damping is about 5×10^{-5}M (Fig. 5-43). The reason for the reduced damping at concentrations in excess of this optimum is often maintained to be in the faster

TABLE 5-VII

Damping in presence of Monolayers.

System	Frequency (cycles sec.$^{-1}$)	Observed Δ (cm.$^{-1}$)	Calculated Δ_i (cm.$^{-1}$) from eq. (5.31)
Water with spread film of docosoic alcohol (very high C_s^{-1}) [105].	50	0.24	0.26
Water with spread film of protein (moderate C_s^{-1}, high η_s)[105].	50	0.17	0.19
Water with spread monolayer of oleic acid ($C_s^{-1} = 78$ dynes cm.$^{-1}$)[113g].	210	0.81	0.82
Water with adsorbed film from 10^{-2}M lauryl sulphate + 0.9×10^{-4}M dodecanol[113b].	150	0.67	0.67
Water with adsorbed film of laurylsulphate ions (from 10^{-5}M solutions; very low C_s^{-1})[105]	50	0.23	0.16
Water with spread film of hexadecanol (high C_s^{-1}), Vines (1960)[112].	49.4	0.31	0.23
Water with spread film of ethylstearate (low C_s^{-1}), Vines (1960)[112].	49.4	0.12–0.16	0.16
Water with adsorbed sodium lauryl sulphate from 10^{-3}M solution[113b].	150	0.75	0.47
ibid., with 10^{-2}M solution[113b].	150	0.96	0.59
ibid., but with 10^{-3}M di-(ethylhexyl)-sulphosuccinate[113b].	150	0.70	0.69
ibid., but with 0.1% solution of nonylphenol condensed with 15 moles of ethylene oxide[113b].	150	0.76	0.66
Water with adsorbed film of laurylsulphate ions[113], 10^{-5}M	150	0.19	0.40
ibid., 3×10^{-5}M	150	0.24	0.40
ibid., 5×10^{-5}M	150	0.56	0.45
ibid., 10^{-4}M	150	0.50	0.46
ibid., 3×10^{-4}M	150	0.64	0.56
ibid., 10^{-3}M	150	0.74	0.71
ibid., 10^{-2}M	150	0.81	0.79

adsorption and desorption rates from and to the more concentrated solutions, thus short-circuiting the surface-pressure gradients in the plane of the surface[114]. There are difficulties in this theory, however, and a calculation of the relaxation-time (for adsorption and desorption) of a monolayer (page 183) shows that this should be fairly short ($\sim$5 milliseconds) for lauryl sulphate ions, and not vary greatly with concentration.

We believe that the high maximum damping is a fundamental property of the surface film; that it should not vary greatly with frequency and that it is due to localised reversals in the surface stresses as the waves pass, caused by the marked compressional elastic behaviour of surface films. This view is also consistent with Dorrestein's analysis (see below).

Values of Δ measured for film-covered surfaces are summarized in Table 5-VII. Apart from the first four values, certain experimental values are *higher* than according to theory (eq. 5.31), particularly for the soluble surface-active agents such as sodium lauryl sulphate; here "short-circuiting" of the surface-compressional stresses might have been expected to reduce the experimental values *below* the theoretical for an immobile surface.

In a mathematical analysis of the problem of an *insoluble surface film* on the surface, Dorrestein[113e] found that the surface behaviour should approximate, as $C_s^{-1} \rightarrow 0$, to that for a fully mobile interface, and for $C_s^{-1} \rightarrow \infty$ to eq. (5.31). In general, however, the theory[113e] predicts that at some intermediate value of C_s^{-1}, Δ will pass through a maximum, of the order twice that for a completely rigid film (which would correspond to $C_s^{-1} \rightarrow \infty$). For capillary waves (of $2 > \lambda > 0.1$) the value of C_s^{-1} at which damping coefficient should just exceed Δ_i is 4.3 dynes cm.$^{-1}$; and when the film is compressed to a C_s^{-1} of about 17 dynes cm.$^{-1}$, the damping coefficient reaches the maximum (of the order $2\Delta_i$). When C_s^{-1} is increased to about 80 dynes cm.$^{-1}$, the damping coefficient should decrease to a value only 20% higher than Δ_i, which is approached in the limit of $C_s^{-1} \rightarrow \infty$.

Dorrestein's analysis[113e] indicates that the *surface viscosity* η_s of the film should also influence wave-damping. Surface-viscosity alone, however, cannot cause Δ to pass through a maximum: if $C_s^{-1} = 0$, Δ increases steadily from Δ_c to Δ_i as η_s is increased. Calculations show[113e] that 80% of the change from Δ_c to Δ_i occurs (for $\lambda = 1$ cm.) when $\eta_s = 45 \times 10^{-3}$ surface poise; or for $\lambda = 0.1$ cm., when $\eta_s = 2.3 \times 10^{-3}$ surface poise. In practice one never finds non-zero values of η_s unless C_s^{-1} is also appreciable, and when both factors are acting simultaneously, a non-zero value of η_s depresses the maximum through which Δ passes between Δ_c and Δ_i. At low values of $\lambda (< 0.01$ cm.) surface viscosity should become very important in damping (the critical value of η_s depending approximately on $\lambda^{5/4}$); values as low as 10^{-4} s.p. or 10^{-5} s.p. should have a marked effect.

From the above theory, Δ should exceed Δ_i when C_s^{-1} is of the order 4 dynes cm.$^{-1}$ and η_s is small. These conditions are satisfied for solutions of pure sodium lauryl sulphate below 10^{-3}M, and it is such solutions particularly that give surface films with a high damping coefficient (Table 5-VII). When the concentration is higher, and when there is some dodecanol present, both C_s^{-1} and η_s are higher and, as expected from theory, Δ tends to Δ_i (Table 5-VII). The maximum in Δ has been observed also by Tailby and Portalski[113c] for various dissolved surface-active agents at concentrations of the order 10^{-4}–10^{-3}M.

At frequencies in the megacycle range[113f], damping by surface-active agents seems largely due to surface viscosity—a value of the order 10^{-5}–10^{-4} surface poise is deduced from the experiments for solutions of surface-active agents in the range 10^{-2}–1% (wt). There is also for "Hostapal C. V." an amplitude-dependent phase angle between the capillary wave and the excitation, suggesting that C_s^{-1} is of some importance. The importance of surface viscosity for these very short waves is predicted by Dorrestein's analysis.

With soluble surface-active agents, the damping is complicated by the possible adsorption and desorption of some of the surface film during the passage of the wave. The alternating tendency (Lamb[112]) for expansion (in the troughs) and for contraction (at the crests) still causes surface pressure fluctuations, though the latter may now be reduced by adsorption and

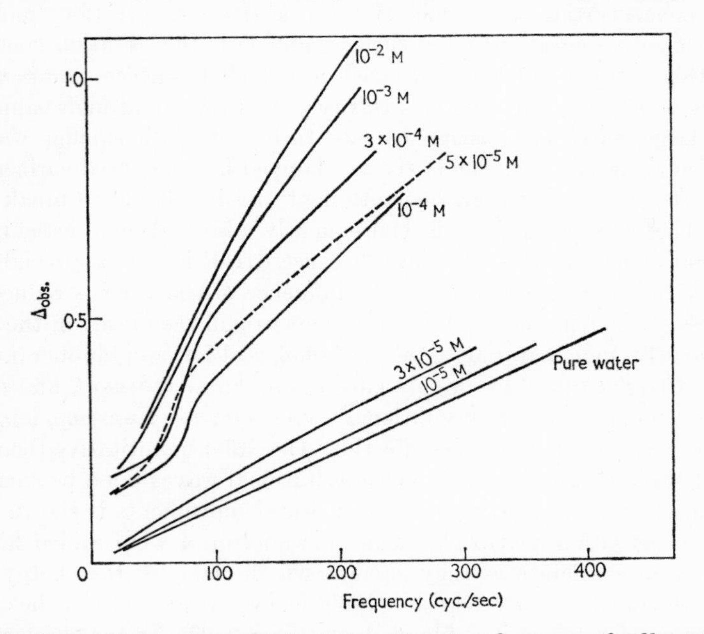

Fig. 5-45. Observed damping coefficients (cm.$^{-1}$) against frequency, for films of sodium lauryl sulphate on water, adsorbed from solutions of various concentrations[113].

desorption from the underlying bulk solution during the passage of the wave. Such "short-circuiting" of the surface stresses will, if it occurs, reduce Δ relative to the value for an insoluble but otherwise similar surface film. Since the calculated relaxation times of films of lauryl-sulphate ions are of the order of 5 milliseconds (pp. 182–3), a change in the characteristics of the wave-damping by such films is predicted when each expansion (and each contraction) occurs over a period of 5 milliseconds, i.e. when the frequency of the waves is 100 cycles sec.$^{-1}$. This effect has now been found experimentally[113], and some of the results are shown in Fig. 5-45. Values of Δ_i for certain of these systems are quoted in Table 5-VII. As eq. (5.31) shows, the plot of Δ_i against frequency should be nearly rectilinear; hence the relatively sudden decreases in the observed Δ as the frequency is decreased below about 100 cycles sec.$^{-1}$ indicates that, in accord with theory, at the low frequencies the surface stresses are reduced. In this region appreciable adsorption and desorption can occur during the passage of each wave.

A different theory, due to Levich[114], predicts that the group C_s^{-1}/γ should be responsible for the damping of short waves, and that when this group approaches unity, the damping should be very considerable. For soluble films, Levich allows for the "short-circuiting" effect by a term $D(\partial c/\partial x)_{x=0}$, which is included in the differential equations expressing the variation of surface concentration with time. Hence c is the concentration (moles l^{-1}) of surface-active agent near the surface, and D is the diffusion coefficient. Using this concept, Levich[114] derived a modified surface compressional modulus which, according to this theory, tends to zero at high values of c.

The **large waves on a rough sea** are turbulent, with smaller waves or ripples on their surfaces. The latter are damped by condensed surface films, making the waves smoother, less turbulent and less liable to break at the crests. The drag coefficient is consequently also reduced, lessening the formation of new ripples. Further, because $d\Pi/dl$ is now appreciable, the tendency to wave-breaking under the influence of the wind is reduced, the net stress of the wind now being insufficient to pull the crests off the waves. Consequently, such materials as cetyl alcohol, seal-fat, and blubber (i.e. fatty acids and triglycerides) are highly effective in damping waves, and only an "oily" swell of the longer gravitational waves remains. Kerosene, long-chain acetamides and phenols are less effective. On either quantitative theory, the effect of surface films on the longer gravitational waves must be small.

The marked effect of surface films on water movements is shown by the striking observation that slicks of hexadecanol on a wind-ruffled lake (see Frontispiece) accelerate as they move down the lake[114a]: the energy of the wind is all transferred to the water lying below the area of the slick, which therefore accelerates as it is blown down the lake[113a]. In the absence of the film, turbulent ripples dissipate most of the energy transferred to the water.

SHEAR ELASTIC MODULI OF MONOLAYERS

When a monolayer is sheared, at constant area, between two concentric rings lying in the plane of the surface (Fig. 5-46), the applied stress on any element of the film is $L/2\pi r^2$, where L is the applied torque, which is constant throughout the annulus. The corresponding shear strain is defined as $r d\boldsymbol{\theta}/dr$. The shear elastic modulus $\mathbf{E}_s$ is defined as the ratio of these quantities[101,115,116]:

$$\mathbf{E}_s = \frac{L}{2\pi r^3} \cdot \frac{dr}{d\boldsymbol{\theta}}$$

Integration gives

$$\mathbf{E}_s = \frac{L}{4\pi\boldsymbol{\theta}} \left(\frac{1}{r_1^2} - \frac{1}{r_2^2} \right) \tag{5.32}$$

IP—K

and substitution for L (if one ring is stationary while the other, attached to a torsion wire and having a total moment of inertia $\mathbf{I}$, is allowed to oscillate) gives the relation[111,116]:

$$\mathbf{E}_s = \pi \mathbf{I} \left\{ \frac{1}{t^2} - \frac{1}{t_0^2} \right\} \left\{ \frac{1}{r_1^2} - \frac{1}{r_2^2} \right\} \tag{5.33}$$

Here t_0 and t are the periods measured respectively for the clean and film-covered surfaces. $\mathbf{E}_s$ has the dimensions of dynes cm.$^{-1}$ or g.sec.$^{-2}$

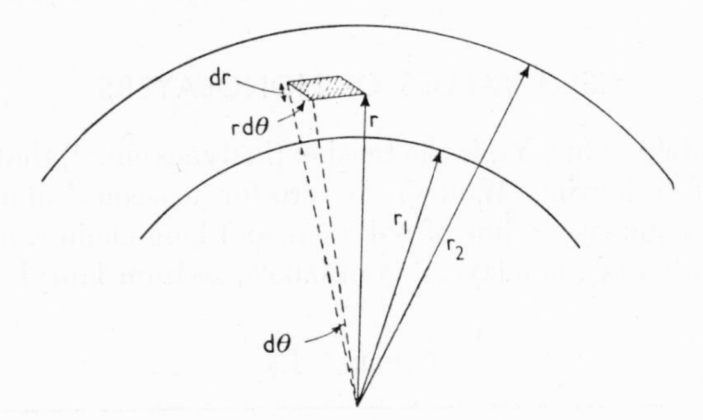

Fig. 5-46. Shear strain of an element of a surface film between two concentric rings lying in the plane of the surface.

The experimental methods[115-119] are based on measuring the movement in the surface of a circular ring, weighted and joined to a torsion wire and head. The film between this ring and the containing circular dish is thus subjected to shear, and the damping is measured. Alternatively[119], a ring in the surface is attached to a torsion head. The torsion head is then rotated through a small

TABLE 5-VIII

Monolayer (Air-water)	E_s(dynes cm.$^{-1}$)
Clean surface	0
"Gaseous"	0
Ovalbumin spread[119] on 0.1 N-HCl at A = 3 m.^{2}mg.$^{-1}$	0.05
As above, A = 2 m.^{2}mg.$^{-1}$	0.13
As above, A = 1.2 m.^{2}mg.$^{-1}$	0.3
Poly-ε-amino-caproic acid at A = 50 Å^2 per residue[119]	0.02
As above[119], A = 25 Å^2 per residue	0.35
Condensed films of palmitic or stearic acid[117]	10
Adsorbed films[118] of saponin (multilayers?) at short times	70
As above, after long times[118]	300

angle, thus applying a torque to the ring and a stress to the monolayer. The consequent strain is followed by noting the positions of the ring, as ascertained using an optical lever, at short time intervals, and applying eq. (5.32). Hysteresis diagrams permit a detailed analysis of the viscous-elastic properties of the monolayer; for proteins the effect of pH is considerable, and is apparently related to cross-linking through NH_2 groups on the side-chains[120] and to the rupturing of hydrogen bonds accompanying the spreading process[121]. Typical values of $\mathbf{E}_s$ are shown in Table 5-VIII.

YIELD VALUES OF MONOLAYERS

The surface yield value, Y_s, is the tension (in dynes cm.$^{-1}$) that a monolayer will sustain before flowing. While Y_s is zero for "gaseous" films, it becomes appreciable for condensed films of polymers and long-chain compounds. The surface yield value of monolayers from 0.05% sodium lauryl sulphate con-

TABLE 5-IX

Monolayer	Y_s(dynes cm.$^{-1}$)
Sodium lauryl sulphate adsorbed from 0.1% solution, with no lauryl alcohol[123]	0
As above[123], with 0.005 g. lauryl alcohol per 100 ml.	0.05
As above[123], with 0.008 g. lauryl alcohol for 100 ml.	0.06
Lauryl alcohol alone, adsorbed from saturated solution[123]	0.12
Spread protein film[102] at pressures above 4.4 dynes cm.$^{-1}$	0.1
Adsorbed film[118] (polymolecular?) of saponin	0.3 to 1

taining 0.005% long-chain alcohol increases linearly with the number of carbon atoms in the alcohol[122].

Typical numerical values of surface yield values are given in Table 5-IX.

DIFFUSION IN MONOLAYERS

To estimate the two-dimensional diffusion coefficient, the surface of the trough is divided into two parts by a very thin glass rod. On one side of this is spread protein, and on the other castor oil or other material, so that the pressure on the two films is equal. The rod is now removed, and the advancing front of protein as a function of time is found from surface potential measurements along the trough[124].

FIBRES FROM MONOLAYERS

When protein monolayers are compressed in a Langmuir trough to occupy a very small area, they may frequently be removed quantitatively from the surface, by means of a fine glass rod, in the form of single elastic fibres, of diameters 10^{-3} to 10^{-4} cm. The readiness with which such fibres can be formed is a rough index of molecular size and of the amount of cross-linking between the molecules in the monolayer, and depends also on whether a lipid is co-adsorbed in the surface[125]. If ovalbumin is treated with heat, ultraviolet radiation, or urea before spreading, its monolayers give either no fibres at all or else fibres much more fragile than those derived from untreated protein[126]. Such fibres, formed by wind blowing over a monolayer, may have been important in the development of the first living cells[125].

To form protein fibres, a species of water-snail from South America is able to use its foot as a Langmuir trough: the edible fibres are formed from protein monolayers as follows. A young snail in the usual resting position, adhering to the wall of the tank or to a floating plant just below the water surface, curves part of its foot to form, either alone or with the wall of the tank, a funnel with its rim in the surface. By rapid undulatory movements of its foot, the snail then draws the protein monolayer towards the mouth of the funnel, compressing the protein film into fibrous masses which are subsequently eaten, thus avoiding a disproportionately high intake of water[127].

The ingestion of protein monolayers seems fairly common among small aquatic animals; a tadpole may ingurgitate in one day its own weight of protein, taken from a monolayer at the air-water interface: ponds, lakes, and rivers usually contain monolayers of protein, presumably from decaying leaves or animal matter[128]. Larger animals are less interested in such monolayers: although adult water-snails sometimes compress the protein monolayer, absent-mindedly as it were, while engaged in eating a lettuce leaf, this latter pursuit is so much more rewarding that the snail quickly abandons any protein fibres it has formed[127].

REFERENCES

1. Franklin (1765), see "The Ingenious Dr. Franklin" (ed. Goodman) (1931); "Famous American Men of Science" (ed. Crowther), Secker and Warburg, London (1937);
   Rayleigh, *Phil. Mag.* **48**, 321 (1899);
   Devaux, *P.V. Soc. Sci. phys. nat. Bordeaux*, 19th Nov., 3rd Dec. (1903); ibid. 7th Jan., 14th April (1904); *J. Phys. Rad.* **3**, 450 (1904);
   Hardy, *Proc. roy. Soc.* **A86**, 610 (1912); **A88**, 303 (1913);
   Langmuir, *J. Amer. chem. Soc.* **38**, 2221 (1916); **39**, 1848 (1917);
   Adam, see ref. 43;
   Rideal, "Surface Chemistry", Cambridge University Press (1930).
2. Fox and Zisman, *Rev sci. Instrum.* **19**, 274 (1948).
3. Ställberg and Teorell, *Trans. Faraday Soc.* **35**, 1413 (1939).
4. Dervichian, *Nature, Lond.* **144**, 629 (1939).
5. Guastalla, *Cah. Phys.* **10**, 30 (1942); *C.R. Acad. Sci., Paris* **206**, 993 (1938).
6. Harkins, "Physical Chemistry of Surface Films", Reinhold, London (1952).
7. Dervichian, *J. Phys. Rad.* **6**, 221, 429 (1935).
8. Dervichian and de Bernard, *Bull. Soc. Chim. Biol., Paris* **37**, 943 (1955);
   Dervichian, *Kolloidzschr.* **126**, 15 (1952).
9. Guastalla, *C.R. Acad. Sci., Paris* **189**, 241 (1929).
10. Harkins and Anderson, T. F., *J. Amer. chem. Soc.* **59**, 2189 (1937).
11. Wilhelmy, *Ann Phys., Lpz.* **119**, 177 (1863).
12. Abribat and Dognon, *C.R. Acad Sci., Paris* **208**, 1881 (1939); *J. Phys. Rad.* **10**, 22 (1939).
13. Cheesman, *Biochem. J.* **50**, 667 (1952).
14. Guastalla, *C.R. Acad Sci., Paris* **206**, 993 (1938); *J. Chim. phys.* **43**, 184 (1946).
15. Ter Minassian-Saraga, *Proc. 2nd Int. Congr. Surf. Act.* **1**, 36, Butterworths (1957).
16. Kalousek, *J. Chem. Soc.* 894 (1949);
   Tvaroha, *Chem. Listy* **48**, 183 (1954).
17. Askew and Danielli, *Proc. roy. Soc.* **A155**, 695 (1936); *Trans. Faraday Soc.* **36**, 785 (1940).
18. Langmuir, *Cold Spr. Harb. Symp. quant. Biol.* **6**, 193 (1938).
19. Alexander, A. E. and Teorell, *Trans. Faraday Soc.* **35**, 727 (1939).
20. Crisp., *J. Colloid Sci.* **1**, 49, 161 (1946).
21. Davies and Llopis, *Proc. roy. Soc.* **A227**, 537 (1955).
22. Cheesman, *Ark. Kemi. Min. Geol.* **22B** (1) and **24B** (4) (1946).
23. Davies, *Biochim. biophys. Acta.* **11**, 165 (1953);
   Davies, *Trans. Faraday Soc.* **48**, 1052 (1952).
24. Michel, *J. Chim. phys.* **54**, 211 (1957).
25. Langmuir, *J. chem. Phys.* **1**, 756 (1933).
26. Frumkin, *Z. phys. Chem.* **116**, 485 (1925).
27. Volmer, *Z. phys. Chem.* **115**, 253 (1925).
28. Haydon and Taylor, F. H., *Phil. Trans..* **252**, 225 (1960).
29. Davies, *J. Colloid Sci.* **11**, 377 (1956).
30. Ter Minassian-Saraga, *J. Chim. phys.* **52**, 80 (1955).
31. Phillips, J. N., Thesis, University of London (1954).
32. Davies, in "Surface Phenomena in Chemistry and Biology" p. 55, Pergamon Press, London and New York (1958).
33. Brady, *J. Colloid Sci.* **4**, 417 (1949).

34. Crisp, in "Surface Chemistry" p. 65, Butterworths (1949).
35. Brady and Brown, A.G., "Monomolecular Layers" p. 33, A.A.A.S. (1954).
36. Hutchinson, E., *J. Colloid Sci.* **3**, 413 (1948).
37. Pethica, *Trans. Faraday Soc.* **50**, 413 (1954).
38. Davies, *Proc. roy. Soc.* **A208**, 224 (1951).
39. Cockbain, *Trans. Faraday Soc.* **50**, 874 (1954).
40. Kling and Lange, *Proc. 2nd Int. Congr. Surf. Act.* **1**, 295, Butterworths (1957).
41. Haydon and Phillips, *Trans. Faraday Soc.* **54**, 698 (1958).
42. Ter Minassian-Saraga, *Proc. 2nd Int. Congr. Surf. Act.* **1**, 36, Butterworths (1957).
43. Adam, "The Physics and Chemistry of Surfaces" (3rd ed.), Oxford University Press (1941).
44. Anderson, P. J. and Pethica, "Biochemical Problems of Lipids" p. 24, Butterworths (1955).
45. Davies, *Trans. Faraday Soc.* **44**, 909 (1948).
46. Cheesman and Davies, *Advanc. Protein Chem.* **9**, 439 (1954).
47. Ekwall and Ekholm, *Proc. 2nd Int. Congr. Surf. Act.* **1**, 23, Butterworths (1957).
48. Parreira and Pethica, ibid. **1**, 44 (1957).
49. Osipow, Snell, and Hickson, ibid. **1**, 50 (1957).
50. Scholberg, Guenthner, and Coon, *J. phys. Chem.* **57**, 932 (1953).
51. Arrington and Patterson, *J. phys. Chem.* **51**, 241 (1953).
52. Klevens and Davies, *Proc. 2nd Int. Congr. Surf. Act.* **1**, 31, Butterworths (1957).
53. Ellison and Zisman, *J. phys. Chem.* **60**, 416 (1956).
54. Banks, *Nature, Lond.* **174**, 365 (1954).
55. Banks, *Proc. 2nd Int. Congr. Surf. Act.* **1**, 16, Butterworths (1957).
56. Fox, H. W., Taylor, P. W., and Zisman, *Industr. Engng Chem.* **39**, 1401 (1947).
57. Spink and Sanders, *Trans. Faraday Soc.* **51**, 1154 (1955).
58. Goodrich, *Proc. 2nd Int. Congr. Surf. Act.* **1**, 85, Butterworths (1957).
59. Dervichian, in "Surface Phenomena in Chemistry and Biology" p. 70, Pergamon Press, London and New York (1958).
60. Davies and Mayers, *Trans. Faraday Soc.* **56**, 690 (1960).
61. Schulman and Cockbain, *Trans. Faraday Soc.* **36**, 651, 661 (1940).
62. Schulman and Rideal, *Proc. roy. Soc.* **A130**, 284 (1931); *Proc. roy. Soc.* **A138** 436 (1932).
63. Betts and Pethica, *Trans. Faraday Soc.* **52**, 1581 (1956).
64. Payens, *Proc. 2nd Int. Congr. Surf. Act.* **1**, 64, Butterworths (1957).
65. Labrouste, *Ann. Phys., Paris* **14**, 164 (1920).
66. Crisp, in "Surface Chemistry" pp. 17, 23, Butterworths (1949).
67. Stenhagen, *Nature, Lond.* **55**, 36 (1945).
68. Dervichian, in "Surface Chemistry" p. 47, Butterworths (1949).
69. Joly, in "Surface Phenomena in Chemistry and Biology" p. 88, Pergamon Press, London and New York (1958); in "Surface Chemistry" p. 37, Butterworths (1949).
70. Cary and Rideal, *Proc roy. Soc.* **A109**, 301, 312 (1925).
71. Boyd and Schubert, *J. phys. Chem.* **61**, 1271 (1957);
Boyd, *J. phys. Chem.* **62**, 536 (1958).
72. Devaux, *P.V. Soc. Sci. phys. nat., Bordeaux*, Nov. (1903);
Gorter and Grendel, *Trans Faraday Soc.* **22**, 477 (1926);
Hughes and Rideal, *Proc. roy. Soc.* **A137**, 62 (1932);
Cheesman and Davies, *Advanc. Protein Chem.* **9**, 439 (1954).

73. Bull, *Advanc. Protein Chem.* **3**, 95 (1947); *J. Amer. chem. Soc.* **67**, 4 (1945).
74. Guastalla, *C.R. Acad. Sci., Paris* **208**, 1078 (1939).
75. Benhamou, *J. Chim. phys.* **53**, 32, 44 (1956);
Dieu, *Bull. Soc. chim. Belg.* **65**, 740, 847, 1035 (1956).
76. Davies, *Biochim biophys. Acta* **11**, 165 (1953).
77. Michel and Benhamou, *C.R. Acad Sci., Paris* **228**, 1577 (1949).
78. Imahori, *Bull. chem. Soc. Japan* **25**, 121 (1952). In English.
79. Singer, *J. chem. Phys.* **16**, 872 (1948).
80. Huggins, *J. phys. Chem.* **46**, 151 (1942).
81. Davies, *J. Colloid Sci. Suppl.* **1**, 9 (1954).
82. Crisp, in "Surface Phenomena in Chemistry and Biology" p. 23, Pergamon Press, London and New York (1958).
83. Cheesman and Schuller, *J. Colloid Sci.* **9**, 113 (1954).
84. Katchalsky and Miller, *J. phys. Chem.* **55**, 1182 (1951).
85. Miller and Katchalsky, *Proc. 2nd Int. Congr. Surf. Act.* **1**, 159, Butterworths (1957).
86. Rideal, *J. Polym. Sci.* **16**, 531 (1955).
87. Allen and Alexander, A. E., *Trans. Faraday Soc.* **46**, 316 (1950).
88. Frisch and Simha, *J. chem. Phys.* **24**, 652 (1956).
89. Llopis and Rebollo, *J. Colloid Sci.* **11**, 543 (1956);
Llopis and Albert, *Arch. Biochem. Biophys.* **81**, 146, 159 (1959).
90. Schulman and Teorell, *Trans. Faraday Soc.* **34**, 1337 (1938);
Crisp, *Trans. Faraday Soc.* **42**, 619 (1946);
Milne-Thompson, "Hydrodynamics" (4th ed.) p. 597, Macmillan, London (1960).
91. Joly, *Kolloidzschr.* **89**, 26 (1939).
92. Brown, A. G., Thuman, and McBain. *J. Colloid Sci.* **8**, 491 (1953).
93. Sanders, Camp, and Durham, *Research Suppl.* **8**, S18 (1955).
94. Bernard, *Proc. 2nd Int. Congr. Surf. Act.* **1**, 7, Butterworths (1957).
95. Kalousek and Vysin, *Coll. Trav. chim. Tchécosl.* **20**, 777 (1955).
96. Ewers and Sack, *Nature, Lond.* **168**, 964 (1951); *Aust. J. Chem.* **7**, 40 (1954).
97. Davies, *Proc. 2nd Int. Congr. Surf. Act.* **1**, 220, Butterworths (1957).
98. Joly, *J. Chim. phys.* **44**, 206 (1947); *J. Colloid Sci.* **5**, 49 (1950); *J. Colloid Sci.* **11**, 519 (1956).
99. Cumper and Alexander, A. E., *Trans. Faraday Soc.* **46**, 235 (1950);
Blakey and Lawrence, *Disc. Faraday Soc.* **18**, 268 (1954).
99a. Adam and Rosenheim, *Proc. roy. Soc.* **A126**, 25 (1929);
Langmuir, Schaefer, V. J. and Sobotka, *J. Amer. chem. Soc.* **59**, 1251 (1937);
Fourt, *J. phys. Chem.* **43**, 887 (1939).
100. Davies and Rideal, *J. Colloid Sci. Suppl.* **1**, 1 (1954).
101. Tschoegl, *J. Colloid Sci.* **13**, 500 (1958).
102. Cheesman and Sten-Knudsen, *Biochem. biophys. Acta* **33**, 158 (1959).
103. Pockels, *Nature, Lond.* **43**, 437 (1891).
104. Brown, R. C., *Proc. phys. Soc. Lond.* **48**, 312, 323 (1936).
105. Davies and Bradley, Research Project, Dept. Chem. Engng, Cambridge (1960).
106. Thomson, *Phil. Mag.* **42**, 368 (1871).
Lamb, "Hydrodynamics", Cambridge University Press (1959).
107. Rayleigh, *Phil. Mag.* **30**, 386 (1890);
Lamb, see ref. 106.
108. Downing and Truesdale, *J. appl. Chem.* **5**, 570 (1955).
109. Jeffreys, *Proc. roy. Soc.* **A107**, 189 (1925);
Stanton, Marshall, and Houghton, *Proc. roy. Soc.* **A137**, 283 (1932);

Keulegan, *J. Res. nat. Bur. Stand.* **46**, 358 (1951);
Van Dorn, *J. Mar. Res.* **12**, 249 (1953);
Ursell, in "Surveys of Mechanics" p. 216, Cambridge University Press (1956);
van Rossum, *Chem. Engng Sci.* **11**, 35 (1959).
110. Bond, J. and Donald, *Chem. Engng Sci.* **6**, 237 (1957);
Kirkbride, *Industr. Engng Chem.* (*Anal.*) **26**, 425 (1934);
Friedman and Miller, *Industr. Engng Chem.* **33**, 885 (1941);
Binnie, *J. Fluid Mech.* **2**, 551 (1957); **5**, 561 (1959);
Benjamin, Brooke *J. Fluid Mech.* **2**, 554 (1957); **3**, 657 (1958).
111. Davies, Bell, and Law, Research Project, Dept. Chem. Engng, Cambridge (1960).
112. Lamb, "Hydrodynamics" Arts. 236, 240, 250, 334a, 348–351, Dover, New York (1945);
Wiegart, *Phys. Z.* **44**, 101 (1943);
Deacon, Sheppard, and Webb, E. K. *Aust. J. Phys.* **9**, 511 (1956);
Mansfield, *Aust. J. appl. Sci.* **10**, 73 (1959);
Vines, *Quart. J. R. met. Soc.* **85**, 159 (1959); *Aust. J. Phys.* **13**, 43 (1960);
Keulegan, *J. Res. nat. Bur. Stand.* **46**, 358 (1951);
Francis, *Proc. roy. Soc.* **A206**, 387 (1951);
Van Dorn, *J. Mar. Res.* **12**, 249 (1953);
Sutton, "Micrometerology" p. 80, McGraw-Hill, New York (1953).
113. Davies and Vose, to be published (1963).
113a. Davies, *Chemistry and Industry* 906 (1962).
113b. van den Tempel, van voorst Vader and Jonkman, to be published (1963).
113c. Tailby and Portalski, *Trans. Inst. Chem. Engrs* **39**, 328 (1961).
113d. Kafesjian, Plank and Gerhard, *A.I.Ch.E.J.* **7**, 463 (1961).
113e. Dorrestein, *Proc. Acad. Sci. Amst.* **B54**, 260, 350 (1951).
113f. Eisenmenger, *Acoustica* **9**, 327 (1959).
113g. Goodrich, *Proc. roy. Soc.* **A260**, 481, 490, 503 (1961).
114. Levich, "Physico-chemical Hydrodynamics" U.S.S.R. Acad. Sci. (1952).
114a. McArthur, *Research* **15**, 230 (1962).
115. Langmuir and Schaefer, *J. Amer. chem. Soc.* **59**, 2400 (1937).
116. Trapeznikov, *C.R. Acad. Sci., U.S.S.R.* **63**, 57 (1948).
117. Mouquin and Rideal, *Proc. roy. Soc.* **A114**, 690 (1927).
118. van Wazer, *J. Colloid Sci.* **2**, 223 (1947).
119. Tachibana and Inokuchi, *J. Colloid Sci.* **8**, 341 (1953).
120. Fourt, *J. phys. Chem.* **43**, 887 (1939).
121. Llopis and Albert, *An. Soc. esp. Fís. Quím.* **558**, 109 (1959).
122. Evans, W. P., quoted in *Proc. 2nd Int. Congr. Surf. Act.* **1**, 228, Butterworths (1957).
123. Brown, A. G., Thuman, and McBain, *J. Colloid Sci.* **8**, 491 (1953).
124. Imahori, *Bull. chem. Soc.* Japan **25**, 13 (1952). In English.
125. Goldacre, in "Surface Phenomena in Chemistry and Biology" p. 278, Pergamon Press, London and New York (1958).
126. Kaplan and Fraser, *Nature, Lond.* **171**, 559 (1953).
127. Cheesman, *Nature, Lond.* **178**, 987 (1956).
128. Goldacre, *J. Anim. Ecol.* **18**, 36 (1949).

Chapter 6

Reactions at Liquid Surfaces

REACTIONS IN MONOLAYERS

What factors distinguish monolayer reactions from those in the bulk phase? Is the energy of activation the same? What of catalysis at liquid interfaces? These are the questions to be answered in this chapter. In particular, there are two main differences[1] between reactions at solid surfaces and those at liquid surfaces: firstly, only the latter are equipotential surfaces, i.e. all the molecules in the surface are at the same chemical potential. Secondly, only at a liquid surface can the orientation and accessibility of the reactant groups in the monolayer be controlled at will; in reactions at solid surfaces no control is possible, while in bulk-phase reactions the orientation of the reactant groups is random.

In a monolayer the orientation of all the molecules (and reactive groups therein) is practically identical, depending only on the surface pressure. Further, this orientation can be determined precisely from studies of the surface dipole moments (Chapter 2). With the film-forming reactant molecules in different orientations and configurations, the corresponding energies of activation can be measured. Complete inhibition can result from applying a pressure high enough to force the reactive groups away from the aqueous surface, and, in photochemical reactions, orientation of all the chromophores can alter the quantum efficiency.

Electrical charges, too, are important, and very large catalytic effects as well as kinetic salt factors can result from the presence of charged groups in the monolayer.

Rate Constants

The energies of activation for reactions in monolayers and in the bulk phase are usually equal[1]; the apparent exceptions to this are provided by reactions carried out at constant surface pressure over the temperature range in which the film "expands" (Chapter 5), this expansion providing an increased accessibility of reactive groups to the reactant in the aqueous phase, superimposed on the normal effect of temperature on a chemical reaction.

For reactions carried out at constant surface area, however, there is a remarkable similarity between the energies of activation and rate constants in the surface and the values of these for similar reactions in bulk phases; deviations which sometimes occur are caused principally by "screening" of the reactive groups by the close packing of the surface. Slow desorption of the products of reaction (e.g. of the soap in the alkaline hydrolysis of trilaurin monolayers[2]) can also retard reaction, particularly in the later stages of the reaction.

Experimental methods

Suppose that a monolayer of a long-chain reactant (e.g. ethyl palmitate) is spread on aqueous alkali. The ester will then be hydrolysed (e.g. to palmitic acid and ethyl alcohol), and if after time t the surface concentration of the remaining reactant is n_r molecules $cm.^{-2}$, the rate of reaction is expressed by

$$\frac{-dn_r}{dt} = k\, n_r \cdot {}_s[OH'] \tag{6.1}$$

Here k is the velocity constant for the hydrolysis and ${}_s[OH']$ is the concentration of hydroxyl ions in the plane of the monolayer. For a reaction such as this it is always valid to assume that the kinetics are pseudo first order, since the amount of reactant taken from the bulk phase by the monolayer will be quite insignificant in affecting the concentration of the bulk. Further, if the film is uncharged, ${}_s[OH'] = {}_b[OH']$, the latter being the concentration of hydroxyl ions in the bulk of the underlying liquid. Hence, by integration of eq. (6.1),

$$\int \frac{dn_r}{n_r} = -k \cdot {}_b[OH']t + \text{constant}$$

i.e.

$$\ln n_r = -k \cdot {}_b[OH']t + \text{constant} \tag{6.2}$$

By plotting $\ln n_r$ against t we should therefore obtain a straight line, the slope giving k. Units of k are conveniently $min.^{-1}\ mole^{-1}$ litre.

In practice the most convenient variable to study is not n_r but ΔV, the surface potential. If the film is at constant area, ΔV after any time t is given by

$$\Delta V = 4\pi n_r \mu_r + 4\pi n_p \mu_p$$

where n_r and n_p are the surface concentrations of reactant and product at time t, and μ_r and μ_p are their (constant) surface dipole moments. Further, if these were initially n molecules of reactant per $cm.^2$, and if no long-chain molecules are lost,

$$n_r + n_p = n$$

and so n_r is given in terms of ΔV by:

$$n_r = \frac{\Delta V - \Delta V_{t=\infty}}{4\pi(\mu_r - \mu_p)} \tag{6.3}$$

where $\Delta V_{t=\infty}$ has been substituted for $4\pi n\mu_p$ on the assumption that the reaction eventually proceeds to completion, with all the surface reactant converted to product.

From eqs. (6.2) and (6.3) it is clear that

$$\ln(\Delta V - \Delta V_{\infty}) = -k._b[OH']t + \text{constant}, \tag{6.4}$$

so that a plot of $\ln(\Delta V - \Delta V_{\infty})$ against t will permit evaluation of k for any given surface area and temperature (see Fig. 6-9 below).

Radioactive tracers can also be used to follow the course of a reaction. The principle is that radiation from an appreciable distance below the surface will be absorbed in the water, and so will not affect a Geiger counter above the surface. The same tracers as are discussed in Chapter 4 may be used: these are listed in Table 4-IX. Tritium is the most effective.

Steric Factors

The first, and still one of the most striking, of the monolayer reactions investigated was the oxidation of a film of oleic acid[3,4] by permanganate. The double bond is oxidized thus:

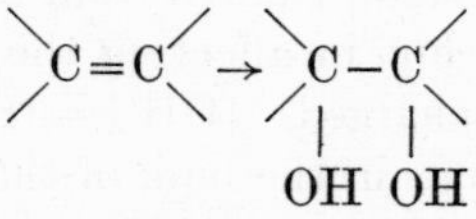

$$\rangle C{=}C\langle \;\rightarrow\; \rangle C(OH){-}C(OH)\langle$$

When the available area is high, the oleic acid molecules lie flat on the water surface (Fig. 6-1), the double bonds being freely accessible to the underlying permanganate solution. If, however, we compress the film, the surface area

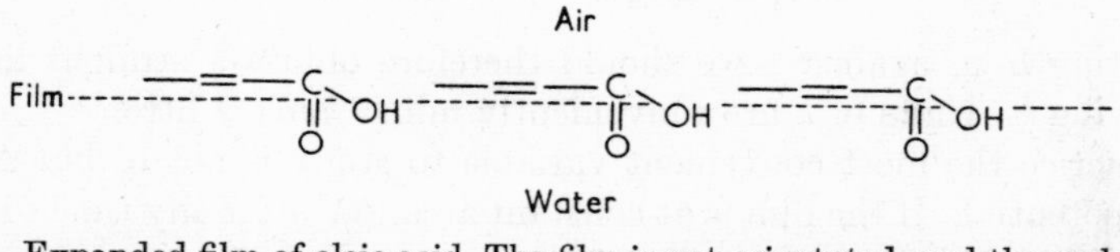

Air

Film ——=——C(=O)OH ——=——C(=O)OH ——=——C(=O)OH

Water

Fig. 6-1. Expanded film of oleic acid. The film is not orientated, and the molecules all lie flat and independently on the surface of the water. The double bonds are all in contact with the water.

available to each chain will eventually become less than the area occupied by the hydrocarbon chain lying flat on the surface, causing some of the chains to stand more nearly vertically, and finally all of them do so at sufficient compression (Fig. 6-2). Those double bonds which have been pushed above the water surface can no longer be oxidized, so that the rate constant for the oxidation process should decrease. This is borne out by experiment (Fig. 6-3). At very high surface pressures the rate of oxidation is very small, practically all the double bonds being separated from the aqueous phase by a "membrane" of hydrocarbon chains. Although this layer is only 10 Å thick, because of the high surface pressure of the orientated film, the permanganate ions cannot break through. Such protection by extremely thin "membranes" is of obvious importance in fields as far apart as corrosion and physiology. Triolein films[5], as well as those of erucic and brassidic acids[6], show similar effects, together with individual characteristics. Oxidation may also be studied with ozone in the gas phase.

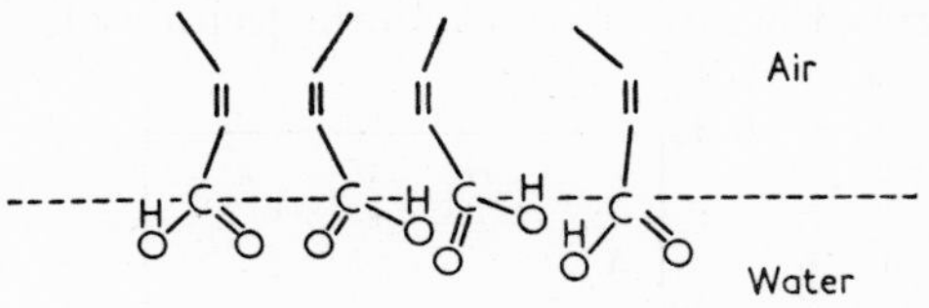

Fig. 6-2. Compressed film of oleic acid, showing that the non-polar hydrocarbon chains are orientated towards the air while the dipoles of the carboxyl group are orientated towards the water. Practically all the double bonds have been forced above the water.

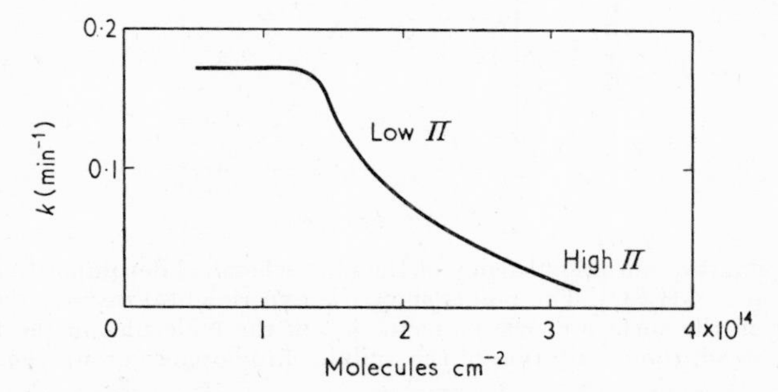

Fig. 6-3. The rate of attack of permanganate on the double bonds of oleic acid is lower when Π is high enough to force some of them off the aqueous surfaces. The oleic acid film is spread on 0.003 % permanganate in 0.01 N H_2SO_4.

Another type of monolayer reaction showing steric factors is the *hydrolysis of long-chain esters* by acid or alkali in the substrate. When, for example, ethyl palmitate molecules are packed tightly in the surface, surface potential studies show that the ethyl chains are forced down into the water below the potentially reactive carbonyl links[7] (Fig. 2-13). This layer of hydrocarbon chains protects the ester groups from attack by the acid or alkali in the water, with the result that hydrolysis is retarded to only 12% of its former rate. These ester hydrolyses confirm the use of eq. (6.4) in interpreting the rate: for the hydrolysis of γ-stearolactone monolayers[8] on alkaline substrates of concentrations between 0.4 N and 2.0 N, the reaction velocity constant at 25°C is $(8.4 \pm 1.0) \times 10^{-2}$ min.$^{-1}$ mole^{-1} litre.

Enzymic hydrolyses at surfaces containing long-chain esters also show strong steric effects, due again to a protective sheath of hydrocarbon chains protecting the ester linkages. Drops of ethyl butyrate, for example, are readily hydrolysed by pancreatin, though an emulsion of ethyl benzoate is not digested at all under the same conditions[9]. This suggests that the bulky benzoate ring hinders the approach of the pancreatin to the surface of the drops. Monolayers of lecithin can be enzymically hydrolysed, giving finally films of lipolecithin: the snake venom enzymes are particularly effective[10]. At higher film pressures, the rate of attack on the lecithin film by the enzymes in

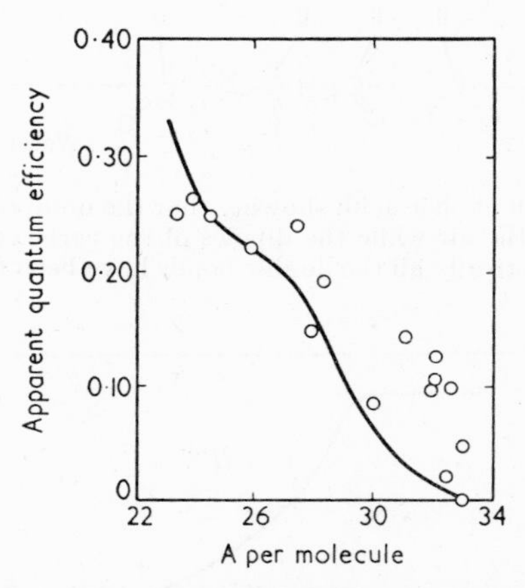

Fig. 6-4. Apparent quantum efficiency of the photochemical decomposition of films of stearanilide on 5 N-H_2SO_4. The points show the experimental results: the full line is calculated from the surface dipole moment, μ_D, of the molecules in the film. As the film is compressed, the orientation of the anilide chromophore group changes[12].

the underlying solution can be reduced fourteen times. This suggests that the enzyme of the venom must require to be coupled not only with the unsaturated hydrocarbon chain (which becomes hydrolysed) but also with some other point in the lecithin molecule. Compression of the lecithin film would then affect the rate by altering the spacing between these essential points of attachment. Protein has a remarkable protective effect on the lecithin monolayer: ovalbumin will reduce the rate of attack of the venom of the black tiger snake by tenfold[10].

Another interesting reaction is that of lecithinase on a monolayer of lecithin, conveniently studied by the radioactive tracer method. Normally this enzyme hydrolyses only lyso-phospholipids, and it only attacks the lecithin monolayer if the latter contains small amounts of dicetyl-phosphoric acid or similar molecules[11]. We believe that an electrical charge on the surface is necessary to attract some part of the enzyme molecule close to the monolayer, added to which there is a weak van der Waals interaction between the enzyme and the lecithin monolayer. Only when these factors act together will the enzyme be attached sufficiently closely to the surface to cause hydrolysis, and the experimental complexities in the dependence of the rate on surface pressure and charge may be due to this additivity of the two energies.

Photochemical reactions in monolayers are markedly affected by the anisotropy of the orientated chromophores in the film. In particular, the apparent quantum efficiency may vary widely according to the orientation

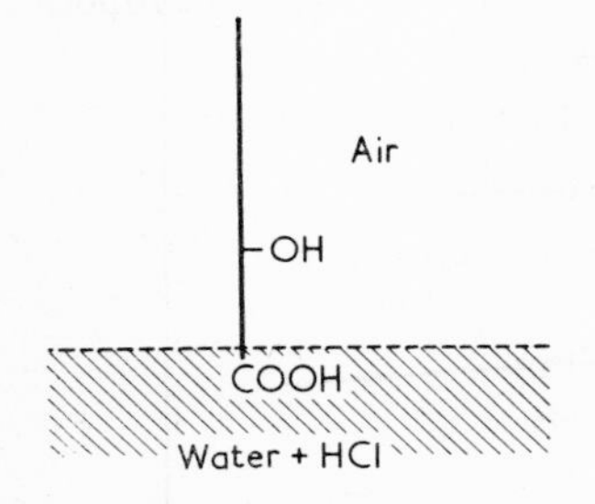

Fig. 6-5. Molecular orientation in a film of γ-hydroxystearic acid at high pressures. The hydroxyl group is forced out of the aqueous surface, and lactonization by aqueous acid is slow.

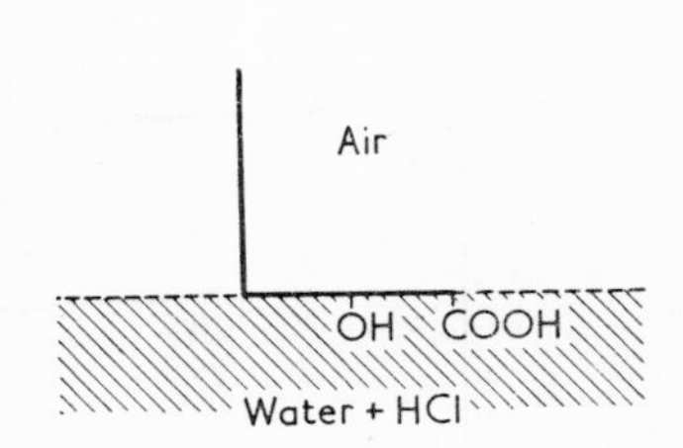

Fig. 6-6. Work must be done against the pressure of the film to push the hydroxyl group of γ-hydroxystearic acid on to the water surface. This occurs more readily if Π is low. If Π is very low, most of the hydroxyl groups are on the aqueous acid, and hence lactonization is rapid.

of the absorbing molecular groups at any particular molecular area in the surface, and this orientation can be exactly deduced from the apparent vertical dipole moment of the molecules in the monolayer. Figure 6-4 shows results on the photochemical decomposition of stearanilide[12], leading ultimately to a film of stearic acid, with the aniline diffusing away into the water as fast as it is formed. Similar results are found for the photolysis of benzylstearylamine and β-phenylethylstearyl-amine, as well as of proteins[13].

Polymerization within a monolayer depends on the relative orientation of the reactive groups at different film pressures. Gee[14] found that polymerization within films of the maleic anhydride compound of β-eleostearin is faster at higher film pressures under certain conditions, although reaction between the same film and an oxidizing agent in the underlying solution is slower at high film pressures. In Russia reactions in monolayers between amines and aldehydes have also been studied[15].

Lactonization, catalysed by acid, can also occur as a monolayer reaction, with marked steric effect. Such reactions as, for example, the lactonization of γ-hydroxystearic acid, are of interest for two reasons: the reaction includes two groups, both held in the film, and also it is amenable to strict mathematical treatment[1]. The retardation at higher film pressures is a direct function of the number of hydroxyl groups forced out of the aqueous interface (Figs. 6-5 and 6-6), and the fraction of hydroxyl groups forced out at any pressure can be calculated from the pressure directly or from the surface dipole moments; this fraction can be compared with the experimental decrease in the rate of hydrolysis (Fig. 6-7).

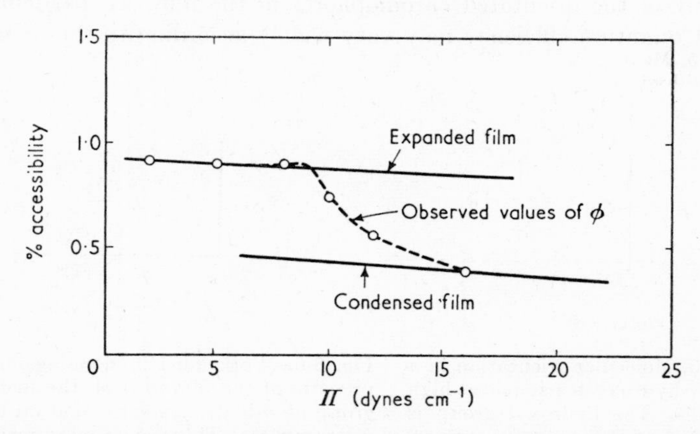

Fig. 6-7. Variation of the accessibility of the hydroxyl groups of γ-hydroxystearic acid with film pressure. Full lines show values calculated from rates of lactonization in the monolayer.

Electrical Factors

At a surface the kinetic effect of an electrical charge is much more pronounced than in bulk. There are two reasons[1] for this; firstly, all the electrical lines of force, instead of radiating spherically as from an isolated point charge, are concentrated into the region immediately below the film (Fig. 2-17) where their high density results in a very steep electrical gradient, locally of the order of 1 million volts cm.$^{-1}$ Further, the effect of the surface charge remains important relative to $\boldsymbol{k}$T, the thermal energy, to distances as great as several hundred angstrom units from the interface. Secondly, the effective concentration of charged groups at the interface may be very high, as the long-chain ions may be unable to desorb into the bulk if the hydrocarbon chain is long enough (e.g. 20 carbon atoms). These ions may therefore be packed more closely than is ever possible in bulk solution.

The electrical field of a charged monolayer alters the concentration of soluble ions near the interface: a film of long-chain anions has but few hydroxyl ions immediately subjacent to it. For example, in the hydrolysis of a monolayer of monocetylsuccinate ions[16], k as calculated from eq. (6.4) using various bulk concentrations of NaOH, is not constant (Table 6-I, column 3). This is because ψ, the potential near the film, is a function of the ionic strength of the NaOH solution; consequently the concentration of hydroxide (or hydrogen) ions at the level of the monolayer is neither equal to nor proportional to the concentration of alkali (or acid) in the bulk phase, but is given by:

$$_{s}[OH'] = {}_{b}[OH']e^{+\varepsilon\psi/kT} \tag{2.26}$$

$$_{s}[H^{+}] = {}_{b}[H^{+}]e^{-\varepsilon\psi/kT} \tag{2.25}$$

where subscripts refer to surface and bulk. For a negatively charged film ψ is negative, so that $_{s}[OH']$ is less than $_{b}[OH']$: the difference between these

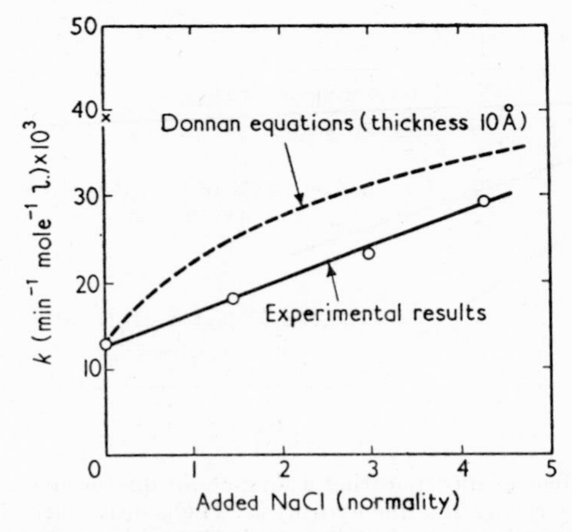

Fig. 6-8. Catalytic effect of neutral salt on the alkaline hydrolysis of monocetyl succinate ions, $C_{16}H_{33}O{\cdot}CO{\cdot}CH_2{\cdot}CH_2{\cdot}CO_2^-$. The monolayer bears a net negative electrical charge. This repels the similarly charged hydrolytic hydroxyl ions. The height of this electrical repulsive barrier is reduced greatly by any "neutral" salts[16].

two concentrations increases when ψ is large, i.e. at low ionic strengths. If a neutral salt (e.g. NaCl) is added to the solution of alkali, on which a monolayer of monocetylsuccinate is being hydrolysed, ψ can be very considerably reduced; the salt acts as a catalyst in that it may increase the velocity of hydrolysis two or three times (Fig. 6-8).

Quantitatively, this means that one must retain $_{s}[OH']$ in eqs. (6.1, 6.2, and 6.4) if the film is charged, or in place of eq. (6.1) one may write, using eq. (2.26),

$$-\frac{dn_r}{dt} = kn_{r \cdot s}[OH'] = kn_{r \cdot b}[OH']e^{\varepsilon\psi/kT} \quad (6.5)$$

For given values of k and $_b[OH']$, this implies that, since ψ is zero in an uncharged film, one may write generally

$$\frac{\text{Rate of reaction in charged film}}{\text{Rate of reaction in neutral film}} = e^{-z_2\varepsilon\psi/kT} \quad (6.6)$$

where z_2 is the valency of the counter-ion (i.e. -1 for OH′): the equation also applies to hydrolysis by H^+ ($z_2 = +1$). Alkaline hydrolysis of mono-cetylsuccinate will therefore be slower (since ψ and z_2 are both negative) than if the film were charged, or than if NaCl were added to reduce ψ to a very low value.

If, however, ψ is positive (and $z_2 = -1$), the reaction will proceed more rapidly on account of the charge: this is found[1] in the alkaline hydrolysis of octadecylacetate monolayers into which a little $C_{18}H_{37}N(CH_3)_3^+$ has been

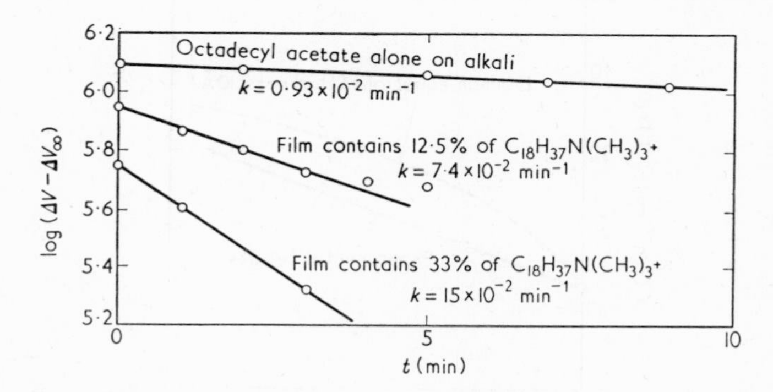

Fig. 6-9. Catalytic effect of incorporating a long-chain quaternary amine in a film of octadecyl acetate undergoing alkaline hydrolysis. The positive electrical charge on the film attracts the negatively charged hydrolytic hydroxyl ions. For octadecyl acetate alone (upper curve), $k = 0.93 \times 10^{-2}$min.$^{-1}$ If one in every eight molecules in the film is replaced by $C_{18}H_{37}N(CH_3)_3^+$, $k = 7.4 \times 10^{-2}$min.$^{-1}$. Replacement of one in every three molecules by $C_{18}H_{37}N(CH_3)_3^+$ increases k to 15×10^{-2} min.$^{-1}$

incorporated (Fig. 6-9). An acceleration of the reaction by the quaternary ions (which may be regarded as catalysts) by seventeen times is easily obtained.

The effect of charge on the film can now be predicted quantitatively if ψ or $_s[OH']$ can be calculated, followed by substitution into eq. (6.5). As explained in Chapter 2, there are equations for achieving this: ψ can be assumed equal to ψ_G (eqs. (2.27), (2.29), or (2.30)), or ψ can be taken equal to $\psi_{Don.}$ (eq. (2.41)). Either procedure may be used to calculate the reaction rate by eq. (6.5), though the assumptions are different in each.

The Gouy equations are based on the model of a uniform charged plane, with the counter-ions represented as point charges, while the Donnan treatment assumes that the insoluble monolayer and the immediately subjacent solution together constitute a surface phase of non-zero thickness, within which all the counter-ions may be considered to be concentrated (Chapter 2). Where the ester grouping and the charged group are separate, treatment as a "surface phase" of appreciable thickness is an advantage: it also allows for the counter-ions occupying a non-zero volume in the "surface phase". We might expect, therefore, that the hydrolysis of monocetylsuccinate monolayers could be represented by the "surface phase" treatment, and the calculations summarized in column 4 of Table 6-I show that this is indeed so, if the thickness is 10 Å. The effect of neutral salt (Fig. 6-8) evidently requires a rather small value of the thickness.

TABLE 6-I

Concentration of NaOH in Bulk	Alkaline Hydrolysis of Monolayer of Stearolactone (neutral film)[a] $k \times 10^2$	Alkaline Hydrolysis of Monolayer of Monocetyl Succinate Ions (negatively charged film)	
		$k \times 10^3$ Calculated from Bulk Concentration	$k \times 10^2$ Calculated from eqs. (6.5) and (2.41) with thickness of "surface phase" of 10 Å
0.4 *N*	8.5	7.9	6.4
0.6 *N*	8.3	9.0	4.7
0.8 *N*	8.3	12.8	5.0
1.0 *N*	8.5	15.0	4.9
1.3 *N*	8.4	17.8	4.9
2.0 *N*	8.4	31.0	6.4

[a] k is expressed in min.$^{-1}$ mol.$^{-1}$ litre units.

In 1953 we tested more precisely the two approaches to ionic reactions: monolayers of cholesterol formate were hydrolysed by 0.1 N-HCl, in the presence of various amounts of long-chain sulphate ions incorporated into the monolayer[1,17]: the negative charge on the monolayer did increase the concentration of H^+ near the surface; and reaction was accelerated. The same result follows from eq. (6.6) since ψ is negative and $z_2 = +1$ for H^+. Quantitative comparison of the Gouy and "surface-phase" treatments is made in Fig. 6-10(a), and similar results for the incorporation of a long-chain quaternary amine into the film (ψ now positive) are shown in Fig. 6-10(b). A logarithmic plot of the reaction velocity for both positive and negative values of ψ_0 (calculated by the Gouy equations) is shown in Fig. 6-11, the slope being the theoretical one predicted by eq. (6.6).

Other examples of the effect of surface charge on reaction kinetics are also known. It was found in 1940 that, whereas the acid-catalysed lactonization of γ-hydroxystearic acid in a monolayer is of the same order of rate as are bulk lactonization reactions, that of β-hydroxyethyl-stearylmalonic acid is very rapid even at pH 5. This latter reaction would occur in bulk only in a strongly acid medium. Here apparently the rate is affected by the negative potential of the surface film, due to the ionization of one of the carboxyl groups of the malonic acid derivative which, unlike simple carboxylic acids,

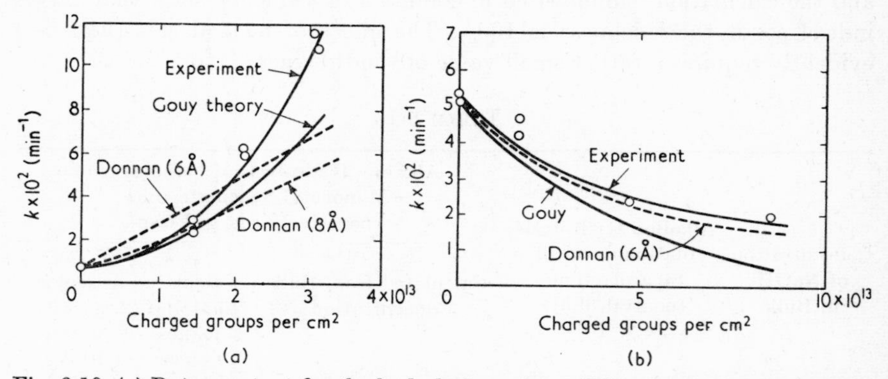

Fig. 6-10. (a) Rate constant for the hydrolysis on 0.1 N-HCI of a cholesterol formate monolayer. The incorporation into the film of a little long-chain sulphate ($C_{22}H_{45}SO_4'$) greatly accelerates the reaction. The calculated increases in reaction rate according to the Gouy and Donnan equations are shown; (b) For hydrolysis on 0.66 N-HCl incorporation of $C_{18}H_{37}N(CH_3)_3^+$ into the film retards reaction because hydrogen ions are repelled from the surface.

is a fairly strong acid. This implies that the negative ψ of the film of the malonic acid derivative attracts hydrogen ions into the surface region, where they act catalytically: this is responsible for the anomalously high rate of lactonization[18].

Havinga[18] in 1954 published data showing that the hydrolysis on 0.1 N alkali of trilaurin films is similarly retarded by stearate ions incorporated in the monolayer but, if the cations of stearylguanidine are incorporated in the film, a small but probably significant increase in reaction rate is observed. That the increase is not large in this system is due to the incomplete ionization of the stearylguanidine: no quantitative treatment has been possible here.

Catalysis by an electrical charge has found application in the degradation of large molecules[19]; proteins are hydrolysed 100 times faster by lauryl sulphonic acid than by HCl of the same bulk concentration, because the lauryl sulphate ions adsorbed on the protein strongly attract hydrogen ions to this region. Further, amide groups ($-CO.NH_2$) are preferentially broken, with the peptide bonds ($-CO-NH-CHR-$) proving more resistant. Selective catalysis of this type promises to be an important tool in the investigation of other large molecules, either in solution if soluble, or otherwise in foams or emulsions if, like Terylene, nylon, and poly-aminoacids, they are insoluble but can be spread at interfaces.

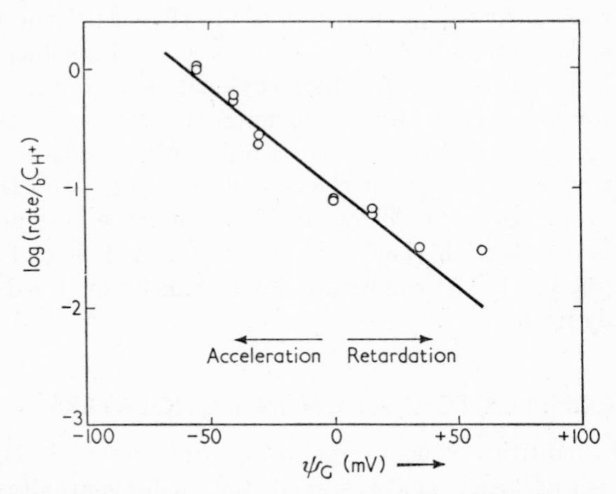

Fig. 6-11. Plot of log (rate/$_bc_{H^+}$) against ψ_G, for the hydrolysis of cholesterol formate monolayers by aqueous HCl. The potential is varied by incorporating different amounts of $C_{22}H_{45}SO_4'$ or $C_{18}H_{37}N(CH_3)_3^+$ into the ester film[1,17]. The slope of the line is the theoretical (—60 mV).

REACTIONS IN EMULSIONS

The large areas of oil-water interface in an emulsion have proved important in both emulsion polymerization and saponification. In emulsion polymerization the intimate contact between the monomer surface and the water-soluble catalyst affects greatly the reaction rate, so that this and the molecular weight of the product are independent variables. Further, because the reaction rate can be high even at room temperature, the formation of the undesirable products of high-temperature operation (particularly branched chain compounds) can be prevented. Block-copolymers can also be conveniently prepared using emulsion methods[20]; if monomer "A" is water-soluble, the polymer radicles initially generated in this phase consist of short chains $-A-A-A-A-A-$. These then add molecules of the water insoluble monomer "B" on reaching the surface of the droplets of "B". Acrylic acid ("A") may, for example, initially polymerize in water under the action of uranyl nitrate as photosensitizer, followed by addition of styrene ("B") to these molecular chains when they come in contact with the surfaces of the styrene drops.

Fats may conveniently be hydrolysed in emulsions, the speed of saponification depending on the fineness of the emulsion.

Emulsion polymerization of styrene does not occur in or on emulsion droplets, but is initiated in styrene solubilized in micelles of soaps[1,21], this process lying outside the scope of the present work.

A spectacular laboratory demonstration of interfacial polymerization may be carried out as follows. Into a clean beaker is poured a solution of 1.5 cc. of sebacoyl chloride in 50 cc. perchlorethylene. On to this is carefully poured a solution of 4 g. sodium carbonate and 2.2 g. hexamethylenediamine in 50 cc. of water. Immediately a milky film begins to form at the interface: one may grasp the film at the centre with tongs, and then pull out continuously a "rope" of nylon. The process can be made fully automatic by setting the beaker on a high shelf, and starting 3 or 4 feet of the "rope" moving over a glass rod. The same experiment can be modified to produce polyesters or polyurethanes[22].

COMPLEX FORMATION IN MONOLAYERS

If very small quantities of heavy metal ions are present in the water on which a monolayer of stearic acid is spread, they will eventually, by normal processes of diffusion and convection, come in contact with the monolayer[23]. One part only of Al^{+3} in 2×10^9 parts of water is sufficient to alter the physical properties of the stearic acid monolayer, and copper ions are likewise very effective in forming salts of stearic acid in a monolayer.

Besides showing the important effect of such small amounts of impurities, these studies cast light on the sensitization of monolayer reactions: the course of a photochemical reaction may be completely altered by traces of heavy metal ions. For the decomposition to proceed in films of α-hydroxystearic acid on 0.01 N-HCl under the action of radiation at 2537 Å, for example, sub-analytical quantities of nickel ions must be present: one molecule of α-hydroxystearic acid must combine with one nickel ion before CO_2 can be split off[24].

The interaction of monolayers of fatty acids and long-chain sulphates with ions of the heavy metals injected beneath them is marked if the pH is such that the basic metal ions are formed in solution[25]. Ions such as Ca^{++}, however, form complexes (insoluble soaps), though the basic ions do not, owing to steric hindrance. Adsorption of long-chain ionic compounds from solution on to the surfaces of solid metals or their minerals takes place under conditions of pH and stereochemistry similar to those required by the monolayer interactions; the latter are therefore a guide to the possibility of "flotation" of the particular minerals with the long-chain compounds under consideration[26].

PENETRATION INTO MONOLAYERS

If, underneath a monolayer, a solution of another molecular species is injected, a variety of phenomena may occur[27]. The simplest interaction between the two types of molecules is mainly electrostatic, leading to changes in surface potential with relatively minor alterations in surface pressure (Fig. 6-12). An example of this is provided by the injection of very dilute solutions of long-chain salts under many monolayers.

Subsequent interaction results from the *penetration* into the monolayer of the injected material (Fig. 6-13). If the energy of desorption is not very high, further compression of the mixed monolayer may eject the additive from the surface, giving again the situation shown in Fig. 6-12. Penetration is characterized by large changes both in surface pressure and surface potential. With some molecular species the contribution of each to the surface pressure and potential are independent, the resulting film being a simple mixture of the two components. Astacene and oleic acid, spread on 0.01 N-HCl, behave in this way at low pressures[28]. This independence of the molecules in the monolayer is not always found, however, and its absence is interpreted as indicating the formation of a molecular complex (Chapter 5) between the two types of molecule in the surface.

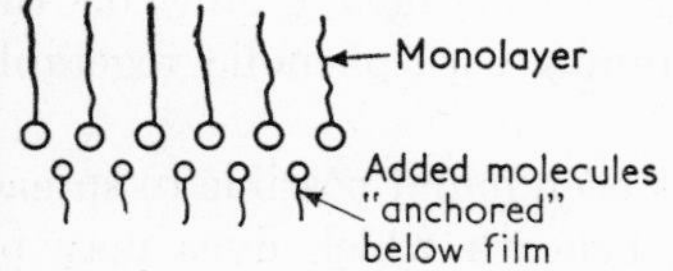

Added molecules
penetrate into film

Fig. 6-12. "Anchoring" of molecules below a film.

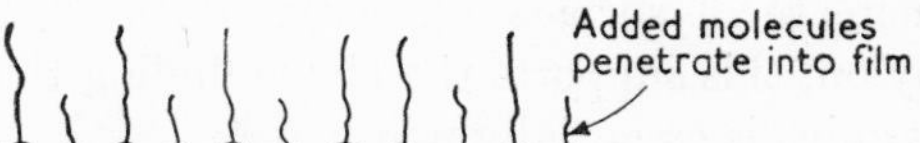

Fig. 6-13. "Penetration" of molecules into a film.

Such complex formation[27,29] is a function of the polarity of the "head-groups" as well as of the lengths and shapes of the hydrocarbon "tails". For example, injection of dilute solutions of substances of the general structure $CH_3.(CH_2)_{11}X$ under monolayers of cholesterol and proteins[27,30] at the air-water surface leads to the following order of reactivities of the group X:

$$NH_3^+ > SO_4' > SO_3' > COO' > N(CH_3)_3^+$$

There is also a strong interaction between the two long saturated hydrocarbon chains[31], which is greatly weakened if there is a double bond in one of the chains. Under certain conditions a stoichiometric complex seems to be formed. An equimolar (1:1) complex is formed between digitonin and cholesterol, or between sodium cetyl sulphate and cholesterol, while digitonin and cetyl alcohol form a 1:2 complex in the surface.

Schulman and Cockbain[32] have concluded from the stability and viscosity of emulsions that similar complexes are formed between long-chain ions and sodium cetyl sulphate, both adsorbed at an oil-water interface.

It is possible for material anchored beneath the film to react chemically with it. Whereas the complexes just discussed do not seem to involve the formation of new chemical bonds, films containing amino-groups react strongly with tannic acid injected below them[27]. A monolayer of protein may, in this way, undergo two-dimensional "tanning", with extraordinary toughening of the monolayer. Ellis and Pankhurst[33] studied this with particular reference to collagen, following the changes in both surface viscosity and surface potential (about 100 mV) during the tanning reaction. Mimosa (catechol) tannins produce a marked condensation of the monolayer, indicating "multi-point" association between the tannin molecules and several protein chains, leading to a compact, cross-linked structure. The change in surface potential suggests that above pH 3 there is some neutralization of the positive charges on the protein, though it is uncertain whether this contributes to the tanning reaction. On the other hand, chromium sulphate produces a slight expansion of the film and an increase of surface potential: these changes are due to an initially electrovalent binding of the chromium cations to the negatively charged groups on the side-chains of the protein, producing a rather open network. Monolayers of nylon (and its methyl-methoxy derivatives) are tanned by mimosa tannin[34], with a large increase of surface viscosity. Since there are no charged groups on the nylon, this supports the above finding that tanning with phenolic vegetable tannins is essentially non-ionic[35].

In studying the mechanism of dyeing, it has been found possible to spread keratin derivatives to monomolecular films, beneath which dyes may be injected[36]. The keratin derivative, prepared by reduction of wool with alkaline thioglycollate, may be spread from water containing about 0.1% amyl alcohol, to assist spreading (Chapter 5). Orange II solutions, injected below these spread films, increase the surface pressure at pH 2 (where the protein carries a strong positive charge); i.e. the dye anions penetrate into the film. At pH 3, however, though the surface pressure is not increased, the viscosity of the keratin film is augmented by injected Orange II, suggesting ionic binding. At pH 4 there is no measurable interaction. Higher ionic strengths favour the interaction of the dye with the protein.

It is now clear that many molecules found in biological systems are capable of forming surface complexes: many natural membranes and interfaces must consist of mixed monolayers of just the types that are prone to complex formation. Indeed, a variety of lytic, agglutinating, and sensitizing agents affect mixed films of cholesterol and gliadin[24], and all agents capable of causing haemolysis either penetrate such films or disperse the protein therefrom. But agents which cause only agglutination or sensitization merely adsorb on the protein of the mixed film, without penetrating the "islands" of sterol in the film.

Another interesting observation[37] is that the addition of haemin lowers the interfacial pressure of adsorbed films of globin at the oil-water interface: this reaction is highly specific. Conversely, a film of stearic acid is penetrated by protein dissolved in the bulk, with a marked rise in surface pressure at the air-water surface[38]. This suggests that there is interaction between the non-polar side chains of the protein and the monolayer, which is supported by quantitative studies of the energy changes, using proteins of different mean side-chain lengths. The energy of penetration is rather low, however (500 to 700 cal. mole^{-1} $-CH_2-$), suggesting that it is more difficult to pull the hydrocarbon side chains into the film on account of their attachment to the peptide "backbone", which may have to be deformed in the process.

Thermodynamics of Penetration

When a molecule in the bulk penetrates into an uncharged monolayer, the following energy terms must be considered:

(i) the energy of removal of the hydrophobic part of the molecule from the water,
(ii) the energy to make a hole in the monolayer (compressing it in the surrounding area),
(iii) the energy of interaction of the penetrating hydrocarbon chain with the molecules of the monolayer (including complex-formation),
(iv) the dipole interactions between the head-groups.

If long-chain ions are penetrating, the electrical factors must also be allowed for, i.e.

(v) the energy of formation of the ionic double layer,
and (vi) the energy (and particularly entropy) of partial dehydration of the ionic head-groups in the surface, because of the lower dielectric constant.

If the original monolayer is ionized, the term (v) must refer to the energy of alteration of the ionic double layer[39].

In practice, because factor (iii) is often appreciable, the energy of desorption of the penetrating compound may be higher than if no monolayer were originally present on the surface: thus the energy of desorption per $-CH_2-$ group may reach 800 cal. mole^{-1} if the original monolayer is close packed[39,40]. Consequently one may compress certain penetrated films to a higher surface pressure than could be attained for a film of the penetrating substance alone or, if a complex is formed, than could be attained with the original monolayer[31,39]. Conversely, if the penetrating molecule has only a short chain (C_4 or C_5), the total lateral cohesion in the monolayer is so reduced[40] through the separation of the long chains of the original monolayer that the film becomes almost "gaseous".

The importance of factor (ii) can be assessed from the experimentally low rates of penetration, and if possible molecular complexes are to be studied in penetrated films, the original monolayer must be expanded to about five times the final area during the penetration process[41] and subsequently compressed again. In this way the energy (ii), which would act as an energy of activation, is lowered enough for the penetration to proceed rapidly to completion. A detailed study, using radioactive tracers, would be of great interest.

The number of adsorbed molecules which has penetrated the original insoluble monolayer of any substance I at a constant total area can be found from the increase in film pressure[42]. The interpretation involves the Gibbs equation (4.52), including a term $-\boldsymbol{k}\mathrm{T}n_{\mathrm{I}}\mathrm{d}\ln a_{\mathrm{I}}$ for the insoluble monolayer. In this expression a_{I} may change during the penetration with the increase of Π, and this is allowed for by the thermodynamic expression

$$\left(\frac{\partial \ln a}{\partial \Pi}\right)_{\mathrm{T}} = \bar{\mathrm{A}}_{\mathrm{I}} \qquad (6.7)$$

where $\bar{\mathrm{A}}_{\mathrm{I}}$ is the partial molar surface area of I, assumed the same as in a monolayer of pure I. In this way the number of lauryl sulphate ions, for example, penetrating a monolayer of cholesterol or cetyl alcohol may be simply deduced[42,43]. Again, experiments using radioactive tracers are required. The heats of adsorption may also be deduced from an equation of the Clausius Clapeyron type[43].

Penetration from the Vapour Phase

Instead of the penetrating molecules entering the monolayer from the underlying liquid phase, they may be adsorbed from the vapour. Dean[44] has studied such behaviour extensively, finding that monolayers of stearic acid are readily penetrated by the vapour of benzene or n-hexane, even at very high film pressures. Over a wide range of molecular areas of stearic acid, the adsorption of n-hexane from saturated vapour remains between 3.6×10^{14} and 4.8×10^{14} molecules cm.$^{-2}$ The result is an increase of surface pressure at constant area, or an increase in surface area at constant pressure.

The Gibbs equation may be used to interpret the results; again the chemical potential of the monolayer (e.g. stearic acid) may itself be altered by the penetrating molecules. The initial heat of adsorption of n-hexane on dilute stearic acid monolayers is high (14 K.cal. mole^{-1}), corresponding to a two-dimensional solution of the stearic acid clusters in hexane, though as the surface becomes covered with hexane the value falls to about 7 K.cal. mole^{-1}, close to the heat of condensation of pure hexane.

REFERENCES

1. Davies, *Advanc. Catalys.* **6**, 1 (1954); in "Surface Phenomena in Chemistry and Biology" p. 55, Pergamon Press, London and New York (1958).
2. Alexander, A. E. and Rideal, *Proc. roy. Soc.* **A163**, 70 (1937).
3. Adam, *Proc. roy. Soc.* **A112**, 362 (1926).
4. Hughes and Rideal, *Proc. roy. Soc.* **A140**, 253 (1933).
5. Mittelmann and Palmer, *Trans. Faraday Soc.* **38**, 506 (1942).
6. Marsden and Rideal, *J. Chem. Soc.* 1163 (1938);
   Nasini and Mattei, *Gazz. chim. ital.* **71**, 302 (1941).
7. Alexander, A. E. and Schulman, *Proc. roy. Soc.* **A161**, 115 (1937).
8. Fosbinder and Rideal, *Proc. roy. Soc.* **A143**, 61 (1933).
9. Schulman, *Trans. Faraday Soc.* **37**, 134 (1941).
10. Hughes, *Biochem. J* **29**, 437 (1935).
11. Bangham and Dawson, *Nature, Lond.* **182**, 1292 (1958); *Biochem. J.* **72**, 493 (1959); ibid. **75**, 133 (1960).
12. Rideal and Mitchell, *Proc. roy. Soc.*, **A159**, 206 (1937).
13. Mitchell and Rideal, *Proc. roy. Soc.* **A167**, 342 (1938);
    Carpenter, *Science* **89**, 251 (1939); *J. Amer. chem. Soc.* **62**, 289 (1940).
14. Gee, *Proc. roy. Soc.* **A153**, 129 (1935);
    Gee and Rideal, *Proc. roy. Soc.* **A153**, 116 (1935).
15. Bresler, Judin, and Talmud, *Acta Phys.-chim. URSS.* **14**, 71 (1941).
16. Davies and Rideal, *Proc. roy. Soc.* **A194**, 417 (1948).
17. Llopis and Davies, *Ann. Soc. esp. Fís. Quím.* **49**, 671 (1953).
18. Havinga, in "Monomolecular Layers" p. 192, (ed. Sobotka), A.A.A.S., Washington (1954).
19. Steinhardt and Fugitt, *J. Res. nat. Bur. Stand.* **29**, 315 (1942);
    Schramm and Primosigh, *Hoppe-Seyl. Z.* **283**, 34 (1948).
20. Dunn and Melville, *Nature, Lond.* **169**, 699 (1952).
21. Bovey, Kolthoff, Medalia, and Meehan, "Emulsion Polymerization" Interscience, New York (1954).
22. Morgan, *Du Pont Mag.* **53** (4), 30 (1959).
23. Langmuir and Schaefer, V. J., *J. Amer. chem. Soc.* **59**, 2400 (1937).
24. Mitchell, Rideal, and Schulman, *Nature Lond.* **139**, 625 (1937).
25. Wolstenholme and Schulman, *Trans. Faraday Soc.* **46**, 475 (1950); **47**, 788 (1951);
    Schulman and Dogan, *Disc. Faraday Soc.* **16**, 158 (1954);
    Spink and Sanders, *Trans. Faraday Soc.* **51**, 1154 (1955).
26. Smith, T. D. and Schulman, *Kolloidzschr.* **126**, 20 (1952);
    Cuming and Schulman, *Symp. Miner. Dressing*, Inst. Miner & Metallurg. (1952).
27. Schulman and Rideal, *Proc. roy. Soc.* **B122**, 29, 46 (1937).
28. Danielli and Fox, D. L., *Biochem. J.* **35**, 1388 (1941).
29. Schulman and Stenhagen, *Proc. roy. Soc.* **B126**, 356 (1938).
30. Marsden and Schulman, *Trans. Faraday Soc.* **34**, 748 (1938).

31. Dervichian, in "Surface Phenomena in Chemistry and Biology" p. 70, Pergamon Press, London and New York (1958).
32. Schulman and Cockbain, *Trans. Faraday Soc.* **36**, 651 (1940).
33. Ellis and Pankhurst, *Disc. Faraday Soc.* **16**, 170 (1954).
34. Pankhurst, *Disc. Faraday Soc.* **16**, 240 (1954).
35. Pankhurst in "Surface Phenomena in Chemistry and Biology" p. 100, Pergamon Press, London and New York (1958).
36. Harrap, *Proc. 2nd Internat. Congr. Surface Activity* **4**, 295, Butterworths, London (1957).
37. Haurowitz et al., *Nature, Lond.* **180**, 437 (1957).
38. Eley and Hedge, *J. Colloid Sci.* **12**, 419 (1957).
39. Crisp in "Surface Chemistry" pp. 23, 65, Butterworths, London (1949).
40. Adam, Askew, and Pankhurst, *Proc. roy. Soc.* **A170**, 485 (1939); Pankhurst, *Proc. roy. Soc.* **A179**, 393 (1942).
41. Joly, *Nature, Lond.* **158**, 26 (1946).
42. Pethica, *Trans. Faraday Soc.* **51**, 1402 (1955).
43. Anderson, P. J. and Pethica, *Trans. Faraday Soc.* **52**, 1080 (1956).
44. Dean and Fa-Si Li, *J. Amer. chem. Soc.* **72**, 3979 (1950);
Dean and Hayes, *J. Amer. chem. Soc.* **73**, 5583 (1951);
Hayes and Dean, *J. Amer. chem. Soc.* **73**, 5584 (1951);
Dean and Hayes, *J. Amer. chem. Soc.* **74**, 5982 (1952);
Dean and McBain, *J. Colloid Sci.* **2**, 383 (1947);
Dean, Hayes and Neville, *J. Colloid Sci.* **8**, 377 (1953).

Chapter 7

Mass Transfer across Interfaces

If a molecule passes across the gas-liquid interface it encounters, in general, a total resistance R which is the sum of three separate diffusional resistances, due respectively to diffusion in the gas phase, across the monomolecular region constituting the interface, and through the liquid below the interface[1] (Fig. 7-1). This may be expressed as:

$$R = R_G + R_I + R_L \tag{7.1}$$

Of these resistances, R_L is usually the highest, corresponding to molecular diffusion of the solute through a non-turbulent liquid layer adjacent to the surface. There are, however, interesting exceptions to this statement; R_L can sometimes be made so low that R_I or R_G may be rate-controlling. Before dealing with specific examples, we shall discuss briefly the orders of magnitude of these separate resistances, and the units in which they may most conveniently be expressed.

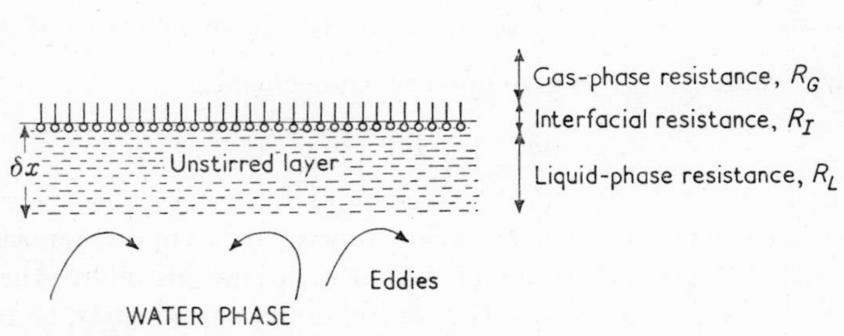

Fig. 7-1. Total resistance $R = R_G + R_I + R_L$

The basic differential equation for transfer across any plane surface is:

$$\frac{dq}{dt} = A.k.\Delta c \tag{7.2}$$

where q is in moles of material transferring, t in seconds, A is the area across which transfer occurs, and Δc is the concentration difference (in moles cm.$^{-3}$) in the region for which the permeability coefficient k (in cm. sec.$^{-1}$) is measured. The reciprocal of k is the resistance R, which is thus expressed in sec. cm.$^{-1}$ Equation (7.2) may alternatively be written in terms of a diffusion coefficient D for the region of thickness Δx under consideration:

hence
$$k = \frac{1}{R} = \frac{D}{\Delta x} \tag{7.3}$$

and
$$\frac{dq}{dt} = A.D.\frac{\Delta c}{\Delta x} \tag{7.4}$$

where $\Delta c/\Delta x$ is the concentration gradient.

The value of R_I for a clean liquid surface is relatively low, and depends on the chance that a molecule in the interfacial region of the liquid has enough energy to overcome the attraction of its neighbours and evaporate into the gas phase. At equilibrium the rate of evaporation must be equal to the number of molecules condensing on to unit area of surface per second: this latter rate is more easily calculated than is the rate of evaporation, since, according to the kinetic theory of gases, the number of molecules arriving per second from the vapour is simply $p(2\pi m\boldsymbol{k}T)^{-\frac{1}{2}}$, where p is the equilibrium vapour pressure (dynes cm.$^{-2}$), m is the mass of a single molecule, and $\boldsymbol{k}$ is the Boltzmann constant. If, of these molecules arriving, a fraction α condenses on to the liquid surface, the rate of condensation (or evaporation) can be expressed as:

$$\frac{1}{A}\frac{dw}{dt} = \frac{p\alpha m}{(2\pi m\boldsymbol{k}T)^{\frac{1}{2}}} = p\alpha\left(\frac{M}{2\pi RT}\right)^{\frac{1}{2}} \tag{7.5}$$

where w is the mass of material condensing (or evaporating), M is the molecular weight, A is the surface area, and R is the gas constant in ergs per degree. This equation can, if required, be rewritten in terms of moles of material q, where $q = \frac{w}{M}$, to give upon rearrangement:

$$\frac{dq}{dt} = A\alpha\left(\frac{RT}{2\pi M}\right)^{\frac{1}{2}}\frac{p}{RT} \tag{7.6}$$

If one now considers the condensation process on a molecular scale, one postulates that the concentration of vapour molecules just above the surface is c_0, while the vapour concentration in the liquid surface may be regarded as zero: the concentration gradient is therefore $c_0 - 0$. Now c_0 is related to the vapour pressure p, being equal to p/RT for an ideal gas, so eq. (7.6) may be written as:

$$\frac{dq}{dt} = A\alpha\left(\frac{RT}{2\pi M}\right)^{\frac{1}{2}}(c_0 - 0)$$

or, by comparison with eq. (7.2),

$$k_I = \frac{1}{R_I} = \alpha\left(\frac{RT}{2\pi M}\right)^{\frac{1}{2}} \tag{7.7}$$

Hence for a clean water surface $R_I(=1/k_I)$ is 0.002 sec. cm.$^{-1}$, using the experimental value for α of 0.034 for a clean water surface at room temperature. This value of α, suggesting that only 3.4% of the molecules of water striking a water surface enter it, shows that a free energy of activation of about 2300 cal. mole.$^{-1}$ is required to produce or find a hole in the liquid surface in which to be accommodated. The presence of certain monolayers can reduce α by as much as 10^4 times, and increase R_I to about 10 sec. cm.$^{-1}$ If there is air above the water surface and the net rate of evaporation is measured, the resistance R_G may be found to be of the order 80 sec. cm.$^{-1}$, though this can be reduced to only a few sec. cm.$^{-1}$ either by reducing the pressure below atmospheric or by intensive stirring of the air.

For a pure liquid R_L is necessarily zero, though if a solute (e.g. CO_2) is diffusing from water to the gas phase, the concentration gradient of the solute below the surface may extend over an appreciable thickness Δx, so reducing k_L according to eq. (7.3). A typical value of R_L in such a system is 500 sec. cm.$^{-1}$: the exact value will depend on D and the thickness Δx of the unstirred region below the surface: Δx depends markedly on the hydrodynamics of the system, and is discussed in detail below.

EVAPORATION

Spread at a **plane air-water surface**, a close-packed monolayer of a long-chain compound can considerably increase R_I. In this system R_L is necessarily zero so that, since $R = R_G + R_I$, the relative effect of the monolayer on the total resistance to evaporation will depend on R_G. The latter may be quite large if there is a stagnant film of gas at atmospheric pressure, situated above the surface[1]. This diffusion barrier R_G can be reduced, however, by evacuating the system, and in this way Rideal[2] was able to reduce R_G to a level comparable with R_I. Indeed, he showed that in such a system certain monolayers can reduce the rate of evaporation by as much as 50%. Langmuir[2] defined the total resistance to evaporation simply as the reciprocal of the measured rate; consequently he was able to express Rideal's results in units of cm.^{2} sec. gm.$^{-1}$, $(R_G + R_I)$ for the clean surface being 770, and $(R_G + R_I)$ for a surface covered with a monolayer of lauric acid being 1340. The effect of the monolayer is therefore to increase R_I by 570 cm.^{2} sec. gm.$^{-1}$ Langmuir also found that monolayers of hexadecanol offer a very high resistance—about 60,000 cm.^{2} sec. gm.$^{-1}$ (of the order 1 sec. cm.$^{-1}$). For more quantitative studies of the evaporation resistance of monolayers, Sebba and Sutin[3] have developed a Langmuir trough enclosed in a box which can be evacuated.

Another method, though less effective than evacuation, of reducing the gas-phase resistance R_G consists in causing a stream of dried air to flow across the water surface, and measuring the amount of moisture taken up. By this method Sebba and Briscoe[4] found that the resistances at 20 dynes cm.$^{-1}$ of films of the straight-chain alcohols increased with the length of the hydrocarbon chain, and that at this pressure the film of the C_{22} alcohol reduces evaporation to 19% of the rate at a clean surface: the same film compressed to 40 dynes cm.$^{-1}$ reduces evaporation to only 12%, while at 48 dynes cm.$^{-1}$ it virtually eliminates all evaporation. Hexadecanol at 40 dynes cm.$^{-1}$ reduces the evaporation to 60% of the rate at a clean surface.

The barrier R_I in the presence of long-chain acids and alcohols (found to be about 5 or 10 sec. cm.$^{-1}$) is detectable even in still air at atmospheric pressure (R_G about 80 sec. cm.$^{-1}$): there is an appreciable reduction in the evaporation rate[5], and temperature coefficients can easily be measured in this system. For a film of C_{19} acid these show that the enthalpy of activation for evaporation is 14,600 cal. mole^{-1}, while use of compounds of different chain lengths indicates that each $-CH_2-$ group is responsible for 300 cal. mole^{-1} of this barrier. Consequently the close-packed carboxyl groups in the monolayer must exert the surprisingly high barrier R_I of 9500 cal. mole.$^{-1}$: the resistance of an $-OH$ end-group is even higher. Traces of impurities may greatly reduce the efficacy of the monolayer: if benzene is used as the spreading solvent for the long-chain acids, enough benzene may remain in the surface to cause dislocations in the otherwise tightly packed, condensed monolayer and, since the resistances of different parts of the surface behave as if in parallel, R_I may well be reduced as much as twenty times. If the regions of the surface occupied by impurity have the resistance of a clean surface (0.002 sec. cm.$^{-1}$), one may thus calculate[5] that even 0.01% of these in the monolayer may reduce R_I by 20%. Curiously, cholesterol has no effect on evaporation.

The striking effects of these long-chain acids and alcohols are of great practical interest, since it is most important to be able to reduce the rate of evaporation of water from lakes and reservoirs in hot, arid regions where the amount of water lost by evaporation may exceed the amount usefully used. In other terms, evaporation may lower the level of a reservoir by as much as 10 feet annually. To reduce this evaporation by using only a monolayer of a polar oil[6,7] has not only the advantage that quite small amounts are required, but also that the oxygen necessary to support life can still diffuse into the water, and stagnation of the lake is thereby avoided. The reason that enough oxygen penetrates a quiescent surface covered with a monolayer lies in the high diffusional resistance R_L encountered in the aqueous solution subjacent to the surface: compared with this resistance, the monolayer has a smaller effect[8]. If, however, a thick film of oil were used to retard evaporation, its enormous resistance would become dominant in retarding the entry of oxygen. Under natural conditions, the uptake of oxygen into a lake or reservoir is aided by the wind and by convection currents, which stir the liquid near the surface: the surface viscosity and the resistance to local compression of a monolayer will reduce this stirring and so decrease the uptake of oxygen towards the rate for a quiescent surface. The effect of this in practice is that the monolayer doubles R_L, and this in turn doubles the oxygen deficit[8]; consequently, the effect of placing a monolayer of hexadecanol on a reservoir containing water which is 90% saturated is to reduce the oxygen content to 80% of saturation. This has little effect on the living organisms in the water.

To make sure that the monolayer is always present, in spite of some local crumpling by dust, rain, and winds in excess of 5 m.p.h., a material is required that spreads readily (see p. 29). A high final surface pressure is desirable not only to squeeze from the film impurities that may be permeable to water-vapour[5,9], but also to pack tightly the long chains so that they offer the greatest possible resistance to evaporation[4,5].

Hexadecanol is very suitable in practice, spreading spontaneously and sufficiently rapidly from solid beads in small, gauze-covered "rafts" in the surface to give a monolayer of $\Pi = 40$ dynes cm.$^{-1}$: this retards evaporation into the atmosphere by about 50%[8]. Spreading from solution in kerosene is also feasible[7], and is preferable if the reservoir is dirty. Trials reported in America describe the pumping of stearyl alcohol suspensions through a perforated plastic hose to give a spray application at the windward end of the reservoir[7].

Commercial "cetyl alcohol" may be only about 45% pure, the remainder being myristyl alcohol and stearyl alcohol. This mixed product, however, fortunately gives a higher resistance than pure hexadecanol, the figures for a quiescent water surface being respectively 14.7×10^4 and 5.3×10^4 cm.^{2} sec. gm.$^{-1}$ Addition of 6% oleyl alcohol to the commercial "cetyl alcohol" reduces the film resistance at a quiescent surface slightly, to 10×10^4 cm.^{2} sec. gm.$^{-1}$, corresponding to about 20% reduction in the rate of evaporation[10]: with a wind of 7.5 m.p.h. the film reduces the evaporation by 60% in the laboratory. The oleyl alcohol is beneficial under practical conditions, however, as it increases both the rate of spreading of the cetyl alcohol and the recovery of the monolayer after collapse by waves[10]. Under the conditions obtaining in a natural reservoir, the spreading of a cetyl alcohol monolayer can now be carried out so as to effect a 30% to 35% reduction in the rate of evaporation[7,10]. Longer hydrocarbon chains improve the efficacy, though the spreading properties of the alcohols then become less favourable. To overcome this difficulty, one may add an ethylene-oxide group at the polar end of the molecule[10a].

Another application of the effect of higher alcohols in reducing evaporation lies in the temperature control of certain insects. The wax cuticle, of the order of microns in thickness, becomes appreciably permeable to water around 30°C, thus permitting more rapid evaporation and consequent cooling[12].

The amount of **evaporation from droplets of liquid** or from spheres of solid into still air obeys experimentally the equation[11]:

$$\frac{\mathrm{d}A}{\mathrm{dt}} = \text{constant} \tag{7.8}$$

where A ($=4\pi a^2$) is the surface area of the droplet. For example, for water the constant is 3×10^{-5} cm.^{2} sec.$^{-1}$ and for iodine it is 5.4×10^{-4} cm.^{2} sec.$^{-1}$ These constants are most easily determined experimentally for very small drops ($a \sim 10^{-4}$ cm.) by observing microscopically their rate of fall, which, by the Stokes equation, should vary with a^2. One usually finds a linear relation between the velocity of fall and the time t which has elapsed, so that da^2/dt is constant, leading to equation (7.8). For larger drops ($a \sim 1$ mm.) the course of evaporation may be followed directly, placing the drop on the pan (coated with carbon-black to prevent wetting) of a silica spiral micro-balance[11].

From eq. (7.8) it follows that the loss of weight from the droplet varies as a, and that the evaporation from unit area of surface $\left(\text{i.e. } \frac{1}{A}\frac{\mathrm{d}w}{\mathrm{dt}}\right)$ varies as $\frac{1}{a}$, and therefore increases steeply as the radius of the droplet becomes very small. This result, at first rather surprising, is found for water, aniline, p-cresol, benzyl benzoate, and other substances, over ranges of a from 5×10^{-5}cm. to 0.2 cm.

Equation (7.8) is consistent with evaporation being a diffusion-controlled process, i.e. R_G is rate-controlling: the following expression may be obtained from diffusion theory[13]:

$$\frac{\mathrm{d}w}{\mathrm{dt}} = \frac{4\pi \mathrm{MD\,pa}}{\mathrm{RT}} \tag{7.9}$$

where w is the weight of material evaporating, and M, D, and p are the molecular weight, diffusion coefficient, and saturation vapour pressure of the vapour under consideration. It is assumed that the surrounding air is completely unsaturated. According to this equation, $-\frac{1}{w}\frac{\mathrm{d}w}{\mathrm{dt}}$, the fractional loss of weight of a droplet, varies as $1/a^2$: the high fractional losses are illustrated by the following calculation. A drop of water of initial radius of 0.1 cm. and at 18°C, evaporating into still, dry air, would take 11 minutes to disappear completely, though a drop of initial radius 10 microns would require only 60 milliseconds. If the air had a relative humidity of 80%, the latter time would become 2.5 sec.

Equation (7.9) may be written in terms of surface area either as

$$\frac{1}{A}\frac{dw}{dt} = \frac{MDp}{RTa} \tag{7.10}$$

or as

$$-\frac{dA}{dt} = \frac{8\pi MDp}{\rho_l RT} \tag{7.11}$$

where ρ_l is the density of the liquid in the evaporating sphere; eq. (7.11) is in accord with the experimental equation (7.8). The high rates of loss of mass per unit area of surface when a is small are seen, on this diffusion theory, to be due to the greater ease of diffusion in a radial diffusion field (where geometry requires that R_G is reduced because the concentration falls off rather steeply away from the surface). Numerical agreement of eq. (7.11) with experiment is quite good, the calculated value for iodine being 4.4×10^{-4} cm.^{2}sec.$^{-1}$, compared with 5.4×10^{-4}cm.^{2}sec.$^{-1}$ from experiment. Consequently for these systems the importance of diffusion (i.e. R_G) as the rate-controlling factor is established[11].

If we apply diffusion theory to very small droplets we see that since $\frac{1}{A}\frac{dw}{dt}$, the rate of evaporation from unit area of surface, varies as 1/a by eq. (7.10) this rate must for a limitingly small droplet become infinite. This result is, however, a consequence of assuming that R_G is so large that R_I can be neglected, and this is no longer valid for droplets so small that the geometry of the system reduces R_G very greatly. The term R_I (in sec.cm.$^{-1}$) can, however, be included into eq. (7.10) to give[13]

$$\frac{1}{A}\frac{dw}{dt} = \frac{Mp}{RT(a/D + R_I)} \tag{7.12}$$

This shows that for very small droplets in air, the rate of evaporation tends to that in a vacuum (as given by eq. (7.5) in conjunction with eq. (7.7)). For a drop of radius 10 microns the correction due to R_I increases the calculated life-time by less than 5%, though for a 0.1 micron droplet the use of the corrected equation (7.12) instead of (7.10) increases the calculated life-time sixfold. The cooling of small drops when the rate of evaporation is high must also be taken into account[13], the effect being much more important in the evaporation of toluene than of water. Normal heat-transfer theory may be used to estimate the temperature of the surface of the drop.

The importance of R_I relative to R_G in determining the rate of evaporation of a small droplet (if a is very small in eq. (7.12)) is reflected in the extreme sensitivity to traces of impurities of the rate of evaporation of such systems. Thus, a monolayer of a long-chain alcohol or acid, which at room temperature can increase R_I (from 0.002 sec.cm.$^{-1}$ for the clean water surface) to 10 sec.cm.$^{-1}$, should be able to reduce the rate of evaporation of a very small drop by 5000 times. The life-times in dry air of water drops of 1 micron radius are correspondingly increased from a few milliseconds to about a minute by such a monolayer; in air of 80% relative humidity the life of a 10 micron drop is increased from 2.5 sec. to 1300 sec.[14] Even oleic acid monolayers (making R_I about 0.1 sec.cm.$^{-1}$) considerably reduce the rate of evaporation of water droplets, as Whytlaw-Gray showed[11]. Besides monolayers, traces of dust, non-volatile impurities, and oxidation or decomposition products may similarly reduce the evaporation rate, particularly if the droplets are small. The tarry material in town fogs is thus responsible for delaying dispersal of the fog by "isothermal distillation" (Chapter 8). Mercury droplets, if a skin of oxidized products is allowed to form, evaporate at least 1000 times more slowly than calculated.

If the drop is exposed to a turbulent stream of air, eq. (7.9) must be modified by multiplication by the calculated correction factor $(1+b(\mathrm{Re})^{1/2})$, where the second term in the bracket allows for eddy diffusion of the vapour away from the surface in the air-flow of Reynolds number Re. The constant b is related to the diffusion coefficient of the solute and the kinematic viscosity of the gas. This correction factor is in good accord with experiment. Self-cooling may become more important under these conditions: a full discussion of the experimental findings for droplets moving relative to the gas is given by Green and Lane[13].

An interesting application of small water drops is in the binding of dust in coal mines. It is, however, necessary to prevent the evaporation of the aerosol of water by previously dispersing in the water a little cetyl alcohol. The spray of water drops, each about 12 microns in radius, is then relatively stable, though without the cetyl alcohol complete evaporation occurs in a few seconds[14].

SOLUTE TRANSFER AT THE GAS-LIQUID SURFACE

If ammonia or CO_2 is being absorbed from the gas phase into solution in water, R_L becomes relatively important in controlling the rate of absorption. This is also true of the desorption of a gas from solution into the gas phase. Usually R_L is of the order 10^2 or 10^3 sec.cm.$^{-1}$, though the exact value is a function of the hydrodynamics of the system: consequently various hydrodynamic conditions give a variety of equations relating R_L to the Reynolds number and other physical variables in the system. For the simplest system, where the liquid is infinite in extent and completely stagnant, one can solve the diffusion equation

$$\frac{\partial c}{\partial t}=D\left(\frac{\partial^2 c}{\partial x^2}\right) \tag{7.13}$$

in which x is the distance from the interface and D the diffusion coefficient.

Hence, for absorption controlled by R_L in a stagnant liquid

$$\frac{dq}{dt} = A(c_s - c_\infty)\left(\frac{D}{\pi t}\right)^{1/2} \tag{7.14}$$

where t is the time for which the surface has been exposed to the gas, and c refers to concentration. Subscript s refers to the constant (saturation) concentration of the gas at the surface, and subscript ∞ refers to the concentration of dissolved gas in the liquid far from the surface: c_∞ is usually made zero. By comparison with eqs. (7.2) and (7.3) we see that

$$k_L = \frac{1}{R_L} = \left(\frac{D}{\pi t}\right)^{1/2} \tag{7.15}$$

so that R_L is a function of the time of exposure. If, for example, $D = 10^{-5}$ cm.^{2}sec.$^{-1}$, R_L is about 500 sec.cm.$^{-1}$ when t = 1 second, increasing to 5000 sec.cm.$^{-1}$ after 100 seconds of exposure. For the desorption of a gas the same equations apply, c_s usually being zero if the desorbing gas is rapidly removed, for example, by a stream of inert gas or by evacuation of the vessel.

The total number of moles that have transferred in a time t is found by integrating eq. (7.14):

$$q = 2A(c_s - c_\infty)\left(\frac{Dt}{\pi}\right)^{1/2} \tag{7.16}$$

and it is this quantity q which is often conveniently measured and compared with theory. If the liquid phase is not infinite in extent, the formulae become more complicated[15].

In general, when the system is subject to stirring by mechanical means or by density or temperature variations during absorption, R_L is difficult to calculate from fundamental principles, and each system has to be considered separately. Among the complicating factors is "surface turbulence", which may reduce R_L by as much as five times, and, before discussing the relation of R_L to the external hydrodynamics, we shall outline the conditions under which spontaneous surface turbulence may occur during mass transfer.

Surface Instability

During the transfer of a surface-active solute across the surface, unstable surface-tension gradients may occur in the plane of the surface. A good example is furnished by Langmuir's experiment[2] on the evaporation of ether from a saturated (5.5%) solution in water: talc sprinkled on the surface shows abrupt local movements, caused by differences in surface tension in different regions. The mechanism is that convection currents, either in the air or in the water, cause more ether to be present at some point in the surface than in neighbouring regions. Consequently, at this point the pressure Π of the adsorbed monolayer of ether is higher, and so this ether film will therefore spread further over the surface, as shown in Fig. 7-2. In doing so it will necessarily drag some of the subjacent liquid with it[16], so that an eddy of solution as yet undepleted in ether is brought to the point in question. The consequent local increase in concentration of ether rapidly amplifies the interfacial movement to visible dimensions. Such "surface turbulence" greatly reduces R_L, and we should therefore expect the ether to evaporate very rapidly from water, since R_I and R_G are small. This is confirmed by experiment: the ether escapes sufficiently rapidly to burn actively when ignited. A monolayer of oleic acid (or other material—the effect is non-specific) on the surface is sufficiently viscous in the plane of the surface and sufficiently resistant to local compression to reduce considerably the "surface

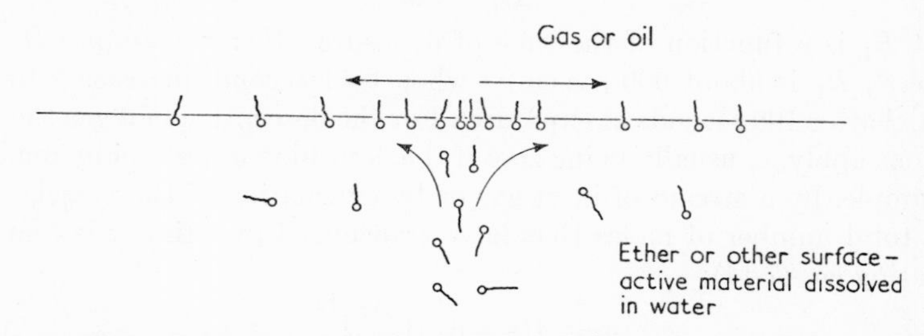

Fig. 7-2. Ether molecules (symbolically shown as ⦚) spreading from a locally high concentration in the surface, carry some of the underlying water with them, bringing up more ether and so amplifying the disturbance.

turbulence"; a thicker, unstirred layer of solution immediately subjacent to the surface is set up as a result of the presence of the monolayer. Because of the necessity for the ether to diffuse through this layer, R_L and hence R are increased (cf. eq. (7.1)), and the ether no longer escapes from the surface rapidly enough to burn[2]. Stirring with a glass rod will temporarily offset the effect of the monolayer. A similar effect was found by Groothuis and Kramers[17] for the absorption of SO_2 into drops of n-heptane. The surface of the drop becomes violently agitated, and R_L is thereby reduced (see below).

Turbulence of this kind depends on the solute transferring across the surface: if eddies of fluid rich in solute coming to the surface raise Π locally, they may cause redistribution by spreading in the surface before the solute is removed into the other phase. The effect must therefore depend[18] both on the distribution coefficient of the solute and on the sign and magnitude of $(d\gamma/dc)$. This is discussed more fully on p. 324.

In a mathematical treatment of the hydrodynamics of surface and interfacial turbulence, Sternling and Scriven[19] have discussed the conditions under which a molecular fluctuation in surface tension during mass-transfer can build up into a macroscopic eddy. They point out that their equations predict that surface eddying will be promoted if the solute transfers from the phase of higher viscosity and lower diffusivity, if there are large differences in D and also in kinematic viscosity $\nu(=\eta/\rho_l)$ between the two phases, if there are steep concentration differences near the interface, if $d\gamma/dc$ is large negatively, if surface-active agents are absent, and if the interface is large in extent. The heat of transfer can produce interfacial forces only about 0.1% of those due to concentration fluctuations. That there is no mention here[19] of the critical concentration of solute required just to produce surface turbulence, nor of the distribution coefficient of the solute between the phases, distinguishes this theory from that of Haydon[18], and shows that neither treatment is comprehensive in explaining the causative and hydrodynamic behaviour of systems showing spontaneous surface turbulence.

Theoretical Values of R_G

In the passage of (say) CO_2 from water to air, the simple diffusion theory (eqs. (7.14 and 7.15)) will apply if the air is completely unstirred: D for a gas is of the order of 0.2 cm.^{2}sec.$^{-1}$ at 1 atmosphere pressure, so R_G is only about 1% of R_L, having values of about 4 sec.cm.$^{-1}$ when t=1 sec. or 40 sec.cm.$^{-1}$ when t=100 secs. Usually, however, the air will be in turbulent motion, and then R_G will depend on the resistance of a layer of air near the surface: typical values of R_G are 5 to 80 sec.cm.$^{-1}$ If pure gas is absorbing into liquid, and if the gas is not highly soluble in the liquid, then usually diffusion through the liquid is rate-determining.

Theoretical Values of R_L

In a *plane static system*, eq. (7.15) above gives exactly the value of R_L. In general, however, thermal convection currents or stirring can reduce R_L much below the value calculated from eq. (7.15).

Lewis and Whitman[1] proposed that *turbulent stirring* maintains the composition of the solute constant within the liquid, up to a distance Δx below the surface, though within this distance the liquid flow is laminar and parallel to the surface (see Fig. 7-1). Through this "laminar sublayer" transfer occurs at a steady rate as if the layer were stagnant, and is therefore given by eqs. (7.2) and (7.3). For this mechanism R_L is thus constant for a given value of Δx or, if the rate of stirring is varied, R_L can be calculated from the variation of Δx near a solid surface with the Reynolds number characterizing the turbulence in the bulk of the liquid[20]:

$$\Delta x = \text{const.} \times (\text{Re})^{-0.67} \tag{7.17}$$

where $\text{Re} = NL^2/\nu$ with N the speed of stirring, L the tip-to-tip length of the stirrer blades, and ν the kinematic viscosity.

Further, if the concentration difference across the laminar sublayer does not vary appreciably with time, eqs. (7.2) to (7.4) may be integrated directly to give linear relations between q and $kt\Delta c$, between k and D, and between q and $Dt(\Delta c/\Delta x)$.

In the limit of extreme turbulence, when eddies of fresh solution are rapidly swept into the immediate vicinity of the interface, neither the laminar sublayer nor a stationary surface can exist: the diffusion path becomes so short that diffusion is no longer rate-controlling, and consequently for such liquid-phase transfer

$$\frac{dq}{dt} = A.\Delta c.\bar{v}_n \tag{7.18}$$

where $\bar{v}_n$ is the mean velocity of the liquid normal to the interface. By eq. (7.2) $\bar{v}_n$ is equal to k_L, and the latter is now independent of D: this limit of extreme turbulence is never reached in practice.

The steady-state theory of Lewis and Whitman cannot be valid, however, at short times of contact of the gas with a turbulent liquid, when diffusion according to eq. (7.14) must be important. This condition may be of practical importance in the flow of a liquid over packing in a gas-absorption column, when the flow may be laminar for short times as the liquid runs over each piece of packing, though complete mixing may occur momentarily as the liquid passes from one piece of packing to the next. For the absorption of a gas in such a column, *Higbie*[21] proposed that moderate liquid movements would have no effect on the diffusion rate for a very short time of exposure of the liquid surface, and that accordingly one should use eqs. (7.14) and (7.16). This is confirmed by experimental studies on the absorption of CO_2 into water with times of exposure t up to 0.12 sec.; k_L and q vary closely with $t^{1/2}$. Further k_L should vary with $D^{1/2}$: this is indeed confirmed by experiments on packed towers[22].

The Higbie theory can, however, be valid only for times of exposure so short and for turbulence so low that the diffusing gas does not penetrate into those parts of the liquid which have a velocity appreciably different from that at the surface. At longer times, eddies may be pictured as continually exposing fresh liquid surfaces to the gas, while at the same time sweeping into the bulk those parts of the surface which have been in contact with the gas. *Kishinevskii*[23] pointed out that, if this occurred, both eq. (7.14) and eq. (7.18) must be simultaneously applied, the two mechanisms acting in parallel. *Danckwerts*[24] suggested that elements of liquid had definite residence times in the surface region, before being swept back into the bulk, and that during this residence time the rate of diffusion of the gas into the element of liquid should obey eq. (7.14): this is equivalent to applying eqs. (7.14) and (7.18) consecutively (i.e. in series). If now s is the fractional rate of turbulent replacement of liquid elements (of any age) in the surface, k_L and R_L are given[24] by

$$\frac{dq}{dt}=A(c_s-c_\infty)(Ds)^{1/2} \tag{7.19}$$

$$k_L=\frac{1}{R_L}=(Ds)^{1/2} \tag{7.20}$$

and

$$q=A(c_s-c_\infty)(Ds)^{1/2}t \tag{7.21}$$

Since k_L here varies with $D^{1/2}$, as in the simple diffusion or Higbie theories, a precise knowledge of s is required to distinguish them. *Toor and Marchello*[25] consider that at very low rates of stirring the Lewis and Whitman model is valid, though at moderate turbulence the rate-controlling step becomes the replacement of elements of liquid according to eq. (7.19).

Experiments on Static Systems

In the rapid absorption of O_2 and CO_2 into unstirred, chemically reactive solutions[27], monolayers of stearyl alcohol may decrease the uptake of gas by 30%. In such studies, R_G can be made negligibly low by using pure gases, and R_L is fairly small because of the simultaneous reaction and diffusion, at least in the terms of the few minutes required for measurements. Hence R_I may readily be determined. Oleic acid, proteins[26] and cholesterol are rather ineffective, but the values of R_I for oxygen diffusing through films of the C_{16} and C_{18} alcohols are much higher than found in evaporation studies, being respectively about 80 and 290 sec.cm.$^{-1}$; the values for CO_2 are similar to these figures[27].

Another technique for studying the absorption of CO_2 into water uses an interferometer to obtain the concentration gradients as close as 0.01 cm. to the surface[28]: a ciné-camera permits results to be obtained within 5 sec. of the admission of the CO_2. Though various corrections are required, it is claimed that this method eliminates convection difficulties and that resistances as low as 0.25 sec.cm.$^{-1}$ can be detected. Experimental results for CO_2 into distilled water show no detectable interfacial resistance, though, when surface-active agents (Lissapol, Teepol) are dissolved in the water, the values of R_I are about 35 sec.cm.$^{-1}$

That these figures are higher than the evaporation studies would indicate for such expanded films suggests that the blocking effect of the "head-groups" is again important; a layer of molecules of (bound) liquid water is always present around these "head-groups" (which therefore oppose evaporation moderately), and this bound water can appreciably retard the transfer of solute molecules (cf. p. 369). More precise studies of the effects of chain length and of the density of packing of the "head-groups" would be of great interest.

Experiments on Dynamic Systems

In the steady-state absorption of oxygen into water *in stirred vessels*, R_L may vary between 120 sec.cm.$^{-1}$ and 12,000 sec.cm.$^{-1}$, the lower figure corresponding to the most rapid mechanical stirring, and the latter to stirring by convection only[29]. Application of eq. (7.3) to the experimental data shows that, since D is of the order 10^{-5}cm.^{2}sec.$^{-1}$, Δx would be 0.1 cm. in water stirred by convection only, 26.6×10^{-3}cm. in water stirred at 50 r.p.m., and 5.8×10^{-3}cm. for stirring at 270 r.p.m. Comparison of the ratio of these two thicknesses with the ratio of the stirring speeds, gives a power of -0.9 instead of the -0.67 required by eq. (7.17): this may well be due to the interface being able to move, instead of being static as in the data on which eq. (7.17) is based. At the lower stirring speeds, and with no wind, a monolayer of hexadecanol (effective in reducing evaporation by 25%—see above) does not alter significantly the rate of oxygen absorption, since R_I is relatively low (about 10 sec.cm.$^{-1}$), and since also R_L is unaffected by the presence of the film. At high stirring rates, however, the monolayer may increase R_L fivefold, by partially damping the eddies of liquid approaching the surface. This hydrodynamic effect of the viscosity and incompressibility of a monolayer is discussed in more detail below. At the lower stirring speeds the monolayer is effective only if the hexadecanol monolayer is compressed on part of the surface by a stream of air; in this way R_L may be approximately doubled. The effect of a monolayer of hexadecanol on the oxygen level of a wind-blown reservoir may be found from eq. (7.2). Thus, for a given oxygen demand dq/dt by the living material in the reservoir, halving k must double Δc, where Δc is the difference between the oxygen concentrations at the surface ($c_{sat.}$) and in the bulk of the water (c), i.e. it is the oxygen deficit ($c_{sat.} - c$). In practice c is about 90% of $c_{sat.}$, so that Δc is 10% of $c_{sat.}$; consequently doubling the deficit c will now increase ($c_{sat.} - c$) to 20% of $c_{sat.}$; or halving k will reduce c from 90% to 80% of $c_{sat.}$. This change in oxygen level is not important to the life in the reservoir. Quantitative measurements[31] show that at moderate stirring speeds k_L varies as $D^{0.25}$. Experiments with CO_2 absorbing into alkali are claimed[30] to show that the surface flow must be more complicated than envisaged in the Danckwerts theory.

Fig. 7-3 shows some recent quantitative results: a protein film makes the interface less subject to renewal from the bulk.

As a model for an industrial packed column, and to study the rate of absorption of a gas at short times, a *wetted-wall column* is very convenient[33]. This consists of a sheet of liquid running down the surface of a tube, usually a few centimeters in diameter and made of steel. To ensure that the surface is completely wetted by the sheet of liquid, the tube may be sand-blasted to roughen it (cf. Chapter 1) if it is of stainless steel or, if mild steel, the tube should be "blued" by heating to redness and quenching in water. The latter treatment both burns away organic contaminants and also produces a hydrophilic oxide coating. Channelling is more likely to occur if there are appreciable ($\sim$5 dynes $cm.^{-1}$) changes in surface tension during gas absorption (cf. p. 28).

The flow of liquid down the surface of such a tube is essentially laminar if $Re < 1200$, where Re is defined as $4u/\nu l$, u being the volume flow rate of liquid, l the perimeter of the tube, and ν the kinematic viscosity of the solvent. Under these conditions, if there are no surface forces acting, the velocity of

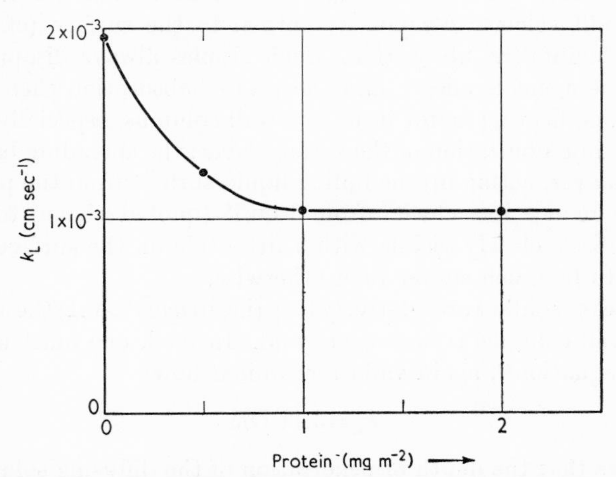

Fig. 7-3. A spread monolayer of protein decreases k_L for the absorption of CO_2 (from the gas phase at 1 atm. partial pressure) into water[32]. The surface is cleaned thoroughly before the experiment, and the contra-rotating stirrers in the liquid are running at 437 r.p.m. The surface is non-rotating in all the experiments described here, and when k_L is reduced to 1.1×10^{-3}cm.sec.$^{-1}$, all random surface movements are also eliminated.

the air-water surface of the falling film is $1.5u/l\delta$, where δ is the thickness of the falling liquid sheet. The residence time t_r of an element of surface in contact with gas is therefore given by $t_r = hl\delta/1.5u$, if it is in contact with the gas while it falls through a height h. If absorption may be regarded as occurring into a semi-infinite mass of liquid, this calculated value of t_r may be substituted into eq. (7.14) or, more strictly, this is valid provided that $hD\nu/g\delta^4 < 0.1$, where g is the acceleration due to gravity[24]. If, however, this inequality is not satisfied, the liquid cannot be treated as semi-infinite, nor can the effect of the velocity gradient in the liquid be neglected. To test the applicability of the use of this value of t_r in eq. (7.15), i.e. of the expression

$$k_L = \frac{1}{R_L} = \left(\frac{D}{\pi t_r}\right)^{1/2} \tag{7.22}$$

Kramers[33] et al. studied the absorption of SO_2 into water. This is a comparatively simple process, R_G being negligibly small and any hydrolysis reactions between CO_2 or SO_2 and water occurring very rapidly relative to the times of contact involved. At low flow rates agreement with theory is satisfactory only if a surface-active agent is present in the system, though without such an agent the experimental results can be as much as 100% too high[33]. At high flow rates, the absorption of SO_2 is independent of the presence of a surface-active agent. These effects are due to rippling of the water sheet which occurs in the absence of surface-active agents: a ripple causes a small velocity component normal to the surface (cf. Chapter 5, p. 266), so facilitating absorption. Such ripples always disappear when a surface-active agent is added, and the rate of absorption then agrees with theory. A complicating factor in wetted-wall columns, especially if these are short, is the immobilization of the surface layer: the spreading back-pressure of the monolayer, acting up the falling liquid surface from the pool of liquid below, may reduce its velocity from 1.5u/lδ (quoted above) to practically zero. This effect, clearly visible with a little talc on the surface, causes gas absorption to be much slower than otherwise.

If the times of contact are relatively long (i.e. $hD\nu/g\delta^4 > 0.1$) the above theory of wetted-wall columns is no longer valid. Instead, one must use Pigford's theoretical equation[22], again valid for laminar flow

$$k_L = 3.41\ D/\delta \tag{7.23}$$

This assumes that the depth of penetration of the diffusing solute molecules is at least equal to δ, i.e. that $3.6(Dt)^{1/2} > \delta$. Equation (7.23) is based on normal diffusion theory, and is of the general form of eq. (7.3). Again, the ripples formed on the liquid surface (see p. 266 and Fig. 5-41) may cause results to be several times greater than calculated by diffusion theory if no surface-active agent is present. Now R_L is at least several hundred sec.cm.$^{-1}$ in typical wetted-wall experiments, and since the surface-active agents used would not increase R_I beyond 30 sec.cm.$^{-1}$ at the most, the effect of the surface-active agents is entirely hydrodynamic: the ripples when present decrease R_L below the calculated values.

Absorption of a gaseous solute into a moving liquid at short times can also be conveniently measured using a *falling vertical jet* of liquid, formed in such a way that the liquid velocity is uniform across every section of the jet. Results for CO_2 absorbing into water and dekalin are in close agreement with an equation of the form of (7.16), modified to allow for the form of the jet during absorption. This accord shows that R_L is dominant, R_I being negligible[34,35]. If surface-active agents are added, the same effects are found as for short wetted-wall columns, an immobile monolayer forming on the lower part of the jet. This monolayer retains powder sprinkled on the surface, and extends to such a distance up the jet that the spreading tendency (forcing it further up the jet) is balanced by the shear stress between the monolayer and the moving liquid of the jet, i.e. $\Delta\Pi = h\tau$, where h is the height the monolayer extends up the jet, and τ is the calculated mean shear stress. Typical values[35] for 0.2% Teepol are $\tau = 20$ dynes cm.$^{-2}$ and h = 1.4 cm., whence $\Delta\Pi = 28$ dynes cm.$^{-1}$, a quite reasonable value.

Liquid flowing slowly over the surface of a solid sphere is in essentially laminar flow, except that there must be a radial component due to the

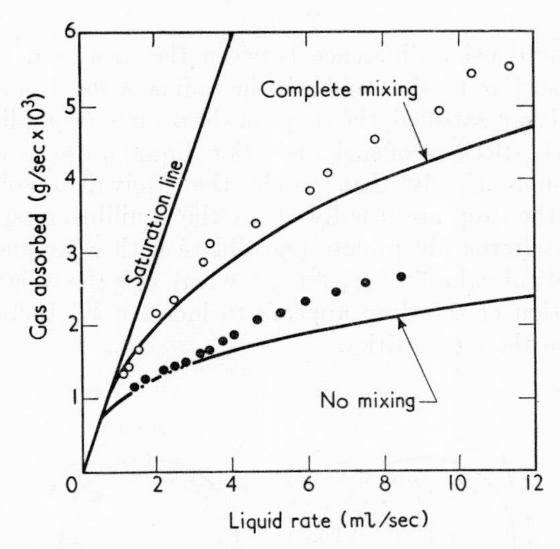

Fig. 7-4. Carbon dioxide absorption rate for 5 spheres at 25°C. O, experimental results for pure water; ●, Lissapol solution. (Ref. 37).

increase of the amount of surface at the "equator" relative to that at the "poles". This system, useful as a model of packing in an absorption column, has been studied mathematically[36,37]. Further, if a vertical "string" of touching spheres is studied, one finds that this laminar flow is interrupted at each meniscus between the spheres: complete mixing of the liquid occurs at these points. This is confirmed by measurements and calculations of R_L for the absorption of CO_2 (Fig. 7-4). If now a surface-active agent is added to the water flowing over the spheres, the uptake of gas is reduced by about 40%, because the adsorbed monolayer reduces k_L (by reducing both the eddy mixing at the menisci and the rippling over the spheres[37]). This rippling of the clean surface occurs in the system of Fig. 7-4 at flow rates greater than 3 cm.^{3}sec.$^{-1}$, and deviations from the calculated curves at high flow rates are ascribed to ripples and to inertial effects in the liquid flow, not allowed for in the calculations[37]. There is no evidence that R_I is important in any of these studies.

The value of R_L *in a falling drop of liquid* is of interest in view of the applications of spray absorbers[38]. A wind-tunnel[39] for the study of individual liquid drops, balanced in a stream of gas, has shown[40] that R_L for a drop depends on its shape, velocity, oscillations, and internal circulation. The drop will remain roughly spherical only if[41]

$$\frac{\Delta\rho a^2 g}{\gamma} < 0.1 \tag{7.24}$$

where $\Delta\rho$ is the density difference between the drop and the continuous phase, γ is the surface tension, and a is the radius of the drop. In general, if this inequality is not satisfied, the shape of the drop is a complicated function of the variables[42]. Because of such distortion, liquid drops over about 2 mm. diameter fall more slowly than would the equivalent solid sphere[41,42]. Oscillations of the drop are usually about the equilibrium spherical shape, which becomes alternately prolate and oblate with a frequency in accord with the theoretical value[43] of $(8\gamma/3\pi w)^{1/2}$, where w is the weight of the drop. Though oscillation of the drop appears to increase k_L, there is no simple relation between these quantities.

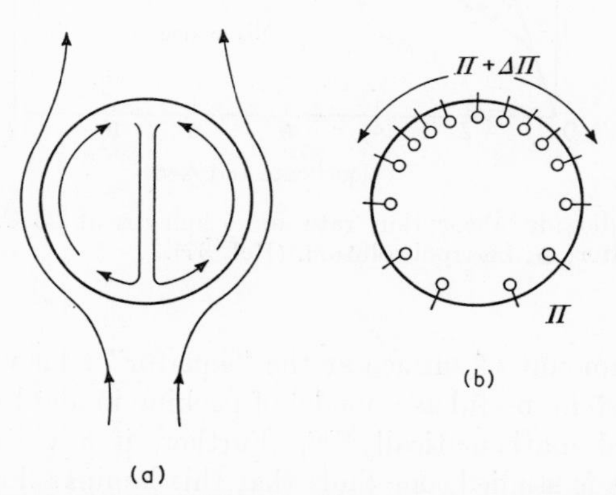

Fig. 7-5. Circulation within a falling liquid drop, and opposing effect of surface pressure gradient.

During the first few seconds after the formation of a drop, k_L for gas absorption may be as much as sixty times greater than predicted by diffusion into a static sphere[40,44]: this is related to the high initial rate of circulation in the drop, caused by the break-away of the drop from the nozzle. As this initial rapid circulation dies away, k_L falls to about 2.5 times that predicted for a static sphere: this factor of 2.5 is in accord with calculation[45], assuming that the natural circulation inside the falling drop is in streamline flow (Fig. 7-5a).

This natural circulation occurs by a direct transfer of momentum across the interface, and the presence of a monolayer at the interface will affect it in two ways. Firstly, the surface viscosity of the monolayer may cause a dissipation of energy and momentum at the surface, so that the drop behaves rather more as a solid than as a liquid, i.e. the internal circulation is reduced. Secondly, momentum transfer across the surface is reduced by the incompressibility of the film, which the moving stream of gas will tend to sweep to the rear of the drop (Fig. 7-5b) whence, by its back-spreading pressure Π, it resists further compression and so damps the movement of the surface and hence the transfer of momentum into the drop[45a]. This is discussed quantitatively on p. 336, where eq. (7.42) should apply equally well to drops of liquid in a gas.

If this natural circulation within the drop is thus reduced by adsorption at the surface, k_L and the rate of gas absorption fall to the values calculated for molecular diffusion into a stagnant sphere[40].

Any surface turbulence will greatly increase k_L: this is found in the absorption of SO_2 into drops of n-heptane, which is ten times faster than expected according to Groothuis and Kramers[17].

Transfer of gas from a rising bubble into a liquid is becoming increasingly important: bubble and foam columns are often more efficient than are packed towers for gas absorption (cf. p. 395).

Swarms of rising bubbles may be produced in pure liquids by the methods described on p. 393, and, in the absorption of gas from such a bubble swarm, *small bubbles* (<0.2 cm. diameter) often behave as rigid spheres, because they are very sensitive to surface films (see eq. (7.42) below): k_L is independent both of the mechanical agitation and of bubble size. However, k_L does vary linearly with $D^{2/3}$: this is all as required by theory[46,47,48]. Surface films reduce the momentum transfer at the bubble surface and hence the rate of rise[47,48]. *Large bubbles* no longer behave as rigid spheres, though k_L is still independent of the bubble size and of the liquid agitation; the values of k_L, however, are considerably higher than for small bubbles[46-49], and k_L varies with $D^{0.84}$. The movements of the surface of a large rising bubble are very complex and defy exact hydrodynamic analysis.

SOLUTE TRANSFER AT THE LIQUID-LIQUID INTERFACE

If a solute such as acetic acid is diffusing from water to benzene, the sum of the two R_L terms is usually much greater than the interfacial resistance R_I. In the simplest, unstirred system, with both liquids taken as infinite in extent, one must solve the diffusion equations[50]

$$\frac{\partial c}{\partial t} = D_1\left(\frac{\partial^2 c}{\partial x^2}\right) \quad \text{for } x<0 \tag{7.25}$$

$$\frac{\partial c}{\partial t} = D_2\left(\frac{\partial^2 c}{\partial x^2}\right) \quad \text{for } x>0 \tag{7.26}$$

referring respectively to the lower and upper liquid (Fig. 7-6a). The initial conditions are typically that, at $t=0$, $c=c_0$ for $x<0$, and $c=0$ for $x>0$. If there is a negligible interfacial resistance R_I, then $c_{2I}/c_{1I}=B$ for all $t>0$, where c_{1I} and c_{2I} are the respective concentrations of the solute immediately

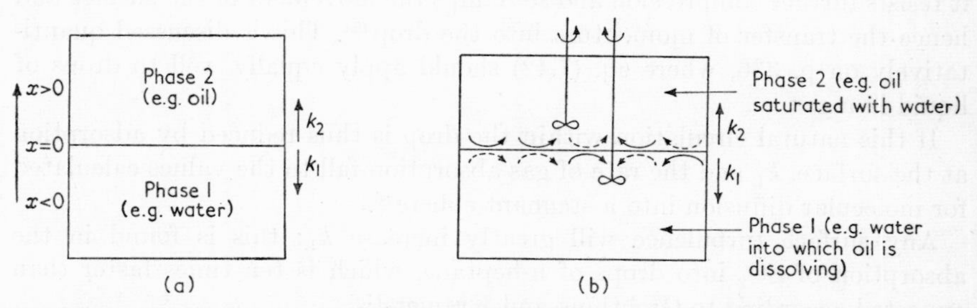

Fig. 7-6. (a) Unstirred cell. (b) Stirred cell. Full arrows show eddies produced directly by stirrer, broken arrows show eddies induced by momentum transfer across the turbulent liquid interface.

on each side of the interface, and B is the distribution coefficient. The continuity of mass-transfer across the interface requires that at $x=0$ the condition $D_1(dc_1/dx)=D_2(dc_2/dx)$ is always satisfied, and with these boundary conditions the solutions to the diffusion equations above (if $x=0$ and $t>0$) are:

$$c_{1I} = \frac{c_0}{1+B(D_2/D_1)^{1/2}}$$

and

$$c_{2I} = \frac{Bc_0}{1+B(D_2/D_1)^{1/2}}$$

These equations imply that the interfacial concentrations must remain constant as long as the diffusing material has not reached the ends of the cell: for cells which cannot be considered infinite the treatment is more difficult[15].

The total number of moles q of solute which has diffused across the interface after any time t is given by $\int_0^\infty c_2 dx$, which may be integrated to[51]

$$q=\left(\frac{2AB\mathrm{c}_0}{1+\mathrm{B}(\mathrm{D}_2/\mathrm{D}_1)^{1/2}}\right)\left(\frac{\mathrm{D}_2\mathrm{t}}{\pi}\right)^{1/2} \tag{7.27}$$

from which

$$q=2A\mathrm{c}_{2\mathrm{I}}\left(\frac{\mathrm{D}_2\mathrm{t}}{\pi}\right)^{1/2} \tag{7.28}$$

whence the value of k_{L} is given with respect to $\mathrm{c}_{2\mathrm{I}}$ with these initial conditions by

$$k_{\mathrm{L}}=\frac{1}{R_{\mathrm{L}}}=\left(\frac{\mathrm{D}_2}{\pi\mathrm{t}}\right)^{1/2} \tag{7.29}$$

If there is an interfacial resistance, the equations become much more complicated[51,52], and both q and the ratio $\mathrm{c}_{2\mathrm{I}}/\mathrm{c}_{1\mathrm{I}}$ are altered.

These equations, referring to completely unstirred systems, are not usually valid in practice; complications such as spontaneous interfacial turbulence and spontaneous emulsification often arise during transfer, while, if external stirring or agitation is applied to decrease R_{L}, the hydrodynamics become complicated and each system must be considered separately. The testing of the above equations will be discussed below, after a consideration of overall coefficients and of interfacial turbulence.

In practice, whether the cell is stirred or not, transfer is conveniently referred to "overall" rather than interfacial concentrations. The overall transfer coefficient $\mathbf{K}_{\mathrm{L}}$ is related to the individual liquid coefficients by the relations

$$\frac{1}{A}\frac{\mathrm{d}q}{\mathrm{dt}}=k_{\mathrm{L}_1}(\mathrm{c}_1-\mathrm{c}_{1\mathrm{I}})=k_{\mathrm{L}_2}(\mathrm{c}_{2\mathrm{I}}-\mathrm{c}_2) \tag{7.30}$$

and

$$\frac{1}{A}\frac{\mathrm{d}q}{\mathrm{dt}}=\mathbf{K}_{\mathrm{L}}(\mathrm{c}_1-\mathrm{c}_2/\mathrm{B}) \tag{7.31}$$

implying that the material diffusing into the interface must also diffuse away from it. They are applicable at any time t. The factor B must be introduced in the latter equation since the transfer rate will be zero at equilibrium, i.e. when c_2 finally reaches $\mathrm{c}_1\mathrm{B}$. Elimination of the rates from eqs. (7.30) and (7.31) gives

$$\frac{1}{\mathbf{K}_{\mathrm{L}}}=\frac{1}{k_{\mathrm{L}_1}}+\frac{1}{\mathrm{B}k_{\mathrm{L}_2}} \tag{7.32}$$

or

$$R_{\mathrm{L(overall)}}=R_{\mathrm{L}_1}+R_{\mathrm{L}_2}/\mathrm{B} \tag{7.33}$$

If there is also an interfacial resistance, it will be given by the measured resistance less the $R_{\mathrm{L(overall)}}$ value calculated by eq. (7.33).

Interfacial Instability

When the system contains only two components, such as ethyl acetate and water, the rates of mutual saturation are generally uncomplicated. If, however, a third component is diffusing across the oil-water interface, two complicating effects may occur. The first, spontaneous emulsification, is seen when, for example, 1.9 N-acetic acid, initially dissolved in water, is allowed to diffuse into benzene: fine droplets of benzene appear on the aqueous side of the interface. The mechanism of this spontaneous emulsification is that of "diffusion and stranding", discussed in detail in Chapter 8. No system has yet been found where, under close observation, spontaneous emulsion did not form during the transfer of the third component. Although surface-active agents, including protein monolayers, may reduce the amount of emulsion by reducing the interfacial turbulence, as discussed below, they never eliminate it completely.

The second complicating factor is interfacial turbulence[18,53], very similar to the surface turbulence discussed above. It is readily seen when a solution of 4% acetone dissolved in toluene is quietly placed in contact with water: talc particles sprinkled on to the plane oil surface fall to the interface, where they undergo rapid, jerky movements. If a pendant drop of water is formed in the toluene-acetone, this drop will undergo violent, erratic pulsations or "kicks", each of which is rapidly damped out by viscous drag. The frequency of kicking of the drop diminishes with time, and ceases when all the acetone is distributed between the oil and water phases in accordance with the partition coefficient. Aluminium powder, suspended in the liquids, shows that "kicking" of a drop is associated with greatly enhanced flowing of liquid near the interface, which in turn leads to mass-transfer of the acetone or other third component at rates greater than expected.

The molecular mechanism of interfacial turbulence is that at some point on the interface an eddy of oil brings up rather more acetone than is present at other points on the interface. Such eddies may originate either in the formation of the drop, in thermal inequalities in the system, or, once the process has started, in movements associated with previous kicks. At the interface, the locally high concentration of acetone results in increased adsorption over a small region, and the accompanying higher value of Π (lower γ) results in a momentary change in the pressure inside the drop (since the excess pressure P_e is $2\ \gamma/a$) and hence causes it to "kick" (Fig. 7-7). There is subsequent rapid spreading of the acetone across the surface (in a few milliseconds), followed by rapid desorption into the water. The spreading of the monolayer, however, causes liquid movements (as shown in Fig. 7-7) with the result that more of the toluene-acetone solution is brought up to the same point, thus enlarging what was originally a very small effect into a pronounced one. When the oscillations have nearly ceased, and much of the

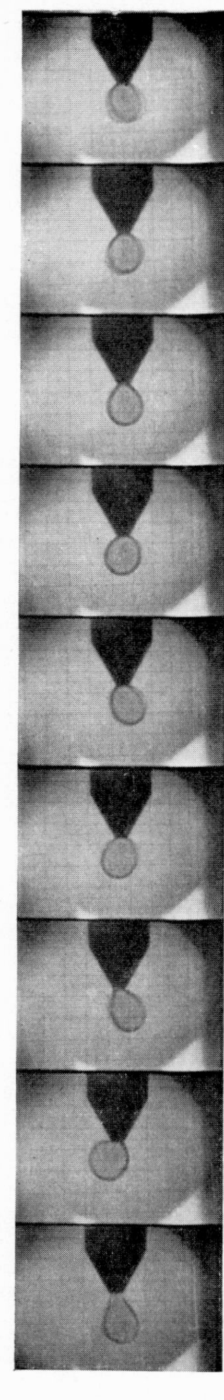

Fig. 7-8. A typical film sequence of a "kicking" drop.

[*To face p. 323*

acetone from the initial interfacial region has diffused into the drop, the system is ready to show another kick, induced by another eddy containing fresh 4% acetone in toluene. A typical sequence is shown in Fig. 7-8.

This mechanism requires a relatively uncontaminated interface and the possibility of appreciable depletion of the third component near the inter-

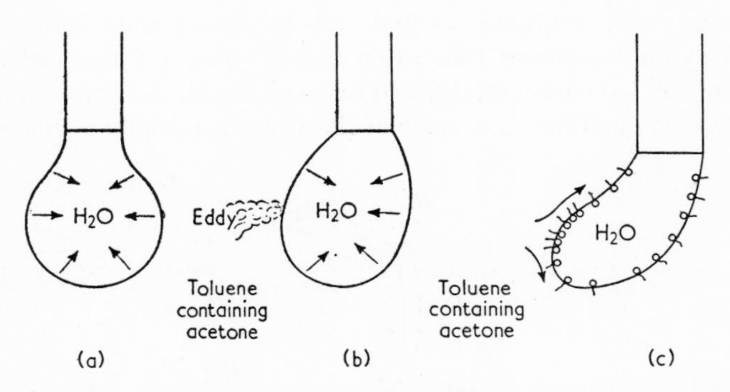

Fig. 7-7. (a) Arrows represent pressures exerted inwards by the interfacial tension. (b) Interfacial tension is reduced by an eddy of toluene undepleted in acetone, and the local change in interfacial tension causes the drop to "kick". (c) Spreading of monolayer after adsorption from the eddy causes liquid movements which bring up further undepleted solution of acetone in toluene, so amplifying the whole process.

face due to mass transfer, and is consistent with the following experimental observations[18].

(i) If the solubility of acetone in water is reduced by substituting 1 M-NaCl solution to form the drop, practically no kicking occurs.

(ii) An air-bubble, which can take up relatively little acetone, does not kick in the acetone-toluene solution.

(iii) A 4 mM solution of dodecyltrimethylammonium chloride inhibits the kicking, though it does not affect the partitition coefficient. Evidently the rather strongly held monolayer of long-chain ions is important, for the presence of the acetone now scarcely affects the interfacial tension.

(iv) Studies with a ciné-camera show that, if an eddy of acetone is squirted from a fine syringe on to the "equator" of a drop of water formed in pure toluene, the drop gives a sharp kick as soon as the jet of acetone reaches it, the first movement being towards the jet. An air bubble behaves similarly. The kicks are, as far as observable, simultaneous with the arrival of the solute at the surface, showing that temperature changes due to redistribution of the solute between the bulk phases cannot primarily be responsible for the kicks.

To investigate the effects quantitatively, it is necessary to find the concentration c_k of solute at which kicking becomes just appreciable. By magnifying the image of the pendant drop on a screen, one can determine the amplitude of small kicks, 0.08 radian being a convenient standard. Since this amplitude increases steeply with solute concentration, c_k can be found to an accuracy of $\pm 5\%$.

In an analytical treatment to find $\Delta\Pi$ due to an eddy arriving at the surface, Haydon[53] assumes that after a short time (of the order 3 sec.) equilibrium will have been established between the oil and water phases close to the interface by diffusion of some of the acetone, initially of concentration

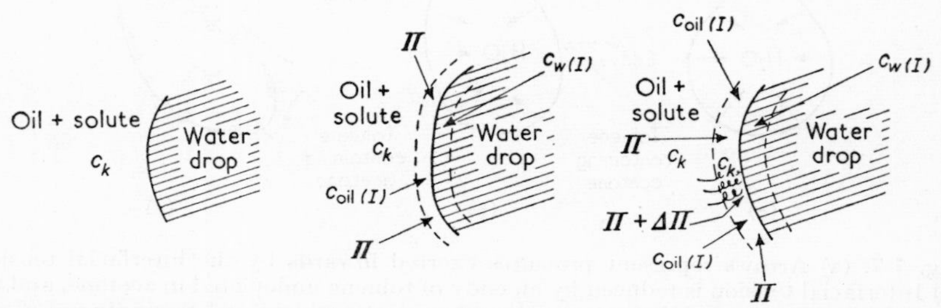

Fig. 7-9. Sequence of events leading to "kicking" of drop; the water drop is just formed in the first diagram, then some of the solute near the interface is distributed between the phases, and finally an eddy of oil which is as yet undepleted in solute arrives at the interface and locally increases the surface pressure.

c_k, from the oil immediately adjacent to the interface, into the water adjacent to the interface. In symbols, if B is the distribution coefficient,

$$c_{oil(I)}/c_{w(I)} = B \tag{7.34}$$

and

$$c_{w(I)} + c_{oil(I)} = c_k \tag{7.35}$$

where $c_{w(I)}$ and $c_{oil(I)}$ refer respectively to the concentrations of solute (acetone) in the water and oil close to the interface, as in Fig. 7-9. The concentration difference when an eddy of oil, undepleted in acetone (the concentration still being c_k), arrives at the interface is therefore

$$c_k - c_{oil(I)} = c_k\left(\frac{1}{1+B}\right)$$

whence, knowing c_k, we can deduce $c_{oil(I)}$ and hence $c_{w(I)}$. These three quantities are respectively 0.135 M, 0.0175 M, and 0.118 M for the first system of Table 7-I. From a curve such as that of Fig. 7-10 one can then find the interfacial tensions corresponding to c_k and c_{oil}, here 41.9 dynes cm.$^{-1}$ and 46 dynes cm.$^{-1}$ Consequently the eddy of acetone momentarily raises the surface pressure over a small area A_1 of the interface by a quantity $\Delta\Pi$, which is here 4.1 dynes cm.$^{-1}$

TABLE 7-I

Continuous phase	Drop	$\frac{1}{B}$(20°C)	c_k (moles $l.^{-1}$)	γ_0 (dynes cm.$^{-1}$)	Π (dynes cm.$^{-1}$)	$\Delta\Pi$ (dynes cm.$^{-1}$)	$\mathfrak{W}_1/A_1$ (ergs cm.$^{-2}$)	$10^3\mathfrak{W}_2(0.08/\theta)^2$ (ergs)
Petroleum ether + acetone	water	6.8	0.135	49.7	3.7	4.1	5.8	1.14
Petroleum ether + acetone	water + 0.3mM-$C_{12}H_{25}N(CH_3)_3Br$	6.8	0.290	49.7	9.3	4.5	5.4	0.93
Petroleum ether + acetone	water + 1.36mM-$C_{12}H_{25}N(CH_3)_3Br$	6.8	0.525	49.7	15.8	4.5	4.9	0.82
Petroleum ether + ethyl alcohol	water	18.2	0.169	49.7	2.4	3.4	5.7	1.61
Petroleum ether + ethyl alcohol	water + 1.36mM-$C_{12}H_{25}N(CH_3)_3Br$	18.2	0.658	49.7	16.1	4.3	4.8	0.81
Toluene + acetic acid	water	23.6	0.087	36.0	2.3	2.5	3.5	0.87
Toluene + diethylamine	water	1.7	0.463	36.0	14.8	5.5	6.4	0.38
Benzene + acetone	water	1.1	0.200	35.0	4.2	2.3	2.8	0.67
Ethyl acetate + acetone	water	0.81	0.650	7.8	1.0	0.24	0.26	0.075
Ethyl acetate + acetic acid	water	1.19	0.167	7.8	0.3	0.50	0.71	0.11
Butyl acetate + acetone	water	1.23	0.260	12.6	2.1	0.36	0.38	0.31

The work done by the additional adsorption from the eddy can now be calculated. If the surface film has a pressure Π before the eddy arrives (here $\Pi = 49.7 - 46.0 = 3.7$ dynes cm.$^{-1}$) one can show that the total energy $\mathfrak{W}_1$ imparted to the interface is given by:

$$\mathfrak{W}_1 = (\Pi + \Delta\Pi) A_1 \ln\left\{\frac{\Pi + \Delta\Pi}{\Pi}\right\} \tag{7.36}$$

where the quantities refer to a deflection of the drop during kicking of about 0.08 radian, and A_1 refers to the initial area of contact of the convection current and the surface of the drop. Figures for $\mathfrak{W}_1/A_1$ are shown in Table 7-I.

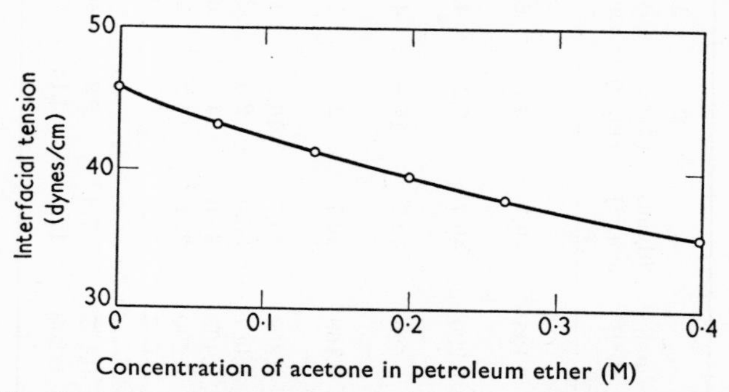

Fig. 7-10. The dependence of the interfacial tension, between petroleum ether and water, on the acetone concentration in the oil. The acetone in the water phase was maintained at 0.118 M.

It is now of interest to compare the energy of adsorption (eq. 7.36) with the energy dissipated during a single kick and the subsequent damped oscillations of a drop. The latter energy can be obtained by a combination of ciné-records of drop movement and a complete hydrodynamic theory of the movements, put forward by T. V. Davies and Haydon[53]. This is possible in practice only if the drop can be made to give a single kick, by a technique similar to that of (iv) above: one squirts the acetone solution from a horizontal jet (0.07 cm. diameter) immersed in the oil and situated about 0.25 cm. from the equator of the drop. The concentrations of acetone or other solute in the oil and in the water drop are chosen to be those given by relations (7.34) and (7.35) above, while the eddy squirted from the jet is of oil containing a concentration c_k of solute: in this way the conditions for kicking are as close as possible to those in the spontaneously kicking systems described above, with periods of oscillation of the pendant drop of the order 0.2 sec. The single, induced kick and the subsequent damped oscillations of a single pendant drop of a given size dissipate energy $\mathfrak{W}_2$ which is proportional to θ^2, where θ is the deflection of the initial kick. The energy $\mathfrak{W}_2$ can therefore be converted to the equivalent energy had the kick been through only 0.08 radian, i.e. $\mathfrak{W}_2(0.08/\theta)^2$.

Comparison of the energy $\mathfrak{W}_1$ supplied to the drop by adsorption with the equivalent energy dissipated hydrodynamically is made in Fig. 7-11: with the exception of the system toluene + diethylamine against water, the plot confirms the validity of the basic assumptions. Further, since $\mathfrak{W}_2$ should equal $\mathfrak{W}_1$ when $\theta = 0.08$, A_1 must follow from the plot of Fig. 7-11. The mean value of A_1 thus found is 2×10^{-4}cm.^{2}, a reasonable value for the initial area affected by the eddy on the surface of a spontaneously kicking drop.

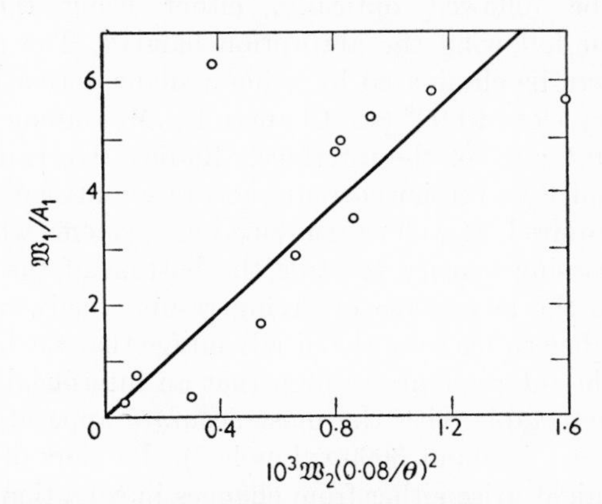

Fig. 7-11. Plot of the energy per unit area producing one series of oscillations, against the energy dissipated by the drops against viscous forces. Each point represents a different system.

If the drop is of oil-solute solution, with the continuous phase water, kicking still occurs, accompanied by strong circulation of the liquid inside the drop: these movements are too complicated to permit detailed analysis. A hydrodynamic theory[19] of the eddying in each phase accompanying spontaneous interfacial turbulence has been attempted (see p. 310): so far it has not been possible to introduce the distribution coefficient, the concentration c_k, or the properties of the adsorbed monolayer into this theory.

Theoretical Values of R_L

For plane, unstirred cells the above equations ((7.29) and (7.33)) are accurate, while in plane stirred cells the theories of Lewis and Whitman, Kishinevskii, and Danckwerts may be tested against the experimental results. In particular, if eqs. (7.14) and (7.18) represent processes occurring in parallel, q should depend on some power of D between 0.5 and 0, whereas if they apply in series (to consecutive exposure of new surface, and its later removal), eqs. (7.19-7.21) should apply.

Experiments on Static Systems

For completely unstirred systems, one may test whether results agree with theory (assuming no interfacial resistance) in one of two ways. Firstly, one may find experimentally c_{2I} and c_{1I} and test whether their ratio is B. Secondly, one may find q at different times and compare the result with eq. (7.28). If equilibrium does not prevail across the interface (i.e. if interfacial concentrations are not in the ratio of B), or if the experimental q is less than calculated, then an interfacial resistance must be operative. In systems which are completely unstirred, the amount of material diffusing across the interface may be followed optically, either using the Lamm scale method[51,52,54,55] or following the absorption bands[56]. The meniscus at the sides of the cell can be eliminated by using a silane derivative to make the contact angle very close to 90° (see Chapter 1). Measurements can then be made to within 0.1 mm. of the interface. Radioactive tracers can also be used[57,58]. To minimize convection currents, accurate control of the temperature to ± 0.001°C is required, as well as studying only systems which maintain at all times an increasing density towards the bottom of the cell. While unstirred cells have the advantage of giving results easily comparable with diffusion theory, they suffer from the disadvantage that the liquid resistances on each side of the interface are so high that an interfacial resistance R_I is experimentally detectable with the most accurate apparatus[51,52] only if it exceeds 1000 sec.cm.$^{-1}$ (about 3000 cal.mole^{-1}). Barriers of this magnitude at the interface might arise either from changes in solvation of the diffusing species in passing between the phases, or from a polymolecular adsorbed film.

When acetic acid is diffusing from a 1.9 N solution in water into benzene, spontaneous emulsion forms on the aqueous side of the interface, accompanied by a little interfacial turbulence. Results can be obtained with this system, however, if in analysing the refractive index gradient near the surface a correction is made for the spontaneous emulsion: the rate of transfer is then in excellent agreement[52] with eq. (7.28) (Fig. 7-12). Consequently there is no appreciable energy barrier due to re-solvation of the acetic acid molecules at the interface, nor does the spontaneous emulsion affect the transfer. With a monolayer of sodium lauryl sulphate or protein at the interface, spontaneous emulsification is virtually eliminated, but the monolayer has no effect on the rate of transfer of the acetic acid (Fig. 7-12). If, however, a polymolecular "skin" of sorbitan tetrastearate is allowed to form in the interface by dissolving this agent in the benzene, the rate of transfer is considerably reduced[52], the value of R_I being 3000 to 6000 sec.cm.$^{-1}$ If acetic acid transfers from water to toluene, spontaneous emulsion and turbulence are again visible[52], and the amount transferred is slightly faster than calculated, presumably on account of the latter factor. Propionic acid transfers from water to toluene in accord with diffusion theory[55,59], as determined both by comparing the concentrations on each side of the interface with the partition coefficient and by considering the total amount transferred. Butyric and valeric acids, too, give concentration ratios close to the interface in accord with the normal bulk partition coefficients, showing that there is no measurable interfacial barrier: the amounts transferring cannot be determined because of the interfacial turbulence[55].

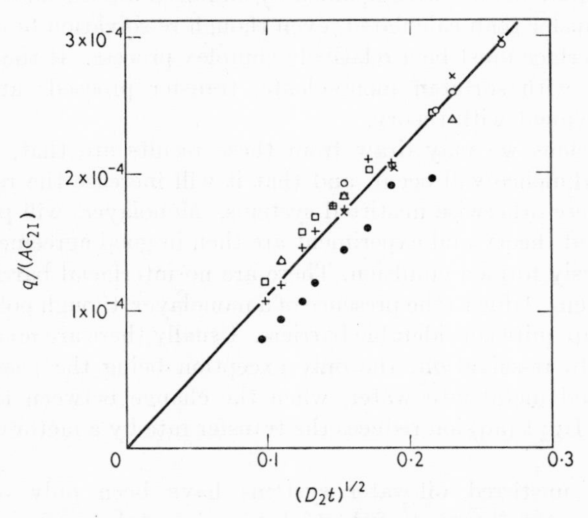

Fig. 7-12. Experimental data for system of Fig. 7-6 (a), plotted and compared with calculation (eq. (7.28)) on the basis of no interfacial resistance to the diffusion of acetic acid from water to benzene[52]. Points are: +, clean interface; ○, 0.00125 M pure sodium dodecylsulphate; △, 0.00250 M pure sodium dodecylsulphate: □, 0.00250 M sodium dodecylsulphate + 2.4% lauryl alcohol; +, spread protein; ●, 5000 p.p.m. of sorbitan tetrastearate in the benzene. Units are: q in moles cm.$^{-2}$; c in g. moles l.$^{-1}$ and $(Dt)^{1/2}$ in cm.

Diethylamine, transferring from 0.16 M solution in toluene to water, shows such strong spontaneous emulsification and turbulence that quantitative results are precluded[51]. If, however, 1.25 mM-sodium lauryl sulphate is present in the water, the interfacial turbulence is completely eliminated, and transfer then is calculable, if allowance is made for some spontaneous emulsion which is still formed, though less heavily than before. This transfer is in accord with diffusion theory.

The amount of sulphuric acid transferring between water and phenol shows no interfacial resistance for the first 100 minutes[57,58], though subsequently the amount transferring is lower than calculated, suggesting an interfacial resistance of about 4×10^5 sec.cm.$^{-1}$ Sulphur transferring between certain organic liquids also seems abnormally slow. In view of the long times involved and the possible impurities forming in the system, further experiments would be desirable. In the binary system of n-heptane and liquid SO_2, interfacial resistances of 25,000 to 100,000 sec.cm.$^{-1}$ are reported[57,58], though the difficulty of keeping the interface clean in the cells containing rubber gaskets may have been responsible for these extraordinarily high figures.

The extraction of uranyl nitrate from 1 M aqueous solution into 30% tributylphosphate in oil is accompanied by an initial interfacial turbulence[56], with more transfer than calculated, even though re-solvation of each uranyl ion at the interface must be a relatively complex process. If the turbulence is suppressed with sorbitan mono-oleate, transfer proceeds at a rate in excellent agreement with theory.

The conclusions we may draw from these results are that, in general, interfacial turbulence will occur, and that it will increase the rate of mass transfer in these otherwise unstirred systems. Monolayers will prevent this turbulence, and theory and experiment are then in good agreement, in spite of spontaneously formed emulsion. There are no interfacial barriers greater than 1000 sec.cm.$^{-1}$ due to the presence of a monolayer, though polymolecular films can set up quite considerable barriers. Usually there are no appreciable barriers due to re-solvation: the only exception being the passage of Hg from the liquid metal into water, when the change between the metallic state and the Hg_2^{++} (aq) ion reduces the transfer rate by a factor of the order 1000.

Completely unstirred oil-water systems have been only occasionally studied, however, on account of the high experimental accuracy required to obtain reliable results. Much early work concerned slightly stirred systems, studied by direct analysis of samples: again, however, monolayers of protein and other material at the interface do not affect the diffusion rates of solutes which are inappreciably adsorbed. These include various inorganic electrolytes and benzoic acid[60]. If the solute is surface-active, however, the strong spontaneous interfacial turbulence accompanying transfer greatly increases the rate of transfer relative to that calculated from diffusion coefficients, especially in the early stages of the diffusion. Since this turbulence can be completely eliminated by surface films, it is not surprising that these reduce the observed high rate of transfer by as much as four times: typical examples of this are found in the transfer of propanol or butanol from water to benzene[61], of phenol from sulphuric acid to water[59], and of acetic acid from CCl_4 to water[62].

Experiments on Dynamic Systems

The use of vigorously stirred cells in studying interfacial resistances has the advantage that the resistances of each bulk phase adjacent to the interface can be made quite small; typical values are 10,000 sec. cm.$^{-1}$ to 100 sec.cm.$^{-1}$, depending on the rate of stirring. Against this advantage must be set the complications of the turbulent flow: the eddies near the interface are likely to be affected by the proximity of a monolayer.

As an empirical correlation for clean surfaces, J. B. Lewis[63] found that his results on systems of the type shown in Fig. 7-6(b) obeyed the relation

$$k_1 = 1.13 \times 10^{-7} \nu_1 (\mathrm{Re}_1 + \mathrm{Re}_2 \eta_2/\eta_1)^{1.65} + 0.0167 \nu_1 \qquad (7.37)$$

where subscripts 1 and 2 refer to the two liquid phases, where η refers to viscosity, ν to kinematic viscosity ($=\eta/\rho_l$, in cm.^{2}sec.$^{-1}$), and where Re is the Reynolds number of either phase (defined by L^2N/ν, L being the tip-to-tip length of the stirrer blades and N the number of revolutions of the stirrer per second). A similar equation applies to k_2 by interchanging the subscripts.

Various objections have been raised to eq. (7.37): though numerically satisfactory to $\pm 40\%$ as written, it requires a further length term on the left side for dimensional uniformity. This could easily be effected using the constant term L. A more piquant criticism[64] is that, since there are no terms in D_1, this is a form of eq. (7.18). It is, however, difficult to see how the eddy diffusion could be completely dominant at the moderate stirring speeds used in the experimental work. Further, the implied proportionality of k_1 and η_1 at low stirring speeds is physically unrealistic, as is the complete cancellation of the term in η_2 in eq. (7.37). A different empirical correlation[65] of the transfer across a clean interface in a stirred cell is:

$$k_1 = 0.00316(\mathrm{D}_1/L)(\mathrm{Re}_1\mathrm{Re}_2)^{0.5}(\eta_2/\eta_1)^{1.9}(0.6 + \eta_2/\eta_1)^{-2.4}(\mathrm{Sch.}_1)^{5/6} \qquad (7.38)$$

where D_1 is the diffusion coefficient of the diffusing species in phase 1 and Sch. is the Schmidt number, defined as ν/D. Consequently k_1 varies with $\mathrm{D}_1^{1/6}$, this dependence lying between that of the Danckwerts equation (7.20) and that of equation (7.37) of J. B. Lewis. This correlation, which is accurate in predicting k_1 to $\pm 40\%$ for different systems, suggests that when the interface is uncontaminated the replacement by turbulent flow of elements of liquid in the surface is indeed very important, and that molecular diffusion from these elements has then to occur over only a very short distance.

This continual replacement of liquid is readily visible with talc particles sprinkled on to the interface: though stationary on the average (if the stirrers in phases 1 and 2 are contra-rotated at appropriate relative speeds) they make occasional sudden, apparently random, local movements, which indicate that considerable replacement of the interface is occurring by liquid impelled into the interface from the bulk. Spontaneous interfacial turbulence, associated with such processes as the transfer of acetone from solvent to water, may further increase the rate of transfer by a factor of two or three times[59,63,66]. Other systems[63], such as benzoic acid transferring (in either direction) between water and toluene, give transfer rates only about 50% of those calculated by eq. (7.37), suggesting that either there is an interfacial barrier or that eq. (7.37) no longer holds. This is discussed further below.

Study in a stirred cell of the transfer of uranyl ions from water to organic solvents confirms the result for unstirred cells that transfer is faster than theoretical when interfacial turbulence is visible[67]. After long times, in systems showing no visible turbulence, the transfer coefficients decrease, becoming less than those calculated from eqs. (7.33) and (7.37): this retardation is a function of time only and is not due to contamination of the interface. The explanation may lie either in there being some slight spontaneous interfacial turbulence which initially offsets a small error in eq. (7.37), or in the slow re-solvation of the uranyl ions at the interface.

A technique for studying extraction after times of 10^{-3} sec. to 1 sec. has been developed[67], though the hydrodynamic conditions are not exactly known.

For a surface covered with a monolayer (e.g. a protein), the value of k_1 is reduced by as much as 80%: the effect of a compressed protein monolayer at different stirring speeds can be expressed[63] by:

$$k_1 = 1.13 \times 10^{-7} \nu_1 \mathrm{Re}_1^{1.65} + 0.0167 \nu_1 \qquad (7.39)$$

This is eq. (7.37) with the term in Re_2 omitted; physically this means that the momentum of the turbulent eddies of oil near the interface is damped out by the monolayer, so that the stirring in the oil (phase 2) can no longer contribute to increasing k_1.

The observed rates of transfer are lower than those calculated by the correlation (7.37) for organic molecules which themselves are surface-active, without specifically added long-chain molecules: thus in the transference of $(C_4H_9)_4NI$ from water to nitrobenzene, of benzoic acid from toluene to water and the reverse, of diethylamine between butyl acetate and water, of n-butanol from water to benzene, and of propionic acid between toluene and water, the rates[59,63] are of the order one-quarter to one-half those calculated by eqs. (7.33) and (7.37). Since with these systems the solute itself is interfacially active, and therefore its monolayers should reduce the transfer of momentum, we interpret these findings as indicative that R_{L_1} and R_{L_2} are increased in this way. This is confirmed by the experimental rate of transfer of propionic acid from water to toluene in the stirred cell being lower than calculated, while in the unstirred cell there is no interfacial resistance R_I which would explain this. Conversely, sulphuric acid, which is not surface-active, transfers from water to phenol even faster than calculated (although in the unstirred cell a resistance, presumably due to an interfacial skin of impurities, is reported)[57,59].

The effect of protein and other monolayers on mass-transfer rates depends quantitatively[65] on the surface compressional modulus (C_s^{-1}, see p. 265): this quantity correlates the effect of monolayers, both spread and adsorbed, on $\mathbf{K}_L$. No simple modification of eq. (7.38) is possible to cover this phenomenon and, as one may show qualitatively with talc particles, the eddy velocity at the interface is greatly reduced by the monolayer. The latter restrains fresh liquid from being swept into the surface, i.e. there is less "clearing" of the old surface. If now $\Delta\Pi$ is the surface pressure resisting the eddy due to its

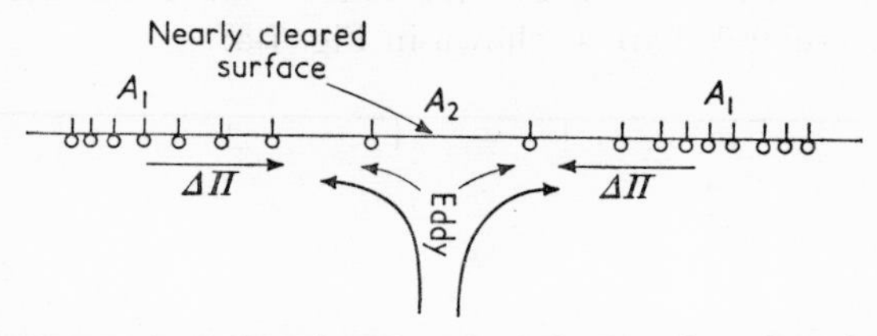

Fig. 7-13. Eddy brings fresh liquid surface into the interface, but this is opposed by the back pressure $\Delta\Pi$ of the spread film (cf. also Fig. 5-31).

partly clearing an area (Fig. 7-13) in the interface (and consequently changing the available area per molecule from A_1 to A_2), then

$$\Delta\Pi = \left(\frac{\partial\Pi}{\partial A}\right)(A_2 - A_1)$$

whence, if A_2/A_1 is defined as j,

$$\Delta\Pi = C_s^{-1}(j-1) \tag{7.40}$$

Since this differential spreading pressure $\Delta\Pi$ will oppose the movement of the eddy at the interface, it will also oppose surface renewal and hence mass transfer: a plot of $\mathbf{K}$ vs. C_s^{-1} is shown in Fig. 7-14: it represents the transition from a general "surface-renewal" mechanism towards the "stagnant-layer" process of Lewis and Whitman. Similarly, in the solution of oil in water (Fig. 7-6b), k_1 is reduced to only 20% of the value with a clean surface, when bovine plasma albumin is spread at the interface between ethyl acetate and water[65].

As with wave-damping, (p. 273), solubility of the film will "short-circuit" the compressional modulus: a minimum in the transfer coefficient is often observed[65,68] (Fig. 7-15). From this figure it is also clear that, whereas monolayers affect greatly the stirring near the surface, and so reduce k_L and the transfer rates in stirred cells, they have no measurable effect in unstirred cells.

Diffusion from single drops is easily measured: the process of formation of a drop of organic liquid, containing a solvent to be extracted into water, induces internal circulation which in turn so promotes transfer that up to 50% of the extraction may occur during the period of formation of the drop[69]. Even after release of the drop, as it rises freely through the water, the rate of extraction is often as much as twenty or forty times higher than that calculated from diffusion alone, suggesting that the liquid in the drop must still be circulating rapidly[69,72]. Indeed, this internal circulation[41], often accompanied by oscillation, may cause removal of each element of liquid at the interface after a residence only 10% of the period required for the drop to rise through one diameter: empirically[68] one finds that **K** varies as $D^{0.38}$. The reason for this circulation of liquid within the drop lies in the drag exerted along the surface by the relative motion of the continuous phase: the circulation patterns[41,71,72] are as shown in Fig. 7-5.

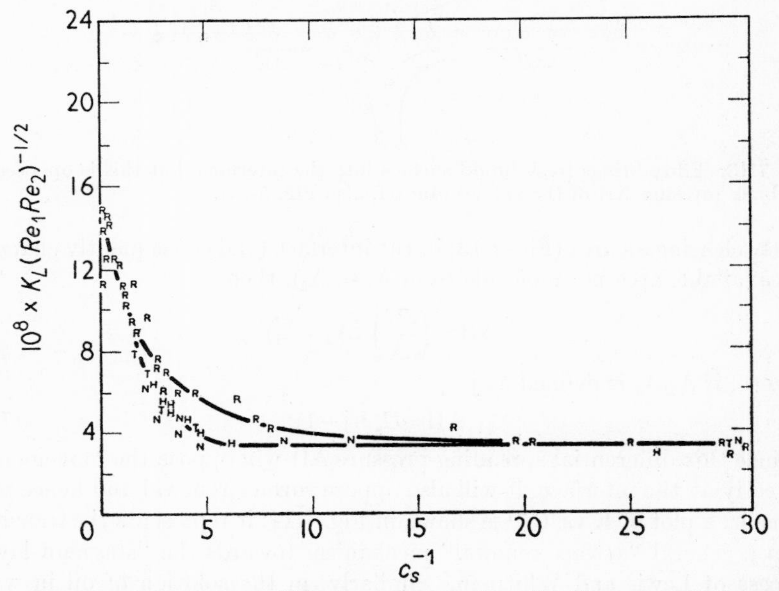

Fig. 7-14. Plot of $\mathbf{K}_L(Re_1Re_2)^{-1/2}$ (in cm.sec.$^{-1}$) against C_s^{-1} (dynes cm.$^{-1}$) for spread monolayers of bovine plasma albumin. $\mathbf{K}_L$ refers to the overall mass-transfer coefficient for isopropanol transferring from water to benzene, and various stirring speeds are employed in the apparatus of Fig. 7-6 (b). R refers to the runs using redistilled water, H to those using 0.01 N-HCl, N to those using 0.01 NaOH, and T to those using tap water[65].

The effects of interfacial monolayers on extraction from drops are particularly striking. Early work showed that traces of either impurity or surface-active additives can drastically reduce extraction rates: even plasticiser, in sub-analytical quantities dissolved from plastic tubing by benzene, reduces the mass-transfer rate by about ten times by retarding the hydrodynamic renewal of elements of liquid at the interface[73]. Further, more polar solvents are particularly liable to give the high mass-transfer coefficients associated with circulation[74], presumably because on their surfaces the energy of adsorption of surface-active impurities is relatively low.

The mechanism by which the surface film inhibits internal circulation is that the fluid flow will drive the adsorbed material towards the rear of the drop[45a]: consequently the surface concentration and surface pressure will be higher here, and the monolayer will tend to resist further local compression (Fig. 7-5b). This resistance to flow in the surface damps down circulation inside the drop by reducing the movement of the interface and hence reduces the transfer of momentum across it (and also the rate of rise or fall of the drop[45a,73], by perhaps 12%). Again, as for stirred liquids separated by a

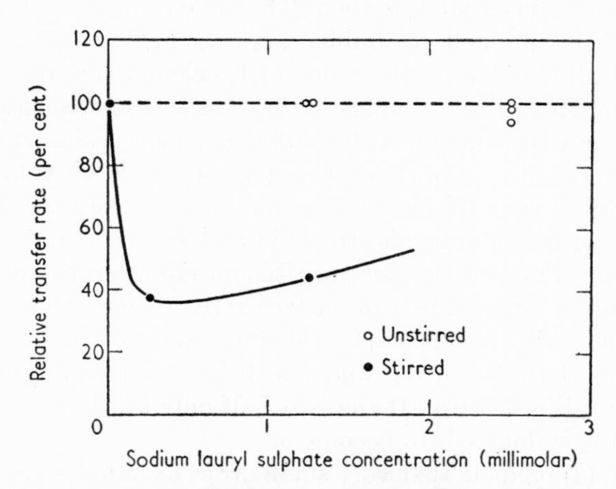

Fig. 7-15. Comparison of effect of sodium lauryl sulphate on the transfer of acetic acid from water to benzene at 25°C in an unstirred[51,52] and in a stirred cell[65].

plane interface (p. 333), the reciprocal compressibility modulus C_s^{-1} of the interfacial film is often the determining quantity, though with a highly viscous monolayer the interfacial viscosity must also play a part. The resistance to the circulation of a drop will depend on the drop radius a, since the smaller the drop the larger will be the surface pressure gradient between the front and the rear, and the smaller will be the tangential frictional stress. In terms of a dimensionless group, the circulation will be reduced by some function of $(C_s^{-1}/a^2.g.\Delta\rho_l)$ where a is the radius of the drop, $\Delta\rho_l$ is the difference in density of the liquids in the drop and in the continuous phase, and g is the gravitational acceleration[75]. The ratio of the bulk viscosities of the outer and inner liquids, $(\eta_{outer}/\eta_{inner})$ must also have an effect on the circulation[41]. In general,

$$\text{Degree of circulation} = f_1\left[\frac{\eta_{outer}}{\eta_{inner}}\right] \times f_2\left[\frac{C_s^{-1}}{a^2 g|\Delta\rho_l|}\right] \qquad (7.41)$$

where f_1 and f_2 are functions to be determined. If this expression is interpreted in terms of the velocity of circulation within the drop and of the fraction of the liquid in the drop which is still circulating for any given value of C_s^{-1}, then one may write[75a]

$$\% \text{ circulation} = \left[\frac{100}{1+1.5(\eta_{\text{inner}}/\eta_{\text{outer}})}\right]\left[1-\frac{f_3 C_s^{-1}}{a^2 g|\Delta\rho_l|}\right] \quad (7.42)$$

where f_3 is a numerical coefficient. For drops free of contamination (i.e. C_s^{-1} zero), circulation will always occur, at a rate dependent on the viscosity ratio, acccrding to this expression. From the experiments of Linton and Sutherland[72] we deduce that f_3 is about 0.6, while from the published data[75b] on the rate of rise of small gas bubbles, f_3 is found[75a] to be about 14. For elimination of all circulation, the value of f_3 calculated[75a] from the surface stresses on a moving drop or bubble is 1.5. These discrepancies may be due partly to certain experimental surface films not being completely insoluble: only if the fractional rate of desorption (eq. (4.21)) is very low compared with the reciprocal time for the drop or bubble to move through one diameter[75a] will the surface film be effectively insoluble, as is assumed in the foregoing treatment. Further, the rates of rise and fall of drops and bubbles are rather more sensitive to traces of surface-active agents than is the mass-transfer coefficient[68]: this is explained[75a] by the turbulent wake, which largely controls the rates of rise and fall, being most readily affected by a surface film (see Fig. 7-5(b)). If the rear half only of the drop is stagnant, the calculated f_3 is doubled, to become 3.

Equation (7.42) predicts that very small drops or bubbles are highly sensitive to minute traces of surface-active impurity: a benzene drop of radius 0.01 cm. should have its circulation prevented by a "gaseous" monolayer covering only 0.02% of the surface. It is found experimentally that gas bubbles of radius below 0.03 cm. are always non-circulating under practical conditions, and that even for bubbles of radius 0.07 cm, circulation is inhibited by only 10^{-6} M sodium lauryl sulphate[75b].

Many commercial solvents, particularly those which are non-polar, are very liable to have poor circulation, due to traces of strongly adsorbed impurities: the commercial polar oils form circulating drops because the impurities are less strongly adsorbed, and so C_s^{-1} is low. Further, according to eq. (7.42), the smaller the drop the more sensitive it should become to traces of surface-active impurity, as is confirmed by Linton and Sutherland's experiments[72]. These show that 1 mm. benzene drops can be made to circulate only by rigorous attention to the removal of adventitious impurities; even large drops of commercial benzene are always stagnant. Protein in concentrations as low as 0.0005% will reduce the circulation of drops (5 mm. in diameter) of oils of interfacial tension greater than 30 dynes cm.$^{-1}$

One can displace the film of impurity from the interface by adding a sufficient amount of a short-chain alcohol. The films of short-chain alcohols, are, however, so readily desorbed that they do not greatly inhibit the circulation of the liquid, though they may produce other circulation patterns due to the spontaneous interfacial turbulence associated with their redistribution. In this way the reduction (by a surface film) of k_L for any component being extracted can be offset by the addition of a few per cent of short-chain alcohol or acetic acid to the oil drop[71-73]: this increases the extraction efficiency as much as ten times, bringing it back to the value for a circulating drop[71]. In the absence of a surface film the spontaneous interfacial turbulence, on the addition of a little alcohol or otherwise, increases k_L still further[66]. Since interfacial turbulence is reduced or nullified by strongly adsorbed monolayers, it is likely that only the effects of rather weakly adsorbed surface films can be offset by the addition of a short-chain alcohol.

Practical Extraction Columns

In liquid-liquid extraction using wetted-wall columns, analysis is possible only by dimensionless groups[74]: for the core fluid, flowing up inside the tube, k_c varies as approximately $D_c^{0.67}$ and for the fluid falling down the inner walls, k_w varies as $D_w^{0.38}$. Systems studied include phenol-kerosene-water, acetic acid-methylisobutylketone-water, and uranyl nitrate between water and organic solvents[74,76-78]: interfacial resistances of the order 100 sec.cm^{-1} are observed in the last system. These resistances are interpreted as being caused by a rather slow third-order interfacial exchange of solvent molecules (S) co-ordinated about each UO_2^{++} ion:

$$UO_2(NO_3)_2\,6H_2O + 2S \rightleftharpoons UO_2(NO_3)_2.2S.4H_2O + 2H_2O$$

The square of the uranyl nitrate concentration is required in the kinetic analysis, suggesting a slow process of de-ionization before the solvent exchange. Mass transfer with chemical reaction[22,30] is, however, too specialized a subject to be discussed fully here.

Study of the efficiency of packed columns in liquid-liquid extraction has shown that spontaneous interfacial turbulence or emulsification can increase mass-transfer rates by as much as three times when, for example, acetone is extracted from water to an organic solvent[79]. Another factor associated with the fluctuations of interfacial tension is the coalescence rate: if between two adjacent drops of the phase into which extraction is occurring, there arrives an eddy of undepleted continuous phase, the drops must "kick" towards each other (cf. Fig. 7-7), this process assisting coalescence. This increase in the coalescence rate will occur if (as is usual) the fluid component being extracted causes an appreciable decrease in interfacial tension. An alternative explanation[79] of the more ready coalescence under certain conditions is that, *when transfer is occurring from the drops* to the continuous phase, the region between the drops is more concentrated in solute than is a region away from the line of approach of the drops, and adsorption of the solute is therefore enhanced between the drops. This adsorption in turn leads to spreading of the surface film away from this region, some of the intervening bulk liquid also being carried away with the surface film (cf. p. 252): the film of liquid separating the drops is thus thinned, and coalescence of the drops is consequently promoted. However, the "kicking" mechanism leads to the same conclusion, that the drops are brought into close proximity if the liquid between them is more concentrated in solute, and so far the two explanations have not been subjected to rigorous experimental tests. The important experimental difference is that in the former explanation a local thinning of the liquid film between the drops is responsible (leading presumably to

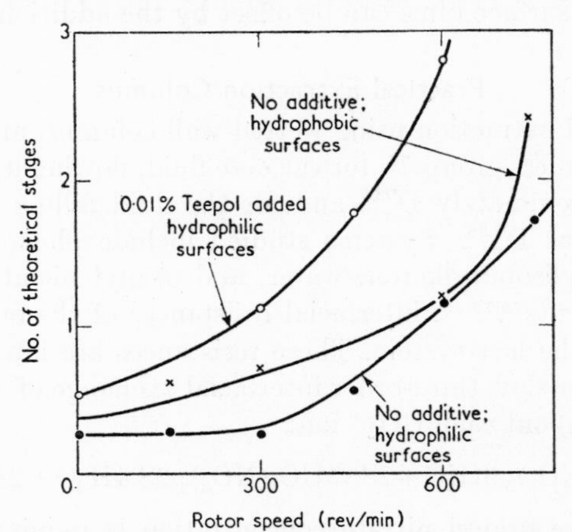

Fig. 7-16. Efficiency of a rotating disc contactor of given height under different conditions[81].

protuberances on the surfaces of the drops), whereas in the latter it is the excess pressure inside the drops, far from the point of approach, that deforms the whole drops, driving them together. Possibly high-speed cine-photography or variations in the viscosity of the phases might lead to a clarification of the mechanism.

If in laminar flow, *transfer is occurring into the drops*, the region between two of the latter becomes preferentially depleted in solute. The lowered adsorption in this region then causes spreading of the surface film and associated liquid into this region (or possibly the drops "kick" away from each other), and coalescence becomes unlikely. As pointed out above, however, if an eddy of the undepleted solution happens to be swept between the drops, this will favour coalescence.

In practical plants for liquid-liquid extraction the effect of surface-active agents is usually to increase the rate of extraction: presumably the smaller droplet sizes associated with the lower interfacial tension more than compensate for the reduction in the stirring by momentum transfer at the interface. Thus, in a column packed with Raschig rings, the rate of extraction increases linearly (to about 50%) with the decrease in interfacial tension[80], though at higher concentrations of surface-active additive the rate passes through a maximum, due either to adsorption of rather impermeable multilayers or to the back-mixing associated with very small drop sizes. In a rotating-disc contactor, addition of 0.01% Teepol similarly increases the extraction efficiency (Fig. 7-16)[81]. Further, if the column is operated with oil as the continuous phase, the dispersed drops of water coalesce on, and subsequently run down, the glass and metal surfaces. This reduces the efficiency of extraction by a "by-passing" effect, which can be avoided by rendering these surface hydrophobic with silicones (Fig. 7-16).

Distillation

When the diffusion of a component from a vapour bubble to the liquid is measured, one finds that the mass-transfer coefficient is larger when the surface tension is increased by the mass transfer[82-5]. Further, when the surface tension of the liquid in a distillation column is higher towards the bottom of the column, the plate efficiencies are relatively high because thin sheets of liquid, such as stabilize slightly the bubbles of vapour, are fairly stable, leading to a longer time of contact of vapour and liquid on each plate. With a surface tension decreasing down the column, however, the bubbles are highly unstable, and the plate efficiency is lowered. Another explanation is that drops of liquid coming into contact with the bubbles will break these if the surface tension is lowered locally, causing spreading in the bubble surface away from the affected region, and so destabilizing the liquid lamellae (cf. p. 415). This subject is treated more fully in ref. 75a.

REFERENCES

1. Lewis, W. K., and Whitman, *Industr. Engng Chem.* **16**, 1215 (1924).
2. Rideal, *J. phys. Chem.* **29**, 1585 (1925);
   Langmuir, I. and Langmuir, D. B., *J. phys. Chem.* **31**, 1719 (1927).
3. Sebba and Sutin, *J. chem. Soc.* 2513 (1952).
4. Sebba and Briscoe, *J. chem. Soc.* 106 (1940).
5. Archer and La Mer, *J. phys. Chem.* **59**, 200 (1955).
6. Mansfield, *Nature, Lond.* **175**, 246 (1955); *Aust. J. Appl. Sci.* **9**, 245 (1958); **10**, 65, 73 (1959).
7. Grundy, *Proc. 2nd Internat. Congr. Surface Activity* **1**, 270, Butterworths (1957);
   Dressler and Johanson, *Chem. Engng Progr.* **54**, 66 (1958).
8. Linton and Sutherland, *Aust. J. Appl. Sci.* **9**, 18 (1958).
9. La Mer, *Proc. 2nd Internat. Congr. Surface Activity* **1**, 259, Butterworths (1957).
10. McArthur and Durham, ibid. 262;
    McArthur and Jones, T. G., *5th Colloq. Surface-Active Agents*, Paris (1959).

10a. Deo, Sanjana, Kulkarni, Gharpurey and Biswas, *Nature, Lond.* **187**, 870 (1960).

11. Morse, *Proc. Acad. Arts Sci.* **45**, 363 (1910);
    Rie, *Ann. Phys., Paris* **63**, 759 (1920);
    Whytlaw-Gray and Patterson, "Smoke", Edward Arnold, London (1932);
    Twort, Baker, Finn, and Powell, *J. Hyg., Camb.* **40**, 17 (1940);
    Bradley, R. S., *Proc. roy. Soc.* **A205**, 553 (1951); **A206**, 65 (1951);
    Monchick and Reiss, *J. chem. Phys.* **22**, 831 (1954).
12. Beament, *J. exp. Biol.* **32**, 514 (1955); **35**, 494 (1958).
13. Maxwell, "Scientific Papers" **2**, 639, Cambridge University Press (1890);
    Langmuir, *Phys. Rev.* **12**, 368 (1918);
    Fuchs, *Phys. Z. Sowjet* **6**, 224 (1934); "Evaporation and Droplet Growth in Gaseous Media", Pergamon Press, London and New York (1959);
    Bradley, R. S., Evans, M. G. and Whytlaw-Gray, *Proc. roy. Soc.* **A186**, 368 (1946);
    Luchak and Langstroth, *Canad. J. Res.* **A28**, 574 (1950);
    Johnson, J. C., *J. appl. Phys.* **21**, 22 (1950);
    Green and Lane, "Particulate Clouds, Dusts, Smokes and Mists", E. and F. N. Spon, London (1957);
    Langstroth, Diehl, and Winhold, *Canad. J. Res.* **A28**, 580 (1950).
14. Bradley, R. S., *J. Colloid Sci.* **10**, 571 (1955);
    Eisner, Brookes, and Quince, *Nature, Lond.* **192**, 1724 (1958);
    Eisner, Quince, and Slack, "Experiments on Dust Binding with Aerosols" Minist. Power S.M.R.E. Rep. **180** (1959).
15. Scott, Tung, and Drickamer, *J. chem. Phys.* **19**, 1075 (1951).
16. Schulman and Teorell, *Trans. Faraday Soc.* **34**, 1337 (1938).
17. Groothuis and Kramers, *Chem. Engng Sci.* **4**, 17 (1955).
18. Haydon, *Nature, Lond.* **176**, 839 (1955); *Proc. roy. Soc.* **A243**, 483 (1958).
19. Sternling and Scriven, *Amer. Inst. chem. Engrs J.* **5**, 514 (1959).
20. For summary see Rideal, "An Introduction to Surface Chemistry", pp. 283–6 Cambridge University Press (1930);
    Vielstich, *Z. Elektrochem.* **57**, 646 (1953);
    Levich, *J. phys. Chem., Moscou* **22**, 575, 711 (1948).
21. Higbie, *Trans. Amer. Inst. chem. Engrs* **31**, 365 (1935).
22. Sherwood and Pigford, "Absorption and Extraction" pp. 20, 22, 265–7 McGraw-Hill, New York (1952).
23. Kishinevskii and Pamfilov, *Zh. Prikl. Khim., Leningr.* **22**, 1173 (1949).
24. Danckwerts, *Industr. Engng Chem.* (*Anal.*) **43**, 1460 (1951).
25. Toor and Marchello, *Amer. Inst. chem. Engrs J.* **4**, 97 (1958).
26. Sebba and Rideal, *Trans. Faraday Soc.* **37**, 273 (1941).
27. Blank and Roughton, *Trans. Faraday Soc.* **56**, 1832 (1960).
28. Harvey and Smith, W., *Chem. Engng Sci.* **10**, 274 (1959).
29. Downing and Truesdale, *J. appl. Chem.* **5**, 570 (1955); "Water Pollution Research 1956" H.M.S.O., London;
    Linton and Sutherland, *Aust. J. appl. Sci.* **9**, 18 (1958).
30. Kishinevskii, *Zh. Prikl. Khim., Leningr.* **27**, 382, 450 (1954).
31. Hutchinson, M. H., and Sherwood, *Industr. Engng Chem.* **29**, 836 (1937).
32. Davies and Kilner, Unpublished Results (1960).
33. Ternovskaya and Belopolskii, *J. phys. Chem., Moscou.* **24**, 43, 981 (1950); **26**, 1090, 1097 (1952);
    Emmert and Pigford, *Chem. Engng Progr.* **50** 87 (1954);
    Lynn, Straatemeir and Kramers, *Chem. Engng Sci.* **4**, 49, 58 (1955);
    Bond J. and Donald, *Chem. Engng Sci.* **6**, 237 (1957);

Nysing and Kramers, *Chem. Engng Sci.* **8**, 81 (1958);
Haselden and Malaty, *Trans. Instn. chem. Engrs, Lond.* **37**, 137 (1959).
34. Vielstich, *Chem.-Ing.-Tech.* **28**, 543 (1956).
35. Cullen and Davidson, *Trans. Faraday Soc.* **53**, 113 (1957).
36. Lynn, Straatemeier, and Kramers, *Chem. Engng Sci.* **4**, 63 (1955);
Davidson and Cullen, *Trans. Instn. chem. Engrs, Lond.* **35**, 51 (1957);
Cullen and Davidson, *Chem. Engng Sci.* **8**, 49 (1956);
Davidson, *Trans. Instn. chem. Engrs, Lond.* **37**, 131 (1959).
37. Davidson, Cullen, Hanson and Roberts, *Trans. Instn. Chem. Engrs, Lond.* **37**, 122 (1959).
38. Jowitt, *Industr. Chem. Mfr.* **26**, 489 (1950).
39. Garner and Kendrick, *Trans. Instn. Chem. Engrs, Lond.* **37**, 155 (1959).
40. Garner and Hale, *Chem. Engng Sci.* **2**, 157 (1953);
Garner and Lane, *Trans. Instn. Chem. Engrs, Lond.* **37**, 162 (1959).
41. Hadamard, *C.R. Acad. Sci., Paris* **152**, 1735 (1911); **154**, 109 (1912);
Rybczynski, *Bull. Acad. Sci. Crac.* **1**, 40 (1911);
Bond, W. N., *Phil. Mag.* **4**, 890 (1927);
Bond, W. N. and Newton, *Phil. Mag.* **5**, 794 (1928);
Boussinesq, *Ann. Chim. phys.* **29**, 349 (1913);
Robinson, *J. phys. Chem.* **51**, 431 (1947);
Nutt, *Bgham. Univ. chem. Engr.* **3** (3), 16 (1952).
Davies, C. N. *Proc. phys. Soc., Lond.* **57**, 259 (1945);
Bartok and Mason, *J. Colloid Sci.* **13**, 293 (1958).
42. McDonald, *J. Met.* **11**, 478 (1954); Finlay, quoted in ref. 40.
43. Lamb, "Hydrodynamics" Cambridge (1916).
44. Groothuis and Kramers, *Chem. Engng Sci.* **4**, 17 (1955);
Hatta and Baba, *J. Soc. chem. Ind. Japan* **38**, 544B (1935).
Calderbank and Korchinski, *Chem. Engng Sci.* **6**, 65 (1956).
45. Kronig and Brink, *Appl. sci. Res., Hague* **A2**, 142 (1950).
45a. Frumkin and Levich, *Zh. fiz. Khim.* **21**, 1183 (1947).
46. Calderbank, *Trans. Instn. Chem. Engrs, Lond.* **36**, 443 (1958); **37**, 173 (1959).
47. Frossling, *Beitr. Geophys.* **32**, 170 (1938).
48. Stuke, *Naturwissenschaften* **39**, 325 (1952).
49. Hammerton and Garner, *Trans. Instn. Chem. Engrs, Lond.* **32**, S18 (1954).
50. Crank "The Mathematics of Diffusion" Oxford University Press, London (1956).
51. Wiggill, J.B., Thesis, University of Cambridge (1958).
52. Davies and Wiggill, J. B., *Proc. roy. Soc.* **A255**, 277 (1960).
53. Quincke, *Ann. Phys., Lpz.* **35**, 593 (1888);
Lewis, J.B., and Pratt, *Nature, Lond.* **171**, 1155 (1953);
Garner, Nutt, and Mohtadi, *Nature, Lond.* **175**, 603 (1955);
Haydon, *Proc. roy. Soc.* **A243**, 483 (1958);
Davies, T. V., and Haydon, *Proc. roy. Soc.* **A243**, 492 (1958);
Sigwart and Nassenstein, *Naturwissenschaften* **42**, 458 (1955);
Nassenstein and Kraus, *Chem.-Ing.-Tech.* **28**, 220 (1956);
Davies, Boothroyd, and Palmer, Research Project in Department of Chemical Engineering, Cambridge (1956).
54. Lamm "The Ultracentrifuge" (ed. Svedberg and Pedersen), Oxford University Press (1940);
Lamm and Polson, *Biochem. J.* **30**, 528 (1936).
55. Ward and Brooks, L. H., *Trans. Faraday Soc.* **48**, 1124 (1952).

56. Hahn, Report HW.32626 Richland, U.S. Atomic Energy Commission, Wash.
57. Tung and Drickamer, *J. chem. Phys.* **20**, 6, 10 (1952).
58. Sinfelt and Drickamer, *J. chem. Phys.* **23**, 1095 (1955).
59. Blokker, *Proc. 2nd Internat. Congr. Surface Activity* **1**, 503, Butterworths (1957).
60. Sjölin, *Acta. physiol. scand.* **4**, 365 (1942);
Davies, *J. phys. Chem.* **54**, 185 (1950).
61. Hutchinson, E., *J. phys. Chem.* **52**, 897 (1948).
62. Sigwart and Nassenstein, *Naturwissenschaften* **42**, 458 (1955).
63. Lewis, J. B., *Chem. Engng Sci.* **3**, 248, 260 (1954).
64. Sherwood, *Chem. Engng Sci.* **4**, 290 (1955).
65. Davies and Mayers, *Chem. Engng Sci.* **16**, 55 (1961).
66. Sherwood and Wei, *Industr. Engng Chem.* **49**, 1030 (1957).
67. Lewis, J. B. *Chem. Engng Sci.* **8**, 295 (1958);
Martin, *Nature, Lond.* **183**, 312 (1959).
68. Holm and Terjesen, *Chem. Engng Sci.* **4**, 265 (1955);
Boye-Christensen and Terjesen, *Chem. Engng Sci.* **7**, 222 (1958), **9**, 225 (1959).
Thorsen and Terjesen, *Chem. Engng Sci.* **17**, 137 (1962).
69. Sherwood, Evans, J. E. and Longcor, *Amer. Inst. chem. Engrs* **35**, 597 (1939);
Licht and Conway, *Industr. Engng Chem.* **42**, 1151 (1950);
Licht and Pansing, *Industr. Engng Chem.* **45**, 1885 (1953).
70. Hughes and Gilliland, *Chem. Engng Progr.* **48**, 497 (1952);
Coulson and Skinner, *Chem. Engng Sci.* **1**, 197 (1952);
Garner and Skelland, *Trans. Instn. chem. Engrs, Lond.* **29**, 315 (1951);
Industr. Engng Chem. **46**, 1255 (1954).
71. Levich *J. Exp. theor. Phys.* **19**, 18 (1949); *J. phys. Chem., Moscou* **21**, 1183 (1947);
Garner and Skelland, *Chem. Engng Sci.* **4**, 149 (1955).
72. Linton and Sutherland, *Proc. 2nd Int. Congr. Surf. Act.* **1**, 494, Butterworths (1957).
73. West et al., *Industr. Engng Chem.* **43**, 234 (1951); **44**, 625 (1952);
Garner and Hale, *Chem. Engng Sci.* **2**, 157 (1953);
Lindland and Terjesen, *Chem. Engng Sci.* **5**, 1 (1956).
74. Pratt, *Ind. Chemist* **31**, 63 (1955).
75. Davies, *Trans. Instn. Chem. Engrs, Lond.* **38**, 289 (1960).
75a. Davies, *Adv. Chem. Engng.* **4** (1963).
75b. Okazaki, Miyazaki and Sasaki, *Proc. 3rd. Int. Congr. Surf. Act.* (Univ. Press, Mainz) **2**, 549 (1961).
76. Hunter, Nash, et al., *J. Soc. chem. Ind., Lond.* **54**, 49T (1935); **56**, 50T (1937).
77. Brinsmade and Bliss, *Trans. Amer. Instn. Chem. Engrs* **39**, 679 (1943).
78. Murdoch and Pratt, *Trans. Inst. chem. Engrs, Lond.* **31** 307 (1953).
79. Gayler and Pratt, *Trans. Inst. chem. Engrs, Lond.* **35**, 273 (1957);
Groothuis and Zuiderweg, *Chem. Engng Sci.* **12**, 288 (1960).
80. Chu, Taylor, C. C. and Levy, *Industr. Engng Chem.* **42**, 1157 (1950).
81. Davies, Ritchie and Southward, *Trans. Inst. chem. Engrs, Lond.* **38**, 331 (1960).
82. Zuiderweg and Harmens, *Chem. Engng Sci.* **9**, 89 (1958).
83. Frank, Diss. E. Technische Hochschule No. 2827, Zürich (1958).
84. Grassmann and Anderès, *Chem.-Ing.-Tech.* **31**, 154 (1959).
85. Danckwerts, Smith, W. and Sawistowski, "Internat. Symp. Distillation", 1. Instn. chem. Engrs., London (1960).

Chapter 8

Disperse Systems and Adhesion

Disperse systems are often thermodynamically unstable, because any free energy associated with the large interfacial area between the dispersed phase and the continuous phase can decrease by the aggregation or coalescence of the dispersed phase. Such systems, in which the interfacial free energy is positive, are typical *lyophobic dispersions*, meaning that, rather than remain in contact with the solvent, the particles or bubbles ultimately aggregate with or coalesce with each other. If one dimension of the dispersed particles is in the range 10^{-7}cm. to 10^{-4}cm., these particles constitute a "colloidal" solution.

The formation of lyophobic colloids may be brought about by a process either of dispersion or of condensation. The former implies that when two phases are present, one of these is physically torn asunder, to be dispersed in the other. Condensation methods involve the supersaturation of a phase with solute: the latter is then precipitated in the form of very small particles or droplets.

Lyophobic colloids with concentrations of particles in excess of about 10^{10} per cm.^{3} coagulate rapidly, provided that every collision leads to adhesion between the colloidal particles. The exact collision frequency has been calculated by Smoluchowski and, as shown below, agreement with experiment is good. Often, however, it is possible to stabilize the dispersion by electrical or solvation barriers, so that dispersions of relatively high concentrations (10^{14} particles per cm.^{3}) can persist for long periods. In these systems only one collision in 10^6 or 10^8 can lead to coagulation, and very stable dispersions are now quite common. Though these may last for several years, mists and smokes (aerosols) cannot be so stabilized. Foams, however, are also stabilized by electrical forces, and may likewise persist for many years, though, because of the usually very high phase volume of gas in the liquid, the coalescence theory of Smoluchowski cannot be applied to foams: their stability depends on the drainage of liquid from the foam and other factors.

Often it is required to "break" colloidal dispersions quickly: if we know what is the stabilizing factor, appropriate counter-measures against it can be taken. We shall see below that induced rapid coagulation of liquid systems, of mists and smokes, and of emulsions and foams is possible, each by a different approach. Before discussing the individual properties of aerosols, emulsions and foams, however, we shall outline the general theory of Smoluchowski to obtain the collision rate of colloidal particles.

COLLISION RATES IN DISPERSE SYSTEMS

According to the theory of the coagulation of colloids advanced in 1916 by Smoluchowski[1], the rate of disappearance of primary spherical particles, due to collision following diffusion of the particles towards each other, is related to the diffusion coefficient D of the particles through the medium, to the number $\mathbf{n}_1$ of *primary particles* present per cm.^{3}, and to the collision radius of the particles, $\boldsymbol{R}$, according to the expression

$$-d\mathbf{n}_1/dt = 8\pi D\boldsymbol{R}\mathbf{n}_1^2 \tag{8.1}$$

This assumes that every collision is effective in removing two of the primary particles from the system, if the centres of the particles approach to within

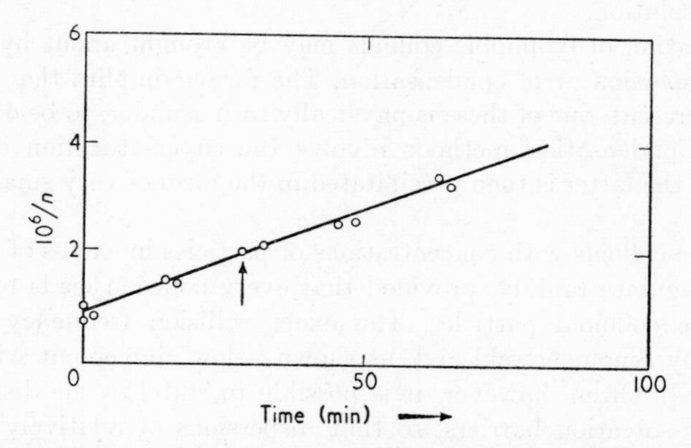

Fig. 8-1. Coagulation of a stearic acid smoke, plotted as $10^6/\mathbf{n}$ against time[8]. Here $\mathbf{n}$ is the total number of particles, of whatever size, remaining in the system after time t: $\mathbf{n}$ decreases as particles coagulate with each other. The time for half coagulation is that at which $10^6/\mathbf{n}$ has increases from 1 to 2, i.e. 28 min.

the collision radius $\boldsymbol{R}$. Usually, however, we are concerned with the *total* number of colloidal particles in the system, and each collision will remove in effect only one particle from the system, since there were two particles before collision and only one afterwards. Hence, if by $\mathbf{n}$ we denote the *total number of particles in the system*, eq. (8.1) must be rewritten with a factor four:

$$-d\mathbf{n}/dt = 4\pi D\boldsymbol{R}\mathbf{n}^2 \tag{8.2}$$

This integrates to

$$1/\mathbf{n} = 4\pi D\boldsymbol{R}t + \text{constant}$$

and if, when $t=0$, there are $\mathbf{n}_0$ particles per cm.[3] in the dispersion,

$$1/\mathbf{n} - 1/\mathbf{n}_0 = 4\pi D\boldsymbol{R}t \tag{8.3}$$

There are two principal ways of testing this equation by comparison with experiment. One consists in plotting $1/\mathbf{n}$ against t, when a straight line should result (Fig. 8-1). The other consists in finding the time ($t_{1/2}$) for the number of particles to be reduced to half the initial value. This is easily obtained from (8.3) by substituting $\mathbf{n} = \mathbf{n}_0/2$, $t = t_{1/2}$. This leads to

$$t_{1/2} = 1/4\pi D\boldsymbol{R}\mathbf{n}_0 \tag{8.4}$$

As for a second-order chemical reaction, therefore, the time for half-coagulation varies inversely with the number of particles originally present.

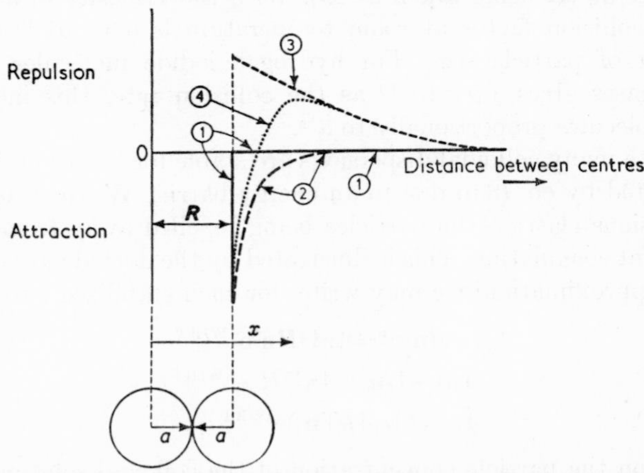

Fig. 8-2. Interaction energies between two equal spherical particles. The full curve 1 shows the infinitely deep, vertically walled energy well assumed in the basic theory of Smoluchowski (eqs. 8.1 to 8.7). Curve 2 shows how the energy of attraction should, more correctly, be represented, as a longer range effect (eq. (8.57) below). Curve 3 shows a repulsive barrier, as in eq. (8.54) below, and curve 4 is the summation of curves 2 and 3. The lower part of the figure shows that, at the simplest, $\boldsymbol{R} = 2a$.

The diffusion coefficient D is given by the Einstein formula

$$D = \boldsymbol{k}T/6\pi\eta a \tag{8.5}$$

where $\boldsymbol{k}$ is the Boltzmann constant, η is the viscosity of the medium, and a is the radius of the particle. The Einstein relation is valid provided the colloidal particles are large compared with the mean free path of the molecules in the continuous medium.

Usually, to a close approximation, the collision radius $\boldsymbol{R}$ is 2a (Fig. 8-2): i.e., the particles must just touch before adhesion can occur, there being no long range attraction between them. With this approximation, equations (8.4) and (8.5) can be combined to give

$$t_{1/2} = 3\eta/4\boldsymbol{k}T\mathbf{n}_0 \tag{8.6}$$

For water, η is 0.01 poise, so at 25°C and with c.g.s. units throughout,

$$t_{1/2} = 2 \times 10^{11}/\mathbf{n}_0 \text{ seconds} \tag{8.7}$$

This last equation is valid only for aqueous dispersions which are not stabilized in any way. For a concentrated aqueous dispersion ($n \approx 10^{14}$ particles cm.$^{-3}$), $t_{1/2}$ is about 1/500 sec., though for normal concentrations ($n = 10^9$ to 10^{11}) $t_{1/2}$ is of the order of a second or a minute.

It is of interest to note that the collision factor $8\pi D\boldsymbol{R}$ of eq. (8.1) is of the same order of magnitude as that between gas molecules. Thus, if D from the Einstein formula (8.5) is substituted into eq. (8.1), this collision factor becomes $8\boldsymbol{k}T/3\eta$ ($\boldsymbol{R}$ being taken as 2a). With the viscosity of air 1.7×10^{-4} poise, this collision factor at room temperature is 6.3×10^{-10} c.g.s. units, independent of particle size. For hydrogen iodide molecules the kinetic theory of gases gives 1.5×10^{-10} as the collision rate, this increasing for different molecules proportionally to $a^{1/2}$.

In practice many colloidal dispersions are stable for longer periods of time than predicted by eq. (8.6) due to an energy barrier **W** which makes many of the collisions elastic—the particles being repelled away from each other again without coagulating. This is illustrated by the dotted curve in Fig. 8-2. To a first approximation[1] we may write, for such stabilized systems:

$$-d\mathbf{n}/dt = 4\pi D\boldsymbol{R}\mathbf{n}^2 e^{-W/RT}$$

$$1/\mathbf{n} - 1/\mathbf{n}_0 = 4\pi D\boldsymbol{R} t e^{-W/RT} \tag{8.8}$$

and

$$t_{1/2} = (3\eta/4\boldsymbol{k}T\mathbf{n}_0)e^{W/RT} \tag{8.9}$$

Depending on the particle concentration of the colloidal solution, values of **W** from 15RT to 25RT are necessary if it is to be considered stable, i.e. if the time of half-coagulation is to be in excess of 10^6 seconds (about 11 days).

In equation (8.2), the potential well responsible for cohesion is assumed to be infinitely deep (Fig. 8-2). This implies that two particles, having touched, never separate again. Though approximately valid for such suspended particles as those of colloidal metals, this is not true for many systems, which exhibit lesser degrees of "stickiness". Into the coagulation equations one must therefore insert a term for the declumping (i.e. redispersal) of the coagulated particles:

$$-\frac{d\mathbf{n}}{dt} = 4\pi D\boldsymbol{R}\mathbf{n}^2 e^{-W/RT} - k' \text{ (concn. of clumps)} \tag{8.10}$$

At equilibrium, $d\mathbf{n}/dt = 0$, and the concentration of clumps is determined by the ratio $4\pi D\boldsymbol{R}\mathbf{n}^2/k'$. The application of this equation is discussed on p. 429.

AEROSOLS

As with colloidal dispersions in general, there are two chief ways of forming an "aerosol", i.e. a colloidal dispersion of a solid or liquid in gas: one may adopt either dispersion or condensation methods[2].

Dispersion Methods

The simplest dispersion method of producing an aerosol involves introducing a powder or a solution of the solid to be dispersed into a strong air stream, which then expands into a large chamber. Alternatively, a solvent can be used if it is readily volatile, so that only the solid suspension remains: Freon, CCl_2F_2, is therefore often used. If a mist is to be formed, an

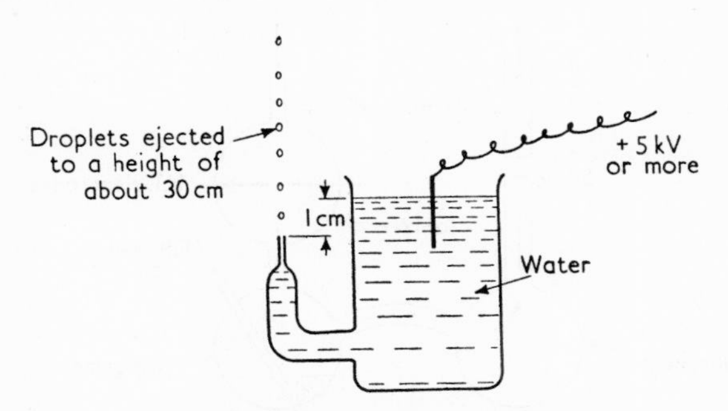

Fig. 8-3. The electrostatic fountain[3].

"atomizer" of the scent-spray type usually suffices, though it is not easy to obtain a very fine dispersion in this way. An electrical charge[3] assists the formation of an aerosol and, if a capillary containing a liquid of low conductivity is connected to a d.c. supply of electricity at a potential of 5-10 kV (positive) or higher, a mist of droplets, 1 micron or less in diameter, is expelled: the electrical repulsion offsets the contractile effect of the surface tension, with the net result that the surface tends to expand till the electrical potential is reduced, through the charge becoming spread over a sufficiently large area of surface and through a large volume of space. In this way aerosols of ethanol, distilled water or oil may be produced. An interesting experimental variation of the method is the electrostatic fountain[3] (Fig. 8-3): with tap water, the droplets may reach 30 cm. in height from a capillary a few tenths of a millimetre in diameter. These droplets are large enough (0.1 mm. diameter) to be easily visible.

Condensation Methods

In these methods the vapour of the material to be dispersed as an aerosol is supercooled. The liquid or solid particles then form by condensation on to very small units, called nuclei, of the liquid or solid respectively. If the supercooling is sufficient, these nuclei form spontaneously, as explained below: this often occurs following a chemical reaction, as between hydrogen chloride and ammonia. Similarly, a smoke of stearic acid forms spontaneously if a stream of hot air is passed over stearic acid and subsequently cooled; a smoke of silver is likewise formed by passing air through an electric arc between silver electrodes, again followed by cooling of the air. If, however, the supersaturation of the vapour is only slight, foreign nuclei[2] must be present on which the mist or smoke may condense. This is the principle of the Sinclair-La Mer apparatus[2] for producing relatively monodisperse aerosols; if the condensation of supersaturated vapour on foreign nuclei is slow and

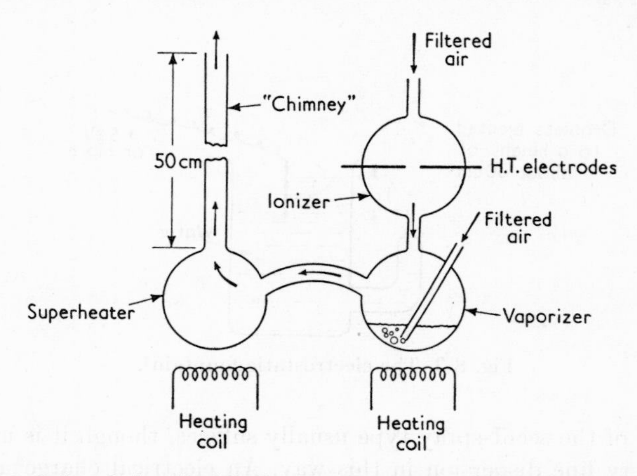

Fig. 8-4. Principle of the Sinclair-La Mer apparatus for producing monodisperse aerosols.

uniform, the variation in the particle radius of the resultant aerosol may be only 10% about the mean. For forming aerosols from liquids of B.Pt. in the range 300-500°C, the apparatus shown in Fig. 8-4 is used: the vapour from the liquid (heated in the vaporizer to 100-200°C) is mixed with a stream of foreign nuclei from the ionizer. These nuclei may result from a high-voltage electric spark between the electrodes, or the latter may be replaced by a heated coil of wire covered with NaCl. To achieve uniform mixing, the stream is fed into a superheater, maintained at about 300°C, which also evaporates any spray carried over from the vaporizer. Finally the gaseous stream is passed through a long chimney in which it gradually cools, condensation of the vapour then occurring slowly and uniformly on to the foreign nuclei. The size of the aerosol particles will depend on the number of foreign nuclei relative to the amount of vapour present.

The kinetics of condensation processes are of great interest: if foreign nuclei are present, only slight supercooling of the vapour will lead to quite rapid condensation to form aerosol droplets, the rate of condensation being simply proportional to the vapour concentration at any particular temperature, and the process is therefore like a first-order chemical reaction. Nuclei of NaCl of radius about 10^{-6}cm. cause water vapour to condense when the vapour pressure is only about 1.1 to 1.6 times the saturation vapour pressure, although surface-active agents may alter this[4]: formaldehyde adsorbed on the NaCl surface increases the required super-saturation to 2.4, though a silicone film lowers it from 1.3 to 1.1.

If, however, no foreign nuclei are present, supersaturation to four or five times the normal saturation vapour pressure is required to condense the vapour. The rate of condensation will now depend on the rate of spontaneous formation of nuclei from the supersaturated vapour, as well as on the rate at which condensation on to the nuclei proceeds. The rate of spontaneous nucleus formation depends, as shown in detail below, on the temperature, the entropy of condensation, and on the surface tension between the liquid or solid and the vapour.

Kinetics of Nucleation of a Supercooled Vapour

In general, the rate at which a new phase forms depends on the free energy of formation of an initial very small critical mass (the nucleus) of the new phase[4,5,6]: the requisite energy for this depends appreciably on the interfacial properties. The exact relation for the kinetics of phase transitions is of the form

$$\text{Rate} \propto e^{-\Delta G_n/RT}$$

where ΔG_n, the free energy of nucleus formation, depends not only on the free energy to form the bulk of the nucleus of the new phase, but also on the surface energy of this nucleus.

Given a sufficiently long time interval, any vapour should condense to drops of liquid at the boiling point, if heat is removed from the system. In practice, however, quite considerable supercooling of a vapour is usually required before even small amounts of condensation will occur, but on slight further cooling below this temperature, condensation becomes very rapid. We suspect, therefore, that ΔG_n in the above relation must decrease markedly as the temperature of the system is lowered. That this is indeed so is shown rigorously below.

In a supercooled vapour, small liquid droplets of molecular dimensions will be formed by random fluctuations. These liquid droplets, containing only a few molecules, are known as embryos. Will such embryos grow, or will they evaporate again? On the answer to this question depends the rate of condensation of the vapour. In a small embryo, the interface and the bulk are of comparable energies; to the free energy of condensation of the material in the bulk of the embryo must be added a term for the surface free energy of the embryo, since the molecules at the surface of the embryo possess an additional free energy because of their unsymmetrical valency environment. The formation of this new interface is therefore accompanied by a positive ΔH. Now the surface layer is partly ordered, i.e. ΔS is negative, so that, since $\Delta G = \Delta H - T\Delta S$, ΔG for the formation of the surface of the embryo is positive; consequently, the embryo surface tends to disappear by re-evaporation. The formation of the liquid interior of the embryo, however, is accompanied by a negative ΔG, since the vapour is supercooled, and the balance of these free energies of the surface and of the bulk determines the total free energy and hence the fate of the embryo.

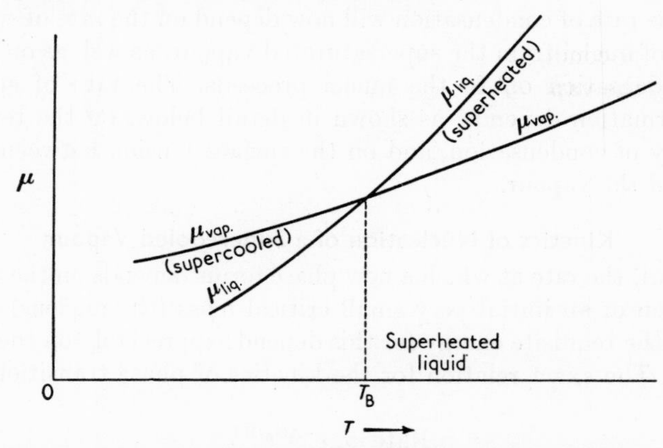

Fig. 8-5. Variation of chemical potentials with temperature.

If the embryo is large enough, the bulk term in the free energy of its formation will equal the surface energy term, and the embryo will just be stable. It is then called a *nucleus*. On this nucleus more molecules of vapour may condense; the bigger the nucleus grows the more stable it is, the interfacial term becoming relatively less important.

The *free energy of formation* of an embryo is found as follows. Consider that in the supercooled vapour a small spherical liquid embryo, of radius a cm., has formed. The total free energy of formation of this embryo may be expressed[5] as:

$$\Delta G = \text{free energy to liquify bulk of embryo} + \text{surface free energy of embryo,}$$

or
$$\Delta G = \frac{1}{v}\left(\frac{4}{3}\pi a^3\right)(\mu_{\text{liq.}} - \mu_{\text{vap.}}) + 4\pi a^2 \gamma_0 \quad (8.11)$$

Here $\mu_{\text{liq.}}$ and $\mu_{\text{vap.}}$ are the chemical potentials of the bulk liquid and the bulk vapour phases, and may be conveniently expressed in ergs per molecule. The volume occupied by a molecule in the liquid is v cm.^{3}, and γ_0 is the surface energy (or tension) of the vapour-liquid surface, in ergs cm.$^{-2}$ Thus ΔG is expressed in ergs, and refers to the formation of a liquid embryo of radius a from the vapour. Below the boiling point T_B, $\mu_{liq.} < \mu_{vap.}$, and the first term in eq. (8.11) is therefore negative, favouring condensation. This is opposed by the second, positive, energy term. We may note here that, as expected, ΔG is indeed strongly temperature-dependent, since $\mu_{liq.}$, $\mu_{vap.}$, and their difference depend very markedly on temperature (Fig. 8-5).

Will the embryo grow? Any process will occur spontaneously if the free energy ΔG is thereby decreased, and this principle may be applied to the possible growth of an embryo. If we plot the ΔG function of eq. (8.11) for

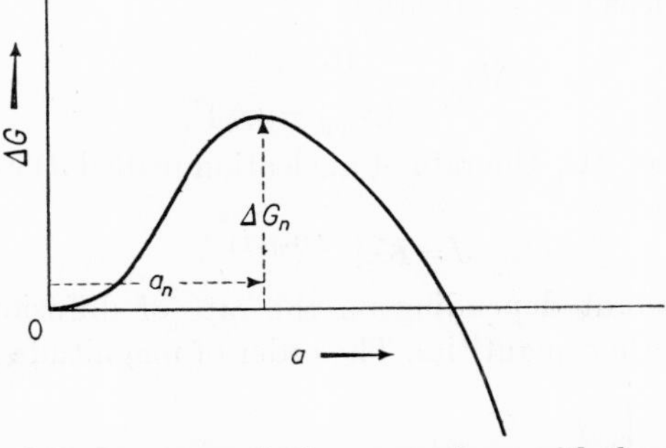

Fig. 8-6. Variation of the free energy to form an embryo with the radius of the embryo. If the radius of the latter exceeds a_n, the free energy is decreased by further growth, which is therefore spontaneous.

embryos of different radii (Fig. 8-6), we see that at low values of a the free energy is positive, since the first term (negative and involving a^3) is then negligible compared with the second (positive and involving a^2). For larger embryos, however, the ratio of volume to surface increases, and ΔG reaches a maximum, decreases, and becomes negative for very large embryos. Clearly, ΔG is decreased with further growth after the embryo has reached a radius a_n (Fig. 8-6), and an embryo of this critical size, which will continue to grow in the supercooled vapour, is a spontaneously formed nucleus. This will continue to grow indefinitely, retaining the vapour molecules which strike it; even a single nucleus will eventually lead to liquefaction of all the supercooled vapour. The latter process, however, would be very slow and, in practice, numbers of nuclei of the order 10 to 10^7 must be formed per cm.^{3} per second for condensation to proceed at a measurable rate.

Since **the radius of the nucleus** is that at which ΔG reaches a maximum (Fig. 8-6), it may be calculated from

$$\frac{\partial(\Delta G)}{\partial a}=0 \tag{8.12}$$

or, using eq. (8.11),

$$\frac{4\pi a_n^2}{v}(\mu_{liq.}-\mu_{vap.})+8\pi a\gamma_0=0$$

whence

$$a_n=\frac{2v\gamma_0}{(\mu_{vap.}-\mu_{liq.})} \tag{8.13}$$

Here a_n is the radius of the embryo for which $\frac{\partial(\Delta G)}{\partial a}=0$ i.e. the radius of the nucleus. It is always positive since $\mu_{liq.}<\mu_{vap.}$ for supercooled liquids. From a_n we may obtain ΔG_n by substituting into eq. (8.11). This energy is the free energy of formation of an embryo just large enough to grow, i.e. **the energy of formation of a nucleus** and is given by

$$\Delta G_n=\frac{16\pi\gamma_0^3 v^2}{3(\mu_{vap.}-\mu_{liq.})^2} \tag{8.14}$$

With a knowledge of ΔG_n the rate of nucleation (called $\boldsymbol{J}$) can now be found, since

$$\boldsymbol{J}=\boldsymbol{K}e^{(-\Delta G_n/\boldsymbol{k}T)} \tag{8.15}$$

where $\boldsymbol{K}$ is a constant depending on the rate of diffusion of the vapour molecules and on other quantities. The order of magnitude of $\boldsymbol{K}$ is 10^{24}sec.$^{-1}$

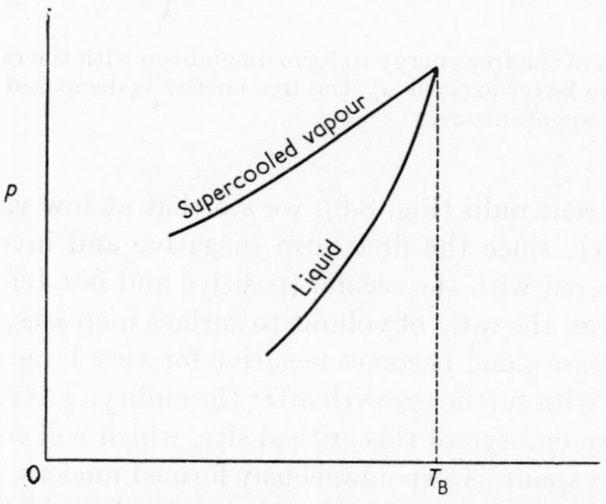

Fig. 8-7. Vapour pressures of supercooled vapour, compared with those over the equilibrated liquid, assuming a plane surface.

In calculating ΔG_n in practice, however, it is convenient to convert the terms in eq. (8.14) into vapour pressures, as follows. For the supercooled vapour, of vapour pressure p,

$$\mu_{vap.}=\mu_{vap.}^0+\boldsymbol{k}T\ln p \tag{8.16}$$

while, after condensation has occurred at the temperature T, the vapour pressure will have decreased to $p_{equil.}$, that of vapour in equilibrium with liquid water (Fig. 8-7). When this has occurred,

$$\mu_{liq.} = \mu_{vap.(equil.)} = \mu^0_{vap.} + \boldsymbol{k}T \ln p_{equil.} \tag{8.17}$$

so that, from eqs. (8.16) and (8.17),

$$\mu_{vap.} - \mu_{liq.} = \boldsymbol{k}T \ln\left(\frac{p}{p_{equil.}}\right) \tag{8.18}$$

Thus, from eqs. (8.14) and (8.18) we obtain

$$\Delta G_n = \frac{16\pi\gamma_0^3 v^2}{3\boldsymbol{k}^2T^2\left\{\ln\left(\frac{p}{p_{equil.}}\right)\right\}^2} \tag{8.19}$$

and from eqs. (8.13) and (8.18),

$$a_n = 2v\gamma_0/\boldsymbol{k}T\ln(p/p_{equil.}) \tag{8.20}$$

Thus, from eq. (8.15),

$$J = \boldsymbol{K}\exp\left\{\frac{-16\pi\gamma_0^3 v^2}{3\boldsymbol{k}^3T^3\left\{\ln\left(\frac{p}{p_{equil.}}\right)\right\}^2}\right\} \tag{8.21}$$

The form of (8.21) is such that J remains negligibly small till, on sufficient supersaturation, $p/p_{equil.}$ reaches a certain limit, at which J begins to increase very steeply (Fig. 8-8). As an example[6], air at 2.2°C containing water vapour

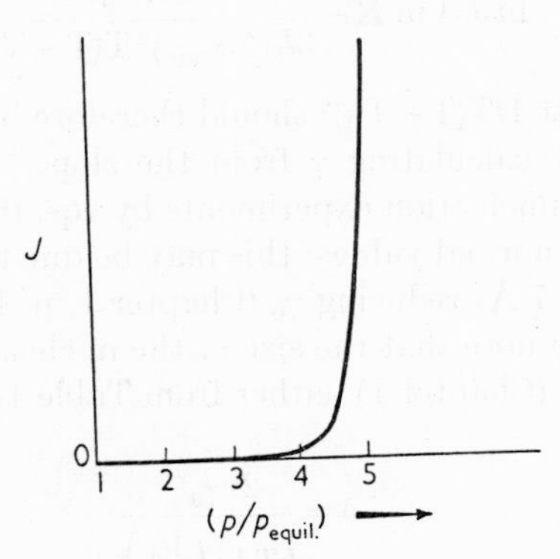

Fig. 8-8. Typical example of the plot of eq. (8.21).

becomes misty only when $p/p_{equil.}$ is 4.21. This compares very well with the predicted value of 4.16 from eq. (8.21), using a reasonable estimate of J (to which the calculation of $p/p_{equil.}$ is very insensitive). The calculated size of the nucleus under these conditions is obtained by substituting in (8.20):

$$a_n = \frac{2 \times (18/N) \times 70}{(4 \times 10^{-14}) \times 1.4} = 6.5 \times 10^{-8} \text{cm.} = 6.5 \text{ Å}$$

The liquid nucleus, just large enough to continue to grow, has a radius of 6.5 Å under these conditions: it contains about 40 water molecules.

The effect of temperature on the rate of condensation can be easily seen using the following approximate treatments. If it is assumed that the entropy of vaporization of the liquid ($\Delta S_{vap.}$) is constant over small temperature ranges (it increases by only 8% for water between 100°C and 80°C), we can integrate the general thermodynamic equation

$$\frac{d(\mu_{liq.} - \mu_{vap.})}{dT} = \Delta S_{vap.}$$

to give

$$\mu_{liq.} - \mu_{vap.} = (\Delta S_{vap.})(T - T_B)$$

where T and T_B are respectively the temperature of the supercooled vapour, and the temperature at which it would have condensed had it been nucleated, i.e. the boiling point. The units of $\Delta S_{vap.}$ are ergs per molecule; for water the value is about 0.2×10^{-14} erg.

The expression (8.14) for ΔG_n now becomes

$$\Delta G_n = \frac{16\pi\gamma_0^3 v^2}{3(\Delta S_{vap.})^2 (T - T_B)^2}$$

and so for J we have, taking logarithms of (8.15) and substituting,

$$\ln J = \ln \boldsymbol{K} - \frac{16\pi\gamma_0^3 v^2}{3\boldsymbol{k}(\Delta S_{vap.})^2 \, T(T - T_B)^2} \tag{8.22}$$

A plot of $\ln J$ against $1/T(T - T_B)^2$ should therefore be linear, and the theory may be checked by calculating γ from the slope. Values of γ_0 for water calculated from the nucleation experiments by eqs. (8.21) or (8.22) are about 15% lower than the normal values: this may be due to the high curvature of the interface ($a_n = 6$–7 Å) reducing γ_0 (Chapter 1, p. 11).

It is instructive to note that the size of the nucleus can be predicted from the Kelvin equation (Chapter 1), either from Table 1-II or from (1.9), which may be rewritten as

$$a = \frac{2v\gamma_0}{\boldsymbol{k}T \ln\left(\frac{p_a}{p_\infty}\right)} \tag{1.9}$$

Here both p_a and p_∞ are equilibrium pressures, over the droplet of radius a and over a flat liquid surface, respectively. The equation indicates that the vapour pressure over a small droplet must be equal to the prevailing vapour pressure if the droplet is to be able just to grow and to constitute a nucleus. Smaller droplets will evaporate again since their vapour pressure will be higher than that of the vapour with which they are in contact. Equation (1.9) is thus exactly equivalent to (8.20) above, and the critical nucleus size calculated for condensing water vapour at 4.2 times the saturation pressure (found to be 6.5 Å by the above treatment) could have been predicted from Table 1-II.

The nucleation of crystals from the vapour is very similar to the above process, although the exact value of the free energy of nucleus formation will now depend on the shape and irregularity of the solid crystalline nucleus; the edges and corners of microcrystals on the nucleus may be expected to be particularly important. Little is known, however, about the surface energies of such edges and corners, and it is therefore convenient to assume, as a first approximation, that the solid nucleus is spherical and that its surface is uniform. The theory of nucleation then leads to a form exactly analogous to eq. (8.14):

$$\Delta G_n = \frac{16\pi\gamma^3 v^2}{3(\mu_{vap.} - \mu_{solid})^2} \tag{8.23}$$

where γ is now the solid-vapour interfacial energy, and v is the volume (in cm.^{3}) of a molecule in the solid. Experiments[7] on the formation of clouds from supersaturated water vapour below -62°C suggest that ice nuclei are being formed; there is fair agreement with theory if the surface energy of the ice crystals is taken as 70 ergs cm.$^{-2}$ with a temperature coefficient of -0.06.

The Stability of Aerosols

That fogs and smokes appear to be very stable compared with aqueous dispersions is common knowledge. Although the explanation of this is apparently quite simple, various complicated reasons have been advanced from time to time for the persistence of fogs and smokes. For the coagulation of an unstabilized colloidal system (there being no energy barrier to the close approach of the particles or drops), the Smoluchowski theory gives:

$$1/\mathbf{n} - 1/\mathbf{n}_0 = 4\pi D \boldsymbol{R} t \tag{8.3}$$

and

$$t_{1/2} = 1/4\pi D \boldsymbol{R} \mathbf{n}_0 \tag{8.4}$$

in which D is expressible in terms of a ($=\boldsymbol{R}/2$) and the viscosity of the continuous medium. However, when the mean free path l_f of the molecules of the continuous phase is of the same order of magnitude as the radius of the colloidal particles, the resistance to movement is no longer given exactly by $6\pi\eta a$: a correction by $(a/(a+0.9l_f))$ is required[8]. This indicates that, if the mean free path is of the same order as the radius of the particle, the latter can no longer be regarded as moving through a continuous medium of viscosity η; for some of the time it will be moving unopposed through vacuum till it happens to strike a molecule of the phase in which it is dispersed. The viscous drag is correspondingly reduced, and the viscosity η must be replaced by $\eta\left(\frac{a}{a+0.9l_f}\right)$. With this correction, eqs. (8.4) and (8.5) lead to the following formula for the time of half-coagulation:

$$t_{1/2} = (3\eta/4\boldsymbol{k}T\mathbf{n}_0)(a/(a+0.9l_f)) \tag{8.24}$$

Though in liquids l_f is negligibly small, in air at atmospheric pressure the mean free path is about 1×10^{-5} cm., so that, with particles of aerosol of radii 1×10^{-5} cm., and with 1.7×10^{-4} poise for the viscosity of air, the time for the number of particles to be reduced by half is given by

$$t_{1/2} = \frac{1.6\times10^9}{\mathbf{n}_0} \text{ seconds} \tag{8.25}$$

This suggests that aerosols should have smaller times of coagulation than aqueous sols (cf. eq. (8.7)), but this is only true for equal numbers of particles per cm.$^{-3}$ Now it so happens that the aqueous colloidal dispersions of common occurrence have particle concentrations of the order of 10^{12} cm.$^{-3}$, while the fogs and smokes commonly encountered may have only 10^5 particles cm.$^{-3}$ and yet appear very dense: a fog with only 10^4 particles cm.$^{-3}$ will reduce visibility to 20 metres! In this low concentration of aerosols lies the principle reason for their stability: though the viscosity of air is only about 1/50 that of water, the low concentration of the aerosols lends them great stability. Aqueous dispersions of equally low concentrations are even more stable than are fogs and smokes.

As an example of the application of the theory, consider a fog of 10^5 particles cm.$^{-3}$, and of particle radius 10^{-5}cm. The time for the number of particles to be reduced to one-half, according to eq. (8.25), is 1.6×10^4 seconds, i.e. about $4\frac{1}{2}$ hours. After this time, therefore, the particle concentration is reduced to 5×10^4 particles cm.$^{-3}$, and to reduce this to 2.5×10^4 takes a further 9 hours. The fog will thus still be dense (visibility less than 20 metres) even after $13\frac{1}{2}$ hours, and during this time many new particles may have been added to the smoke or fog. The methods by which the coagulation of fog and mists may be accelerated are thus of considerable interest, and are discussed below.

For a smoke of particle radius 10^{-5}cm. with 10^6 particles cm.$^{-3}$, the time for half-coagulation would be ten times less than above, i.e. 1.6×10^3 seconds (27 minutes). Comparison with experimental data on stearic acid smoke (Fig. 8-1) shows that eq. (8.3) is obeyed extraordinarily well, the slope of the plot of $1/\mathbf{n}$ against t being quite constant even over long time intervals. Further, the time for the number of particles cm.$^{-3}$ to be reduced by half, from 10^6 to 0.5×10^6, is found to be 28 minutes, in agreement with the figure of 27 minutes calculated from the theory of Smoluchowski.

Whereas *an electric charge* on aqueous sols may stabilize them almost indefinitely, with $e^{-W/RT}$ as low as 10^{-8}, aerosols are never stabilized as much as this. The smaller particles of smokes and fogs normally carry a slight electrical charge (about 1ε positive or negative), often caused by collisions with naturally occurring air ions. Such charges on small particles, however, have no appreciable effect on the rate of aggregation.

An alteration in the stability of an aerosol is, however, found with larger particles, such as those in blown silica aerosols or in MgO smoke[9]: here the electrical charges may be appreciable. The system described in Table 8-I is electrically neutral as a whole, with positive and negative charges present in

TABLE 8-I

System	Mean charge, positive or negative	$e^{-W/RT}$
Theoretical	0	1
SiO_2, 10 lb. in.$^{-2}$ air blast	7 ε	17
SiO_2, 25 lb. in.$^{-2}$ air blast	12 ε	40

about the same quantities. For similarly charged particles **W** will be very high ($e^{-W/RT}$ close to zero); for oppositely charged particles **W** will be negative ($e^{-W/RT}$ greater than unity). The mean values of $e^{-W/RT}$ for the whole system will therefore be about one-half the latter values, and it is these that are recorded in Table 8-I. The net result of this electrical charge is that the aerosol coagulates more quickly than if it were completely uncharged.

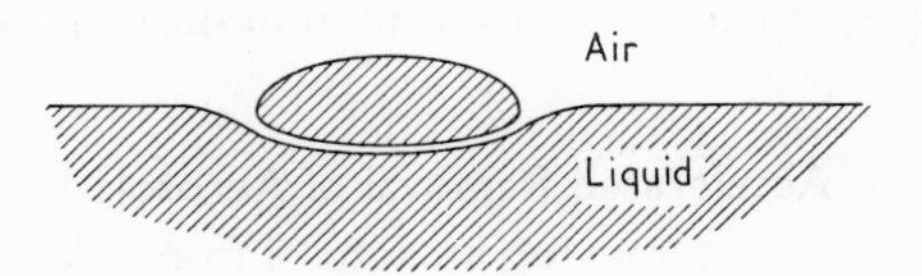

Fig. 8-9. Drop of a liquid floating on a large surface of the same liquid.

Whether an aerosol can be produced in which all the particles are similarly charged is a matter of some interest. However, the low electrical capacity of an aerosol particle has the result that not many electronic charges can be accumulated on it; air ions will therefore rapidly remove any net charge from the system.

A possible model system for aerosol droplet coalescence rates is shown in Fig. 8-9. It has long been known[10] that *floating drops* of a liquid may remain for some seconds on a surface of the same liquid, and in more recent years the presence of ionic surface-active agents has been shown to increase the life of these floating drops. If the model is a good one, the life of an aerosol formed from a solution of a surface-active agent should be likewise increased. Direct tests of this suggestion, however, give entirely negative results[11], the aerosols being no more stable than those from pure water. The model system is therefore not analogous to the approaching droplets of aerosol: because of the large areas of near contact in the former, the entrapped air can flow out only slowly, especially if the surface layers are rendered immobile by a viscous or incompressible monolayer. Besides this, over large areas the electrical repulsion between similar charges on the drops may become very important as the air gap narrows, though this is not so in the close approach of two spheres, which will touch only in point contact. Further, the system is very sensitive to traces of dust, and to chance mechanical shocks; an attractive electrical potential causes rapid coalescence. In many respects, we conclude, the model is unsatisfactory, particularly in that aerosols, unlike large liquid drops, cannot be stabilized with adsorbed monolayers.

Evaporation of Aerosol Droplets

Evaporation of a volatile liquid may cause an aerosol to disappear rapidly. Even water, dispersed in the form of droplets of 10μ radius, evaporates completely, if the surface is clean, in 2.4 sec. into an environment of air at 20°C and 80% relative humidity. The vapour pressure is higher over the smaller drops than over the larger drops (eq. (1.9)), so that liquid may undergo *isothermal distillation* from the small drops to the larger, thus eventually reducing the number of aerosol droplets[2,12]. A coating of tarry material on the drops (as in town fogs) reduces the isothermal distillation. The reduction of evaporation in aerosols used in dust-binding is discussed on p. 308.

Accelerated Removal of Aerosols

The simplest method[2] of removing an aerosol from a large volume of gas is to filter it through a paper, cloth or other filter, just as the domestic vacuum cleaner removes suspended dust by passing the air through a cloth bag. This process, however, requires a considerable pressure drop across the very fine filters, and in recent years other methods have been developed in which no pressure drop occurs. Electrical precipitators[13], in which a brush discharge passes across the high tension electrodes (with voltages up to 75 kV) are very effective in removing the larger aerosol particles. The small particles ($<3\times10^{-4}$cm.), however, tend to remain in the gas. The efficiency of the process for the large particles is very high, and about 94% recovery of a smoke is possible with a time of passage of only 1 second for the gases between the electrodes. Sonic agglomerators[13] are particularly useful in removing small particles in the size range 50 Å to 5000 Å. They thus find application in collecting carbon-black, cement dust, and sulphuric acid fogs. The mechanism of their action is complex, but the process, like electrical precipitation, is rapid: only 3 seconds are required with an intensity of 160 decibels. Centrifugal methods[13a] are also useful, particularly on a laboratory scale: a rapid estimation of particle sizes is also possible.

Often the water in a cloud of supercooled water drops will precipitate quickly through the air on to the ground only if ice crystals can grow: solid CO_2 (which lowers the temperature to the point at which spontaneous nucleation can occur) and AgI smoke (the particles of which have a lattice spacing very similar to that of ice) are both effective in initiating the formation of the ice crystals. Once the crystallization has started, ice particles quickly increase at the expense of the water droplets, and cause precipitation[14], possibly melting again or evaporating during their passage through the lower atmosphere.

EMULSIONS

Like other colloidal systems, emulsions may be formed by methods of either dispersion or condensation. Among the former the use of the classical "colloid-mill" or homogenizer is well known. The principle behind such machines is that shear energy is supplied to a mixture of oil and water, tearing one or both phases into smaller drops. Clayton[15] and Becher[16] give detailed descriptions of many different emulsifying machines and homogenizers. To reduce the amount of mechanical work required to effect emulsification, the interfacial tension is lowered by the adsorption of a surface-active agent; this is also important in stabilizing the newly-formed emulsion, as explained below, and may also alter the type of the emulsion[17]. Condensation of an aerosol of the dispersed phase (which may be produced electrically) into the continuous phase gives an emulsion of uniform particle size[18].

Condensation methods may be illustrated by the laboratory method of preparing an emulsion of (say) toluene in water. This may be achieved by first dissolving the oil in ethyl alcohol, then pouring this mixture into water; the consequent dilution of the alcohol leaves the oil "stranded" in the water, forming emulsion drops about 1 micron in size. In general, this and other systems which emulsify without any stirring whatever are of considerable fundamental interest. In such *spontaneous emulsification* the entire energy required for the emulsification comes from the redistribution of materials within the system. We shall see that spontaneous emulsification may occur by either the condensation or dispersion mechanisms, the visible effect being the same—if the oil is placed quietly on the water, the interfacial region gradually becomes cloudy due to spontaneously formed emulsion. It is convenient to consider separately the subject of the spontaneous formation of emulsions, and to give rules for elucidating the mechanism by which this occurs in any particular system.

Spontaneous Emulsification

If oil and water (with certain additives) are brought quietly into contact, spontaneous emulsification may occur on one or both sides of the interface. Usually the requisite energy comes from the free energy release as the additive is redistributed to its equilibrium state in the two phases, but occasionally electrical energy rather than chemical energy can be responsible for spontaneous emulsification, as explained below. In practice, particular care is necessary to distinguish emulsification occurring truly spontaneously from emulsification occurring easily: this easy emulsification, made possible by a very low interfacial tension, is often referred to commercially as spontaneous emulsification. However, in true spontaneous emulsification no external mechanical work whatever is required.

Three mechanisms have been suggested to account for spontaneous emulsification:

(*i*) *Interfacial Turbulence*

This mechanism was originally proposed by Quincke in 1888: it had previously been noticed that, when solutions of lauric acid in oil are placed very gently on aqueous sodium hydroxide, an emulsion is formed in the water phase[19]. Quincke suggested that the spontaneous emulsification is caused by localized interfacial tension lowerings, due to the unequal formation of soap at different points in the interface. The ensuing violent spreading may tear droplets of oil away from the interface, which are then stabilized by the soap produced.

(*ii*) *Diffusion and Stranding*

Spontaneous emulsification can occur by diffusion alone[20,21,22] when, as in the example mentioned above, a solution of ethyl alcohol and toluene is placed gently in contact with water. The alcohol, as it diffuses from the oil into the water, carries with it some oil (forming a three-component phase in the immediate vicinity of the interface). As the alcohol diffuses further into the water, the associated oil becomes thrown out of solution, and is "stranded" in the water in the form of fine emulsion drops. An emulsion of water in oil may also be formed on the oil side of the interface, since the alcohol in the oil may permit some water to dissolve. As the alcohol passes into the aqueous phase, the water becomes "stranded" in the oil.

(*iii*) *Negative Interfacial Tension*

If the interfacial tension is locally negative, the area of the interface tends to increase spontaneously. This occurs when an adsorbed film is present under conditions such that $\Pi > \gamma_i$, but a zero interfacial tension ($\Pi = \gamma_i$) will also be

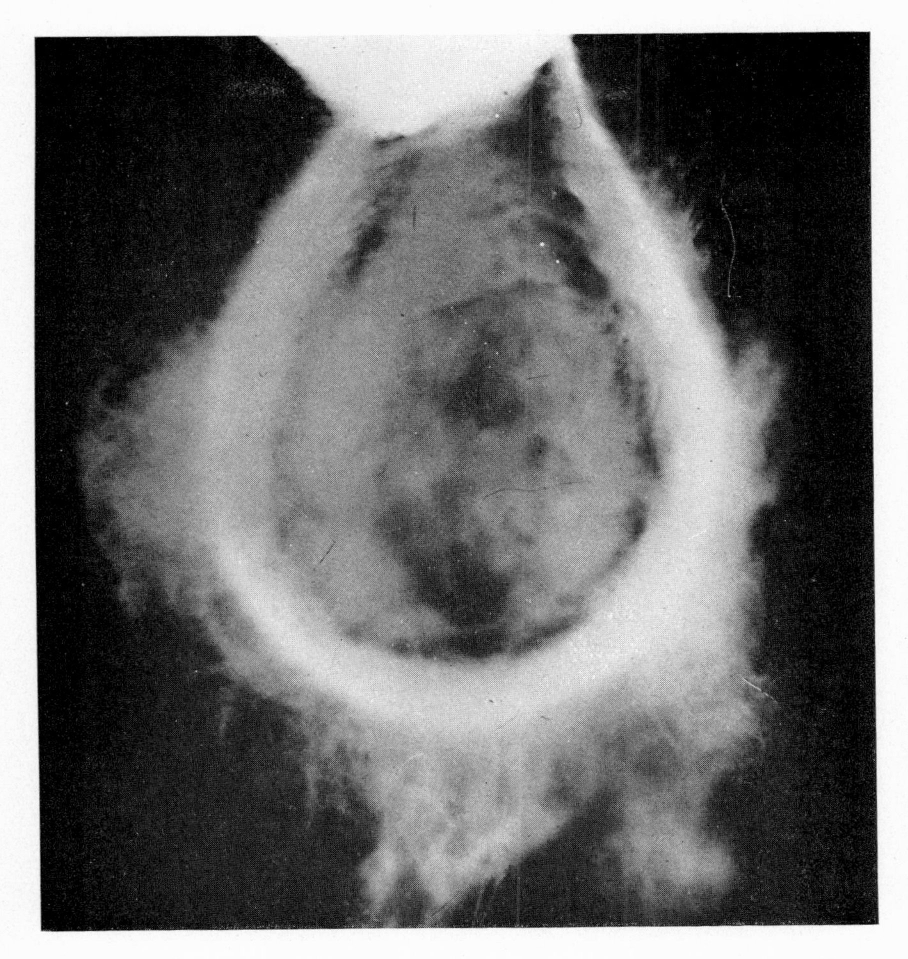

Fig. 8-10. Flash photograph of a drop of water in toluene containing 14% ethanol and saturated with water, after 7 sec. from formation. The spontaneously formed emulsion is seen both inside and outside the drop which is "kicking" strongly[24].

[*To face p. 360*

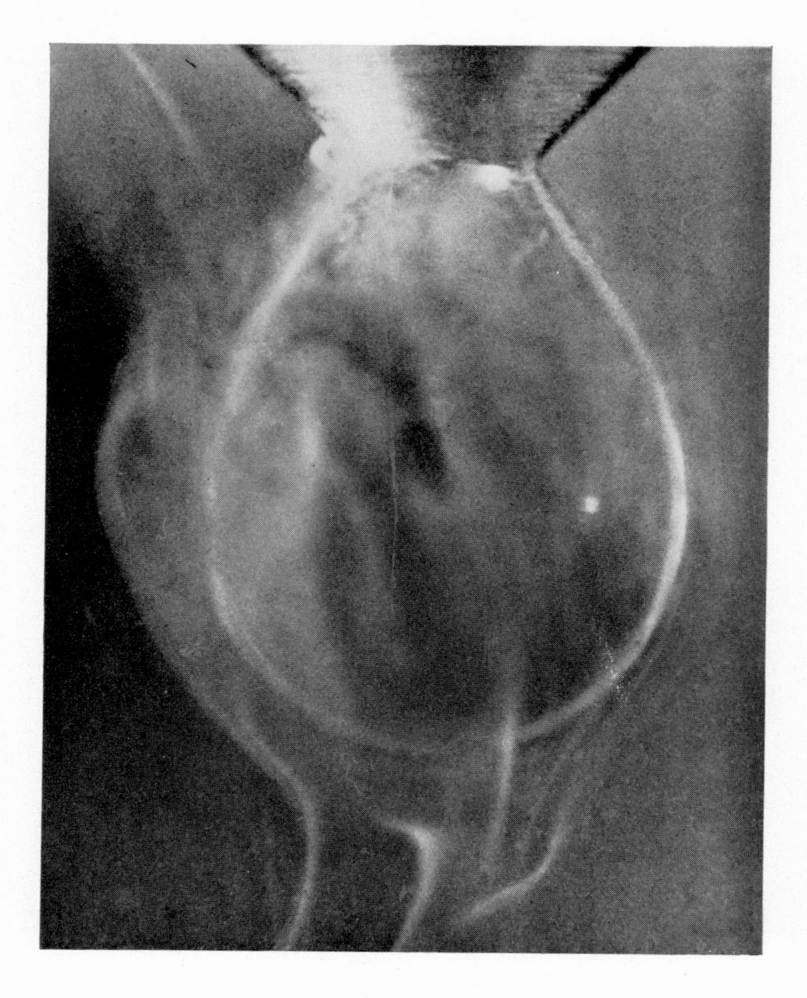

Fig. 8-11. If the drop is prevented from kicking by forming it of 1 N-NaCl, instead of pure water as in the previous figure, emulsion still forms spontaneously, showing that the "diffusion and stranding" mechanism is operative[24].

unstable, since chance vibrations and thermal fluctuations (i.e. entropy effects) will tend to break up the interface. As the area of the interface increases, the interfacial tension will become less negative, the surface pressure of the adsorbed film decreasing till finally Π is a little less than γ_i, i.e. the interfacial tension is low and positive. This appears to be the mechanism operative when oleic acid dissolved in oil is placed on aqueous alkali[23].

To choose between these three possible explanations is not easy, and sometimes more than one of the processes may be occurring. However, it is possible to devise tests[24] to apply to any given experimental system to determine which mechanism is operative.

Examples of Mechanism (i)

No example of this is definitely established, though it is perhaps the most widely accepted mechanism[20]. The obvious system to test for this mechanism is that of methyl or ethyl alcohol in toluene in contact with water[25]. This shows strong spontaneous emulsification (Fig. 8-10), and it also shows marked interfacial turbulence[26], as is discussed in detail in Chapter 7. Surprisingly, however, the emulsification in these systems is accounted for by the second mechanism, since the interfacial turbulence can be completely suppressed by adding a little detergent to the water, by dissolving salt in the water, or by spreading a protein film at the interface[24], while the spontaneous emulsion is still produced (Fig. 8-11). The only effect of the elimination of the interfacial turbulence is that the emulsion is no longer thrown violently off from the interface, but instead streams off quietly. The turbulence is thus *not* responsible, as many had previously thought[20], for the emulsification in this system.

Examples of Mechanism (ii)

We have eliminated from the first mechanism emulsions from solutions of methanol or ethanol in toluene, placed gently in contact with water; prevention of the turbulence has but little effect on the emulsification. Further, the interfacial tension is always positive, of the order 10 dynes cm.$^{-1}$, leaving mechanism (ii) by default. This can be confirmed by presaturating the toluene-alcohol mixture with water; the diffusing alcohol leaves much water stranded in the oil, as well as oil stranded in the water. This mechanism is likely whenever the third component increases considerably the mutual solubility of the oil and the water: thus mixtures of an oil with sulphonated castor oil and sodium oleate, placed in contact with water, emulsify by the diffusing sodium oleate carrying oil with it into the water[27].

The same explanation[24] also fits the spontaneous emulsification seen when a solution in petrol-ether of commercial sodium dodecyl benzene sulphonate is gently placed in contact with pure water[28]. The rapid diffusion of the detergent into the water carries with it some oil, which is then "stranded" in the water as the detergent becomes more diluted. Further, some of the water, which initially becomes solubilized in the oil-detergent solution, apparently becomes "stranded" in the oil as more detergent leaves the oil for the water, and small amounts of W/O emulsion are thus visible[24]. Emulsions may thus be formed on both sides of the interface. In these experiments a temperature of 90°C must be maintained to ensure a sufficient solubility of the detergent in the oil. Even when the interfacial tension is as high as 5 dynes cm.$^{-1}$, the emulsification is still marked.

This "diffusion and stranding" mechanism can occur independently of the value of the interfacial tension, which may be relatively high. However, if sufficient of the third component is added to the oil, the interfacial tension becomes zero or negative, as may be seen by extrapolation of the curves for added ethanol or acetic acid in Fig. 8-12. The same phenomenon occurs with

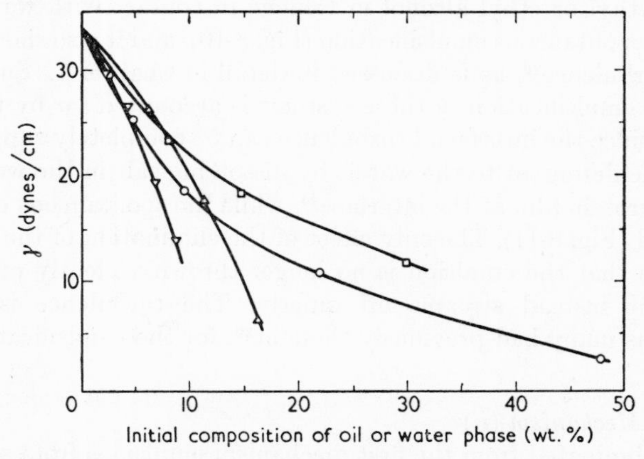

Fig. 8-12. Interfacial tensions (γ) of water-toluene systems, with additives, at 20°C, each measurement taking 8 sec. ▽ Ethanol in the toluene; △ Acetic acid in the toluene; O Acetone in the toluene; □ Acetic acid in the water[24].

surface-active agents in the oil, and thus the interface is effectively abolished and the phases are locally completely miscible. Spontaneous emulsification due to "stranding" still occurs, but this must be distinguished from that occurring by the mechanism discussed below, where the interface, although tending to increase in area, may remain well defined and there may be no miscibility of the phases (e.g. water and mercury).

Examples of Mechanism (iii)

The most straightforward illustration of the effect of a negative interfacial tension is the spontaneous emulsification of mercury in water. Ilkovič[29] showed that if a negative potential is applied to a mercury drop in an aqueous solution of a quaternary ammonium salt, the interfacial tension can be greatly decreased. The electrocapillarity curve for the system is shown in Fig. 8-13 and is of the typical parabolic form (Chapter 2). The quaternary ammonium ion is so resistant to decomposition at the surface of the mercury that a highly compressed monolayer of these cations is held there, both by adsorption of the hydrocarbon residues and by electrical attraction. At a potential of about -2 volts, the electrocapillarity curve (Fig. 8-13) suggests that the interfacial tension must become negative, i.e. $\Pi > \gamma_i$. The negative interfacial tension at the large applied negative potentials results in the surface of the mercury drop disintegrating into a brown cloud of colloidal mercury in water, and at -8 volts the spontaneous emulsification is very striking. The energy required to form the emulsion is supplied electrically, and the process is rather similar to the electrical formation of aerosols (p. 347).

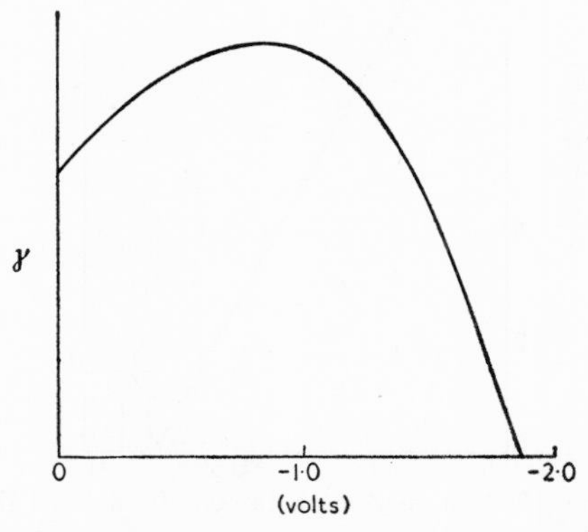

Fig. 8-13. The interfacial tension of mercury in the presence of a quaternary ammonium compound, as a function of applied potential. (After Ilkovič.)

It must be noted that zero or negative interfacial tension will occur whenever an adsorbed film at the interface is so compressed that $\Pi \geqslant \gamma_i$: in two-component systems, however, zero or negative interfacial tension implies miscibility, corresponding to *no* interfacial tension. The above experiment brings out this point; the highly compressed film of quaternary ammonium ions at the mercury-water interface has no effect on the negligibly small miscibility of mercury and water. Another example of a zero interfacial tension was found by Langmuir[30]. He compressed rapidly a film of protein at an oil-water interface, as in Fig. 5-4, eventually reaching the point where $\Pi = \gamma_i$ (p. 222). This is possible because the protein monolayer is unfolded irreversibly by adsorption (Chapter 5) and becomes insoluble. Eventually, however, the protein monolayer may crumple and the surface pressure will again decrease, but this is a rather slow process. In this example, moreover, the very high interfacial viscosity of the monolayer prevents the interface from increasing in area by the spontaneous formation of an emulsion.

The negative interfacial tension mechanism also explains the finding[23] that oil containing 5% to 20% of a fatty acid such as oleic, placed gently on aqueous alkali, leads to spontaneous emulsification. By extrapolation the interfacial tension appears to be negative in the pH range of 9 to 12, in which range the emulsification also occurs.

Another system, which emulsifies because of a negative interfacial tension, is that of solutions of long-chain salts in contact with solutions of cetyl alcohol or cholesterol in oil[24,31,32]. With sodium decyl sulphate in water against cetyl alcohol in toluene, the emulsion can form spontaneously at the

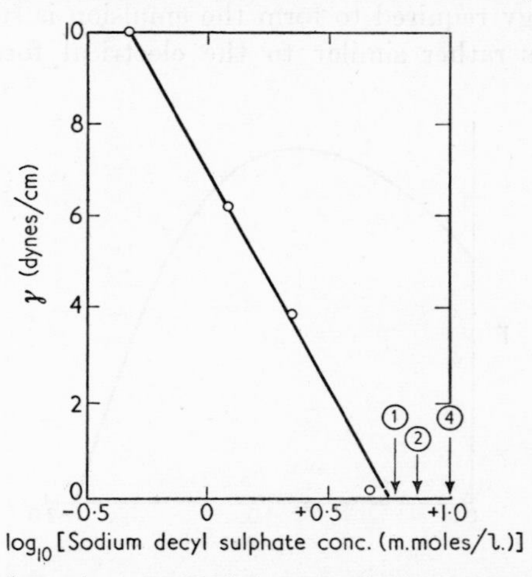

Fig. 8-14. Interfacial tensions at 20°C (measured by the drop-volume method) of sodium decyl sulphate solutions in buffer against 1.0 M-cetyl alcohol in toluene[24].

interface, with concentration limits which are quite sharp for both long-chain ions and alcohol[24]. Further, these concentration limits agree with those at which γ is expected to become negative (according to a short linear extrapolation), as in Figs. 8-14 and 8-15. A semi-quantitative estimate of the intensity of spontaneous emulsification is indicated by the numerals in circles, and it appears that the spontaneous emulsion forms slightly when γ is a few tenths of a dyne cm.$^{-1}$ below zero, and becomes pronounced when $\gamma = -1$ dyne cm.$^{-1}$, i.e. when the pressure of the interfacial monolayer, adsorbed from both the oil and the water phases, exceeds by 1 dyne cm.$^{-1}$ the tension of the clean interface. In general, ionic and non-ionic surface-active agents adsorb almost independently of each other, so this is a general method of obtaining very small or negative interfacial tensions.

Local lowerings of the interfacial tension to values below zero are now believed to be responsible for the spontaneous emulsification of an oil (e.g. xylene) in aqueous dodecylamine hydrochloride solutions. This phenomenon, which, projected from an optical cell, appears very striking, involves only the passage of the oil into the aqueous phase, where it is ultimately solubilized[33]: there is no diffusional reason why excess of oil should appear on the aqueous side of the interface. The explanation[34] is that the dodecylamine ions are strongly adsorbed (possibly with other surface-active impurities[35]) and momentarily reduce the interfacial tension to a negative value. Rapidly the interface increases in area by spontaneous emulsification, while the passage of the oil into the aqueous phase in this form, as well as solubilized, reduces the concentration of free dodecylamine ions near the surface to a level from

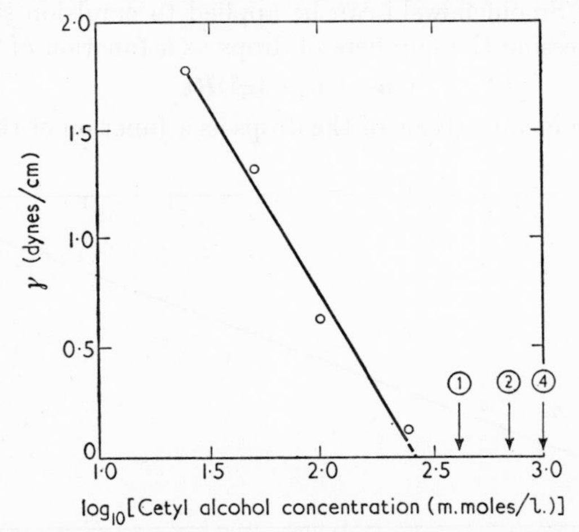

Fig. 8-15. The interfacial tensions at 20°C of toluene solutions of cetyl alcohol (measured by the drop-volume method) against 0.01 M-sodium decyl sulphate in buffer[24].

which adsorption is no longer sufficient to make the interfacial tension negative. At this stage a drop of the oil may be formed on a very fine tip and photographed; it "kicks" (cf. p. 323), unless the emulsion blanket is too thick, because stray convection currents or density differences sweep fresh, undepleted solution of the dodecylamine up to the interface through the enveloping emulsion, thus momentarily lowering the interfacial tension locally to below zero[34]. This fluctuation of the interfacial tension causes both the kicking and the spontaneous emulsification.

Tests of Mechanism of Spontaneous Emulsification[24]

From the foregoing, it is clear that if the interfacial tension is negative, this is a sufficient explanation, provided the monolayer is not too highly viscous, and provided also that there is not complete miscibility at the interface. Emulsification by this mechanism occurs at sharp concentration limits.

If, however, the interfacial tension is appreciably positive and there is no interfacial turbulence, then the "diffusion and stranding" mechanism must be operative.

If the interfacial tension is positive and there is interfacial turbulence, further investigation (e.g. by inhibiting the turbulence with a monolayer of adsorbed protein) is necessary.

The Stability of Emulsions

The theory of Smoluchowski can be applied to emulsion stability, either directly by expressing the numbers of drops as a function of time:

$$1/\mathbf{n} - 1/\mathbf{n}_0 = 4\pi D\boldsymbol{R}t \quad (8.3)$$

or by finding the mean volume of the drops as a function of time, as follows.

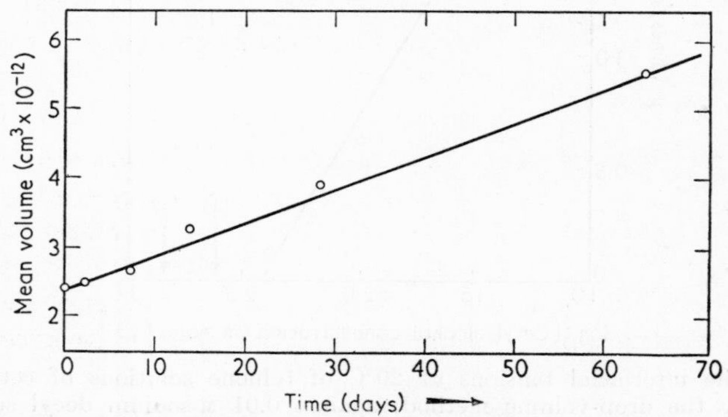

Fig. 8-16. Variation of mean volume with time; emulsion of oil in water stabilized with 1% sodium oleate, 1% phase volume at 25°C[36].

Let φ be the phase-volume of oil in the O/W emulsion, and let there be $\mathbf{n}$ drops, of mean volume $\bar{V}$, after time t. Then by definition

$$\bar{V} = \varphi/\mathbf{n}$$

or from eq. (8.3) above,

$$\bar{V} = \varphi/\mathbf{n}_0 + 4\pi D\boldsymbol{R}\varphi t$$

Since the phase-volume of the oil remains unchanged during coalescence, we have also $\varphi = \mathbf{n}\bar{V} = \mathbf{n}_0 V_0$, where V_0 is the drop volume at t=0. Hence

$$\bar{V} = V_0 + 4\pi D\boldsymbol{R}\varphi t \quad (8.26)$$

The result of testing this equation for emulsions with added stabilizer is shown in Fig. 8-16. The coalescence rate of the emulsion stabilized with adsorbed sodium oleate monolayers may be compared with that predicted by the Smoluchowski theory, by substituting for D from the Einstein equation (8.5) and putting $\boldsymbol{R}=2a$ as before. The time required for the mean volume of the droplets of oil in water to be doubled by coalescence is thus calculated to be 43 seconds, clearly a very much shorter time than is found by experiment (Fig. 8-16). This shows that diffusion alone cannot explain the slow coalescence of the emulsion, but that an energy barrier **W** must also be operative. This may be taken into account by rewriting eq. (8.26)

$$\overline{V}=V_0+4\pi D\boldsymbol{R}\varphi t e^{-\mathbf{W}/\boldsymbol{k}T} \tag{8.27}$$

From the actual variation of $\overline{V}$ with time we can now find **W**: it is about 11 $\boldsymbol{k}$T for an oil in water (O/W) emulsion stabilized with sodium oleate.

Differentiation of eq. (8.27) gives an expression for the coalescence rate of an O/W emulsion. This rate, called Rate 1, is given by

$$\text{Rate 1}=d\overline{V}/dt=(4\varphi\boldsymbol{k}t/3\eta_w)e^{-\mathbf{W}_1/RT}=\mathbf{C}_1 e^{-\mathbf{W}_1/RT} \tag{8.28}$$

where the Einstein equation (8.5) has been used to substitute for D. As before, φ is the phase volume of the oil, and η_w is the viscosity of the continuous phase, here water. $\mathbf{C}_1$ is the collision factor, defined by the equation. If the same system were emulsified to form a W/O emulsion, the corresponding coalescence rate would be

$$\text{Rate 2}=d\overline{V}/dt=(4(1-\varphi)\boldsymbol{k}T/3\eta_0)e^{-\mathbf{W}_2/RT}=\mathbf{C}_2 e^{-\mathbf{W}_2/RT} \tag{8.29}$$

where subscript 0 denotes oil, and $\mathbf{C}_2$ is the corresponding collision factor. In simple, shaken systems the emulsion type (i.e. whether an O/W or a W/O emulsion forms) as well as the stabilities are affected by Rate 1 and Rate 2; apart from the physical method of dispersal[15,16,37-42], the phase volume[16,41-44] and the viscosities of the phases[16,41,42] can, it is well known, greatly influence emulsion type and stability; this is to be expected from the above equations.

The energy barriers are calculated below on the assumption that they remain constant as two drops approach, though this may not be true unless the interfacial viscosity or interfacial compressional modulus (Chapter 5) is high[45]. Quantitatively an interfacial viscosity of at least 10^{-2}s.p. seems to be required if the interfacial films are not to flow aside (allowing coalescence) during the close approach of two drops. The effect of the compressional modulus in preventing displacement of the stabilizing films should be as in eq. (7.41), i.e. it should vary as C_s^{-1}/a^2 where a is the radius of the drop. Clearly drops of small a should be less liable to suffer displacement of the stabilizing films than are large ones.

Electrical Barriers

Consider first two drops of a *paraffinic oil in water* and let the electrical potential on the surfaces of these drops be ψ_0, as before, relative to the bulk of the water far away. Now when two charged surfaces approach through a solution of electrolyte, the repulsive potential varies approximately as ψ_0^2 (see eq. 8.54) whether the surfaces are flat or curved[46,47]. The constant of proportionality depends on the radius of curvature of the approaching surfaces, but since liquid drops nearly in contact may be expected to flatten considerably or even become slightly concave[48,49], the radius of curvature to be used for emulsion drops may be far from the actual drop radius. Under these circumstances we assume[50] that the radius of curvature of the adjacent charged surfaces is constant for emulsion drops of the order 1 micron diameter, and this constant is determined empirically, so that for the coalescence of charged emulsion drops (with no Stern electrical layer):

$$\mathbf{W}_1 = B\psi_0^2$$

with the constant B obtained empirically from coalescence experiments. For emulsions stabilized with sodium oleate, and with $\mathbf{C}_1$ given by the Smoluchowski theory, eq. (8.27) and Fig. 8-16 suggest that $(-B\psi_0^2/\mathrm{RT}) = -11.2$. If there is one electronic charge per 45 Å^2 in the monolayer of oleate on the oil drops, the Gouy equation (Chapter 2) gives $\psi_0 = 165$ mV, whence $B = 0.24$. From experiments on unstable emulsions[51] we have similarly $-B\psi_0^2/\mathrm{RT} = -6$, with $\psi_0 = -123$ mV in the presence of 0.4 N-NaCl according to the Gouy theory; hence $B = 0.23$. Experiments on oil drops in water without additives show[36] that the system is about 750 times more stable than according to the Smoluchowski theory. The ζ-potential of oil drops in water is about -70 mV, due to adsorption of hydroxyl ions, and hence from the relation $\zeta = 0.55\psi_0$ (Chapter 3), ψ_0 is -127 mV. Consequently, for the oil drops in water, $-B \times 127^2/\mathrm{RT} = -6.6$, whence $B = 0.24$. Here we shall use a value of 0.24 for B. In general, a tenfold increase in uni-univalent electrolyte concentration will decrease ψ_0 numerically by 60 mV, and so will greatly decrease the stability of an O/W emulsion.

Consider now two *drops of water in oil:* as shown in Chapter 3, quite high electrical potentials may possibly build up in the oil phase consequent upon adsorption or ionic redistribution. If the thickness of the oil is sufficient to accommodate the necessary counter-ions, most of the potential drop will occur in the oil phase[47]. One may calculate from purely coulombic repulsion that emulsions stabilized by oleates of polyvalent metals should be stable if $\psi_0 \gg 25$ mV, though in practice one finds that some emulsions of water in benzene are stable with $\zeta \approx 10$ mV, while oleates of other metals may give unstable emulsions even when $\zeta > 100$ mV measured in the oil[52]. Clearly other factors must be important here, including interfacial viscosity and the formation of thick layers of oil-wetted hydrolysis products of the oleates: further experiments along these lines are required. An investigation of the effect of the length of the hydrocarbon chain would also be of interest, since on a purely electrostatic theory of stability this variable should not be important. Another difficulty of electrostatic theories is that in practice the usual stabilizing agents for water-in-oil emulsions are non-ionic derivatives of high molecular weight and with hydrocarbon side-chains of considerable length, suggesting a steric mechanism of stabilization. Whether both electrical and steric factors operate together in practice has not been elucidated, nor has the difference in behaviour of water drops in benzene and in petrol-ether; as shown in Chapter 2, water dissolving in the benzene considerably increases the ionic concentration and so reduces $1/\varkappa$, though this does not occur with petrol-ether.

Hydration barriers—"Deep Surfaces"

At a liquid surface there is evidence that several layers of water may be orientated to form a rather rigid layer of "soft ice" (of viscosity about that of toffee or butter): as an illustration, we may imagine a layer 10 Å thick and of mean viscosity about 10^4 poises, the exact value decreasing away from the surface. Thus the dipole moments of the hydroxyl groups of long-chain alcohols at the air-water surface are considerably less than for isolated $-CH_2OH$ groups (Chapter 2 and Fig. 2-12); in Chapter 3 it is shown that the ratio of ζ/ψ_0 depends on the reduced dielectric constant and enhanced viscosity near the surface, and also that flow through very narrow capillaries may be slower than that calculated from electrical retardation alone. Further, the interfacial viscosity of adsorbed films of cetyl alcohol and long-chain sulphates (Chapter 5) suggests a resistance to flow in the aqueous part of the surface, while the rather high resistances to the passage of CO_2 and other gases through surface films (compared with those to evaporation, see Chapter 7) can also be explained by such a thin layer of this "soft ice" associated with the surface film.

Vand's studies[52a] on the viscosity and density of aqueous sucrose solutions show that at 0°C there are about 11 molecules of firmly bound water of hydration around each sucrose molecule, corresponding to about one molecule of water of hydration on every $-OH$ and $-O-$ group of the sucrose molecule. The density of the water of hydration is 1.1. At higher temperatures there is less hydration, because the energy of dehydration is 2.58 K.cal.mole^{-1}. The latter figure is in fair agreement with the value calculated in the next section. Bernal[52b] also cites evidence that the density of orientated water near a surface is higher than normal.

Apparently the dipoles of the film-forming molecules may orientate the dipoles of the water between and below the head-groups, behaving, as McBain suggested[20], rather as a magnet which will pick up several nails. Evidence in favour of this hypothesis comes from a variety of phenomena[53,54]. There is also evidence for strong hydrogen bonding within films of long-chain amides[55]. At charged surfaces the powerful field close to the ionic groups would be expected to orientate a considerable number of water molecules: exact energy and entropy studies are of great importance in this context[56].

Coalescence of Drops—Stability

If two oil drops are to coalesce, this water of hydration must be displaced: the total energy barrier ΣE_h required for this displacement will depend on the total number and type of the hydrated groups on each molecule of the surface-active agent and on θ, the fraction of the interface covered; as a first approximation[50]

$$\mathbf{W}_1 = \theta \Sigma E_h$$

A long-chain, strongly hydrated compound is therefore required to stabilize oil-in-water emulsions: the long chain promotes adsorption and increases θ, while the hydration hinders close approach of the oil surfaces. In general, however, both charge and hydration may be operative in decreasing Rate 1, and the two energy barriers act simultaneously:

$$\text{Rate } 1 = \mathbf{C}_1 \exp\left\{\frac{-0.24\psi_0^2 - \theta \Sigma E_h}{RT}\right\} \tag{8.30}$$

This can explain quantitatively the finding[57] that the addition of a long-chain alcohol to a detergent-stabilized O/W emulsion may greatly enhance its stability, even though the alcohol alone would favour a W/O emulsion: both electrical and hydration factors are now operative. According to eq. (8.30), with $\psi_0 = -175$ mV for the emulsion stabilized by detergent alone, Rate $1 = 10^{-5}\mathbf{C}_1$. If, however, cetyl alcohol is present, the potential ψ_0 will be effectively unaltered, but for this alcohol $\theta = 0.5$ and E_h is 2400 cal. (see below). Hence, by equation (8.30), Rate $1 = 10^{-6}\mathbf{C}_1$, so the emulsion is now, according to the calculation, ten times more stable than without the long-chain alcohol.

The inverse system of two water drops separated by non-polar oil will coalesce by the water bridging the oil gap when the drops approach as close as possible. The rate of coalescence will then depend both on hydrodynamic factors (included in the term $\mathbf{C}_2$), on the number m of $-CH_2-$ groups in each hydrocarbon chain, and on θ, the fraction of surface actually covered with surface-active emulsifier[58-60]. The barrier of each $-CH_2-$ group to the passage of water[61] is about 300 cal. and, since the water will have to bridge the gap across 2m $-CH_2-$ groups,

$$\mathbf{W}_2 = 2m\theta \times 300$$

and

$$\text{Rate } 2 = \mathbf{C}_2 \exp(-600m\theta/RT) \tag{8.31}$$

A high surface coverage of long-chains is therefore required to stabilize a W/O emulsion. The ratio of the rates of coalescence is thus given by:

$$\frac{\text{Rate 2}}{\text{Rate 1}} = \frac{C_2}{C_1} \exp\left\{\frac{+0.24\psi_0^2 + \theta\Sigma E_h - 600m\theta}{RT}\right\} \tag{8.32}$$

As with aerosols, model systems involving the coalescence of **floating drops** at plane oil-water interfaces give only poor correlations with emulsion stability and type: the sensitivity of the interface (small compared with that of an emulsion) to traces of impurity and small electrical potentials, and the importance of draining times, introduce additional complications in such models[62]. The presence of a surface film at the interface hinders the coalescence of the drop with the bulk liquid by increasing the drainage time because the surface film is resistant to flow.

Emulsion Type

The Bancroft rule[63] states that the phase in which the stabilizing agent is more soluble will be the continuous phase: it is a purely empirical correlation.

Another theory, that the different stabilizers act by causing *preferential curvature of the oil-water interface* was first suggested under the title of the "oriented wedge" theory[15]. The latter is now of historical interest only; the radii of curvature of ordinary emulsion drops are of the order one thousand times molecular dimensions, and this would seem to minimize the practical effect of any molecular tendency to preferential curvature. Only if the drops are very small (a few hundred Å in diameter) is this effect important[64].

Recently, emulsifying agents have been classified numerically on the "*H.L.B.*" *scale*[65-67]; this refers to the hydrophilic-lipophilic balance of the emulsifier molecule: the numerical correlation of H.L.B. values and emulsion type must, however, be based on direct experimental tests.

For *shaken mixtures* we believe that the *coalescence kinetics* are responsible not only for the stability but also for the type of the emulsion formed. The kinetics of emulsion breaking have long been of interest, and Cheesman and King[38] and McBain[20] pointed out the importance of such kinetics in determining emulsion type.

We shall now assume that in **shaken systems** (i.e. apart from mechanical factors) emulsion type depends only on the relative rates of coalescence of oil-in-water and water-in-oil systems. From this we may show[50] that:

(i) the logarithm of the relative coalescence rates should vary directly with the H.L.B. values for different emulsifiers,

(ii) the Bancroft rule follows from a correlation between the relative coalescence rates and the preferential solubility of different emulsifiers in one of the phases, under certain conditions, and

(iii) the empirical H.L.B. values have a fundamental significance in terms of free energies, and can be related to the distribution of the surface-active agent between oil and water under certain conditions.

If we shake a mixture of oil and water with an emulsifying agent, we shall form initially a mixture of systems shown (on the left-hand sides) in Figs. 8-17 and 8-18; the relative coalescence rates will determine the emulsion type as observed shortly after the agitation is stopped. Hence follow the relations:

$$\left.\begin{array}{ll} \text{O/W emulsion preferentially stable if} & \dfrac{\text{Rate 2}}{\text{Rate 1}} \gg 1 \\ \text{W/O emulsion preferentially stable if} & \dfrac{\text{Rate 2}}{\text{Rate 1}} \ll 1 \end{array}\right\} \quad (8.33)$$

In practice, emulsions will be *stable* (e.g. for a year) only if Rate 1 or Rate 2 is smaller than about $10^{-6}\mathbf{C}_1$ or $10^{-6}\mathbf{C}_2$.

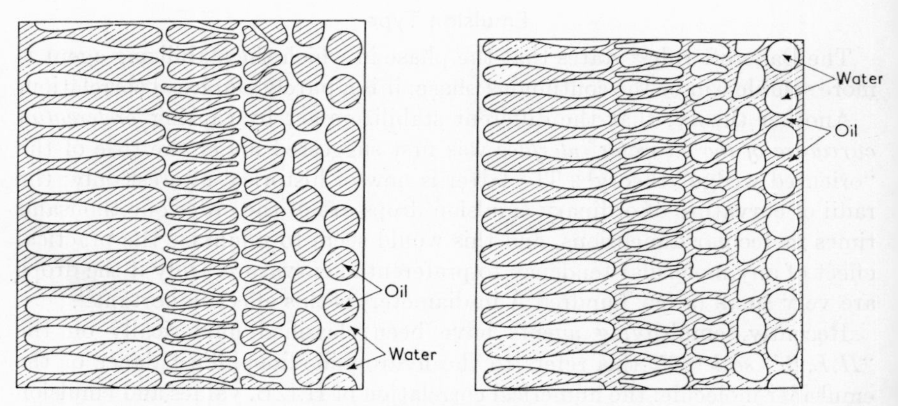

Figs. 8-17 and 8-18. Shaking a mixture of oil and water will lead to the forms on the left-hand sides of the figures: the emulsion type observed will therefore depend on the relative coalescence rates in the two systems.

The *H.L.B. values* as first reported represented an empirical numerical correlation of the emulsifying and solubilizing properties of different surface-active agents. Table 8-II illustrates the various ranges of H.L.B. values, and the values found experimentally for some different surface-active agents are shown in Table 8-III.

However, one can now calculate the H.L.B. values for surface-active agents directly from the chemical formulae, using empirically determined *group numbers*[50]. Table 8-IV shows these group numbers, from which the H.L.B. values in the last column of Table 8-III have been calculated, using the relation:

$$\text{H.L.B.} = \Sigma(\text{hydrophilic group numbers}) - m(\text{group number per } -CH_2- \text{ group}) + 7 \quad (8.34)$$

It must be emphasized that the H.L.B. system is not concerned with the stability of the emulsion once formed: it is a correlation of function (Table 8-II), not of efficacy.

In logarithmic form eq. (8.32) is:

$$\mathrm{RT}\ln\frac{(\mathbf{C}_1\ \mathrm{Rate}\ 2)}{(\mathbf{C}_2\ \mathrm{Rate}\ 1)}=0.24\psi_0^2+\theta\Sigma E_{\mathrm{h}}-600\mathrm{m}\theta \tag{8.35}$$

TABLE 8-II

Classification of Emulsifiers according to H.L.B. values

Range of H.L.B. values	Application
3.5–6	W/O emulsifier
7–9	Wetting agent
8–18	O/W emulsifier
13–15	detergent
15–18	solubilizing agent

TABLE 8-III

Surface Active Agent	H.L.B. from expt.[65,66,67]	H.L.B. from group numbers[50]
Na Lauryl Sulphate	40	(40)
K Oleate	20	(20)
Na Oleate	18	(18)
Tween 80 (Sorbitan "mono-oleate" $+20(CH_2$–CH_2–O) groups)	15	16.5
Alkyl aryl sulphonate	11.7	—
Tween 81 (Sorbitan "mono-oleate" $+6(CH_2$–CH_2–O) groups)	10	11.9
Sorbitan monolaurate	8.6	8.5
Methanol	—	8.3
Ethanol	7.9	7.9
n-Propanol	—	7.4
n-Butanol	7.0	7.0
Sorbitan monpalmitate	6.7	6.6
Sorbitan monostearate	5.9	5.7
Span 80 (Sorbitan "Mono-oleate")	4.3?	5.7
Propyleneglycol monolaurate	4.5	4.6
Glycerol monostearate	3.8	3.7
Propylene glycol monostearate	3.4?	1.8
Sorbitan tristearate	2.1	2.1
Cetyl alcohol	1	1.3
Oleic acid	1	(1)
Sorbitan tetrastearate	~0.5	0.3

This may be compared with eq. (8.34) into which the figure of 0.475 from Table 8-IV has been substituted:

$$(\text{H.L.B.}-7) = \Sigma(\text{hydrophilic group numbers}) - \text{m}\times 0.475 \quad (8.36)$$

By rewriting eqs. (8.35) and (8.36) we obtain:

$$\frac{RT}{600\,\theta}\ln\frac{(\mathbf{C}_1 \text{ Rate } 2)}{(\mathbf{C}_2 \text{ Rate } 1)} = \frac{0.24\psi_0^2}{600\,\theta} + \frac{\Sigma E_\text{h}}{600} - \text{m} \quad (8.37)$$

$$\frac{\text{H.L.B.}-7}{0.475} = \frac{\Sigma(\text{hydrophilic group numbers})}{0.475} - \text{m} \quad (8.38)$$

From these equations follows the first set of three relations showing that the empirical H.L.B. system in fact rests on a kinetic basis:

$$\ln\frac{(\mathbf{C}_1 \text{ Rate } 2)}{(\mathbf{C}_2 \text{ Rate } 1)} = 2.2\theta(\text{H.L.B.}-7) \quad (8.39)$$

$$\text{hydrophilic group number for single hydrated group} = \frac{E_\text{h}}{1260} \quad (8.40)$$

$$\text{hydrophilic group number for charged group} = \frac{1.9\times 10^{-4}\psi_0^2}{\theta} \quad (8.41)$$

TABLE 8-IV

H.L.B. group numbers

Hydrophilic groups	group number
$—SO_4' \, Na^+$	38.7
$—COO' \, K^+$	21.1
$—COO' \, Na^+$	19.1
Sulphonate	about 11
Ester (sorbitan ring)	6.8
Ester (free)	2.4
—COOH	2.1
Hydroxyl (free)	1.9
—O—	1.3
Hydroxyl (sorbitan ring)	0.5
Lipophilic groups	
—CH— $—CH_2—$ $—CH_3$ =CH—	0.475
Derived group	
$-(CH_2-CH_2-O)-$	hydrophilic group number = 0.33

According to eq. (8.39), if $\mathbf{C}_1=\mathbf{C}_2$ (i.e. if the water and oil have equal viscosities and phase volumes) neither a W/O nor an O/W emulsion is favoured if either H.L.B. = 7, or θ, the surface coverage by the adsorbed monolayer, is very small for an emulsifying agent of any H.L.B. For a typically effective O/W emulsifying agent, H.L.B. = 11, $\theta \approx 1$, and hence Rate 1 = 10^{-4} Rate 2. Comparison of eq. (8.39) with experiment for n-alcohols is made in Table 8-V, columns 3 and 4.

TABLE 8-V

Emulsion Stabilizer	H.L.B.	$\frac{\text{Rate 2}}{\text{Rate 1}}$ (calc. from eq. (39) with $\mathbf{C}_1 = \mathbf{C}_2$, $\theta = 1$)	$\frac{\text{Rate 2}}{\text{Rate 1}}$ (observed)	$\left(\frac{c_w}{c_o}\right)^{0.75}$ (data from ref. 68)	$7 + 0.36 \ln \frac{c_w}{c_o}$
Methanol	8.4	23	30	22	8.5
Ethanol	7.9	8	11	8	8.0
n-propanol	7.4	2.7	4	2.8	7.4
n-butanol	7.0	1	1	1	7.0

Emulsions were made by hand-shaking test-tubes containing petrol-ether and water and stabilizer. Rate measurements were made by timing with a stop-watch the breaking of the unstable emulsions on each side of the interface[50].

Equation (8.40) shows that for hydrated groups the H.L.B. group number (Table 8-IV) should be proportional to the energy barrier to coalescence set up by the water which is firmly bound to hydroxyl or ester groups on the surface-active agent. Thus, for a single hydroxyl group, of group number 1.9, E_h is calculated from eq. (8.40) to be $1260 \times 1.9 = 2400$ cal. The relation of H.L.B. to hydration also explains the observed correlations[66,67,69] of H.L.B. with the solubilizing power and with the cloud point (Fig. 8-19). The cloud point is the temperature to which a solution of surface-active agent may be heated before it becomes turbid: this temperature is related to the dehydration energy of the material.

According to eq. (8.41) the H.L.B. group number of a charged group (e.g. sulphate or sulphonate) depends on ψ_0 and θ: the potential ψ_0 is itself determined by the total electrolyte concentration and θ. Thus the contribution of a charged group to the H.L.B. value of an anionic or cationic surface-active agent is not strictly constant. For sodium lauryl sulphate the hydrophilic group contribution is calculated from eq. (8.41) to be 35 if $\psi_0 = 230$ mV and $\theta = 0.29$. Hence the total H.L.B. value for sodium lauryl sulphate is 36 according to eq. (8.36): the experimental value[66] lies between 35 and 40.

The distribution of (say) an alcohol between oil and water is determined by the free energy of transfer of the molecule from water to oil, $\Delta G_{w\to o}$:

$$\Delta G_{w\to o} = -RT\ln\left(\frac{c_o}{c_w}\right) \tag{8.42}$$

where c denotes emulsifier concentration, and subscripts o and w refer to oil and water. The work of transfer, $\Delta G_{w\to o}$, is made up of terms from the hydrophilic and lipophilic parts of the molecule, the former energy being

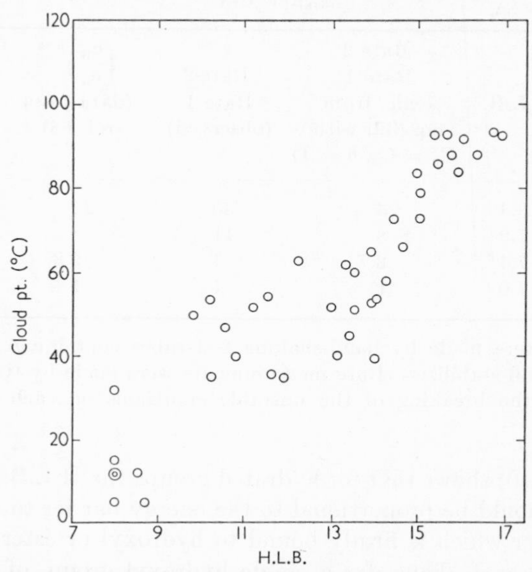

Fig. 8-19. Correlation of cloud points of non-ionic agents with experimental H.L.B. numbers.

$+3200$ cal. mole^{-1} for an $-OH$ group, and the latter about -800 cal. mole^{-1} for a $-CH_2-$ group in a hydrocarbon oil. Thus, for an alcohol,

$$\Delta G_{w\to o} = +3200 - 800m$$

where, as before, m is the number of $-CH_2-$ groups in the molecule.

In general,

$$\Delta G_{w\to o} = \Sigma\Delta G_{w\to o}(\text{hydrophilic groups}) - 800\text{ m} \tag{8.43}$$

Combination of eqs. (8.42) and (8.43) gives:

$$\frac{RT}{800}\ln\left(\frac{c_w}{c_o}\right) = \frac{\Sigma\Delta G_{w\to o}(\text{hydrophilic groups})}{800} - m \tag{8.44}$$

This may be compared with the result of the kinetic treatment (eq. (8.37)) to give:

$$\frac{\mathbf{C}_1 \text{ Rate } 2}{\mathbf{C}_2 \text{ Rate } 1} = \left(\frac{c_w}{c_o}\right)^{0.75\,\theta} \tag{8.45}$$

$$E_h = 0.75\,\Delta G_{w\to o}(\text{uncharged hydrophilic group}) \tag{8.46}$$

$$\frac{0.32\psi_0^2}{\theta} = \Delta G_{w\to o}(\text{charged hydrophilic group}) \tag{8.47}$$

The Bancroft rule is implied in eq. (8.45): additives preferentially soluble in water stabilize O/W systems (cf. relation (8.33) above), and vice versa. This rule therefore rests on a kinetic basis, and will be obeyed if eqs. (8.46) and/or (8.47) are true. Eq. (8.45) is tested for n-alcohols in Table 8-V (columns 4 and 5).

Equation (8.46) predicts that E_h, the work of removing the water from (say) a hydroxyl group, must be of the same magnitude as the total work done in taking the group from water to oil: this is indeed true for non-polar oils such as decane, in which the interaction between oil and a hydroxyl group is small. Thus for non-ionic emulsifiers ($\psi_0 = 0$) the Bancroft rule should always be obeyed if $\mathbf{C}_1 = \mathbf{C}_2$ and the oil is non-polar.

The Bancroft rule (as implied in eq. (8.45)) will also be valid for ionic emulsifying agents such as sodium lauryl sulphate if eq. (8.47) is correct. This, however, holds only as regards orders of magnitude: when ψ_0 is low (small concentrations of emulsifier, high salt concentration) eq. (8.47) will no longer be accurate—and so neither will the Bancroft rule.

If we compare eq. (8.44) with the H.L.B. equation (8.38) above, we obtain the third set of fundamental relations:

$$(\text{H.L.B.} - 7) = 0.36 \ln\left(\frac{c_w}{c_o}\right) \tag{8.48}$$

$$\begin{matrix}\text{hydrophilic group number} \\ \text{for single hydrated group}\end{matrix} = \frac{\Delta G_{w\to o}(\text{uncharged hydrophilic group})}{1680} \tag{8.49}$$

$$\begin{matrix}\text{hydrophilic group number} \\ \text{for charged group}\end{matrix} = \frac{\Delta G_{w\to o}(\text{charged hydrophilic group})}{1680} \tag{8.50}$$

If equations (8.49) and (8.50) are correct, then eq. (8.48) will hold: i.e. the H.L.B. and distribution characteristics of emulsifiers can always be related. The Bancroft rule and H.L.B. are thus equivalent, according to (8.48), so that if the emulsifier is preferentially soluble in water, H.L.B.$>$7, and vice-versa. We can see at once that eq. (8.49) is valid for the simple alcohols, since $\Delta G_{w\to o}/1680$ is $3200/1680 = 1.9$ for an OH group, which is precisely its group number (Table 8-IV).

Equation (8.48) is tested in Table 8-V, columns 2 and 6: it is important that the emulsifying agent should be dissolved in the oil under conditions such that the oil remains non-polar, i.e. no dissolved or solubilized water must be present. Similarly in the aqueous phase, there should be no micelles nor dissolved or solubilized oil. This was carefully observed in the data reported in Fig. 8-20.

Equation (8.50) will be only an approximation, since it follows from eqs. (8.41) and (8.47), the latter being correct only as regards order of magnitude.

In general, the method of dispersal will affect the emulsion type, and the theory above for simple shaken systems has to be modified accordingly[17,42,70]. In practical **emulsifying machines** oil and water are forced between two shearing plates. A laboratory apparatus of this type is shown in Fig. 8-21, the tendency of the mixture to form an emulsion of a certain type being measured by the phase volume φ_i of oil at which the emulsion inverts from

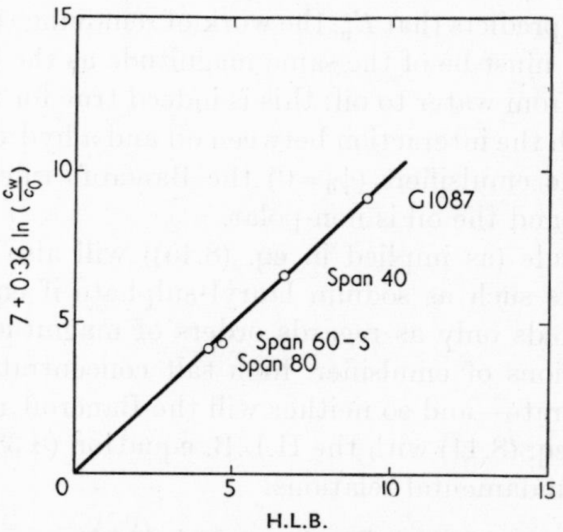

Fig. 8-20. Check of eq. (8.48) for non-ionic stabilizing agents. The analytical figures (unpublished preliminary results of Dr. B. O. G. Schüler, Cambridge) are only approximate, because of the difficulty of obtaining pure polymers.

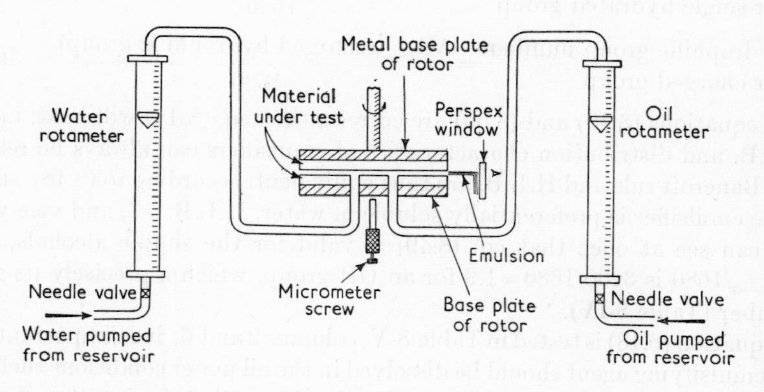

Fig. 8-21. Principle of the laboratory continuous-flow emulsifying machine for investigating the effects of hydrodynamic factors on emulsion type[17].

oil-continuous to water-continuous or vice versa[17]. When φ is near unity the emulsion will always be oil-continuous, irrespective of the other factors, and when φ is small the emulsion will always be water-continuous. Between these two extremes lies the inversion point φ_i: Fig. 8-22 shows a general locus

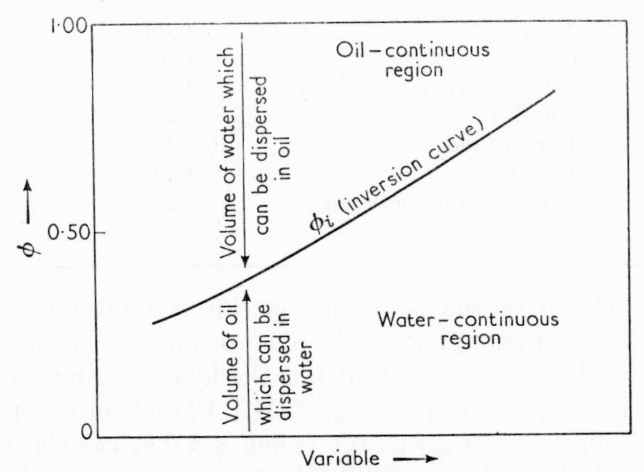

Fig. 8-22. Generalized locus of φ_i with any variable, with the machine of the previous figure.

of φ_i when another quantity (e.g. viscosity of the oil, H.L.B. of the additive, or wettability of the plates) is varied. These plates (Fig. 8-21) of the material under test exert in practice a very marked influence upon emulsion type, oil-wetted plates (Perspex (Lucite) or hard rubber) strongly favouring oil-continuous systems.

The apparatus of Fig. 8-21 runs continuously, water and oil (containing the stabilizing additive) being fed between the central shearing plates through flow-gauges, the readings of which at inversion give a direct measure of φ_i. The emulsion type is easily detected optically by allowing the stream of emulsion to impinge on a Perspex (Lucite) window (Fig. 8-21). The appearance of this shows immediately the emulsion type: if the emulsion forms a continuous white film across the window, it is oil-continuous, whereas if it forms white drops on the window it is water-continuous. This inversion point may be approached from either side in practice, by varying the oil or water flow rates by the needle valves.

Results are shown in Figs. 8-23, 8-24, and 8-25: the difference between glass plates and plastic plates is clearly seen from Fig. 8-24. The rise in the φ_i curve at low H.L.B. values (in Fig. 8-24) is due to the release of free oil, incompletely emulsified. Use of rougher, oil-wetted plates, by enhancing the oil wettability (see p. 37), reduces φ_i still further. Table 8-VI shows typical data for test plates of different materials.

TABLE 8-VI

Material of plates	φ_i (Tween 80, H.L.B. = 15)
Nylon	0.56
Glass	0.47
Sulphonated polystyrene	0.35
Stainless steel	0.33
Hard rubber	0.32
Polystyrene	0.29
Poly-tetrafluoroethylene	0.25
Perspex (Lucite)	0.21

The effect of H.L.B. on the emulsion type (Fig. 8-23) is not in accord with eqs. (8.32), (8.33), and (8.39) for simple mixtures: in the machine the rate of coalescence is no longer so important but the hydrodynamics of formation are dominant; the slope of the plot of φ_i vs. H.L.B. is of the order of only 0.01 (Fig. 8-23), compared with about 0.7 (when $\varphi_i \approx 0.5$) calculated from eqs. (8.28), (8.29), and (8.39). Consequently the two types of emulsion cannot be formed equally in this apparatus; indeed, the marked dependence of φ_i on the materials of the plates confirms the importance of unequal hydrodynamic effects to which the surface-active agents contribute by altering possibly both the relative energies and the relative rates of wetting of the solid plates by the oil and by the water[17]. That the wettability of the solid surfaces influences emulsion type has been repeatedly suggested[42, 43, 71-73], though only with the machine described above have quantitative results been obtained.

Emulsification in general must therefore be considered in terms of rates of

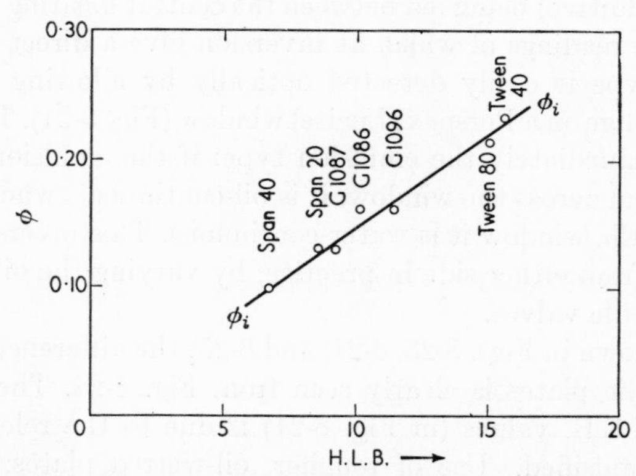

Fig. 8-23. Results of using machine of Fig. 8-21 to find the inversion curves (φ_i vs. H.L.B.) for various surface-active agents 0.5% (wt.wt.) in petrol-ether, emulsified against distilled water at 1425 r.p.m. using smooth Perspex plates[17].

formation of the two types as well as their rates of breaking once formed. In continuous-flow machines the former influence is dominant, and the mechanism may consist in forming "fingers" of one liquid between the plates, these fingers then breaking up spontaneously[74] or because of the high shear gradient. Such "fingers" are readily formed between stationary surfaces wet by liquid (1) when a liquid (2) of lower viscosity flows into it. The "fingers" of liquid (2) thus formed in the liquid (1) have been studied both theoretically

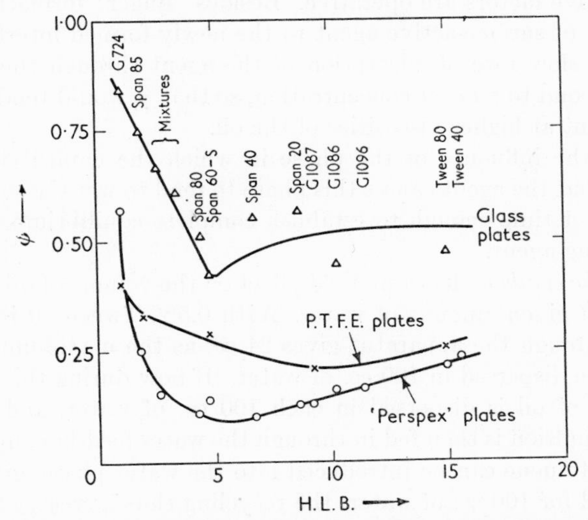

Fig. 8-24. Circles show data of Fig. 8-23, replotted for comparison, and extended to low H.L.B. values. Replacement of the smooth Perspex by poly-tetrafluoroethylene (x) increases the tendency to form water-continuous emulsions, as does the use of glass-plates (Δ). The emulsifying agent is 0.5% in petrol-ether throughout, with a rotor speed of 1425 r.p.m.

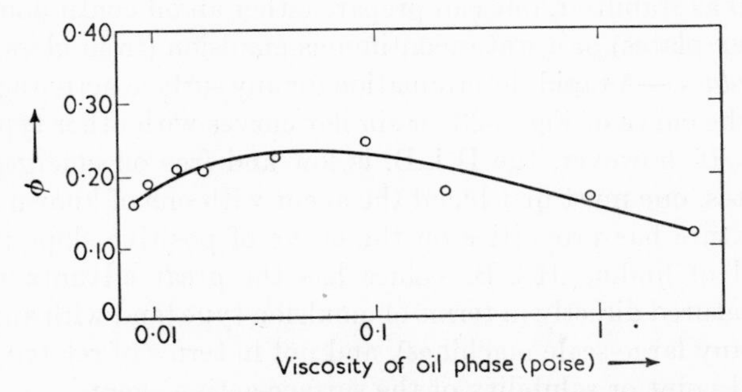

Fig. 8-25. Inversion curve for 0.5% G1096 dissolved in the oil phase, emulsified with distilled water using smooth Perspex plates at 1425 r.p.m. The oil consists of mixtures of petrol-ether and Liquid Paraffin.

and practically[74,75]. This viscosity dependence would cause water to become increasingly dispersed in oils as the viscosity of the latter is increased, and explains the fall in φ_i at high viscosities of the oil (1) (Fig. 8-25). If, however, the water phase (2) preferentially wets the solid walls, it will tend to surround liquid (1) forming "islands" of the latter, which would then be dispersed into small drops in the shear gradient. The effect of viscosity on φ_i (Fig. 8-25) is surprisingly low on any theory, however, and the maximum suggests that two competitive factors are operative. Besides "finger" formation, the rate of adsorption of surface-active agent to the newly-formed interface may be important: a slow rate of adsorption of the agent through the viscous oil would correspond to a lower concentration, so that φ_i would tend to increase on this account at higher viscosities of the oil.

Similarly, the influence of the phase in which the emulsifying agent is dissolved is that the agent causes this phase to tend to wet the solid surfaces, there being not time enough to establish complete equilibrium partition of the emulsifying agent.

Recycling the emulsion has a marked effect on the volume of oil that can be dispersed in a given amount of water. With 0.5% Tween 80 in toluene, a single pass through the apparatus gives 24 cc. as the maximum amount of oil that can be dispersed in 100 cc. of water. If now during this initial pass only 17.3 cc. of oil is dispersed in each 100 cc. of water, and the water-continuous emulsion is then fed in through the water feed-line, up to 26.7 cc. of additional toluene can be introduced into the water phase, giving a total of 44 cc. of oil for 100 cc. of water: the recycling thus increases the possible uptake of oil into the aqueous continuum by 84%[17].

From the results in Table 8-VI and Fig. 8-24 it is clear that by controlling the wettability factors one can prepare with the continuous-flow machine, given any particular oil and stabilizing agent, *a wide variety of emulsions.* For example, Fig. 8-24 shows that from 30 cc. petrol-ether and 70 cc. water, with Span 20 as stabilizer, one can prepare either an oil-continuous emulsion (from Perspex plates) or a water-continuous emulsion (from glass plates).

H.L.B. Values.—A rapid determination for any surface-active agent can be found from the curve of Fig. 8-23, or similar curves with other types of plate (e.g. nylon). If, however, the H.L.B. is low and free oil emerges from the shearing plates, one must first blend the agent with one of known H.L.B., so that the mixture has properties on the curve of positive slope (Fig. 8-23). This method of finding H.L.B. values has the great advantage that the H.L.B. is measured directly in terms of emulsion type (and with an apparatus similar to many large-scale machines), and not in terms of related properties such as cloud point or solubility of the surface-active agent.

The accuracy with which the phase ratio at inversion can be read depends on the *efficiency of emulsification;* any disperse phase emerging incompletely broken up into microscopic droplets will obscure the sharp change in wetting behaviour at the inversion point, as will also rapid coalescence of the emulsion on the window. The most reproducible results are obtained when the emulsion is both readily formed and stable; consequently the optimum H.L.B. for an agent to emulsify any given oil can be found from studying the reproducibility of the phase ratios for each of several stabilizing agents[17].

Clumping of Emulsion Droplets

In an emulsion of oil in water the drops of oil frequently clump together or flocculate without immediate coalescence. The rate of coalescence in such a clump may be 10^{-3}sec.$^{-1}$ in an unstable emulsion, down to 10^{-7}sec.$^{-1}$ in a very stable one. Clumping is discussed further on pp. 346 and 429.

As a model system for a clump, one may press together two drops of oil formed at the ends of fine tubes, and separated by the aqueous phase[76]. In this way it is found that, with 0.01 M-sodium lauryl sulphate as the stabilizing agent, the oil drops may be pushed together without immediate coalescence, the rate of drainage of the intervening water controlling the separation until the oil surfaces are between 100 and 200 Å apart. At this stage electrical repulsion becomes important and further approach of the drops, even over local protuberances on the surfaces, occurs only when some shock, thermal, mechanical, or chemical, disturbs the system. It is noteworthy that only electrical forces are required to explain the results, which obey closely our empirical simplified relation (page 408 below) that the separation is $(6/\varkappa)+60$ Å. Indeed, invoking long-range attractive forces produces disagreement with the observed findings[76]. The model system suffers from the disadvantages that the curvature is much lower than in the much smaller drops of an emulsion, so that drainage times assume undue importance, and coalescence occurs between unspecified protuberances on the surfaces of these large drops.

A recent study of clumping of water drops in oils may be found in ref. 76a.

Breaking of Emulsions

An O/W emulsion will *break* or *invert* whenever Rate 1 becomes large. Whether breaking or inversion occurs will depend on the stability of the inverse system: if this is unstable and Rate 2 is large, the emulsion will break. If, however, Rate 2 is small, the emulsion will invert, as is illustrated in Fig. 8-26. Practical limits to the terms large and small are $10^{-2}\mathbf{C}_1$ or $10^{-2}\mathbf{C}_2$ (coalescence occurring within an hour), and $10^{-5}\mathbf{C}_1$ or $10^{-5}\mathbf{C}_2$ (emulsion stable for several months).

Consider as an example the *chemical breaking* of an emulsion of paraffin oil in water stabilized by a long-chain sulphate or by laurate. Initially we might have $\psi_0=-175$ mV and $\theta=0.4$, so that Rate $1=10^{-5}\mathbf{C}_1$ and Rate

$2 = 10^{-2}\mathbf{C}_2$. After the addition of 0.3 M-NaCl, however, ψ_0 is reduced to about -100 mV, and θ is increased slightly to 0.45 because of the reduced repulsion; Rate 1 is now increased to the large value of $2 \times 10^{-2}\mathbf{C}_1$, whereas Rate 2 is still quite large (because m is only 11 or 12), being about $10^{-2}\mathbf{C}_2$. The conditions for *emulsion breaking* should now be satisfied (sequence a, b, c, d, of Fig. 8-26): this is supported by experiment[51]. We may also note that further additions of sodium chloride should not invert the emulsion: with a hydrocarbon chain of 11 $-CH_2-$ groups, Rate 2 will never become sufficiently small. This is also in accord with experiment[78].

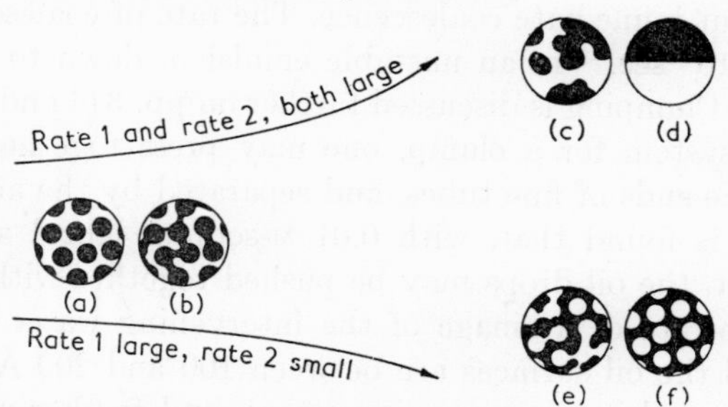

Fig. 8-26. Diagrammatic representation of breaking and inversion of emulsions. The dark liquid is oil, and the white liquid is water. The breaking of this O/W emulsion occurs because Rate 1 is large, and continues to phase separation if Rate 2 is also large, as in the sequence (a,b,c,d). If, however, Rate 2 is small, the sequence leads to inversion (a,b,e,f). This method of representation has been adapted from Sutheim[77] and from Schulman and Cockbain[57], and can be equally applied to W/O emulsions.

The chemical breaking of W/O emulsions with sodium oleate or Teepol[79,80] is also predicted by the theory. A W/O emulsion will break when Rate 2 becomes great, while Rate 1 also remains large. If the surface is well covered ($\theta \sim 1$) with a monolayer of (say) calcium oleate, α-mono-stearin[81], or asphaltic material[79,80], detergent or soap must be added in sufficient quantity to displace this: because of electrical repulsion, θ is lower in this new interfacial film and so Rate 2 will now be greater, perhaps $10^{-2}\mathbf{C}_2$ instead of (say) $10^{-7}\mathbf{C}_2$ as in the original emulsion. At the same time Rate 1 will remain large, changing from about $\mathbf{C}_1$ to about $10^{-2}\mathbf{C}_1$ (assuming there is 0.4 N salt in the aqueous phase); breaking should therefore occur, as is found.

Inversion of an O/W emulsion will occur when Rate 1 becomes large while Rate 2 is small. Consider, for example, an emulsion of oil in water stabilized with a longer-chain compound than before, such as sodium stearate. Initially we might have $\theta = 0.4$ and $\psi_0 = -175$ mV, so that again Rate $1 = 10^{-5}\mathbf{C}_1$. Rate 2 is now $10^{-3}\mathbf{C}_2$. After the addition of 0.5 M-NaCl, however, $\theta = 0.6$ and $\psi_0 = -90$ mV, so that Rate 1 is increased to $0.5 \times 10^{-2}\mathbf{C}_1$; but, because m is now as high as 17, Rate 2 has the low value of $3 \times 10^{-5}\mathbf{C}_2$. Since in this system Rate 2 is so small, inversion will occur[78] (sequence a, b, e, f, of Fig. 8-26) in accord with the general principles above. We may observe that $\frac{\text{Rate 2}}{\text{Rate 1}} \ll 1$ in the system containing added salt.

The inversion will be more complete, of course, if calcium ions or other ions forming undissociated soaps[15] are present. If all the soap in the system is calcium soap, the electrostatic potential ψ_0 in a stearate-stabilized emulsion will be practically zero, so that Rate $1 \sim \mathbf{C}_1$ although, since $\theta \sim 1$ for the condensed interfacial film of calcium soap, Rate $2 = 10^{-7}\mathbf{C}_2$. Here we have the possibility of a stable W/O emulsion, as is widely known from practical experience.

Emulsion breaking or spontaneous inversion can thus be effected by adding sufficient of any surface-active agent to displace to adsorbed film and stabilize, weakly or strongly respectively, the opposite type of emulsion.

Physical Methods of breaking.—Centrifuging, filtration through media whose pores are preferentially wetted by the continuous phase, gentle shaking or stirring, and low intensity ultrasonic vibrations are all effective in breaking emulsions[15,82].

Heating.—This is also effective, since the large energy barriers to coalescence in stable emulsions imply that the retarding term $e^{-W/RT}$ will increase markedly as T increases. In practice, heating to about 70°C will rapidly break most emulsions.

Electrical methods.—These are the most widely used on a large scale, and have proved of great value in breaking oil-field emulsions: about 25% of all crude petroleum reaches the surface containing appreciable amounts of emulsified water. This must be removed before the oil is distilled, or corrosion of the stills will occur. As with the removal of aerosols, a high voltage (about 20 kV.) ensures that the coalescence of the entrained water droplets is rapid: the primary coalescence occurs in a few milliseconds. The mechanism involves the deformation of the drops of water into long streamers in the electrical field, and these form chains by bridging between the separate drops. The emulsion is thus completely broken[82].

Electrical methods can also be used to break O/W emulsions: the process then is essentially an electrophoretic migration of the charged drops to one of the electrodes. This process is rather slow and inefficient, though it has been used for removing traces of lubricating oil emulsified in condenser water.

"Creaming" of Emulsions

If the droplets of dispersed liquid in an emulsion are larger than about 1μ in radius, they will settle preferentially to the top (or bottom) of a vessel under the influence of gravity. If the vessel has a height l cm., it is convenient to use the condition that no appreciable "creaming" will occur provided that:

$$\frac{4}{3}\pi a^3.\Delta\rho_l.g.l \ll kT \quad (8.51)$$

If this is satisfied, thermal movements will off-set the "creaming" under the influence of gravity, a being the radius of the drops, $\Delta\rho_l$ the density difference of the two liquids, and g the gravitational acceleration.

Since the phase volume φ is higher in a "cream" of oil droplets overlying clear water than if the oil droplets were uniformly dispersed, C_1 and hence Rate 1 are increased by such creaming, i.e. the emulsion breaks more readily if it creams first. Creaming may be prevented, if required, by homogenization, i.e. by passing the emulsion through a fine-clearance valve under high pressure, causing the larger drops to be broken up. As may be seen from the above relation, making $\Delta\rho_l$ smaller will also reduce the creaming: this is achieved by producing a poly-phase emulsion consisting of (say) small water drops in larger oil drops, which are in turn dispersed in the continuous aqueous phase. Such a poly-phase emulsion can be made by first adding so little water that, though the emulsifying agent favours an O/W emulsion, a W/O emulsion is first produced. On further addition of water this inverts to an O/W emulsion, though the oil drops have smaller water drops in them, thus reducing $\Delta\rho_l$.

Sometimes creaming is desirable, e.g. in separating rubber particles from the latex emulsion. Sodium alginate[83] or polyvinyl alcohol may be used, the long chains of these polymers adhering to the droplets sufficiently to stick several together (and hence increase the effective radius), while not interacting strongly with the stabilizing agent. The emulsion thus creams but does not necessarily coalesce. Stickiness is discussed further on p. 429.

SOLIDS IN LIQUIDS

Suspensions of solids in liquids may be produced by dispersal or condensation methods.

Certain metals may be dispersed by causing an electric arc to pass between them while immersed in the continuous phase, but the usual dispersive methods are crushing and grinding. As might be expected, one often finds that the work done in grinding up a solid is roughly proportional to the new surface area created. Further, since the liquid-solid interfacial energy is less than the gas-solid energy, wet grinding requires less work than dry grinding, and the work is reduced still further when surface-active agents are present. During any grinding process, however, most of the energy is dissipated as heat, and only a little (e.g. 10%) is available to increase the interfacial area[84].

The interfacial energy between a solid and a liquid may be obtained by comparing the heats of solution of a large crystal and of an equal mass of powder: the difference represents the interfacial enthalpy of the powder, and can be conveniently expressed per unit area of powder surface[85].

Solid dispersions are frequently obtained by condensation (i.e. precipitation), as by pouring 12 cm.^{3} of a 32% ferric chloride solution into 750 cm.^{3} of boiling distilled water: very rapid hydrolysis occurs to give colloidal ferric hydroxide, as a deep red solution. This and similar preparations are listed by Jirgensons and Straumanis[86] and by McBain[20]. The number of particles will depend on the rate of *homogeneous nucleation*, which in turn depends on the supersaturation, as expressed by eqs. (8.14) and (8.15), modified[5] in that γ

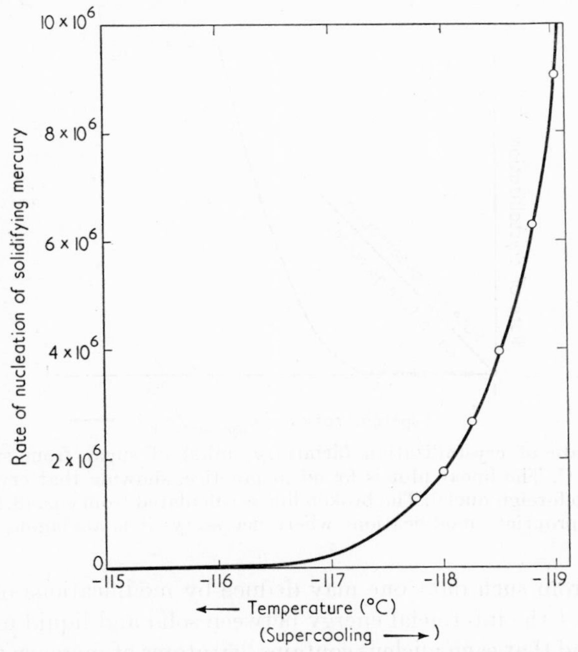

Fig. 8-27. Rate of spontaneous nucleation (in nuclei sec.$^{-1}$cm.$^{-3}$) in supercooled mercury[87]. Melting point of mercury is -38.87°C.

is now the solid-liquid interfacial energy, v is now the volume of a molecule in the solid, and that the relevant difference of chemical potentials is $\mu_{\text{liq.}} - \mu_{\text{solid}}$. This quantitative treatment can be applied to the freezing of pure water (spontaneous nucleation occurs around -35°C), and to the solidification of metals; the μ terms may best be evaluated by the entropy method mentioned on p. 354. In liquid metals $\boldsymbol{K}$ of eq. (8.15) is very high, of the order 10^{40}sec.$^{-1}$. For supersaturated solutions the appropriate chemical potential terms is $(\mu_{\text{supersat.}} - \mu_{\text{sat.}})$ in eq. (8.14), where $\mu_{\text{supersat.}}$ refers to the chemical potential of the solute in the supersaturated solution, and $\mu_{\text{sat.}}$ to the chemical potential of a saturated solution at the same temperature. They are related by:

$$\mu_{\text{supersat.}} - \mu_{\text{sat.}} = \boldsymbol{k}\text{T} \ln(\text{c}_{\text{supersat.}}/\text{c}_{\text{sat.}}) \tag{8.52}$$

In all experimental studies, the most difficult task is to remove traces of foreign nuclei, on which condensation may otherwise occur. For mercury, this may be most easily achieved by dividing the liquid mass into very small droplets. If this subdivision is carried far enough, only a few of these will contain a nucleus of foreign origin, whose influence on the solidification process is thus limited and observable. Thus one finds that metals dispersed as small emulsion droplets may be supercooled as much as 80°C, as shown in

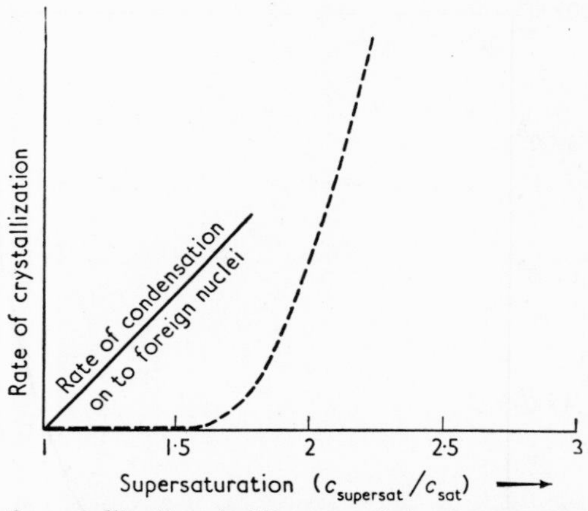

Fig. 8-28. Rate of crystallization (arbitrary units) of sugar from supersaturated solution at 25°C. The linear plot is found in practice, showing that crystallization is occurring on to foreign nuclei. The broken line is calculated from eqs. (8.14), (8.15) and (8.52) with appropriate modifications where necessary: it is analagous to eq. (8.21).

Fig. 8-27. From such data one may deduce by modifications of eqs. (8.14) and (8.15) that the interfacial energy between solid and liquid mercury is 32 ergs $cm.^{-2}$, and that each nucleus contains 260 atoms of mercury (cf. eq. 8.13). In bulk, however, metals are difficult to supercool by more than a few degrees.

To assist the freezing of water in clouds, and so promote precipitation, one may use a smoke of AgI, whose crystals have dimensions very similar to those of ice crystals. Only about 2.5°C supercooling is required for the ice to begin forming on the AgI crystals. Volcanic ash dust may function naturally in the same way. Severe supercooling of a cloud by solid CO_2 also achieves precipitation by causing such strong supercooling that $(\mu_{liq.} - \mu_{solid})$ becomes very large (cf. eqs. (8.14) and (8.15)). In the sugar-house, spontaneous nucleation is rare[5], both because foreign nuclei are common in the liquor and because $\boldsymbol{K}$ of eq. (8.15) is anomalously low in this system, of the order 10^4 instead of the expected 10^{30}. This suggests that the sucrose molecules must strike the nucleus in a very special orientation and on a particular site if they are to remain on the solid. A typical experimental plot is shown in Fig. 8-28.

The size of the particles finally found in the solid depends on the ratio of the rate of formation of nuclei and the rate of growth of the micro-crystals. The particles are of colloidal dimensions (10^{-7} to 10^{-4} cm.) if they are precipitated from reagents at either very high or very low concentrations[88]: very high concentrations of reagents increase the viscosity and so reduce the rate of growth of the nuclei, and very low concentrations also reduce the rate of growth. Intermediate concentrations lead to coarse precipitates, the first nuclei formed having grown rapidly.

Heterogeneous nucleation is often important when solid surfaces are present: if a large volume of a supersaturated solution of sodium thiosulphate is seeded with a single crystal of the solid, rapid solidification occurs through the whole volume. In general, if the transition of a liquid to a solid embryo occurs against *any* plane solid surface, eq. (8.14) must be modified to:

$$\Delta G_n = \left\{\frac{16\pi\gamma_0^3 v^2}{3(\mu_{\text{supersat.}} - \mu_{\text{sat.}})^2}\right\} \left\{\frac{(2-\cos\theta)(1+\cos\theta)^2}{4}\right\}$$

where θ is the contact angle (measured in the liquid) at the junction of the liquid with the plane solid surface having an embryo on it. If θ is zero, the liquid wets the solid preferentially (relative to the embryo), and the correction factor in the second large bracket is unity. If $\theta = 180°$, the embryo "wets" the solid preferentially relative to the liquid, and the correction factor becomes zero, implying that nucleation should occur very readily on the solid. If now instead of a plane solid surface one considers a very narrow V-shaped solid pore, with a strong attraction for embryos (i.e. θ is high) the embryos at the bottom of the pore may be so stabilized that they are retained there above the normal melting-point of the liquid[88a].

The formation of vapour "snakes" during the crystallization of certain liquids is both spectacular and interesting[88b]. If, for example, a sealed glass tube, half-filled with thoroughly degassed liquid cyclohexane, is immersed in a mixture of solid CO_2 and acetone for a short time, a crust of solid cyclohexane forms on the surface. A few seconds later a bubble of vapour penetrates the crust and, elongating into the liquid, induces further crystallization. The bubble now elongates continuously at a rate of several centimeters per second, to form a snake-like tube of vapour, the advancing front of which moves through the whole volume of the liquid, though avoiding close contact with the walls of the vessel or with other coils of the tube. The tube of vapour is coated with solid, and this crust slowly thickens. Other liquids besides cyclohexane exhibit "snakes", though less clearly. They include neopentane, tert-butanol, benzene, cyclohexanol and carbon tetrachloride. The "snake" does not form if even a small amount of air is present: crystallization then occurs throughout the mass of the liquid. Possibly for these substances the embryos of the solid phases are strongly held at the vapour-liquid interface, so that θ is fairly high, and the free energy of nucleation is therefore reduced at the vapour interface: the contraction in the liquid which the solidification causes at the "mouth" of the snake causes the snake to extend further into the liquid[118]. If, however, the liquid contains dissolved gas this is released on freezing, and nucleation then occurs on all the gas bubbles through the liquid mass, and no snake forms.

Electrical Barriers

When two similarly charged particles approach, there is always a repulsive force between them. If the charge on the particles is due to a weak specific adsorption (e.g. of I′ ions on AgI particles), the potential ψ_0 on each surface will remain constant, as the electrical double layers overlap during the approach, by desorption of some I′ ions from the surface into the solution. On the other hand, if the surface charges are immobile and strongly bound (e.g. ionogenic groups held by co-valent forces, or long-chain ions), the surface potentials will rise during the approach of the particles under these conditions of constant surface charge.

The theory of the overlapping double layers—each assumed to be of the Gouy type—has now been worked out[89,90]. In particular, if $\varkappa a < 5$, i.e. the double-layer thickness $1/\varkappa$ is larger than one-fifth the particle radius, the repulsive energy between two charged particles, approaching at constant potential ψ_0 on each surface, is given approximately by

$$\mathbf{W}_R \approx 0.7 D a \psi_0^2 \left(\frac{e^{-\varkappa a(x/a - 2)}}{x/a} \right) \tag{8.53}$$

A similar expression but with a numerical factor of about 0.8 (instead of 0.7) may be used if the surface charge remains constant. Here D is the dielectric constant, x is the distance between centres of the particles (assumed spherical): $x = 2a$ at contact (Fig. 8-2). Insertion of this contact condition into eq. (8.53) gives:

$$\mathbf{W}_R = 0.35 D a \psi_0^2 \tag{8.54}$$

If, however, $\varkappa a > 5$, i.e. the double-layer thickness is less than one-fifth of the particle radius,

$$\mathbf{W}_R \approx 0.5 D a \psi_0^2 \ln(1 + e^{-\varkappa a(x/a - 2)}) \tag{8.55}$$

and, when $x = 2a$ (at contact), this reduces to (8.54). For more exact treatment one should refer to the original papers on the subject[89,90,91].

As an example of the application of eq. (8.54), consider that $\psi_0 = 14$ mV, i.e. that $\varepsilon\psi_0/\boldsymbol{k}T = 0.55$. Then $\mathbf{W}_R$ is $0.35 \times 80a(\boldsymbol{k}T/\varepsilon)^2 \times (0.55)^2 = 0.35 \times 80a \times (4 \times 10^{-14}/4.77 \times 10^{-10})^2 \times 0.55^2$, or $\mathbf{W}_R = 6 \times 10^{-8}a$. Hence, if $a = 10^{-5}$cm.,

$\mathbf{W}_R = 6 \times 10^{-13}$ ergs $= 15\boldsymbol{k}T$. To a first approximation this would be the barrier to the coagulation or coalescence of two particles of this radius and charge, if the attractive van der Waals forces did not operate till the particles were just touching. It corresponds to increasing $t_{1/2}$ of the system (eq. 8.7) by a factor of 10^6, leading to high stability. Derjaguin[89] gives the stability criterion as $Da\psi_0^2 > 20\boldsymbol{k}T$, which indicates an interaction of about $7\boldsymbol{k}T$ for reasonable stability. More exhaustive studies of the theory, with numerical solutions of the appropriate differential equations[91], confirm Derjaguin's result. In general, it is clear that small particles require higher surface potentials for stability than do larger ones.

Long-range Attraction

If, instead of the attractive energy well being vertical at $x = 0$ (curve 1 of Fig. 8-2), one now allows for the fall-off with distance (curve 2) the total interaction will follow curve 4. The attractive energy $\mathbf{W}_A$ between two plates whose surfaces are at a distance 2x apart, and which are of thickness great compared with 2x, may be found by integrating over all the possible atomic pairs undergoing van der Waals attraction[90,92], to give:

$$\mathbf{W}_A = -\frac{\boldsymbol{A}}{48\pi x^2} \tag{8.56}$$

Two spheres, fairly close together ($x < 2.4a$) should have an attraction given by

$$\mathbf{W}_A = -\frac{\boldsymbol{A}}{12\left(\frac{x}{a} - 2\right)} = -\frac{\boldsymbol{A}a}{12h} \tag{8.57}$$

where h is the distance between the surfaces (at the nearest points). Earlier estimates of the constant $\boldsymbol{A}$ were about 10^{-12} erg, but later measurements[93] favour a lower value, around 5×10^{-14} erg. These values are found from direct measurements of the attraction between quartz or glass plates in air: the exact modification required if the attractive forces have to be transmitted through (say) water is not known, though it is suggested[90] that $\boldsymbol{A}$ should be divided by the square of the refractive index of the separating medium. The attractions between the plates and the water, and between the water molecules, may further reduce the net value of $\boldsymbol{A}$ for aqueous systems.

At larger distances the attractive forces fall off more steeply[93], and are small compared with $\boldsymbol{k}T$ at separations greater than 100 Å.

Certain phenomena, previously ascribed to long-range attractive forces, can now be explained otherwise. The spontaneous two-phase formation in solutions of tobacco mosaic virus, for example, is a direct result of electrical repulsion and not of attraction; the effective co-volume of the virus rods is so increased by the electrical double layer around each, that the rods can interfere with each other's rotational movements. In such a system the entropy may increase by some of the rods separating out as a more dense phase, leaving the remaining particles free to rotate. Results at different salt concentrations support this view, as does the volume of the virus sediment after ultra-centrifugation, and there is no evidence of long-range attractive forces[94]. In clays also, long-range attraction is unimportant; cross-linking of stacks of parallel layers by a few non-parallel plates best explains the results[94a].

If two oil drops are pressed together[76] in the presence of an ionic stabilizing agent, the thickness of the water layer between them is found to be greater than that calculated with allowance for long-range forces; these may, therefore, be neglected at these distances (100 Å – 250 Å). No long-range forces are necessary in explaining the thickness of liquid lamellae in foams (p. 408).

Criteria of Stability

It follows from eq. (8.54) that, if a barrier of 15 $\boldsymbol{k}$T is sufficient to stabilize an aqueous suspension of particles of radius 10^{-5}cm., then for stability

$$\psi_0^2 > \frac{45\,\boldsymbol{k}\mathrm{T}}{Da} \tag{8.58}$$

Hence a potential of 14 mV will be sufficient for stability, provided that the attractive forces operate only on contact. If, alternatively, one allows for long-range attraction with the value of $\boldsymbol{A}$ at an extreme upper limit of 10^{-12} erg, the barrier, calculated[90] from the curve of the total interaction (e.g. curve 4 of Fig. 8-2), gives 28.5 mV as the required potential ψ_0 for stability, assuming that $\varkappa = 10^6$cm.$^{-1}$ A lower value of $\boldsymbol{A}$ will reduce the necessary potential.

To examine the range and magnitude of the attractive forces in colloidal systems, one needs quantitative data relating stability and ψ_0: one must also know whether the surface charge is depleted (by desorption or lateral flow) immediately preceding contact. The value of $\varkappa$ does not affect $\mathbf{W}_\mathrm{R}$ at $\mathrm{x}=0$ according to eq. (8.54) but, if long-range attraction is important, the value of $\varkappa$ affects the position and height of the maximum of curve 4 (Fig. 8-2) even if ψ_0 is constant. Allowing for appreciable long-range attraction, Derjaguin[89] derived the stability criterion for weakly charged surfaces

$$\psi_0^2 > \frac{\boldsymbol{A}\varkappa}{2D} \tag{8.59}$$

According to this treatment, the stability conditions are independent of particle size: for $\varkappa = 10^6$cm.$^{-1}$ and $\boldsymbol{A} = 10^{-12}$erg, $\psi_0 = 24$ mV, in good agreement with the 28.5 mV calculated above with the same value of $\boldsymbol{A}$. But if $\boldsymbol{A}$ is taken as the accepted figure of 5×10^{-14}erg, then the requisite ψ_0 would be only about 6 mV. This value is lower than the ζ potentials reported in many systems[95], although because of the possible complications of specific adsorption, particularly with polyvalent ions (Table 2-7), only the simplest systems can be compared with theory. Though Eilers and Korff found that $\zeta^2/\varkappa$ is a measure of emulsion stability, quantitative agreement with eq. (8.59) is unsatisfactory if $\boldsymbol{A}=5\times10^{-14}$erg. We believe that further experiments, on systems such as oil drops with adsorbed films of known charge density and ψ_0, and over a wider range of values of a and $\varkappa$, are required to distinguish the different theories. Meanwhile we prefer the treatments not involving long-range attractive forces.

Electrical Effects in Non-Aqueous Systems

In Chapter 3 the high ζ potentials often found in non-aqueous systems have been discussed. If the particles are coarse ($a=10^{-4}$cm.) eq. (8.54) shows that potentials of 30 mV would give a barrier of about 15 $\boldsymbol{k}$T, enough for strong stabilization, though surface potentials of nearly 100 mV would be required if $a=10^{-5}$cm., and smaller particles still could scarcely be stabilized by this mechanism. The conclusion is therefore that only coarse non-aqueous suspensions can be stabilized electrically and experiments on suspended powders (taking ζ as equal to ψ_0) confirm these results quantitatively[96].

Solvation Barriers

Among the stabilizing agents for carbon-black in water are the poly-ethyleneoxide condensates. Here the stabilizing barrier is primarily one of hydration, as discussed on p. 369.

Solvation effects may be equally important in stabilizing carbon-black or titania in hydrocarbon media: thus non-polar alkyl-aromatic compounds are effective in stabilizing carbon-black in n-heptane, being adsorbed by the aromatic groups on to the polar surface (containing always a little bound oxygen) of the carbon particles. The alkyl chains remain mobile in the hydro-carbon liquid, and prevent coalescence[97], both by adhering to the solvent and by mechanically interfering with the approach of the surfaces. Close approach of the surfaces restricts the freedom of the hydrocarbon chains resulting in an entropy barrier. Barriers as high as 100 $\boldsymbol{k}$T may originate in this way.

Weak Aggregation

Some dispersions aggregate so slowly that macromolecules are added to increase the rate. This phenomenon, important in obtaining easily filterable suspensions, is discussed in detail on p. 429.

GASES IN LIQUIDS

Gases may be dispersed in liquids by mechanical dispersal[98], as by forcing them through a bank of capillary tubes (e.g. steel hypodermic needles), by mechanical beaters, or by pumping the gas through a sieve-plate. The last method produces bubbles of a required size, if pulsations of a given frequency are applied to the gas reservoir. Dispersions of gases can also be produced by nucleation. The latter mechanism is well demonstrated by the rather slow evolution of CO_2 from solution in water, when the pressure is suddenly reduced. Similarly, in a pure liquid such n-pentane, superheating at 1 atm. pressure from the equilibrium boiling point (36°C) to about 120°C is possible if the liquid is highly purified. Conversely, one may raise the pressure over n-pentane and heat the liquid to any desired temperature above the normal boiling point, and then suddenly reduce the pressure back to 1 atm. The time interval that elapses before boiling occurs is related[5] to the probability of spontaneous nucleation at the temperature concerned: typical results[99] are shown in Fig. 8-29. When nucleation has set in, the superheated liquid "bumps" with the sudden violent growth of bubbles.

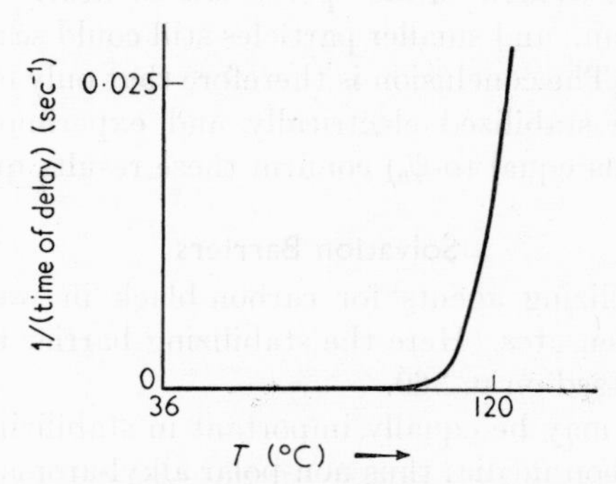

Fig. 8-29. Reciprocal of time elapsed before boiling begins, for liquid n-pentane over which the pressure has suddenly been reduced to 1 atmosphere.

Because radiation, by forming ions which cause repulsion on the surfaces of the embryos, reduces their surface tension and hence the energy of activation required for nucleation, it is utilized in the "bubble chamber" for detecting charged fundamental particles. A vessel containing pure, superheated hydrocarbon constitutes the chamber, the tracks of bubbles in the wake of a charged particle are rapidly photographed stereoscopically.

It has been suggested[100] that natural cosmic radiation is responsible for the finding that pure liquids can be superheated at 1 atm. by only about 90°C, whereas the nucleation theory predicts temperatures very much greater than this. Cosmic radiation has also been cited as the reason why the chemical engineer does not have, in his larger vessels, the troubles with bumping that are so common in laboratory-scale apparatus.

The entrainment of small drops of liquid when a bubble bursts at a liquid surface has been photographed[100].

FOAMS

A foam is a coarse dispersion of a gas in a liquid, most of the phase volume being gas, with the liquid in thin sheets called lamellae between the gas bubbles[101].

The formation of a foam is achieved by agitating the liquid and gas together in the presence of a stabilizing agent; shaking or blowing air through porous carbon or sintered glass discs into the liquid is usually effective. Special rotary pumps[102] which mix intimately the solution of foam-stabilizer and the air, are also used for large-scale production: the solution and the air enter the pump separately, and emerge as a foam. A rapid adsorption of the stabilizing agent is necessary (since the phase volume of gas is high) if immediate breaking of the foam is to be avoided. Adsorption will be fast enough from most solutions of pure surface-active agents (Chapter 4), though traces of lauryl alcohol are necessary to increase the stability of a foam stabilized by detergent and the adsorption of the lauryl alcohol from the very low concentration in the bulk may be rather slow.

Foams find many **industrial uses**: they have become increasingly important in rubber preparations (foamed latex) and in fire-fighting[102], in which they have the double advantage that the density of the aqueous medium is reduced by aeration and that the surface area is greatly increased[103]. This enables the foam to float as a continuous layer on the surface of a burning organic liquid, preventing the evolution of inflammable vapours and presenting a large area for the absorption of radiant heat which assists in cooling the surface and surrounding edges and consequently reduces vapour pressure. Since 1877, when foams were first suggested for fire-fighting, there has been increasing interest in their use, and now, following extensive development during the 1939-45 war to fight oil fires in combat operations, about £100,000 worth of concentrates for producing fire-fighting foams are now produced annually in the U.S.A.

A recent study[104] of mass transfer in foams suggests that there may be future developments in commercial foam columns for gas absorption. In absorption one of the most important rate-controlling factors is the area of contact between the gas and the liquid, and the large interfacial area in a foam, as high as 50 cm.^{2} in 1 cc. of foam, is about fifty times greater than in columns packed with Raschig rings. This advantage may be lost, however, unless the stirring within the foam is specially effective, partly because, as explained below, the effective viscosity of a foam may be rather high. The efficiency of the foam column for mass transfer of gas is shown in comparison with that of other types of column in Fig. 8-30. The height of each transfer unit is required to be as small as possible, so that at low gas flow rates the transfer is more efficient in the foam column than in the packed column.

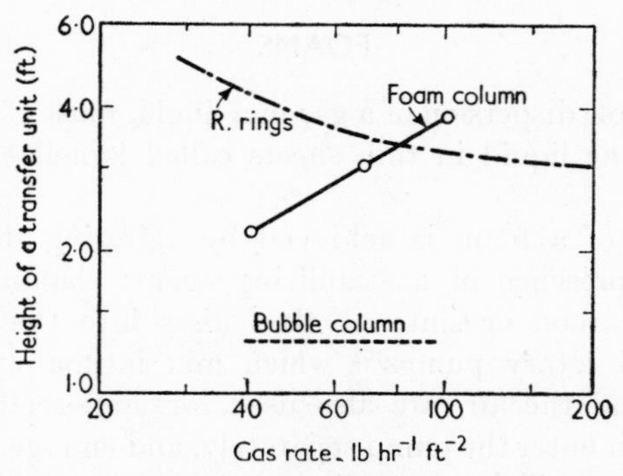

Fig. 8-30. Comparison of mass transfer data for different types of absorption apparatus[104]. Liquid rate = 17,300 lb. hr.$^{-1}$ sq. ft.$^{-1}$

O = foam column

- - - - = data of Shulman and Molstad for gas bubble column; desorption of CO_2 from water (by extrapolation)

—·—·- = data of Cooper, Christl, and Peery for packed column containing 3-inch Raschig rings; absorption of CO_2 in water (by interpolation).

However, the mass transfer is less efficient than in the gas-bubble column, probably because of the high viscosity of the foam.

Separation and purification by foaming is becoming increasingly common. If a stream of gas is passed through a mixed solution, the more surface-active materials are preferentially removed in the foam[105]. It is possible that proteins could be concentrated from biological fluids by this mechanism. Rigorous purification of sodium lauryl sulphate and lauryl sulphonic acid has been made possible by foam fractionation[106]. In these materials slight hydrolysis alters the surface properties drastically, on account of the liberation of lauryl alcohol, which is adsorbed about 1000 times more strongly than the parent compound. Purification to the required level of better than 99.99% can therefore be suitably carried out by foaming, the first fraction of the foam, which contains the lauryl alcohol, being discarded. In the same way foaming of sodium oleate carries off preferentially the oleic acid formed by hydrolysis[107].

Other uses of foam separations are in the flotation of minerals (p. 421) and in the separation of organic and inorganic ions by surface-active agents. Anionic collectors for the latter separations include α-sulpho-lauric acid; while, for separating dyes, cationics such as lauryl pyridinium chloride and quaternary derivatives are used. With successive small additions of the cationic derivative to a mixture of dyes in water, a good separation of the dyes may be achieved in the foam[107a].

Bikerman[108] lists several other applications of separation by foaming: the method may be applied to bile acids, beetroot juice, sugar juice, enzymes, hormones, and even bacteria.

Physical Measurement on Foams

Foams are generally very difficult to pump through pipes, effective viscosities[109] being as much as 100 times that of the liquid from which they are formed. This is due to at least three factors: the necessity for deforming the bubbles (which constitute most of the phase volume) so that they can pass each other, the relative importance of surface viscosity, and the fact that liquid films in large numbers may have to be detached from the walls of the tube. The *viscosity of a foam* depends on the phase volume of gas in the system: in practice, two further quantities are conveniently defined, the *expansion* of the foam and the *density* of the foam. If a volume V_2 of foam is formed from a volume V_1 of liquid, the ratio V_2/V_1 is called the expansion of the foam: fire-fighting foams commonly have expansions between eight and sixteen. The density of the foam is given by $\rho_l V_1/V_2$, where ρ_l is the density of the liquid from which the foam is formed, usually water. Potato juice foam has a very low density, but foams from solutions of anionic detergents may have densities as high as 0.5. White-of-egg foam has a density about 0.2.

Bikerman[108] has defined *a unit of "foaminess"*, Σ, for different foaming solutions. This is based on the finding that the volume V_2 of foam formed at equilibrium is proportional to the volume flow-rate u of air, the constant of proportionality being Σ. Thus we have:

$$V_2 = \Sigma \mathrm{u} \tag{8.60}$$

In practice, air or other gas is blown through a porous septum into the liquid, and the equilibrium height attained by the column of foam gives V_2. A typical value of Σ is 400 sec. for foam stabilized by protein. It is found that Σ is almost independent of the porosity of the septum, probably because the bubble diameter varies but little with the pore size when the pores are near each other.

The *Ross-Miles test*[110] (or "pour test") for "foaminess" consists in running a standard volume (e.g. 200 cc.) of solution through a standard funnel into a cylinder containing another definite volume (e.g. 50 cc.) of the same solution. The volume of froth formed is a measure of the foaming tendency of the solution.

Beating methods[111] are also useful in studying "foaminess", though the amount of foam produced is arbitrarily determined by the characteristics of the beater. The decay of the foam so produced can, however, be accurately studied: if V_2 is the volume of foam present on removal of the beater from the solution, and V the volume of this foam remaining after any time t, the characteristic life-time of the foam t_f is given[112] by $t_f = (1/V_2)\int V \mathrm{d}t$.

Factors determining Foam Stability

A foam can never be thermodynamically stable, since, once the sheet of liquid is ruptured, it must break into drops with a lower total surface area and therefore with a decreased free energy of the system. Nevertheless, although foams from pure liquids and gases are highly unstable (life of less than 1 second), suitable surface-active agents can stabilize a foam almost indefinitely. In considering the action of these stabilizing agents, we must take account of several factors, of which the most important in the early stages is the drainage of liquid from the foam, with consequent weakening of the structure. Also significant is the more rapid growth of the larger bubbles at the expense of the smaller. Some foams, even when most of the surplus liquid has drained away, may still be fairly stable, the air bubbles now being separated by liquid sheets of thickness of the order 200-2000 Å. Stability now depends on the resistance of these thin liquid films to further thinning by chance shocks: if the lamella is thinned to about 50 Å, molecular forces cause rupture. Further, the foam stability is reduced by growth of the larger bubbles at the expense of the smaller, occurring by diffusion of gas through the liquid films, because the excess pressure of a small bubble is higher than that of a large bubble (Chapter 1). We shall now consider in turn each of the factors responsible for foam stability, given a certain mean bubble size.

(*i*) *Drainage Rate*

The drainage rate is the rate at which liquid drains from a foam whose lamellae are not breaking during the process. Drainage is the settling of surplus water to the bottom of the foam column, under the action of gravity or surface tension. Foams of common detergents as well as those of proteins show considerable drainage of liquid, which reduces the stability of the foam considerably by thinning the liquid sheets to such an extent that they are easily broken.

Drainage under the influence of surface tension is illustrated in Fig. 8-31: the pressure within the liquid at point X is lower than elsewhere, because of

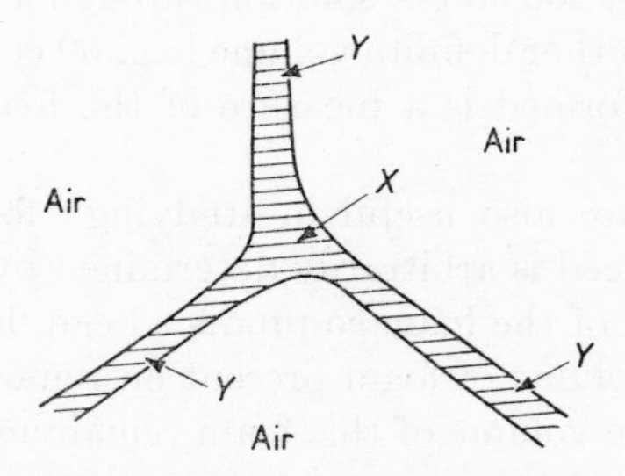

Fig. 8-31. Formation of Plateau border at line of intersection of three gas bubbles. The negative curvature at X makes the pressure there lower than at Y, and liquid flows towards X.

its negative curvature (cf. eq. 1.9), so that liquid is sucked from the lamellae Y into the region at X, which is known as a Plateau border.

In general, the equations

$$V = V_0(1 - e^{-Ct}) \tag{8.61}$$

$$\frac{dV}{dt} = V_0 C e^{-Ct} \tag{8.62}$$

describe the observed drainage rates of diverse foams draining under the influence of gravity, though they are inapplicable to the slow-draining foams made from protein hydrolysates. In these equations V is the volume of liquid (in cc.) drained from the column of foam, the original liquid content being V_0. The time is represented by t, and C is the constant of the drainage process. Eq. (8.61) has the advantage of involving only one empirical constant. As an example of the use of this equation, data for the rapid drainage of detergent-stabilized foam are shown in Fig. 8-32 (as crosses),

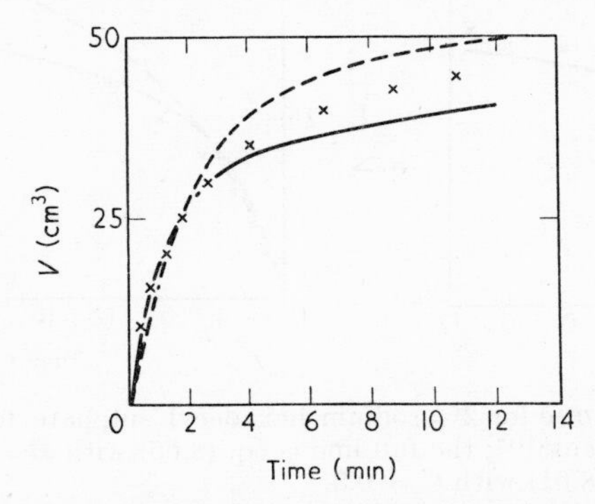

Fig. 8-32. Drainage curve for 2% sodium lauryl sulphate foam of expansion 5: crosses represent the experimental data[113], and the full line is eq. (8.65) with $B = 2.0$, $V_0 = 50$ cm.^{3}, while the broken line is eq. (8.61) with $C = 0.4$.

with eq. (8.61) represented by the broken line, fitted to the early points. The agreement is fair, but the exponential equation clearly makes V approach V_0 too quickly. For slow-draining films (as of sodium hexadecyl sulphate, expansion 9) the agreement with eq. (8.61) is less good (Fig. 8-33), and this is true also of a foam made from a solution of protein hydrolysate containing iron salts (Fig. 8-34). Data for a foam from a commercial non-ionic detergent are shown in Fig. 8-35. Equation (8.61) cannot be fitted to the drainage curves for 0.1% lauryl sulphonic acid foams over both the early and later stages of drainage[114], though, if the drainage for the first one or two minutes is deliberately excluded, eq. (8.61) fits the rest of the drainage curve. Agreement with the drainage of a 0.1% di-alkyl-sulphosuccinate foam is even less satisfactory.

A mathematical study[114] of foam drainage, based on the simple model that the liquid films between the bubbles behave as thin vertical columns of liquid flowing between fixed plates (i.e. the bubble walls are regarded as having infinite viscosity), leads to the drainage equations of Ross:

$$V = (2\mathbf{a}/\mathbf{b})(1-(\mathbf{b}t+1)^{-1/2}) \tag{8.63}$$

$$\frac{dV}{dt} = \mathbf{a}(\mathbf{b}t+1)^{-3/2} \tag{8.64}$$

where V is the volume of liquid drained from the foam after time t, **a** is a constant given by the initial drainage rate (t=0), and the constant **b** represents the time dependence of the drainage rate. Some typical results[113] are shown in Table 8-VII. As expected for an equation containing two

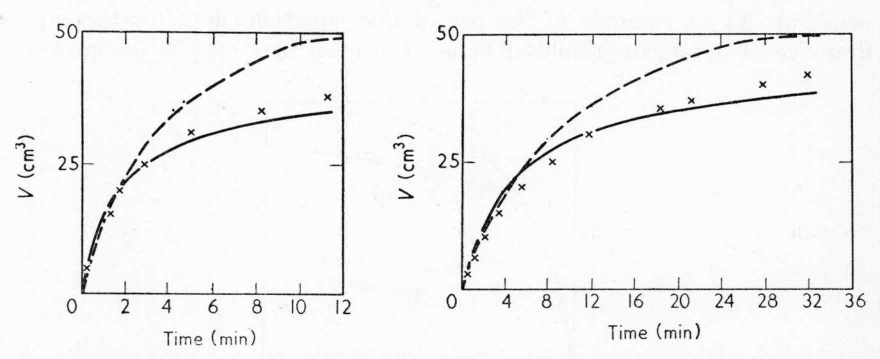

Fig. 8-33. Drainage curve for 2% sodium hexadecyl sulphate foam of expansion 9: the points are experimental[113]; the full line is eq. (8.65) with $\boldsymbol{B}$ = 1.0, V_0 = 50 cm.^{3}; the broken line is eq. (8.61) with C = 0.3.

Fig. 8-34. Drainage curve for 2% Mearlfoam foam of expansion 9.6. Points are experimental[113]: the full line is eq. (8.65) with $\boldsymbol{B}$ = 0.5, V_0 = 50 cm.^{3}, and the broken line is eq, (8.61) with C = 0.1.

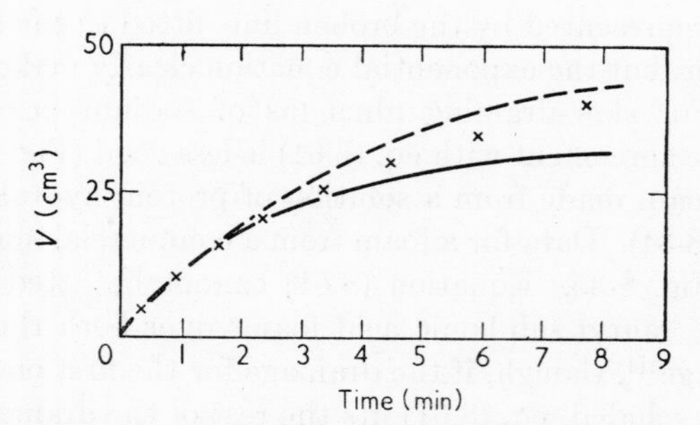

Fig. 8-35. Drainage curve for 2% Triton X-100 foam of expansion 11: crosses show experimental data[113]; the full line is eq. (8.65) with B = 0.8, V_0 = 50 cm.^{3}, while the broken line is eq. (8.61) with C = 0.25.

empirically determined constants, eq. (8.63) correlates well the gravitational drainage of a foam (Fig. 8-36); further, the values of **a** and **b** are reasonable, as the initial drainage rate (**a**) would be expected to decrease with the expansion of the foam; it should also be low for the viscous foams involving protein hydrolysates. The fractional decrease in drainage rate with increasing time, on which the constant **b** depends (cf. eq. (8.64) with t increasing from zero), is also lowest for the films of high viscosity, i.e. very little drainage will occur from the foam even after moderate times. Since in such foams the

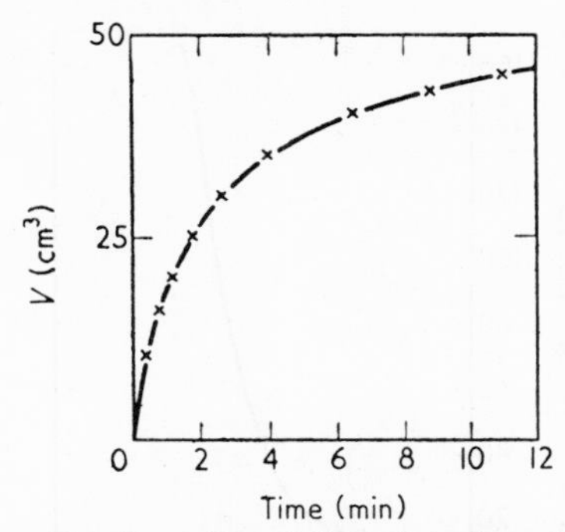

Fig. 8-36. Drainage curve for foam from 2% sodium lauryl sulphate, of expansion 5: crosses, experimental data[113]. The full line is equation (8.63) with **a** = 32, **b** = 1.0.

initial drainage rate (**a**) is also low, the ratio **a**/**b** is approximately constant (Table 8-VII). This suggests that an equation derived from (8.63) but involving only a single constant could be fairly satisfactory. For example, this relation may be modified[114] to:

$$V = V_0(1 - (\boldsymbol{B}\mathrm{t}+1)^{-\frac{1}{2}}) \tag{8.65}$$

or

$$\frac{\mathrm{d}V}{\mathrm{dt}} = \frac{\boldsymbol{B}V_0}{2}(\boldsymbol{B}\mathrm{t}+1)^{-\frac{3}{2}} \tag{8.66}$$

where the amount of liquid drained from the foam should approach V_0 (the total amount of liquid in the foam) as t becomes large: this means that the amount of liquid held in the fully drained lamellae in the foam must be negligibly small. This equation involves only a single arbitrary constant, and fits diverse types of foam reasonably well (Figs. 8-32 to 8-36). The constant $\boldsymbol{B}$ is clearly a measure of the initial drainage rate: values of $\boldsymbol{B}$ found by comparison with experiment decrease markedly with increasing expansion and viscosity of the foam, as expected. Agreement of eq. (8.65) with the drainage from a slower drained foam, for example that from protein hydrolysate (Fig. 8-34), is better than that with the faster drained foams. This is to be expected in view of the assumption of infinite surface viscosity which was required in the derivation.

The variation of drainage time with expansion is, in general, quite marked, as shown in Fig. 8-37. The values will vary with the concentration of stabilizing agent as well as with the ratio of cross-sectional area to depth in the foam column: the results in Figs. 8-34 and 8-37 are thus not directly comparable. Up to an expansion of about 10, Fig. 8-37 shows that the time for 25% drainage is approximately proportional to the expansion of the foam. This result, substituted into eq. (8.65) for V/V_0=constant, implies that $\boldsymbol{B}$ should vary inversely with the expansion of the foam, provided the latter does not

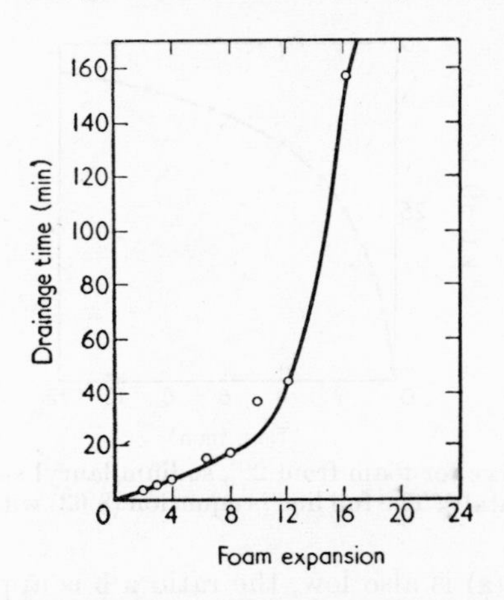

Fig. 8-37. Variation of time for 25% drainage from 6% protein hydrolysate foams of different expansions. Points from data in ref. 102.

exceed 10. That this is approximately true is seen by comparison of the $\boldsymbol{B}$ values of 2 and 1 with the expansions of 5 and 9 for the otherwise very similar foams from sodium tetradecyl sulphate and sodium hexadecyl sulphate (Table 8-VII).

According to the theory used in the derivation of eq. (8.65), the constant $\boldsymbol{B}$ should vary directly with ρ_l/η, the bulk density and viscosity of the liquid. Foams from viscous liquids, for example, should have low values of $\boldsymbol{B}$, i.e. the drainage should be slow: experimental studies[112] of the average life of the liquid in the foam (proportional to $1/\boldsymbol{B}$) as a function of liquid viscosity and density confirm that there is a linear relationship with η/ρ_l for both oil foams and aqueous foams. The effect of the surface viscosity of the stabilizing film is more complex: in deriving eqs. (8.63) and (8.65) it is assumed that (as an approximation) the surface viscosity is always so high that the formula for the flow of liquid between stationary plates is valid, i.e. that the gas-liquid interface is effectively solid. This assumption appears valid for protein films, but not for detergent monolayers. Water layers between the latter will therefore drain more quickly than according to the theory; this is borne out by the fact that the value of $\boldsymbol{B}$ is higher for a foam from sodium hexadecyl sulphate solution than for Mearlfoam of about the same expansion (Table 8-VII).

TABLE 8-VII

Drainage of Foams after relatively short times

Foaming solution	Expansion	C of eq. (8.61) min.$^{-1}$	**a** of eq. (8.63)	**b** of eq. (8.63)	**a/b**	$\boldsymbol{B}$ of eq. (8 65) min.$^{-1}$
Sodium lauryl sulphate	5	0.4	32	1	32	2
Sodium tetradecyl sulphate	5	0.4	30	1	30	2
Sodium hexadecyl sulphate	9	0.3	23	0.8	29	1
Triton X-100 (non-ionic)	11	0.25	13	0.3	43	0.8
Mearlfoam (protein hydrolysate)	10	0.1	7	0.2	35	0.5

The constants **a** and **b** of eq. (8.63) are obtained by fitting this equation to experimental points. All the foams are from 50 cm.^{3} of solutions which are 2% in foaming agent, and times are expressed in minutes[113]. C is the constant of eq. (8.61) which gives the broken curves of Figs. 8-32 to 8-35. The constant $\boldsymbol{B}$ in eq. (8.65) gives the full curves of Figs. 8-32 to 8-35. The constants C and $\boldsymbol{B}$ are obtained by fitting the respective equations to the early experimental points on the drainage curves.

(*ii*) *Diffusion of Air across Liquid Lamellae*

In a column of foam the small bubbles will tend to decrease in size, and the large bubbles will tend to grow, because the gas pressure is higher in the smaller bubbles (eq. (1.9)). The mechanism of transfer of air from the smaller bubbles to the larger is by dissolution in, followed by diffusion through, the liquid lamellae separating the bubbles. A good example is provided by a foam made from Teepol solution: after 15 minutes the total number of bubbles is only 10% of the initial number, although no rupture of the liquid films has occurred[115]. This striking effect is caused by the rapid thinning of the liquid sheets in foams from Teepol and other branched-chain compounds (see below), the rate of diffusion of gas through the liquid sheets varying inversely with the thickness of the sheet. Further, Kitchener and

Cooper[116] point out that the growth of the large bubbles may be followed by mechanical rearrangements of the bubble clusters, with the consequent possibility of mechanical shocks. If a single bubble of air is released beneath the surface, (Fig. 8-38) its diameter decreases with time because of the excess pressure within a bubble[111]. If the radius is a, and if the bubble is half

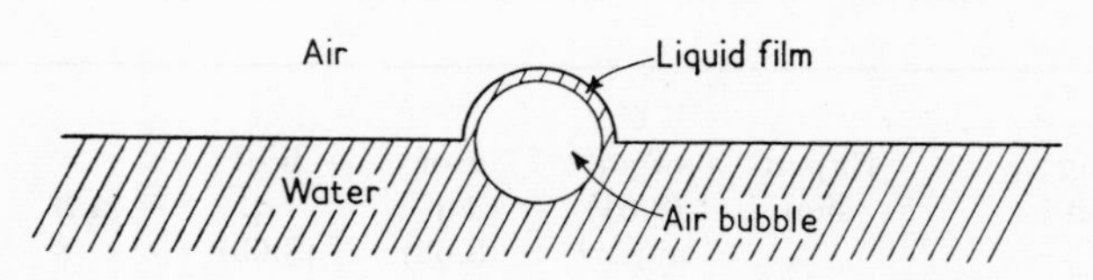

Fig. 8-38. Air bubble injected below an air-water surface. The latter may be covered with a monolayer, or the aqueous phase may contain a surface-active agent.

immersed, the excess pressure P_e in it is $3\gamma/a$ since there are two hemispherical surfaces on the upper half of the bubble and one on the lower. Air therefore passes through the bubble wall into the atmosphere above the bubble. The water must be previously saturated with air, so that diffusion through the water below the bubble is relatively slow, owing to the great length of this diffusion path: the effective area for escape of air from the bubble is thus approximately a hemisphere. The permeability constant of the hemispherical liquid film is k, defined by

$$-\frac{d\mathbf{N}}{dt} = k 2\pi a^2 \Delta c \tag{8.67}$$

where $-d\mathbf{N}/dt$ is the number of moles of gas passing through the bubble wall per second, $2\pi a^2$ is the area (in cm.^{2}) of this film, and Δc is the difference in concentration (in moles cm.$^{-3}$) between the air inside the bubble and in the atmosphere above it. Now $\mathbf{N}$ is related to both the pressure in the bubble and its volume, since the concentration of air will be proportional to its pressure:

$$\frac{\mathbf{N}}{4\pi a^3/3} = \frac{(P+P_e)}{RT}$$

Hence by differentiation $d\mathbf{N}/dt = (4P\pi a^2/RT)(da/dt)$ (since P_e and its differential may be neglected to a first approximation in comparison with P, the atmospheric pressure in dynes cm.$^{-2}$). Similarly,

$$\Delta c = \frac{P_e}{RT} = \frac{3\gamma}{aRT}$$

Substitution for $d\mathbf{N}/dt$ and Δc in eq. (8.67) now leads to

$$P a\, da = -1.5 k \gamma\, dt$$

To express the variation of a with t, this may be integrated to:

$$a^2 = a_0^2 - 3k\gamma t/P \tag{8.68}$$

where a_0 is the radius of the bubble when $t=0$.

A plot of a^2 against time should thus be a straight line of slope $-3k\gamma/P$, from which k may be found. This permeability constant should be independent of the surface tension of the solution and of the size of the bubble being studied. In Fig. 8-39 two typical plots of a^2 against t are shown, and,

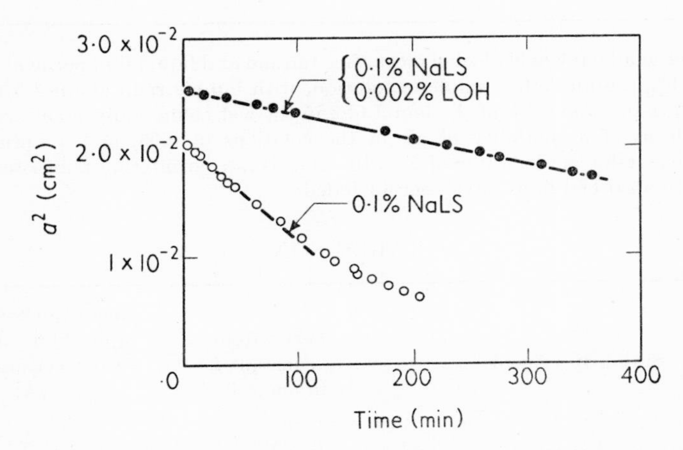

Fig. 8-39. Plot of square of bubble radius against time, in the system of Fig. 8-38, using 0.1% sodium lauryl sulphate (NaLS) solutions. The lower curve was obtained in the absence of lauryl alcohol (LOH), while the upper one shows the effect of 0.002% LOH in reducing the permeability[111].

although the permeability coefficient is constant initially, it decreases when a^2 is reduced to about half its original value. Whether this decrease in k is due to the oxygen of the air escaping preferentially from the bubble in the initial stages of the experiment (it is about twice as soluble as is nitrogen) or whether the diffusion area cannot be simply taken as $2\pi a^2$, has not so far been determined. Separate experiments on bubbles of oxygen and nitrogen would thus be of interest.

Permeabilities of the liquid films are summarized in Tables 8-VIII and 8-IX. The effect of a trace of lauryl alcohol is to decrease the permeability of the lamella: if we take approximate values of the solubility and diffusion coefficient of oxygen, the relative thicknesses are 4600 Å and 12,000 Å. We have assumed that the adsorbed monolayers themselves set up no appreciable barrier to the transfer of oxygen: the resistance of the lamella ($1/k$) is of the order 10^2 sec.cm.$^{-1}$

TABLE 8-VIII

Concentration of lauryl alcohol in g./100 cm.^{3}	Surface tension in dynes cm.$^{-1}$	Permeability constant k in cm. sec.$^{-1}$	Mean thickness h of liquid film calculated from permeability
0.00000	42	1.3×10^{-2}	4,600 Å
0.00025	36	0.5×10^{-2}	12,000 Å
0.002	24	0.5×10^{-2}	12,000 Å
0.008	23	0.6×10^{-2}	10,000 Å

Effect of added lauryl alcohol on surface tension and liquid film permeability of 0.1% sodium lauryl sulphate solution, with bubble radii about 1.5 mm. The mean thickness h of the liquid film in the last column are calculated on the basis of a solubility of air in the solutions of 0.03, and a diffusion coefficient for air in water of 2×10^{-5}cm.^{2}sec.$^{-1}$; no interfacial resistances due to adsorbed films have been included.

TABLE 8-IX

Stabilizing Agent	Permeability constant k in cm. sec.$^{-1}$	Mean thickness h of liquid film calculated from permeability (Å)
Pure sodium lauryl sulphate (anionic)	1.3×10^{-2}	4,600
Triton X-100 (non-ionic)	1.8×10^{-2}	3,300
Quaternary O (cationic)	0.9×10^{-2}	6,600
Potassium laurate (anionic and hydrolysed)	0.2×10^{-2}	30,000
Commercial sodium lauryl sulphate (anionic containing some lauryl alcohol)	0.5×10^{-2}	12,000

Effects of different types of agent in 0.1% solution on permeability of liquid film to air. Thicknesses h of the liquid film are calculated as in Table 8-VIII, and correspond to film drainage after long times.

The results of Table 8-IX suggest that the liquid film is quite thin (3300 Å) for a commercial non-ionic detergent (these often contain 1% or 2% of soap, which may be wholly responsible for the stability of the liquid film), while with potassium laurate (which will be partly hydrolysed to lauric acid) the permeability is very low, corresponding to diffusion of oxygen through a film of water 30,000 Å thick. It is of interest to consider whether or not the film thicknesses thus calculated are pseudo-equilibrium values, determined only by electrical repulsion, or are non-equilibrium values to which the film thins relatively quickly, further drainage being slow. We believe they are due to the latter factor: the higher the viscosity the slower will be the drainage. The effect of small amounts of lauryl alcohol supports this view, as it raises appreciably the rather low surface viscosity of a pure surface-active agent, while the thickness of the layer of solution of potassium laurate is due to the presence of lauric acid formed by hydrolysis penetrating the film. Further, these thicknesses are much greater than h_∞, the "pseudo-equilibrium" values (see below).

Calculation of the rate of thinning of a liquid film may be effected as in drainage theory, neglecting electrical repulsion and taking the surface viscosity to be infinitely high. Thus, for a film 10,000 Å thick, the rate of thinning should be 2000 Å per hour; if, however, the film is only 3000 Å thick, the thinning rate is reduced to only 54 Å per hour. That these figures are so low arises from the absence of Plateau borders and from the low head of the liquid being drained.

These rates for the drainage of liquid from thin films are lower limits, derived from the assumption that the viscosity of the adsorbed monolayers is so great that there is no movement of liquid immediately adjacent to the surface: in practice, the monolayers will certainly be able to move somewhat, and, the lower the surface viscosity, the faster will be the drainage rate of the liquid film. In the upper limit, where there is no film and we have only pure water, the theory will no longer apply, and the bubble will burst as soon as the liquid film can fall away—in perhaps 10^{-2} sec.

(*iii*) *The Thickness of the Electrical Double Layer*

If the film of liquid lies between two charged monolayers, consisting (say) of long-chain sulphate ions, it may be pseudo-stable; i.e. it may resist further thinning, either by shocks or by drainage, after draining for a long time to some thickness h_∞; beyond this, further removal of liquid would bring the charged surfaces closer and would also set up a high osmotic pressure within the liquid layer by accumulating therein many sodium or other counter-ions (Fig. 8-40). Although the free energy of the liquid film will always be greater than if it were broken into drops, the energy barrier to local thinning provided by a charged monolayer (and quite apart from any viscous resistance) is often sufficient to stabilize the film against appreciable shocks: some of Dewar's single lamellae were stable[117] for as long as three years!

Although ψ_0 (the electrical potential of the surfaces due to the charge on the monolayers) will clearly affect the actual pressure required to thin the lamella to any given thickness, we shall assume for the purpose of a simple illustration that $1/\varkappa$, the mean thickness of the ionic double layer, will influence the ultimate thickness when the liquid film is under a relatively low pressure. Let us also assume[118] that each ionic atmosphere extends only to a distance

$3/\varkappa$ into the liquid when the film is under a relatively low excess pressure from the gas in the bubbles; this value corresponds to a repulsive potential of only a few millivolts. Thus, at about 1 atm. pressure,

$$h_\infty = \frac{6}{\varkappa} + 2(\text{monolayer thickness}) \tag{8.69}$$

For charged monolayers adsorbed from 10^{-3} N-sodium oleate, the final total thickness (h_∞) of the aqueous layer should thus be of the order 600 Å $\left(\text{i.e. } \frac{6}{\varkappa} \text{ or } \frac{18}{\sqrt{c}} \text{ Å}\right)$: to this is added 60 Å for the two films of orientated soap molecules (Fig. 8-40), giving a total of 660 Å. The experimental figure[119] is about 700 Å (Fig. 8-41). With NaCl added to a concentration of 10^{-3} N, the estimated thickness is 490 Å (experimental about 420 Å); with 10^{-2} N-NaCl the figures are 240 Å and 280 Å, and with 10^{-1} N-NaCl these become 120 Å and 120 Å. For Aerosol OT (dioctyl sodium sulphosuccinate) of ionic strength approximately 10^{-2} N, the estimated thickness of the liquid film is 240 Å (experimental value = 240 Å) and, with 10^{-2} N-NaCl present, 190 Å, which again agrees with experiment.

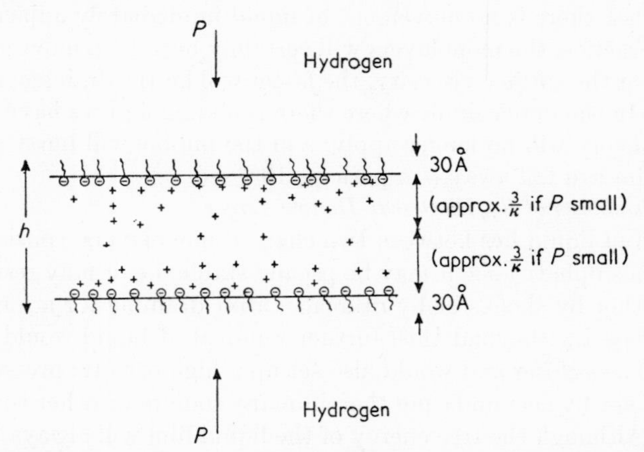

Fig. 8-40. Charged monolayers at the boundaries of a lamella present an electrical barrier to further thinning.

At higher values of the gas pressure P_e (Fig. 8-40), the height decreases below the values given by the above approximation, the h_∞ vs. P_e isotherms being shown in Fig. 8-41: agreement with the results expected from a full analysis[119] of the diffuse electrical double layer (represented by the full lines) is satisfactory. Experimentally one may measure by interference the thickness of the liquid film between two bubbles pressed together, the excess pressure P in them being held at various levels[119].

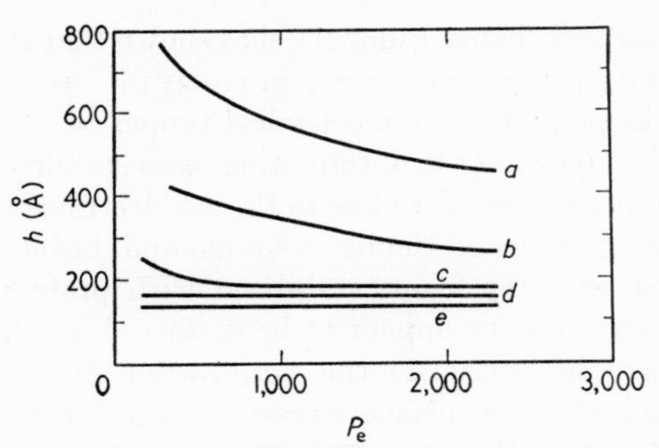

Fig. 8-41. Experimental effects of net pressure (dynes cm.$^{-2}$) in reducing the thickness of the liquid lamella of Fig. 8-40. Curves refer to lamellae found from 10^{-3} N-aqueous sodium oleate, with added: (a) 10^{-4} N-NaCl; (b) 10^{-3} N-NaCl; (c) 10^{-2} N-NaCl; (d) 3×10^{-2} N-NaCl; (e) 10^{-1} N-NaCl.[119]

Bikerman[120], using published results on the electrical conductivity of bubbles, has also reached the conclusion that the theory of the ionic diffuse double layer can explain satisfactorily this pseudo-stability. Evaporation from thin lamellae may also be important[121a].

(*iv*) *Surface viscosity and Yield Values*

In addition to reducing the draining rate, the surface and bulk viscosities cushion the thin liquid films against mechanical, thermal, or chemical shocks. For a given shock, the relative increase in the area where thinning occurs (Fig. 8-42) will be written j, and the above factors reduce this shock area. The highest foam stability is associated with appreciable surface viscosity and yield value, while solutions yielding foams of very poor stability show very low surface viscosity[111]. Teepol and other branched-chain compounds have very low surface cohesion, and their foams show rapid thinning, whereas the materials of higher surface viscosity (potassium laurate and proteins) have longer drainage times. Similar results are found for foam-stabilizing additives, as shown in Figs. 8-43 and 8-44 below.

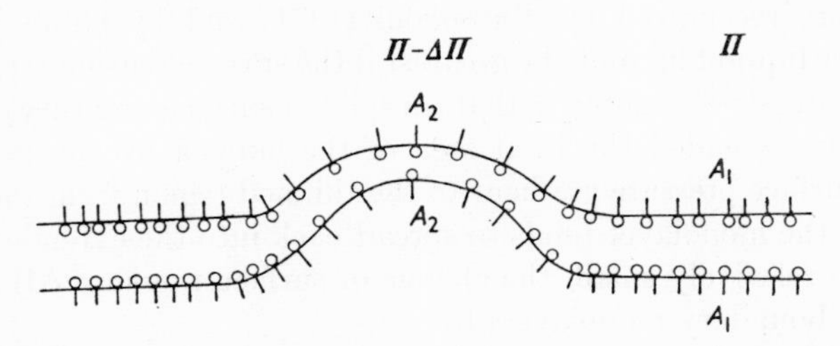

Fig. 8-42. Local thinning of a lamella increases the area available per long chain from A_1 to A_2, and reduces the surface pressure from Π to $\Pi - \Delta\Pi$. The ratio A_2/A_1 is termed j.

For protein-stabilized foams, Cumper[121] has shown that the foam stabilities and surface viscosity pass through a maximum at the same pH for any given protein: this again suggests that mechanical properties of the surface films are of considerable importance in determining foam stability. It is significant that the optimum pH is generally close to the iso-electric point of the protein, when the repulsion is zero, and the net cohesion and, hence, surface viscosity are therefore greatest. For foams stabilized with proteins, therefore, the surface viscosity and rigidity appear to be dominant stabilizing factors.

Surface viscosity also assists in the stabilization of foams from sodium dioctyl-sulphosuccinate, the surface viscosity being 5×10^{-4} surface poises. Surface yield values are also important[122], possibly in addition to surface viscosity, in stabilizing lamellae of solutions of proteins, saponin[123], soaps, and many (slightly impure) detergents.

Whether an enhanced surface viscosity is a *necessary* condition for foam stability and, if so, what is the minimum requisite value, is yet undecided. The sensitive "surface traction" type of surface viscometer (Chapter 5) would be necessary for studying the lower surface viscosity limit. Clearly, however, surface viscosity is not a *sufficient* condition for foam stability: cetyl alcohol gives highly viscous monolayers but does not stabilize foams.

Bikerman[120] points out that surface viscosity may apparently affect foam life through its effect on foam formation if this is carried out by a shaking method. Shaking gives rise to cigar-shaped bubbles of air, which will quickly revert to spheres of the same volume if the surface viscosity is low, but which will break up into a string of small bubbles if the surface viscosity temporarily stabilizes the cigar-shaped foams. Hence a foam of a surface-viscous liquid (formed by shaking) would be expected to have smaller bubbles (which might also be more stable than larger ones) than would a foam from a liquid of lower surface viscosity. Further work on surface viscosity, on the distribution of bubble sizes, and on the persistence of bubbles of different sizes would be welcome.

(*v*) *The restoring effects of surface tension*

These were recognized by Marangoni (1871) and by Gibbs (1878): the stability of a liquid film must be greatest if the surface tension strongly resists deforming forces[124,125]. Suppose that (in spite of surface viscosity) some shock has suddenly extended the local area of the lamella by the factor j. This induces a surface pressure gradient to the thinned region from the rest of the surface, i.e. the monolayer tends to spread back into the extended region. If the latter is relatively small, the change of surface pressure $\Delta\Pi$ is given for each of the boundary monolayers by

$$\Delta\Pi = -\left(\frac{\partial\Pi}{\partial A}\right)(A_2 - A_1)$$

where A_1 and A_2 are respectively the available areas per long-chain in the original and in the extended parts of the surface (Fig. 8-42). Hence

$$\Delta\Pi = -A_1\left(\frac{\partial\Pi}{\partial A}\right)_t\left(\frac{A_2}{A_1}-1\right) = -A_1\left(\frac{\partial\Pi}{\partial A}\right)_t(j-1) = C_s^{-1}(j-1) \qquad (8.70)$$

where j is the area extension factor as before. Note that the term $-A_1\left(\frac{\partial\Pi}{\partial A}\right)$ is the surface compressional modulus of the monolayer, C_s^{-1} (p. 265). For a large restoring pressure $\Delta\Pi$, this modulus should be high. In the extended region the local reduction of the surface pressure to $\Pi - \Delta\Pi$ results in a spreading of molecules from the adjacent parts of the monolayer (at a pressure Π) to the extended region, the rate of increase of Π in this extended region being $\left(\frac{\partial\Pi}{\partial t}\right)^{\text{(spreading)}}$. This differential will be increased by low surface and bulk viscosities. The spreading of the adjacent monolayers into the region extended and thinned by the shock will stabilize the lamella because the monolayers will carry with them a layer of the adjacent liquid (up to 10^{-3}cm. thick cf. p. 252), so opposing the thinning due to the shock. Evidently certain of the factors controlling foam stability oppose each other[126]: a high liquid or surface viscosity makes the drainage rate slower, although, once thin films have been formed, a high viscosity will delay transport of liquid back into a locally extended region. Further, during the time interval Δt the depleted monolayers in the extended region of the liquid film will partially replenish themselves by further adsorption from the adjacent liquid, increasing Π by $\Delta t\left(\frac{\partial\Pi}{\partial t}\right)^{\text{(adsorption)}}$. Though such adsorption will restore the initial value of the surface tension in the thinned region, no liquid is transported to the extended area. Adsorption will therefore not thicken the liquid film again, except possibly by electrical factors (as in section (*iii*) above). Since the term $\left(\frac{\partial\gamma}{\partial t}\right)^{\text{(adsorption)}}$ is greater at high concentrations of surface-active agent (p. 170), the lower foam persistence often observed at the higher detergent concentrations is explicable. An ingenious way of measuring the elastic properties of lamellae consists in forming these across the hair-spring from a pocket watch. The damping of the vibrating spring is a measure of the effects of surface tension changes and of the mechanical properties of the surface films[126a].

Experiments on single bubbles may again be used to illustrate these points; a bubble of air is injected below the surface of a Langmuir trough, as in Fig. 8-38. The bubble rises and floats at the air-liquid surface, and the time required for it to break (as well as the rate of shrinkage due to diffusion) may conveniently be studied. Monolayers, either spread or adsorbed, may be present at the interface.

Sir William Hardy in 1912 carried out the first studies of the stability of single bubbles[127]. His original experiments, as well as those carried out subsequently, show that the persistence time of the bubble increases to a maximum when a monolayer spread on the surface of the trough is compressed to the beginning of the steep part of the Π-A curve: further compression of the film again reduces greatly the stability of the bubble. Later this type of experiment was taken up in Russia[128,129], where again the bubbles were found to be most stable at an available molecular area rather greater than that in a condensed monolayer. For ethyl oleate the maximum bubble life occurs when $A = 120$ Å^2. Protein films, both adsorbed and spread, can be readily studied by this method. Results[121] are rather complicated by the ready coagulation of the material, but for adsorbed films of the protein pepsin at pH 3.7, the following correlation between the single bubble life and the "foaminess" Σ was obtained:

Ionic strength	0.03	0.30	3.00
Single bubble life	20 sec.	50 sec.	160 sec.
Σ	137 sec.	172 sec.	300 sec.

When draining has proceeded till the liquid film is very thin, the effective compressional modulus of the lamella is greatly reduced. This is because there is less liquid free to thicken the film if it has been stretched locally: i.e. the liquid between the surface films no longer has the properties of the liquid in bulk, but, being attached by forces of hydration to the polar groups of the monolayer, becomes more "brittle", and so allows the film to be ruptured by slight shocks or even chance molecular fluctuations. The final stages of the thinning to this state may be rapid, presumably because the van der Waals forces of attraction assist the molecules to lie in orientated sheets, i.e. "liquid crystals", which include a few tightly bound and fully orientated water molecules. Such a liquid crystal, of the order 100 Å thick, constitutes a "black spot"; it ruptures quickly.

Summary of Causes of Foam Stability

The film separating the gas bubbles will break when thinned to some critical thickness of the order 50-100 Å. The thinning may be caused by drainage (under the action of gravity) and by capillary suction at the lines of contact of different bubbles or by mechanical, chemical, or surface shocks. A high liquid or surface viscosity will minimize these effects, as will also a large "pseudo-equilibrium" disjoining pressure of electrical origin, or a compressional modulus of the surface film (or supporting liquid). A simple experimental test of the relative importance of viscous and compressional effects consists in dropping small solid or liquid projectiles through the foam: if only viscosity is important the projectile leaves a hole behind it, as a high viscosity can only retard motion, not reverse its effect.

We may summarize the above discussion in symbols by writing for a foam of given mean bubble size[118]

$$\text{Stability} = f\left(h, j, C_s^{-1}, \left(\frac{\partial \Pi}{\partial t}\right)^{(\text{spreading})}, \left(\frac{\partial \Pi}{\partial t}\right)^{(\text{adsorption})} \right) \tag{8.71}$$

where h depends on the rate of drainage and hence on the surface and bulk viscosities and yield values, and after a long time reaches a limiting value h_∞ dependent on the ionic strength. For stability h should always be as large as possible, to resist shocks and to reduce air diffusion from the small bubbles. The term j is the local extension ratio of the area of a lamella due to a given thermal or mechanical shock; it will be reduced, and the stability thereby increased, by high surface and bulk viscosities and yield values. The compressional modulus C_s^{-1} of the monolayers should, for stability, be as high as possible: this assists rapid flow of the monolayers (and some bulk liquid) into the extended region of the lamella. This monolayer flow-rate is measured by $\left(\frac{\partial \Pi}{\partial t}\right)^{(\text{spreading})}$: it should be as large as possible, which requires low surface and bulk viscosities in addition to a high compressional modulus of the monolayer. The rate of adsorption, measured by $\left(\frac{\partial \Pi}{\partial t}\right)^{(\text{adsorption})}$ should generally be low, so as to allow time for spreading to occur; if, however, the lamella thickness is of the order h_∞ and spreading is very slow, a high rate of adsorption could thicken the extended region.

Foam Stabilizing Additives

If small amounts of lauryl alcohol are added to sodium lauryl sulphate, the foam produced is much more stable. Typical values are 70 min. for the life of a foam from pure sodium lauryl sulphate, and 825 min. with sodium lauryl sulphate containing 1% lauryl alcohol. Lauric-isopropanolamide and other similar substances act in the same way[130].

Whether these foam-stabilizing additives act by altering the surface viscosity and yield point of the adsorbed monolayers is thus of interest. The surface viscosities are low (10^{-4} to 3×10^{-3} surface poises), but they can be studied using the "viscous-traction" instrument, described in detail in Chapter 5; with it, the effect on surface viscosity of very small amounts of additives can be studied in relation to foam life. A convenient basic solution for this purpose is 0.1% sodium laurate at pH 10 in distilled water[131,132]. Unlike laurate at lower pH values, this shows no appreciable foaming without additives, due to the fact that there is no lauric acid produced by hydrolysis at this high pH. The surface viscosity of this solution is too low to be detected, and the foam produced by intense hand-shaking lasts only about 3 seconds.

On the addition of very small amounts of foam-stabilizing additives, in particular lauryl alcohol and lauric-isopropanolamide, the foam life increases drastically, with a concomitant rise in surface viscosity (Figs. 8-43 and 8-44). We see that the increase in foam stability occurs at about the same concentrations of additive as does the rise in surface viscosity. This suggests that at these low concentrations surface viscosity is a primary cause of the

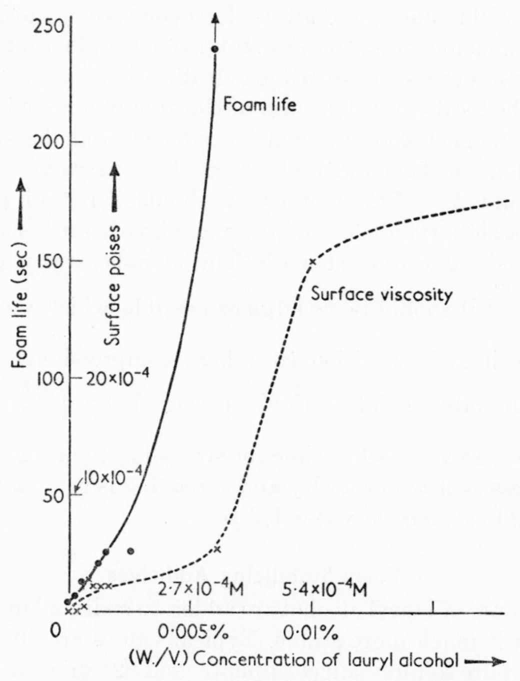

Fig. 8-43. Effect of lauryl alcohol on foaming of 0.1% sodium laurate at pH 10, with the corresponding changes in surface viscosity[132].

increased foam stability. The very great stability at higher additive concentrations is not the result of a corresponding increase in surface viscosity: it may well result from high surface yield values and compressional moduli.

For a given increase in surface viscosity, lauryl alcohol is more effective in stabilizing foam than is the amide, though smaller concentrations of the latter are involved (Figs. 8-43 and 8-44). The isopropanol residue in the long chain amide makes the latter more lipophilic, and hence more strongly adsorbed from solution, so that the amide is effective in molar concentrations only about one-third those of the alcohol; apparently the alkyl-amide group has no *specific* effect.

The stabilizing effect of phosphate ions[131] on foams from laurate at pH 10 remains obscure; it may lie in the increased adsorption of long-chain ions at higher ionic strengths, or in sequestering of traces of polyvalent metal ions.

The rate of attainment of equilibrium is important if foam-stabilizing additives are to increase "foaminess". Thus, from the low concentrations in which lauryl alcohol is present, we may compute by the methods of Chapter 4 and Fig. 4-5 that its rate of penetration into the monolayers of sodium lauryl sulphate is rather low: about 0.5 sec. may be required. During the formation of a foam, therefore, the stabilizer may be unimportant.

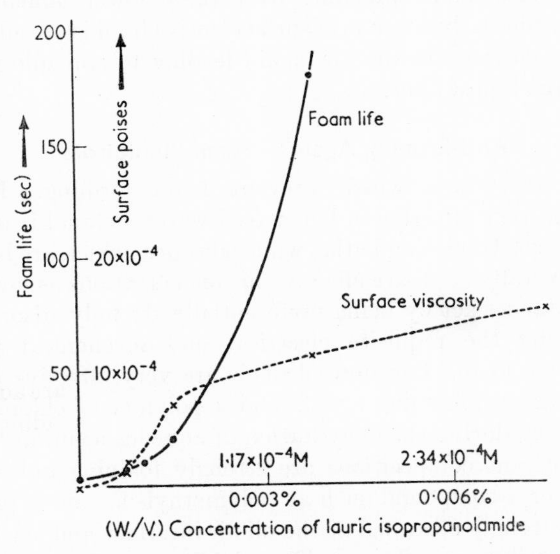

Fig. 8-44. As for previous figure with lauric isopropanolamide as the foam-stabilizing additive.

Destruction of Foam—Foam "Breakers" or "Killers"

These are substances which cause a local rupturing of the liquid film. They act by raising the surface pressure over a small region: from this region spreading therefore occurs, and as the film spreads it carries with it a layer, up to 10^{-3}cm. deep, of the underlying liquid (Chapter 5), thus thinning the film of liquid. This explanation, advanced by Ewers and Sutherland[133], correlates the behaviour of the three groups of foam breakers: solid or liquid surface-active materials other than the substance stabilizing the film, liquid containing the foam stabilizer in higher concentration than it is present in the foam, and vapours of surface-active materials. In each of these groups local adsorption leads to a surface-pressure gradient and, if the spreading is fast enough and can be maintained for long enough, the film is thinned to the critical distance of 50-100 Å, followed by spontaneous breaking by mechanical shocks or thermal or diffusive molecular fluctuations. In general, the faster the molecules of the foam breaker can spread on the liquid film the greater the thinning and the greater the chance of rupture. For example, liquid 1:1-dihydroperfluorobutanol spreads faster on sodium lauryl sulphate solutions (4.6 cm.sec.$^{-1}$) than does n-octanol (3.6 cm.sec.$^{-1}$): the former material is also the better foam breaker. A strong solution of surface-active agent, concentrated by evaporation and drainage at the top of a column of foam, may, as further bubbles break, become sprayed over the lower bubbles, and, spreading over these lower lamellae, may well rupture them. Similarly vapours of surface-active liquids are adsorbed locally by convection currents in the air, again leading to spreading and possible rupturing of the liquid films.

Anti-foaming Agents—Foam "Inhibitors"

There are substances which prevent foam forming. For example, polyamides are very effective in boiler feed water, octanol is used in paper-making and in electroplating baths, while silicones, which are becoming very widely and generally used, are effective in concentrations as low as 10 p.p.m. All these substances act by being preferentially strongly adsorbed, without, however, having the requisite electrical and mechanical properties to produce a stable foam. Perfluoro-alcohols are very effective both as foam breakers and as anti-foaming agents and a polymer of chlorofluoro-olefine prevents foaming during the dehydration of aqueous solutions[108].

Organic lubricant compositions are strongly foaming unless fluorinated hydrocarbons or a compound such as polymethylsiloxane is present. These materials are strongly adsorbed at the oil-air interface and act by displacing the foam-stabilizing monolayers. The rigidity of a monolayer may be destroyed completely by antifoaming agents such as tributylphosphate, with a concomitant decrease in surface viscosity[122]. This effect is also shown by 2-ethylhexanol and 4-methyl-2-pentanol which are effective antifoaming agents for aqueous detergent solutions. The non-ionic reagent Span 20 (sorbitan monolaurate) prevents the foaming of egg-white by similarly being strongly adsorbed but contributing neither to the surface viscosity and rigidity, nor to the electrical double layer which might lead to liquid films of "pseudo-equilibrium" thickness.

CLOSED SHELLS OF FLUIDS

The most common of these is the "soap bubble", in which the shell (A of Fig. 8-45) is a thin layer of water, of the order 1000 Å thick, while the inner and outer phases (B) are air. The stability of such a water shell is similar to that of a foam.

Less well-known are shells of gas (now in position A) separating two phases of water (B of Fig. 8-45). Sometimes known as "inverse bubbles", the closed shells of gas may be produced either by pouring or dropping aqueous solutions of surface-active agents through the surface of a similar solution[134-5], or by bubbling a stream of air through an aqueous solution[136] of (say) 0.5% Lissapol or Tween 80. The stability of such gas shells lies simply in the long drainage times for the gas to escape from between the two almost concentric spheres at the point shown by the arrow: coalescence occurs[136] when thinning

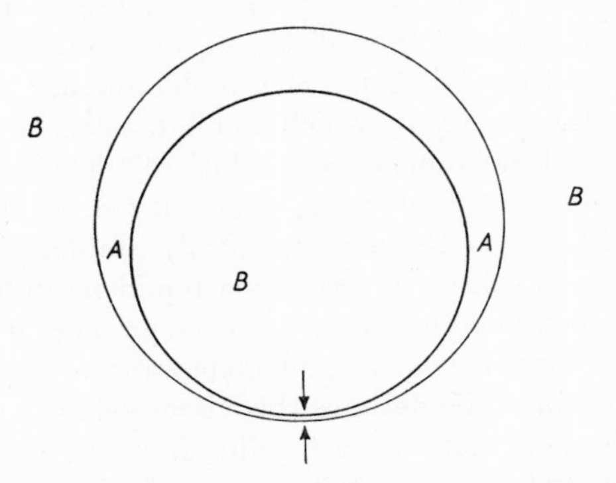

Fig. 8-45. Closed spherical shell of phase A in continuous phase B.

(at the arrows) has proceeded to about 3000 Å, which may take up to about 5 minutes. That coalescence of the water phases may occur across such a great air gap may be due to dust, electrification, or chance thermal, mechanical, or surface-tension fluctuations. The gas shells may be between 0.1 and 2 cm. in diameter; they appear silvery by total reflection of light and rise rather slowly through the liquid.

The mechanism of stabilization of gas shells by surface-active agents is that these so reduce the flow in the surface (compressional modulus and surface viscosity) that the gas has to drain out from between stationary walls, a much slower process than if the boundary layers of liquid can move. This stabilization mechanism is the same as that for *floating drops* (p. 357), and the two phenomena are often found together.

Closed fluid shells are also found in liquid-liquid systems[137]: oil shells (A in Fig. 8-45) in water (B) have been described, as have shells of water (now B) in oil (now A). One may produce the former, for example, by placing a layer of oil over an aqueous phase, and then dropping some of the aqueous phase through the oil film. With protein as the stabilizing agent, such oil shells may be stable for up to 6 days.

DETERGENCY

Detergency is the removal of oil or solid particles ("soil") from a surface or from a cloth, and is markedly dependent on the adsorption of surface-active agents. Soaps have been used as detergents for many centuries, though in the last few decades the long-chain sulphates and sulphonates, nitrogen bases, and poly-ethylene-oxide condensates have largely replaced soaps in Europe and the United States.

The first function of the detergent in cleaning a surface is to remove any greasy material. This may occur by one or more of three mechanisms: solubilization, spontaneous emulsification, or "rolling-up". The second function is to prevent the soil from being re-deposited.

Solubilization is the process by which an oil dissolves in the interior of the micelles of detergent formed in aqueous solution[138]. The micelles may be only 40-500 Å in diameter, even when oil is solubilized, so that the solution is optically clear. It is also thermodynamically stable. Solubilization of a polar soil (e.g. stearic acid) may occur via liquid-crystalline phases[139], and the interface may be broken up by the osmotic pressure if the polar material and the detergent form a nearly solid membrane of the liquid-crystalline complex at the interface. Sometimes this viscous skin at the interface leads to "tentacles" of greasy material protruding from the surface.

Spontaneous emulsification, such as occurs during the solubilization of certain oils (e.g. xylene in contact with an aqueous solution of dodecylamine hydrochloride (p. 365)) may also be important in detergency. Even solid fats may be rapidly spontaneously emulsified at room temperature if the detergent contains some solubilized xylene[139a].

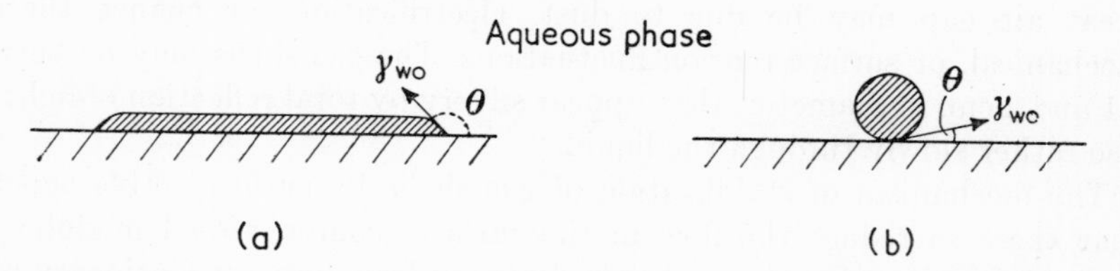

Fig. 8-46. A patch of a non-polar oil (a) becomes rolled up (b) in the presence of a surface-active agent[140].

"*Rolling-up*" *of the oil* (subsequently emulsified) on a surface[140] is illustrated in Fig. 8-46. With the usual subscripts we have:

$$F^{s}_{S/O} = F^{s}_{S/W} + \gamma_{O/W} \cos\theta \qquad (1.42)$$

where θ is measured in the water. For a greasy surface θ may be of the order 150°, but, when a monolayer of detergent is adsorbed, both $F^{s}_{S/W}$ and $\gamma_{O/W}$ are considerably decreased, especially if a branched-chain detergent is used.

This means that to regain equilibrium cos θ must now become more positive i.e. θ becomes smaller. When θ is nearly zero we may regard the detergent monolayer as being preferred energetically to the grease on the surface, and, if $F^s_{S/W}$ and $\gamma_{O/W}$ are reduced sufficiently, cos θ will not be able to become positive enough for the equilibrium condition (1.42) to be satisfied, so that the oil is now completely displaced from the surface.

Unaided, this process may occur only slowly, especially if the solid is non-porous. In practice, therefore, one usually agitates the solution to assist the displacement of the oil, which is then removed as a rather coarse emulsion. Foams may also assist the removal of oil by causing the latter to be sucked away from the surface into the Plateau borders (Fig. 8-31).

Once removed from the surface, the oil must not be redeposited. Provided that depletion of the detergent solution by adsorption is not severe, eq. (1.42) will still be satisfied with a small value θ, the surface being preferentially covered with detergent solution rather than with grease. The oil can remain in solution either as an emulsion or, preferably, solubilized in the micelles.

Solids may be removed by gentle agitation in the detergent solution; monolayers are adsorbed on the solid particles and sometimes on the surface being cleansed, increasing the solid-solid repulsion and so preventing re-deposition of the solid. Foams may again be advantageous here, carrying the solid particles away to the surface of the water, thus reducing the probability of re-deposition.

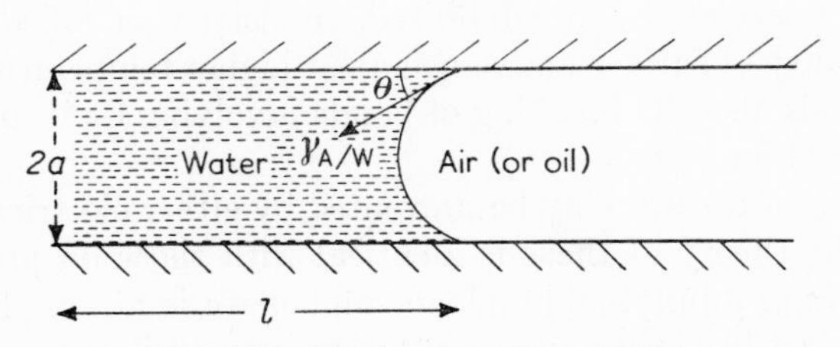

Fig. 8-47. Illustration of eq. (8.72).

In detergency *the rate of wetting* is often important, and as a basis for studying this quantity one may use the Rideal-Washburn equation[141] for the rate of advance of liquid into an air-filled tube (see Fig. 8-47):

$$\frac{dl}{dt} = \frac{a\gamma \cos\theta}{4\eta\, l} \tag{8.72}$$

where a, γ, η, and l refer respectively to the radius of the capillary tube or pore, the surface tension (air-water) of the liquid, the viscosity of the liquid, and the distance the liquid has travelled along the pore. As usual, θ is the contact angle and t is the time. The equation, which neglects gravitational effects, may readily be integrated to

$$t = \frac{2\eta l^2}{a\gamma \cos\theta} \qquad (8.73)$$

In a mass of textile fibres, we are concerned not with one pore but with a series of capillaries of different radii: consideration must therefore be given strictly to these variations in pore size since the resistance to flow in each channel will now be that of a number of capillaries in series[142].

Comparison of eq. (8.73) with direct measurements of the rates of wetting of textiles shows[143] that increasing amounts of sodium dodecyl benzene sulphonate increase t (for a given value of l) by up to three times: this is expected because γ is reduced by the surface-active agent. However, no quantitative theory of the exact value of γ at the advancing liquid front is available, though depletion by adsorption on the walls of the pore, and replenishment by diffusion must clearly be important[144]. Electrolytes increase the adsorption of surface-active agent on the walls of the pore, and so slightly increase γ and decrease t.

With clean cotton fabrics the angle θ is always small, but if they are covered with fatty soil θ becomes great and t of eq. (8.73) is greatly increased. If $\theta > 90°$, no penetration of liquid into the fabric will occur at all. A surface-active agent must then be added to lower θ (cf. eq. (1.42)).

Without a doubt the high *temperature coefficient of detergency* is associated partly with the increased rate of diffusion of surface-active agent into the soil, and partly with the melting of some of the greasy soil which is consequently more readily penetrated and solubilized, emulsified, or rolled up. The much greater solubility of surface-active agents at higher temperatures can also be important, as is also the breaking of bonds involving Ca^{++} and other specifically adsorbed ions (see below).

Redeposition of the soil may be prevented by either electrical or hydration barriers[145]: the theory of these is identical with those on pp. 367 and 369. Although sodium dibutylnaphthalene sulphonate is an excellent dispersing agent for carbon-black in aqueous solution, non-ionic surface-active agents are more effective in preventing redeposition.

Calcium and other polyvalent cations, present in traces on cotton and in hard water, reduce considerably the efficacy of detergency. This is partly because fatty acids in a polar soil may form insoluble hydrophobic soaps from such ions, reducing the rate of wetting of the textile[143], but also because re-deposition of the soil, after it has been removed, is accelerated by traces of ions such as Ca^{++}. The reason for the latter effect is that cotton does not strongly adsorb long-chain anions[146], but has a natural negative charge arising from $-COO'$ groups on the surface, the latter groups being readily neutralized by Ca^{++} because of its double charge and because of its specific adsorption (cf. p. 86). The Ca^{++} in solution should therefore be sequestered into a non-adsorbed complex ion[147], and such materials as $Na_5P_3O_{10}$ and

$Na_4P_2O_7$ (known as "builders") are widely used as additives in detergent formulations.

FLOTATION OF MINERALS

If a particle of a solid is to float, the total upward pull of the meniscus on it must balance the apparent weight of the particle. For a particle of the shape shown in Fig. 8-48a, the vertical force is $(\gamma_{A/W}\cos\theta)\times$ perimeter, whereas for the particle in Fig. 8-48b it is

$$\{\gamma_{A/W}\cos(\theta+\theta_p)\}\times \text{perimeter},$$

where θ_p is the angle the appropriate face of the particle makes with the vertical.

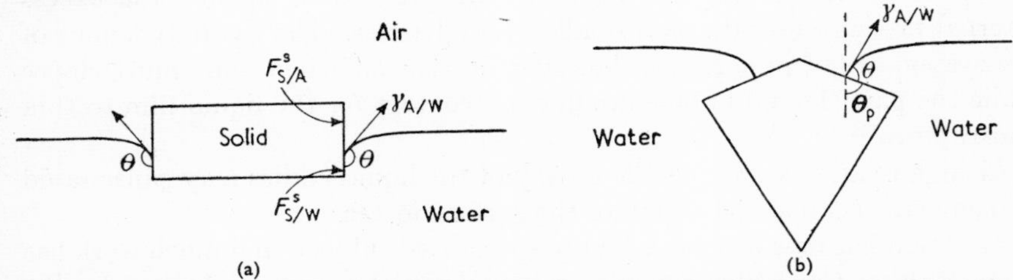

Fig. 8-48 (a) Flotation of a particle by the pull of the meniscus on vertical sides. (b) The flotation of a particle whose face makes an angle θ_p with the verticals.

The value of $\gamma_{A/W}\cos\theta$ depends on $F^s_{S/A}$ and $F^s_{S/W}$ according to the equilibrium relation:

$$\gamma_{A/W}\cos\theta = F^s_{S/A} - F^s_{S/W} \tag{1.35}$$

and, since $F^s_{S/A}$ and $F^s_{S/W}$ are sensitive to surface-active agents, the condition for floating or not can easily be changed. For most natural mineral surfaces $F^s_{S/A}$ is so great that $\cos\theta$ must be great (i.e. θ is of the order zero or else no equilibrium can be established); hence clean water wets the solid completely.

However, the various constituents of many crude ores have different tendencies to float on the surface of water, and these differences can be enhanced by certain additives, known as "collector oils". These adsorb strongly on to the solid (e.g. galena), reducing particularly $F^s_{S/A}$, so that by eq. (1.35) $\cos\theta$ decreases, and θ increases to the point where, as in Fig. 8-48, flotation is possible. Particles, such as sand, on which the additives do not adsorb so strongly, remain wetted by the water.

In practical ore-flotation a foaming agent is also added to stabilize air bubbles, which are formed by forcing air through a sieve-plate at the bottom of a large vessel. The air bubbles, rising through the aqueous suspension of the powdered crude ore, collect those particles that will float into a foam at the top of the vessel, leaving behind the sand. Because of wide variations in shape, size, and roughness of the grains, no single value of θ can be cited that will satisfy the requirements for flotation[84]. However, a value of 50° to 75° is a typical minimum requirement (see Fig. 8-48b): this may be achieved by as little as 5% surface coverage by the adsorbed additive. The organic xanthates and thiophosphates are well-known additives. Often the ore needs pre-treatment before it will adsorb the additive sufficiently, e.g. ZnS must be pre-treated with dilute $CuSO_4$ solution. Similarly, one can increase further the specificity of the flotation process by adding depressants, as for example CN'; this prevents the FeS and ZnS from floating but allows the PbS to float. In this way the components of a mixed ore are readily separated.

There is no energy barrier to the adsorption of a solid particle on to a large bubble, though for small bubbles the work done against the excess internal pressure introduces a small energy barrier. The hydrodynamics of the system are of importance, however, in that sufficient time must elapse while the particle and bubble are in close contact for the liquid film to thin and rupture[148].

At high relative velocities the energy of the liquid eddies may cause rapid detachment, against the action of the surface forces.

Flotation has now become a highly specialized subject, and much work has been done on the hydrodynamics required for the best conditions for the adhesion of the solid particles and the gas bubbles. The "collectors" and "depressants", as well as the foaming agents, have also been extensively studied, and the interested reader is referred to the specialized monographs[149,150] on the subject.

MODIFICATION OF THE HABITS OF CRYSTALS

Sodium chloride and many other substances may be made to alter their crystal habits in the presence of small amounts of surface-active agents[151,152]; this may be shown experimentally as follows. A supersaturated solution of pure NaCl is slowly evaporated in the presence of a crystal "seed" suspended by a nylon thread. The resulting cubic crystals, which may have edges up to 1 cm., are now suspended in supersaturated NaCl solution containing the surface-active agent: the latter both inhibits and alters crystal growth, usually leading to predominant octahedral faces coinciding with the corners of the original cube[151]. Many surface-active agents also produce octahedral NaCl crystals when NaCl solution containing a little of the agent is simply allowed to crystallize very slowly. The explanation of these effects is that, while the (100) plane of the NaCl crystal contains both sodium and chloride ions, the (111) plane contains alternately sodium ions and only chloride ions; the (111) plane, when it becomes positively charged, strongly adsorbs some of the long-chain anions, which protect it from ions arriving from solution. The result is that these (111) planes grow slowly, leading to the octahedral habit, since those faces which have the slowest growth normal to their plane appear largest in the final crystal (Fig. 8-49). The (100) planes, being electrically neutral, adsorb less strongly, and so are still able to grow.

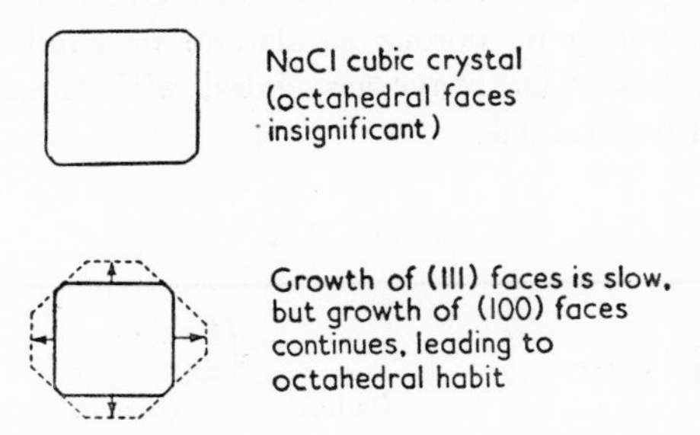

Fig. 8-49. NaCl cubic crystal (octahedral faces insignificant). If growth of (111) faces is slow, but growth of (100) faces continues, an octahedral habit results.

Surface-active agents are not alone in being able to modify crystal habit[153]: dyes and various inorganic salts may act in the same way, and macromolecules having regularly spaced carboxyl groups adsorb so strongly on to micro-crystals of $CaSO_4 2H_2O$ and other salts as to be industrially useful in reducing precipitation in boilers.

LIQUIDS IN FINE PORES

The Rideal-Washburn equation[141],

$$l^2 = \frac{\gamma \text{ at} \cos\theta}{2\eta} \tag{8.73}$$

has been discussed above: it relates the rate of penetration of a liquid into an air-filled horizontal pore of uniform diameter to the surface-tension of the liquid and to the contact angle θ. Table 8-X shows that it applies to a wide variety of liquids. Modifications of this equation have been proposed, to apply when the ore narrows along its length, when there is a network of pores, when one liquid is displacing another from the pore, or when the pores are not horizontal[142].

If eq. (8.73) is to be applied to a plug of a solid powder or to a bundle of capillaries, a tortuosity factor t must be used to correct the value of l, which is replaced in the equation by tl. The factor a/t^2 is evaluated by performing an experiment with a liquid which wets the pores (i.e. $\cos\theta = 1$), so that $\cos\theta$ can then be determined for other liquids. Alternatively, $\cos\theta$ can be obtained for the plug or pad of fibres by measuring directly the pressure resulting from the penetration of the liquid[153a].

In the complete **displacement of oil** from an oil-wetted pore or porous bed, one must reduce θ by adding surface-active agents. This has been extensively investigated both in the laboratory and in the field, in connection with increasing the yield of oil. Natural petroleum occurs under the surface of the earth at depths varying from a few hundred to several thousand feet, filling the interstices of bands of porous sandstone or sand confined between impervious rocks. Water is always associated with the oil, and the latter contains dissolved natural gas.

TABLE 8-X

Liquid	$\left(\frac{\gamma a}{2\eta}\right)^{1/2}$ (calc.)	(observed from $l/t^{1/2}$)
Isobutyl alcohol	3.75	3.70
Isopropyl alcohol	4.10	4.20
Ethyl alcohol	5.52	5.65
Methyl alcohol	8.16	7.90
Chloroform	8.70	8.60
Benzene	8.90	9.90
Ether	11.38	10.95
Water	11.31	11.40

Test of eq. (8.73) for wetting liquids ($\theta = 0$) in a glass capillary of $a = 0.0354$ cm.

Only about 30% of the oil can be obtained from a natural reservoir using the pressure of the associated gas: further extraction of oil is only possible with artificial methods. Chief amongst these is the injection of a water "flood" to drive the oil to the shaft of the well. This technique permits recovery of up to 66% of the oil initially present, the exact figure being dependent on the various interfacial energies. Attempts have been made to add surface-active agents to the flood water to reduce θ, and so to promote increased oil recovery and increased permeability of the porous rock.

The efficiency of the water flooding method depends, *inter alia*, on whether the reservoir rock is preferentially wetted by water or by oil, on the relative proportion of oil and natural water present in the reservoir (determining the relative permeabilities to oil and water), on the spacing and arrangement of the injection and producing wells, on the injection pressures and the rate of movement of the flood, and on the presence of dissolved substances in the flood water.

The reservoir rocks are often preferentially water wetted, being composed of silica, although certain crude oils contain impurities which are adsorbed on to the surfaces of the rock pores, rendering them oil-wetted; metal-porphyrin complexes are particularly effective in this connection. The result is that most reservoirs contain some areas of oil-wet rock, although it is unusual for the whole reservoir to be oil-wetted.

The oil droplets, present in the water-wet areas, will cause a resistance to the flow of water, since the droplets have to be distorted to pass through irregularly shaped pores, and sometimes this resistance may be so high that the oil droplet cannot be displaced. Gas bubbles may similarly retard the liquid flow.

In the oil-wet regions of the rock, however, while it may be possible for the water flood to drive out some of the oil from the wider pores (by "cores" or "fingers" of water[75]), an appreciable amount of oil will remain as a film on the solid surfaces: this film will only be displaced with difficulty or not at all. A slow flooding rate (a few inches per day) is used to help in the displacement of the oil films from the surface without "channelling" of the water through the reservoir. This can be achieved by reducing θ (Fig. 8-47), by adding surface-active agents favouring wetting by water, i.e. reducing $F^s_{S/W}$ and $\gamma_{O/W}$ of eq. (1.42), hence making $\cos\theta$ larger, and so making θ smaller. Since $F^s_{S/O}$ should be but little affected, water-soluble surface active agents should be used[154,155].

The question of whether in practice one should use surface-active agents for oil recovery is purely economic: to give a monolayer coverage to the pores of a typical reservoir would cost about $15 million, and even this estimate may be too low on account of the slow diffusion of surface-active material on to the advancing menisci. Other methods of recovery, such as underground combustion of some of the residual oil (to reduce its viscosity and increase the gas pressure locally) appear more promising.

Resistance to strain and rupture in solids is reduced by adsorption from the surrounding medium. The reason lies in the reduced adhesion of the metal to itself (c.f. eq. (1.20)), so that micro-cracks on the surface can rapidly deepen under any applied stresses, until they extend far into the interior of the metal[156].

Mica crystals, for example, behave in vacuum and in dry air as ideal elastic bodies, though in the presence of water, especially if fatty alcohols or other surface-active agents are present, they show after-effects while the load remains applied. Glass fibres, too, stretched in aqueous solutions of long-chain cationic or anionic agents, show elastic after-effects, such as hysteresis or creep, although in dry air no deviation from elastic behaviour can be observed.

The creep of metals is especially sensitive to adsorption: if a constant stress (below the yield value) is applied to a tin crystal in pure paraffin oil, followed by adding a little oleic acid or cetyl alcohol, the rate of deformation is drastically increased.

The effect of adsorption can be shown to be a maximum when a complete monolayer is formed on the freshly-forming surfaces of the developing defects. The saturated fatty acids and alcohols, adsorbing from hydrocarbon oil on to tin, have an energy of adsorption of 155 cal. mole^{-1}, and a double bond considerably heightens the effect. Metals also fatigue more readily if adsorption occurs at the weak spots. That the effects of adsorption are so striking follows from Rehbinder's theoretical treatment. According to this, the probability P of the formation of rupture cracks at weak spots on the metal surface is given[156] by:

$$P = C_m \exp\left(-\frac{F^s_{S/L}\delta^2_m}{\boldsymbol{k}T}\right) \tag{8.74}$$

where C_m is a constant characteristic of the metal, $F^s_{S/L}$ is the surface energy of the solid-liquid (or solid-air) interface, and δ_m is the distance between the structure defects (about 10^{-6}cm.). That P increases so sharply as $F^s_{S/L}$ is reduced by adsorption of a monolayer is thus comprehensible.

Oils (or "inks") for the **detection of very fine flaws** or leaks in solids contain a fluorescent additive[156a]. The oil is spread over the degreased metal, penetrating into any cracks which extend to the surface. After a rapid wash with petrol-ether or carbon tetrachloride (to remove excess "ink"), the surface is viewed by ultra-violet light. Any cracks are then clearly indicated by a bright green trace. This testing procedure is applicable to metals, plastics, ceremic and clay articles; it is essentially non-destructive. It is also rapid in operation provided that the oil readily penetrates the cracks. To achieve this, the contact angle (measured in the oil) must be low (eq. (8.73)).

For magnetizeable solids such as steel, a "Magnetic Ink" may more conveniently be used. This consists of a fine dispersion of coloured magnetic particles (of diameter less than 5 microns), in an oil base. When some of this "ink" is applied to the surface of a magnetized solid (not necessarily degreased), the magnetic particles become concentrated at the edges of the crack, so rendering visible cracks and pores too small to be seen directly. A water-based magnetic "ink" is sometimes used, the suspension being stabilized with surface-active agents. Instead of pigmenting the magnetic particles in the suspension, one may make them fluorescent, detecting the flaws or leaks by ultra-violet light. This improves the optical contrast of the cracks in the metal surface, and makes detection correspondingly easier.

ADHESION

That adhesive strengths should be very high for perfect surfaces may be illustrated by the following simple calculation. If into eq. (1.32) one inserts

$\theta = 0$ (i.e. the adhesive just wets the solid), and $\gamma = 30$ dynes cm.$^{-1}$, the force required to separate the surfaces by a molecular dimension is $(30 \times 2)/10^{-7}$ dynes cm.$^{-2}$, i.e. 600 kg cm.$^{-2}$ or about 4 tons per square inch. In practice, however, such high adhesive strengths are not attained, usual values being of the order 30 kg cm.$^{-2}$ The reason for the difference lies both in the plastic and viscous flow which may precede fracture, and in the existence of flaws and dislocations in the surface of the solid[157,158].

The adhesion can be studied experimentally as a function of adsorption: the former is expressed as a coefficient of adhesion, which is the ratio of the adhesive force to the load previously applied. The adsorption is measured by direct studies of depletion, by the vibrating-plate potential apparatus modified for the oil-metal interface (Fig. 2-7) or by radioactive tracers.

If two pieces of *glass or hard metal* are carefully cleaned and then pressed together, the adhesion (the force required to separate the surfaces again) is negligible if the air is clean and dry. Under these conditions the surfaces deform elastically at the point of contact, and the stored energy assists separation (Fig. 8-50). In a humid atmosphere, however, the adhesion

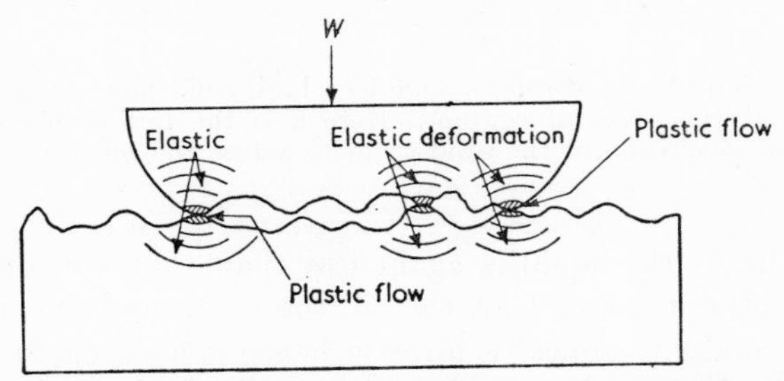

Fig. 8-50. These elastic stresses are released in peeling dry junctions apart[157].

increases towards a limit (reached at saturation vapour pressures). This limit is the adhesion in the presence of a small amount of water between the liquid surfaces. Experimentally, it is convenient to study a sphere in contact with a plane sheet, as in Fig. 8-51. We shall assume that the sphere is smooth and of radius r, and that the liquid collects to form a pool at the tip of the sphere, completely wetting the surface. If the radius of curvature of the profile of the meniscus is a, the pressure inside the liquid is less than atmospheric by approximately γ_0/a (as may be easily shown by a derivation similar to that of eq. (1.10)). The force over the whole liquid pool is thus $\pi s^2 \gamma_0 / a$, and, since as a first approximation the geometrical relation $s^2 = 2r.2a$ is valid (see Fig. 8-51),

$$\text{adhesive force} = 4\pi r \gamma_0 \qquad (8.75)$$

Thus the adhesion should be independent of the thickness of the liquid film and is directly proportional to r. This relation is confirmed quantitatively by experiment, both with thick films of liquid and with films so thin that interference colours cannot be seen[157]. Hence one may deduce that γ_0 is unaffected by the thinning of a liquid below 1000 Å.

Flat, parallel surfaces may also adhere through a film of liquid. Suppose that the surfaces are circular discs, each 5 cm.^{2} in area, and that a film of oil 1000 Å thick separates them. Assuming that the oil wets the metal, the radius of curvature a of the meniscus at the edge of the discs is 500 Å, and

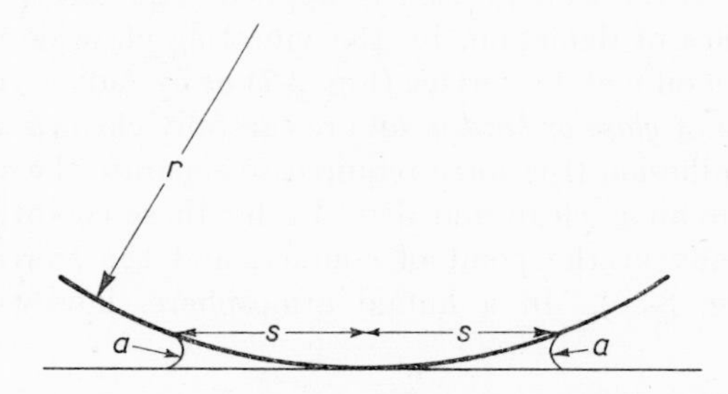

Fig. 8-51. Adhesion of a hard solid sphere to a hard solid plane surface, with a little water present at the point of contact. Here a is the radius of curvature of the meniscus[157], the separation of the solid surfaces across the meniscus being therefore approximately 2a.

the pressure inside the liquid is again less than that outside (1 atm.) by γ_0/a. If γ_0 is 30 dynes cm.$^{-1}$ for the oil, the pressure deficit is thus 6×10^6 dynes cm.$^{-2}$, and the adhesive force is hence calculated to be $5 \times 6 \times 10^6$ dynes $= 31$ kg. Such surface tension forces will buckle thick glass plates[157]; adhesions of the order 20 kg. have been found experimentally. If the thickness of the liquid film is increased, a is also greater, and the adhesion becomes much smaller. The same effect may be found by reducing γ with surface active agents: further, by completely surrounding the solid plates with liquid the surface-tension effect should be zero, as has been confirmed experimentally.

The adhesion of *soft metals* such as indium is very high, even in the complete absence of liquid films, since the adhesion is not opposed by elastic stresses in the solid. The formation of junctions between such surfaces may be lessened by liquid lubricants: if a clean steel ball is tested against a sheet of indium, paraffin oil will reduce its adhesion to about one half. If the indium surface is covered with a monolayer of lauric acid the adhesion is reduced to a negligibly low value; only if heavy loads are used, so that an appreciable indentation is formed in the indium plate, will the steel ball adhere. This occurs when the indentation is large enough to extend the area of the surface covered with the lauric acid monolayer by about 2% in excess of the area of the original plane surface; unless the monolayer is expanded by more than about 2%, it can still prevent appreciable metallic adhesion. The harder metal (steel) of which the ball is made cannot be appreciably deformed and, if this surface, instead of the indium, is covered with the monolayer, the adhesion remains negligible whatever the size of the indentation formed in the indium surface[157].

Lead and tin also show appreciable reduction in adhesion in the presence of paraffin oil: again the adhesion becomes negligible if a monolayer of fatty acid is present.

The adhesion of an oil to a metal, important in lubrication (see below), can be determined directly from the measured value of the spreading coefficient (eqs. (1.41) and (1.43)), or from θ (eq. 1.32): usually experiments are carried out in such a way that the metal/air surface is always covered with an adsorbed film.

If the liquid is allowed to drain out of a tube wetted by it, the thickness of the liquid film adhering to the walls after any time interval is a function of the viscosity of the liquid. Conversely, one may lift a plate continuously out of a liquid and find the steady-state thickness at different points. As a first approximation only density and viscosity terms need be taken into account, but more complete treatments require also the surface tension of the liquid[159,160].

Powders often adhere more strongly if they are damp: this is a direct result of eq. (8.75). With fairly slight condensation of moisture inter-particle adhesion may become comparable with the weight of a small particle, since the former varies directly as r and the latter as r^3. Similarly, inorganic powders dispersed in an oil clump if a little water is present, due to the formation of lenses of water between the particles[161]. For two spherical particles, each of radius r, one can easily show[118] that eq. (8.75) must be modified to give an adhesive force of $2\pi r\gamma_0$.

THE STICKINESS OF PARTICLES, DROPLETS AND CELLS

There are two principal reasons for particles having a limited "stickiness", i.e. an intermediate value of k' in eq. (8.10).

Firstly, *molecules of high molecular weight* such as alginate, methylcellulose, polyelectrolyte or poly-vinylalcohol, may adsorb on to the surfaces of particles or cells, thus increasing $\boldsymbol{R}$ beyond the value where the electrical repulsion between the surfaces is important. Such materials are added to very fine precipitates to make them more readily filtrable, and to emulsions stabilized by electrical charges to cause clumping (and hence creaming) without coalescence. They effectively reduce **W** of eq. (8.10) while at the same time reducing the stickiness: k' is increased from a very small value to around 10^{-4} sec.$^{-1}$ (Table 8-XI). This is a direct result of the macromolecules forming bridges between the surfaces, which prevents intimate contact. The particles in the clump may have their surfaces separated by a distance of the order several times $1/\varkappa$, and the small electrical forces of repulsion at these separations are balanced by the forces of adsorption at each end of the bridging macromolecule. Living cells also may clump in the presence of various polysaccharides or protein molecules, which apparently form bridges between them[161b]. How k' is affected in such systems remains to be determined.

In general, the "bridge" substance must have rather unfolded molecules, and for certain proteins this may involve partial denaturation. Bacterial surfaces may contain polysaccharides[161c] or polypeptides[161d]; proteins also increase the adhesion of certain cells to glass.

The second mechanism of stickiness involves *bridging by polyvalent ions* between charged groups on two surfaces. This is well known in the field of detergency; particles of clay or carbon-black (the latter usually contains COO′ groups on the surface) adhere strongly to anionic groups on the surfaces of fibres in the presence of trace amounts of polyvalent metal ions such as Ca^{++} and Fe^{+3} (p. 420) while rust adheres exceptionally strongly. These ions are specifically adsorbed onto COO′ groups (Table 2-VII); it is noteworthy that Mg^{++} with its weaker specific adsorption, is much less effective than Ca^{++} in binding particles to fibres. Certain living cells, otherwise hydrophilic, can certainly become "sticky" for the same reason; ions of heavy metal such as Pb^{++} strongly promote clumping[161c] and are also highly toxic[161e] (p. 89).

TABLE 8-XI

System	k' (sec.$^{-1}$)	Characteristics of particles
Styrene-butadiene latex + Dupanol M.E.[161a]	~ 0	very strongly sticky
Colloidal Gold[161a]	1.8×10^{-4}	strongly sticky
Styrene-butadiene latex + Dupanol M.E. + methylcellulose[161a]	5×10^{-4}	moderately sticky
Hydrophilic colloids	∞	not sticky

Not only do these polyvalent ions form strong bridges between the anionic surfaces, however; they also reduce strongly the repulsive potentials ψ_0 or ψ_δ, which act as an energy barrier in preventing the close approach of the charged surfaces (cf. eq. (2.44), p. 89 and Fig. 8-2). This type of bridging is characterized by a very close approach of the surfaces, a high temperature coefficient of breaking, and a sensitivity to sequestering and complexing agents (which render ineffective the polyvalent ions concerned). If the surfaces are otherwise strongly hydrophobic, the reduction in the electrical charge by specific adsorption of polyvalent ions will result either in very strong adhesion or in coalescence.

SLIDING FRICTION

Sliding friction is the force retarding the sliding of one surface over another. This involves shearing the "welded" junctions across which adhesion is very strong (Fig. 8-50): there is no recovery of the elastic energy of deformation. The coefficient of friction μ is defined as the ratio of the frictional force to the normal force pressing the surfaces together: under certain conditions μ is constant at different loadings (Amonton's Law). The static friction is measured as the tangential force required to induce sliding: once the surfaces are moving the force, called now the kinetic friction, is less.

If the friction is *high*, there must necessarily be strong adhesion while the surfaces are pressed together, but whether a strong adhesion is *measurable* normal to the surface will depend on the importance of released elastic stresses. If the friction is low, the adhesion must necessarily be low[157] (e.g. for polytetrafluoroethylene).

Lubrication

The sliding friction between perfectly clean surfaces of nickel, tungsten, and copper is very high, though the oxide layers normally present reduce μ, the coefficient of friction, from 5 or more to a value of the order 0.5. A continuous oil layer between the metal surfaces will reduce μ to a much lower value than this, and if the lubricant layer is of sufficient thickness, the resistance to motion is due entirely to the viscosity of the interposed layer, there being no wear on the solid surfaces. Under these conditions of "fluid lubrication", μ is extremely small, of the order 0.002.

Usually, however, it proves impossible to maintain conditions for "fluid lubrication"; the thick lubricant layer breaks down and the metal surfaces are separated by adsorbed films of molecular dimensions, giving rise to "boundary lubrication", for which μ is about 0.1. The exact value of μ now depends on the nature of the metal surface and on the chemical composition of the lubricant, the bulk viscosity of the lubricating oil being of lesser importance. Addition of a little fatty acid to a non-polar oil is especially effective in boundary lubrication in reducing the friction and wear of metal surfaces: cadmium surfaces sliding in contact with the pure oil have $\mu \sim 0.6$, whereas a drop of lauric acid added to the sliding system suddenly reduces μ to 0.07. More quantitative measurements have shown that effective lubrication will occur with a boundary film of only 1 or 2 molecular thicknesses of lauric acid.

Single layers are worn away rapidly, however, and must be readily replaced[157]: clearly the rate of adsorption is important here. Because of the high viscosity of many oils, the diffusion coefficient (which varies inversely as viscosity) may be low, and B_1 (eq. (4.15)) is correspondingly reduced. As an example of such a slow adsorption we may cite the friction of cadmium surfaces in contact with a 0.001% (about 4×10^{-5} M) solution of lauric acid in paraffin oil: μ is 0.45 (comparable with the value for pure paraffin oil) when first measured, but after 12 hours μ has decreased to 0.26. More directly one may follow the rate of adsorption using the apparatus shown in Fig. 2-7; the vibrating plate is situated immediately above the thin oil layer, so that adsorption on to the former cannot occur. Provided adsorption at the oil-air interface is low, the adsorption fatty acid or other additive on to the lower metal surface may be followed by the change in potential with time. If the additives are non-ionic, no compensating electrical double layer can be set up in the oil, and the results are consequently readily interpreted by eq. (2.20).

Adsorption can also be measured with direct analysis (using perhaps a Langmuir trough) for the loss of material from the bulk solution. Longer hydrocarbon chains lead to higher adsorptions on nickel powder, and adsorption at these concentrations is nearly complete after a few minutes. Incidentally, since the acid is adsorbed as a monolayer, one may use adsorption results to estimate the area of the metallic powder, assuming that each adsorbed molecule occupies 20 Å^2 of surface.

With materials other than fatty acids or metallic soaps, more than a monolayer of lubricant is required to reduce μ effectively, as many as 9 layers of cholesterol being required on stainless steel. Chemical bonding of the first layer of a fatty acid may occur on a reactive metal, such as stainless steel or copper, to form metal soap, making the monolayer more adherent than one of an alcohol or ester, although a single layer is usually unable to provide continued protection.

Unreactive surfaces, such as nickel, chromium, platinum, and glass, are scarcely affected by the addition of fatty acid to the paraffinic oil, in contrast to copper, cadmium, and other reactive metals. Consequently, chemical attack of the fatty acids to form soaps is of primary importance in obtaining a strongly coherent film. This is supported by the observation that magnesium laurate (unlike lauric acid) lubricates platinum. Generally, however, soaps formed *in situ,* if this is possible, are more strongly linked to the metal surfaces and provide better lubrication. The chemical attack only occurs if some oxygen and water are present, the former substance forming an oxide layer, which is subsequently attacked by the fatty acid.

Structure of the Lubricating Layer

The first monolayer of a fatty acid film is different from the subsequent layers, in that, while the hydrocarbon chains in the first monolayer lie perpendicular to the surface, the top layers crystallize into the standard crystalline form of the material. Although the structure of the upper layers can be altered by rubbing with a degreased cloth, the first monolayer remains unaffected, due to its strong attachment to the solid surface. Even on non-reactive surfaces the metallic soaps (e.g. barium stearate) form more completely orientated monolayers than do the long-chain acids.

As the temperature is raised the film becomes more disorientated, and the temperature (about 110°C) at which laurate films on zinc, cadmium, and mild steel lose their orientation corresponds roughly to the melting point of the corresponding metallic soap. Lubrication ceases to be effective at about the same temperature (120°C for lauric acid chemically adsorbed on steel, and about 105°C on zinc). The soaps of the perfluoro-acids, however, are effective up to about 250°C[162].

Radioactive tracers have been useful[163] in determining the extent of the chemical adsorption: if a monolayer of a fatty acid is chemi-sorbed on to the surface of a radioactive metal, it will itself show radioactivity when removed. This occurs for fatty acids adsorbed on to zinc or cadmium; but there is no detectible metallic soap formation in the monolayers removed from radioactive platinum or gold.

Mechanism of Boundary Lubrication

The resistance to shear motion lies in intermolecular forces acting strongly at the points of contact of the solids. The lubricant, in causing the formation of monolayers on the surfaces before they come into contact, evidently minimizes intimate metallic contacts; though, since even under very light loads and with the best boundary lubricants μ and the amount of wear are by no means negligible, some metallic adhesion must still occur through the lubricant film (Fig. 8-52). In regions where the "peaks" on the surfaces

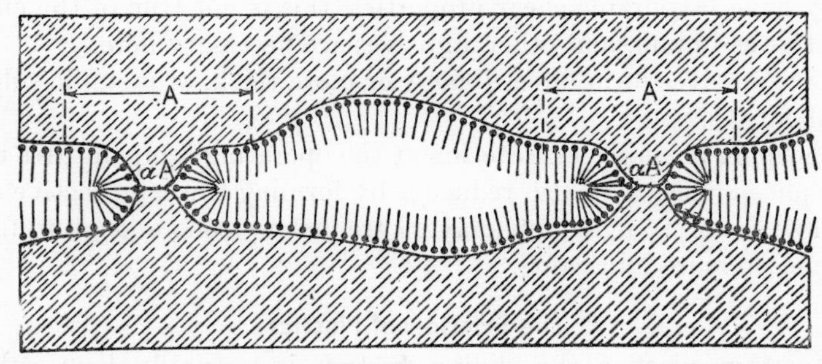

Fig. 8-52. The sliding of solid surfaces under conditions of boundary lubrication Breakdown of the lubricant film occurs at small localized regions. The load is supported over an area A, whilst metallic junctions are formed through the lubricant film over a much smaller area αA. The force to shear these metallic junctions constitutes part of the friction observed; and some friction also arises from shearing the lubricant film[157].

coincide, the pressure will be very high over regions large compared with molecular dimensions, and the resistance to motion will consist partly of the force required to break these intermetallic junctions. There will also be a smaller resistance to sliding due to the lubricant film in the other regions of the surfaces.

The main function of the lubricant film is to reduce the amount of close metallic contacts by interposing monolayers that are strong enough not to be penetrated without difficulty, but have easy shear between them. Firm attachment and strong lateral adhesion are characteristic of monolayers of good lubricants.

The wear of moving parts is enormously reduced by lubricants[164]: a reduction by a factor 10^4 or more may be achieved. That the coefficient of friction decreases only by a factor of about 10 under these conditions reflects the considerable force required to shear a monolayer on a solid surface. This is confirmed directly by experiments on mica[165].

Extreme Pressure Lubricants

Under extreme loads and speeds, monolayers of even the best boundary lubricants of the fatty acid or metallic soap type may break down. Additives that will maintain coherent surface films even under these conditions are now required: they must function at the high temperatures developed under extreme conditions of rubbing. As "extreme pressure" additives, compounds of chlorine, sulphur, and phosphorus are used extensively.

The chlorine compounds, which may include the $\rangle$Se Cl_2 group, react with steel surfaces at the elevated local temperatures (around 200°C) during extreme pressure sliding, giving metallic chloride as surface films many hundreds of angströms thick, and μ values about 0.1[157]. While iron chlorides evidently have favourable shear properties, this is not true of the chlorides of copper or cadmium, and other additives are necessary here.

Sulphur additives form films, 1000 Å or more thick, of metallic sulphides, with μ about 0.5.

The phosphorus additives, reacting at the operating temperatures to form metallic phosphides, may also reduce μ by forming thick protective surface films of low shear strength: monolayers are never strong enough to function as extreme pressure lubricants.

The Friction of Plastics

With certain plastics the sliding friction is naturally low: polytetrafluoroethylene is outstanding in its low friction and adhesion, μ for this material against itself[157,166] being of the order 0.05, close to the value of this plastic against ice[167]. Apparently the large size of the F atoms screens the dipoles of the surface groups, and the molecular cohesion remains very low when two surfaces are pressed together. The interlocking of the macromolecular chains in the bulk, however, ensures a reasonable shear strength for a block of the plastic. The low friction of polytetrafluoroethylene on ice is due to its low adhesion: the contact angle of water on this plastic is about 126°, which (by eq. (1.32)) shows that W is low. Polyethylene has a higher coefficient of friction against itself (order 0.6), but this may be reduced by incorporating oleamide in the material during manufacture: this is exuded during use of the plastic, forming a lubricating monolayer on the surface[166].

ROLLING FRICTION

This is generally less than sliding friction, both because the work of elastic deformation may be recovered[168] and because there is no "ploughing" of one surface by asperities on the other. Such deformation as occurs with a non-recoverable energy is largely responsible for rolling friction: interfacial slip contributes very little resistance. Consequently lubricants, although they may greatly reduce metallic transfer and wear during rolling, are relatively ineffective in reducing rolling friction.

WETTING

A solid is completely wet with a liquid if θ (measured as usual through the liquid) is zero. If an equilibrium is established, the relation of θ to $F^s_{S/A}$ and $F^s_{S/L}$ is:

$$\cos\theta = \frac{F^s_{S/A} - F^s_{S/L}}{\gamma_{L/A}} \tag{1.35}$$

If the right-hand side equals or exceeds unity, the liquid will wet the solid. To attain this, $F^s_{S/A}$ should be large, while $F^s_{S/L}$ and γ should be small. Often $F^s_{S/L}$ is large in practical systems where the surfaces are rough or particulate (e.g. surfaces covered with carbon-black, feathers, certain leaves): this is discussed below. By lowering γ, wetting agents allow the water to penetrate into the crevices on the surface, thereby also reducing $F^s_{S/L}$. Many of the best wetting agents have an irregularly shaped molecule: sodium di-n-octyl-sulphosuccinate is a typical example. Such molecules pack into micelles only with difficulty, with the consequence that the effective concentration of single molecules in the bulk can be exceptionally high, and γ correspondingly low. Even a surface of polytetrafluoroethylene can be wet by aqueous solutions of such materials. Other wetting agents include the sorbitan derivatives in the H.L.B. range 7-9, and non-ionic and ionic fluorinated derivatives.

The free energy terms in eq. (1.35) can be split into enthalpies and entropies: the former can be obtained from heats of immersion at several different temperatures. One may also attempt to calculate the critical wetting temperature (above which $\theta = 0$) from the energy of mixing of molecules of the liquid and of the solid surface[169].

In horticultural spraying, clearly some wetting is desirable, otherwise the spray drops bounce off elastically (p. 440). However, if the wetting is complete ($\theta = 0$) the liquid film on the leaf drains off rapidly and the deposition of fungicide is very low. Between these extremes lies the optimum, when the spray is retained on the leaf in the form of large drops. Non-ionic surface-active agents are very satisfactory in practice[170].

In the boiling of liquids[171], a layer of vapour may spread between the wall of the vessel and the liquid, hindering the conduction of heat: the addition of components which make the solid wall more readily wetted by the liquid can therefore greatly increase the heat transfer. Surface roughness also promotes the heat transfer.

The wetting of fabrics is discussed on p. 442.

NON-WETTING

From the relation

$$\cos \theta = \frac{F^s_{S/A} - F^s_{S/L}}{\gamma_{L/A}} \tag{1.35}$$

it is clear that if θ is to be as large as possible, ($\cos \theta$ to be as small as possible) γ and $F^s_{S/L}$ must be large, and $F^s_{S/A}$ should be low. In practice a contact angle of 90° is often regarded as a sufficient criterion of non-wetting, though the larger the value of θ the less coalescence will there be of the drops on the surface. Mercury, whose γ is high, has a high contact angle (~150°) on many solids, while on rough or hairy solids many liquids have high contact angles because $F^s_{S/L}$ is close to $F^s_{S/A}$. This can also be expressed by:

$$\cos \theta = f_1 \cos \theta_1 - f_2 \tag{1.38}$$

where f_1 and f_2 are the fractions of the composite surface which are respectively liquid-solid and liquid-air. On rough or hairy surfaces, therefore, contact angles of 140° or more can be readily attained. The third method of promoting non-wetting, by making $F^s_{S/A}$ low, has already been mentioned in connection with flotation, and is used in promoting drop-wise condensation of vapours, as discussed in detail below.

Many **natural surfaces** are non-wetted by water: this is partly a consequence of the high γ of pure water. It is well known that swans and ducks sink in water contaminated by oil or surface-active agents, and that a slightly greasy needle cannot be floated under these conditions, though it can on clean water. Aquatic insects and birds have a structure of fine hairs which make them difficult to wet because f_2 is large. In particular duck and swan feathers have a very regular structure (Fig. 8-53), unlike hen feathers or fur where the surface tension of the water is able to pull the fibres together and the animal becomes completely wetted with no pockets of air retained to confer buoyancy. Ordinary insects are resistant to sprays unless the liquid surface tension is lowered: water drops have a contact angle of about 97° on insects, which is reduced to 35° with certain oils, leading to a greater uptake of insecticide[172,173].

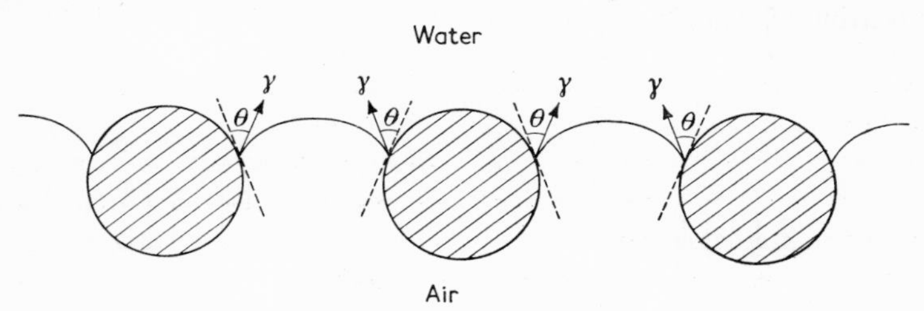

Fig. 8-53. Surface tension prevents water from passing a regular array of hairs or fibres.

While clean surfaces of **metals and glass** are generally completely wetted by clean liquids, it is frequently found that θ is no longer zero if the liquid contains a little surface-active agent[174]. Thus platinum and mica become non-wetted by dilute aqueous solutions of both cationic and anionic agents, though more concentrated solutions of the same agents reduce θ to zero again. Silica and glass, however, are always wetted by anionic agents, though they become readily non-wetted ($\theta \sim 45°$) by very low concentrations of cationic agents, because these adsorb on to the fixed negative charges of the silicate groups at the interface. Solutions of higher concentration again wet the silica or glass: for solutions of $C_{16}H_{33}N(CH_3)_3^+$ this occurs at about 7×10^{-3} M.

When the surface is non-wetted, it is covered by an incomplete monolayer as in Fig. 8-54(a): this film reduces $F^s_{S/A}$ more than it reduces $F^s_{S/L}$, and so $\cos\theta$ is reduced. This is confirmed by the observation[174] that long-chain cations from the air-water surface are deposited on to a glass slide as it is withdrawn through the surface: it follows that there is more adsorption at the glass-air than at the glass-water interface.

At higher concentrations of the surface-active agent, γ is lowered further and also polymolecular adsorption may occur[175], as shown simply by measuring the film thickness by the optical polarization method[176]. In particular, a hydrophilic double layer (Fig. 8-54b) reduces $F^s_{S/L}$ to a very low value, and cos θ increases (θ decreases) to give wetting again.

Radioactive tracers can also be used to measure the adsorption of a monolayer or of a bimolecular layer on glass: an elegant study has shown that strong adsorption of the counter-ions occurs at just the concentration at which θ decreases again (Fig. 8-55). Electrokinetic phenomena support this finding[177].

Solids can also be rendered hydrophobic by adsorption of a monolayer of a silane derivative (p. 36) or by application of a thicker film of wax, silicone, or carbon-black. The silicone is more adherent than the carbon-black, but the latter is more effective (θ approaches 180°) because of the plate-like structure of the coating (f_1 small in eq. (1.38)).

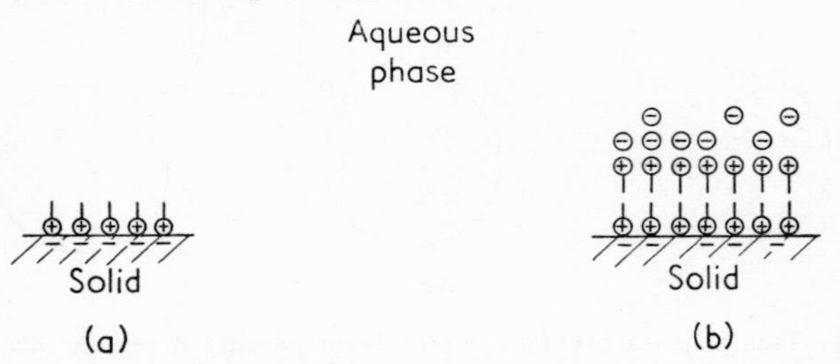

Fig. 8-54. Diagrammatic representation of adsorption of (a) a monolayer and (b) a bimolecular layer at the water-solid interface. In (a) $F^s_{S/A}$ is reduced more than is $F^s_{S/L}$, resulting in non-wetting. In (b) the bimolecular layer, adsorbed from higher concentrations of surface-active agent in the aqueous phase, reduces $F^s_{S/L}$ to a very low value, and the surface becomes wetted.

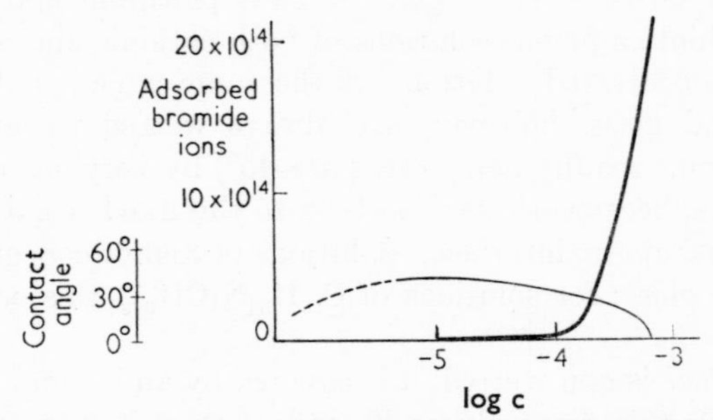

Fig. 8-55. The heavy line shows the variation of the number of adsorbed Br^{82} counter-ions from $C_{16}H_{33}N(CH_3)_3^+Br'$ adsorbed on to glass, as a function of concentration of the long-chain salt[177]: ordinates are ions per cm.^{2} The light line shows the variation of contact angle: the glass plate is wetted ($\theta = 0$) at very low and at high concentrations.

To prevent corrosion or to increase heat transfer across metal surfaces, these are often required to be dewetted on a practical scale. Deposited monolayers of amines and of perfluoro-acids are particularly effective[178] in preventing wetting and corrosion, but are rather fragile. In steam condensers, dropwise condensation of water on to a hydrophobic surface of a metal condenser allows a heat transfer rate about four times greater than if the water condenses into a continuous liquid layer[179]; this is because conduction of heat

(a)

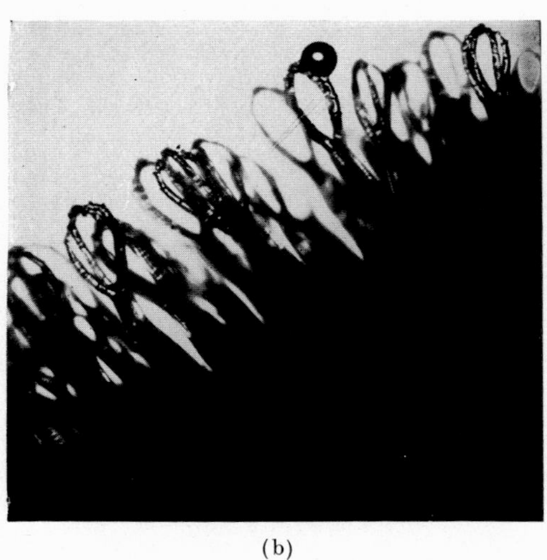

(b)

Fig. 8-57. In (a) the projections on a *Salvinia* leaf are viewed from above. In (b), the projections are seen in profile[173]. (Reproduced by courtesy of Professor N. K. Adam and of *Endeavour*.)

[*To face p. 439*

through the liquid layer can occur only slowly. If, however, discrete drops are formed, these will run off and leave bare metal (Fig. 8-56). In practical condensers the surface film must have a long working life, and sulphur or selenium groups attached to the hydrocarbon chains have proved useful. Monolayers do not last very long, and layers of up to 1000 molecules are required in practice. Sometimes solubilizing groups are included in the molecule, so that traces of the surface-active agent can be added continuously with the steam. Xanthates and pyridine compounds have also been used: the problem is essentially similar to that of obtaining flotation of

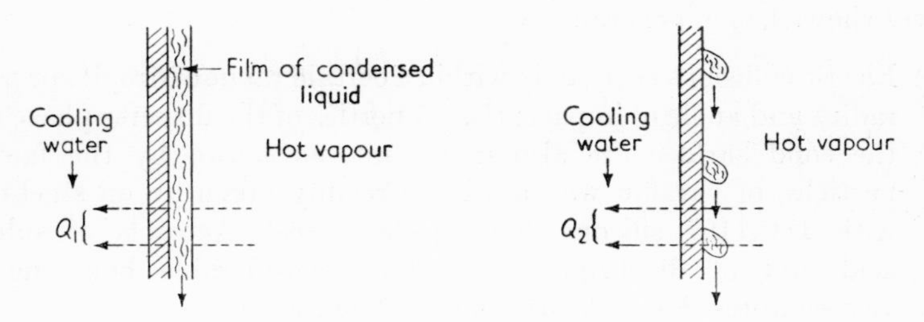

Fig. 8-56. Condensation of the hot vapour into a continuous film compared with dropwise condensation. The heat transfer Q_2 is much greater than Q_1.

a mineral or to that of extreme pressure lubrication. A coating of polytetrafluoroethylene ($F^s_{S/A}$ low) has also been suggested: indeed this material seems essential for the dropwise condensation of organic liquids, whose γ is low. A baked-on film of silicone on mild steel might possibly prove effective.

The leaves of certain plants (e.g. peas and cabbage) consist of plates of waxy materials that confer water-repellency both on account of the wax and of the roughness. Further, the water may make only incomplete contact (f_1 small in eq. (1.38)), the overall result being that water drops on such leaves show contact angles as high as 140°. Other plants (e.g. *Salvinia*) have leaves from which many fine hairs protrude (Fig. 8-57): again f_2 is large and the water forms discrete drops on the leaves[173,180,181]. An interesting observation is that trichloracetic acid and certain other compounds used in agriculture for treatment of the soil before germination are known to increase the susceptibility of some weeds to herbicides: the explanation lies in the fact that, with increasing concentrations of trichloracetic acid, the waxy projections are progressively reduced, with consequent reduction in the contact angle[181].

Reflection of incident drops from a surface showing a high contact angle is important in several practical fields, including horticultural and insecticidal spraying, and the filtration of aerosols.

The use of selective weed killers of the growth-regulator type has focused interest on this type of behaviour of liquids on leaves, since the observed differences in susceptibility to hormonal weed killers may well be due to the fact that certain leaves are able to reflect droplets of spray, so failing to absorb them. Glass surfaces coated with carbon-black or MgO show the same reflective property towards incident water droplets[182], the high contact angle (approaching 180°) again resulting from the heterogeneity of the surface. Drops of mercury, which have a high contact angle even on clean, smooth glass are well known to bounce off readily.

Experimental studies of the collisions of drops and of particles with surfaces show that in general:

(i) Elastic collisions occur only within a certain range of droplet or particle radius and angle of impact: the properties of the disperse phase and of the solid surface are also important, as shown by the fact that particles of paraffin wax are more readily taken up on steel coated with D.C.1107 silicone than on clean steel. Aerosols of sulphuric acid and of dioctylphthalate show considerable bouncing from viscose fibres[183] at velocities above 5 cm.sec.$^{-1}$

(ii) The area of contact between the droplet or particle and the solid depends on the velocity of impact. Shearing or the formation of a "neck" of liquid[184] may then allow most of the particle or droplet to be removed elastically, even if the contact angle is less than 90°.

(iii) Droplet diameter is important: collisions of small water drops ($50\mu > a > 10\mu$) are not elastic, while larger water drops ($a \gg 125\mu$) always bounce off both pea leaves and carbon-black films[185].

(iv) Surface tension is important, as shown by adding methanol or acetic acid to the water spray. The elasticity of the collisions of drops of radius 125μ persists, provided that $\gamma \gg 57$ dynes cm.$^{-1}$ for pea leaves, and $\gamma \gg 34$ dynes cm.$^{-1}$ for carbon-black films. Conversely, for drops of $\gamma \leqslant 39$ dynes cm.$^{-1}$ none of the collisions against pea leaves is elastic; and if $\gamma = 32$ dynes cm.$^{-1}$ only 30% of the incident drops collide elastically with a carbon-black surface.

(v) Lowering the air pressure does not affect the phenomenon, suggesting that a cushion of air is not responsible for the elastic collisions. Further, the drops rebounding from a carbon-black layer carry a little carbon-black with them, showing that close contact occurs on collision. Smoothing a leaf by rubbing with the finger or with solvents destroys the reflecting property.

A theoretical treatment[185] of elastic collisions of drops, relates the energies of the approaching drop and of the drop on the surface to the energy required for complete detachment of the drop. If the drop, while on the solid surface, is in contact with air and with the solid over areas A_1 and A_2 (Fig. 8-58), then the total surface energy of the drop on the surface is

$$F^s_{drop} = A_1\gamma_{L/A} + A_2F^s_{S/L} - A_2F^s_{S/A} \tag{8.76}$$

If θ is the contact angle of the drop on the surface, this equation becomes (by virtue of eq. (1.35))

$$F^s_{drop} = \gamma_{L/A}(A_1 - A_2 \cos\theta)$$

If a is the radius of the drop (assumed spherical in air), A_1 and A_2 are functions of a^2 and θ only. The energy required to detach the drop from the surface is $4\pi a^2\gamma_{L/A} - F^s_{drop}$, i.e.

$$F^s_{detachment} = 4\pi a^2\gamma_{L/A} f(\theta) \tag{8.77}$$

This function is plotted in Fig. 8-59. If the kinetic energy stored in the drop just after it hits the surface exceeds $F^s_{detachment}$, the drop will be reflected.

Consider, for example, a drop of water of radius 50μ; its surface energy ($4\pi a^2\gamma_{L/A}$) is 2.2×10^{-2} erg. Such a drop, moving at 25 cm.sec.$^{-1}$ has a kinetic

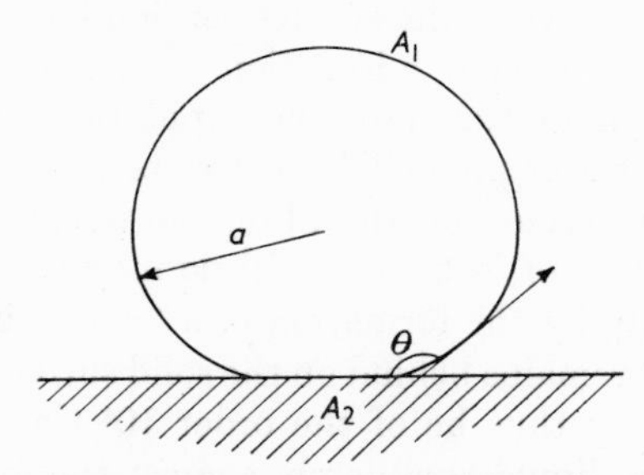

Fig. 8-58. Drop in (temporary) contact with solid surface over a circle of contact of area A_2. The total surface energy of the drop is given by eq. (8.76).

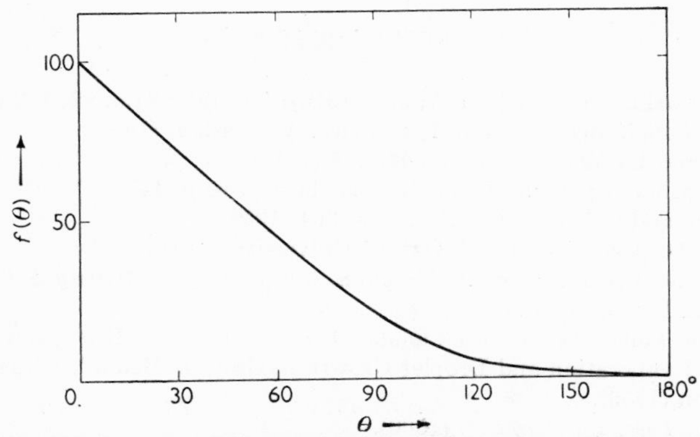

Fig. 8-59. Plot of f(θ) (i.e. energy required to free drop completely from surface, as percentage of $4\pi a^2\gamma_{L/A}$) against θ[185].

energy of 1.5×10^{-4} erg, though, after striking a solid surface, its energy available for detachment will be even less than this, owing to viscous losses during the deformation following collision. Calculation shows that, if 0.5×10^{-4} erg is available for detachment, this energy will be sufficient only if the contact angle is very high, about 179°. This explains why such small drops are always retained on the solid surface under practical conditions. For a drop of radius 125μ, the surface energy in the absence of additives is 13.8×10^{-2} erg and the kinetic energy of the falling drop is about 4×10^{-2} erg, of which we shall assume 2×10^{-2} erg is available for detachment. The ordinate on Fig. 8-59 is therefore $(2 \times 10^{-2}/13.8 \times 10^{-2}) \times 100$, i.e. 14.5, and so reflection of the droplets will occur from a solid surface on which the contact angle exceeds 100°. Lowering $\gamma_{L/A}$ to 40 dynes cm.$^{-1}$ would reduce the surface energy to 8×10^{-2} erg, corresponding to a minimum contact angle of 85° for reflection; this angle is not, however, easily maintained in the presence of surface-active agents, which therefore assist retention of the drops.

Fabrics can be water-proofed if the inter-fibre spacings are uniform and are not reduced by the force of the surface tension between them. It is also necessary, however, to prevent adsorption of water molecules[186], as this could lead to surface migration into the capillaries, ultimately producing wetting. In a fabric water-proofed both by its texture and by adsorbed monolayers of cationic agents on the fibre, the pores remain open for the passage of air. The drops of water on a fabric may readily be shaken off if θ is 90° or 100°, usually by the formation of a "neck" in the liquid drop: a small amount of liquid remains behind on the solid surface. The force required to detach the drop may thus be of the order 10^4 times smaller than that required to remove the liquid completely against the forces of adhesion[184].

REFERENCES

1. Smoluchowski, *Phys. Z.* **17**, 557, 585 (1916); *Z. phys. Chem.* **92**, 129 (1917); Kruyt, "Colloid Science" Vol. I, Elsevier, Amsterdam (1952).
2. Sinclair and La Mer, *Chem. Rev.* **44**, 245 (1949); See also other papers in Chem. Rev. **44** through to p. 417; Burgoyne and Cohen, *J. Colloid Sci.* **8**, 364 (1953); Ingold, "Dispersal in Fungi", Oxford University Press (1953); La Mer and Drozin, *Proc. 2nd Internat. Congr. Surface Activity* **2**, 600, Butterworths, London (1957); Green and Lane, "Particulate Clouds: Dust, Smokes and Mists", London (1957); Fuchs, "Evaporation and Droplet Growth in Gaseous Media", Pergamon Press, London (1959). Dunning, *Disc. Faraday Soc.* (1960).
3. Vonnegut and Neubauer, *J. Colloid Sci.* **7**, 616 (1952); **8**, 551 (1953); Drozin, *J. Colloid Sci.* **10**, 158 (1955).

4. Derjaguin et al., *Proc. 2nd Internat. Congr. Surface Activity* **3**, 594, Butterworths, London (1957).
5. Dunning, in "Chemistry of the Solid State", p. 159, Butterworths, London (1955).
"Nucleation Phenomena", *Industr. Engng Chem.* **44**, p. 1268 ff (1952).
6. Volmer and Weber, *Z. phys. Chem.* **119**, 277 (1925);
Volmer and Flood, *Z. phys. Chem.* **A170**, 273 (1934).
7. Sander and Damköhler, *Naturwissenschaften* **31**, 460 (1943).
8. Whytlaw-Gray and Patterson, "Smoke", Arnold, London (1932).
9. Gillespie, *Proc. roy. Soc.* **A216**, 569 (1953).
10. Rayleigh, *Proc. roy. Soc.* **28**, 406 (1879); **29**, 71 (1879); **34**, 130 (1882); *Phil. Mag.* **48**, 328 (1899);
Mahajan, *Phil. Mag.* **10**, 383 (1930).
Benedicks and Sederholm, *Ark. Mat. Astr. Fys.* **B30** (5) (1944).
11. Linton and Sutherland, *J. Colloid Sci.* **11**, 391 (1956).
12. Elton, Mason, and Picknett, *Trans. Faraday Soc.* **54**, 1724 (1958).
13. Welch in "Colloid Chemistry" VI, 274, (ed. J. Alexander) Reinhold, New York (1946);
Stokes in "Colloid Chemistry" VII, 403, (ed. J. Alexander) Reinhold, New York (1950).
13a. Keith and Derrick, *J. Colloid Sci.* **15**, 340 (1960).
14. Sänger *Proc. 2nd Internat. Congr. Surface Activity* **2**, 583, Butterworths, London (1957).
15. Clayton, "The Theory of Emulsions and their Technical Treatment" (5th ed., Sumner) Churchill, London (1954).
16. Becher, "Emulsions, Theory and Practice", Reinhold, New York (1957).
17. Davies, *Proc. 3rd Internat. Congr. Surface Activity*, Cologne.
18. Nawab and Mason, *J. Colloid Sci.* **13**, 179 (1958).
19. Gad, *Arch. Anat. Physiol., Lpz.* 181 (1878).
Brücke, *Anz. Akad. Wiss. Wien.* **79**, 267 (1879).
Quincke, *Wiedemans Ann.* **35**, 593 (1888).
20. McBain, "Colloid Science", D. C. Heath and Company, Boston (1950);
Kruyt, "Colloid Science" Vol. I, p. 340, Elsevier, Amsterdam (1952).
21. Gurwitsch, "Wissenschaftliche Grundlagen der Erdölbearbeitung (1913) p. 430, (Trans. Moore) Chapman and Hall, London (1932).
22. Raschevsky, *Z. Phys.* **46**, 568 (1928).
23. v. Stackelberg, Klockner, and Mohrhauer, *Kolloidzschr.* **115**, 53 (1949).
24. Davies and Haydon, *Proc. 2nd Internat. Congr. Surface Activity* **1**, 417, 476, Butterworths, London (1957).
25. McBain and Woo, *Proc. roy. Soc.* **A163**, 182 (1937).
Kaminski and McBain, *Proc. roy. Soc.* **A198**, 447 (1949).
26. Haydon, *Nature, Lond.* **176**, 839 (1955);
Lewis, J. B. and Pratt, *Nature, Lond.* **171**, 1155 (1953).
27. Pospelova and Rehbinder, *Acta phys.-chim. URSS.* **16**, 71 (1942).
28. van der Waarden, *J. Colloid Sci.* **7**, 140 (1952).
29. Ilkovič, *Coll. Trav. chim. Tchécosl.* **4**, 480 (1932).
30. Langmuir, *Cold Spr. Harb. Symp.* **6**, 193 (1938).
31. Cockbain and Schulman, *Trans. Faraday Soc.* **36**, 651 (1940).
32. Matalon, *Trans. Faraday Soc.* **46**, 674 (1950).
33. Kaminski and McBain, *Proc. roy. Soc.* **A198**, 447 (1949).

34. Davies, Bell, G. and Law, Research Project in Department of Chemical Engineering, Cambridge (1960).
35. Hartung and Rice, *J. Colloid Sci.* **10**, 436 (1953).
36. Lawrence and Mills, O. S., *Disc. Faraday Soc.* **18**, 98 (1954).
37. Ostwald, *Kolloidzschr.* **6**, 103 (1910).
38. Cheesman and King, A. *Trans. Faraday Soc.* **34**, 594 (1938).
39. Andreas, *J. chem. Educ.* **15**, 523 (1938).
40. Isemura and Kimura, *Mem. Inst. sci. Res. Osaka Univ.* **6**, 54 (1948).
41. Griffin, "Emulsions" in *Encyclopedia of Chemical Technology* **5**, 692, (ed. Kirk and Othmer) New York (1950).
42. Dickinson and Iball, *Research, Lond.* **1**, 614 (1948).
43. Robertson, O., *Kolloidzschr.* **7**, 7 (1910).
44. Stamm, *J. phys. Chem.* **30**, 998 (1926).
45. Robinson, C., *Trans. Faraday Soc.* **32**, 1424 (1936);
Davies and Mayers, *Trans. Faraday Soc.* **56**, 691 (1960).
46. Derjaguin, *Trans. Faraday Soc.* **36**, 203 (1940).
47. Verwey and Overbeek, "Theory of the Stability of Lyophobic Colloids", Elsevier, Amsterdam (1948).
48. Derjaguin and Kussakov, *Acta phys.-chim. URSS.* **10**, 25 (1939).
49. Gillespie and Rideal, *Trans. Faraday Soc.* **52**, 173 (1956).
50. Davies, *Proc. 2nd Internat. Congr. Surface Activity* **1**, 426, Butterworths, London (1957).
51. Pethica and Few, *Disc. Faraday Soc.* **18**, 258 (1954).
52. Albers and Overbeek, *J. Colloid Sci.* **14**, 501, 510 (1959).
52a. Vand, *J. Phys. Coll. Chem.* **52**, 314 (1948).
52b. Bernal, *Trans. Instn. chem. Engrs, Lond.* (1960).
53. Derjaguin, *Proc. 2nd Internat. Congr. Surface Activity* **1**, 477, Butterworths, London (1957).
54. McBain and Henniker, in "Colloid Chemistry" VII, 67, (ed. J. Alexander) Reinhold, New York (1950).
55. Alexander in "Surface Phenomena in Chemistry and Biology" p. 18, Pergamon Press, London (1958).
56. Harkins, *Science* **102**, 292 (1945);
Betts and Pethica, *Trans. Faraday Soc.* **56**, 1515 (1960).
57. Schulman and Cockbain, *Trans. Faraday Soc.* **36**, 651, 661 (1940).
58. van der Waarden, *J. Colloid Sci.* **5**, 317 (1950).
59. van der Waarden, *J. Colloid Sci.* **6**, 443 (1951).
60. Mackor, *J. Colloid Sci.* **6**, 492 (1951);
Mackor and van der Waals, *J. Colloid Sci.* **7**, 535 (1952).
61. Archer and La Mer, *Ann. N.Y. Acad. Sci.* **58**, 807 (1954).
62. Rehbinder and Wenstrom, *Kolloidzschr.* **53**, 145 (1930);
Mahajan, *Phil. Mag.* **10**, 383 (1930);
Benedicks and Sederholm, *Ark. Mat. Astr. Fys.* **30B**, (5) (1944);
Davies, *Perfum. essent. Oil Rec. Yearb.* **43**, 338 (1952);
Cockbain and McRoberts, *J. Colloid Sci.* **8**, 440 (1953);
Nielsen, Wall, and Adams, *J. Colloid Sci.* **13**, 441 (1958);
Watanabe and Kusui, *Bull. chem. Soc. Japan* **31**, 236 (1958).
63. Bancroft, *J. phys. Chem.* **17**, 514 (1913).
64. Bowcott and Schulman, *Z. Elektrochem.* **59**, 283 (1953).
65. Griffin, *J. Soc. cosmet. Chem.* **1**, 311 (1949).

66. Griffin, *J. Soc. cosmet. Chem.* **5**, 4 (1954).
67. Griffin, *Off. Dig. Fed. Paint Varn. Prod. Clubs*, June 1956.
68. Hutchinson, E., *J. phys. Chem.* **52**, 897 (1948).
69. Kanig, Chavkin, and Lerea, *Drug Cosmet. Ind.* (1954);
Griffin, Private communication;
Maclay, *J. Colloid Sci.* **11**, 272 (1956).
Greenwald, Brown, G. L. and Fineman, *Analyt. Chem.* **28**, 1693 (1956);
Tanaka, *Proc. 2nd Internat. Congr. Surface Activity* **4**, 132, Butterworths, London (1957).
70. Clay, *Proc. Acad. Sci. Amst.* **43**, 852 (1940). In English.
71. Dvoretskaya, *Koll. Zhur.* **11**, 311 (1949); **13**, 432 (1951).
72. Stamm, *J. phys. Chem.* **30**, 998 (1926).
73. Cheesman and King, *Trans. Faraday Soc.* **34**, 594 (1938).
74. Rehbinder, *Colloid J. Voronezh* **8**, 157 (1946).
Kremnev and Ravdel, *C. R. Acad. Sci. U.R.S.S.* **90**, 405 (1953).
75. Saffman and Taylor, G. I., *Proc. roy. Soc.* **A245**, 312 (1958).
76. van den Tempel, *Proc. 2nd Internat. Congr. Surface Activity* **1**, 439, Butterworths, London (1957); *J. Colloid Sci.* **13**, 125 (1958).
76a. Davies and Pugh, to be published (1963).
77. Sutheim, "Introduction to Emulsions", New York (1946).
78. Tartar, Lothrop, and Pettengill, *J. phys. Chem.* **34**, 373 (1930).
79. Lawrence, *Chem. & Ind.* (*Rev.*) **39**, 615 (1948).
80. Lawrence and Killner, *J. Inst. Petrol.* **34**, 821 (1948).
81. Jellinek and Anson, *J. Soc. chem. Ind., Lond.* **68**, 108 (1949).
82. Monson and Stenzel, in "Colloid Chemistry" VI, 535, (ed. J. Alexander) Reinhold, New York (1946).
83. Cockbain and McMullen, *Trans. Faraday Soc.* **47**, 322 (1951).
84. Bikerman, "Surface Chemistry" pp. 264, 378–82, Academic Press, N.Y. (1958).
85. Jura and Garland, *J. Amer. chem. Soc.* **74**, 6033 (1952).
Brunauer, Kantro, and Weise, *Canad. J. Chem.* **34**, 729 (1956).
Brunauer, *Proc. 2nd Internat. Congr. Surface Activity* **2**, 17, Butterworths, London (1957).
86. Jirgensons and Straumanis, "A Short Textbook of Colloid Chemistry" e.g. p. 20, Pergamon Press, London (1954).
87. Turnbull, *J. chem. Phys.* **20**, 411 (1952).
88. von Weimarn, *Chem. Rev.* **2**, 217 (1926).
88a. Turnbull, *J. chem. Phys.* **18**, 198 (1950).
88b. Phibbs and Schiff, *J. chem. Phys.* **17**, 843 (1949);
Frank, F. C. *J. chem. Phys.* **18**, 231 (1950);
Seki, *J. chem. Phys.* **18**, 397 (1950).
89. Derjaguin, *Trans. Faraday Soc.* **36**, 203 and 730 (1940); *Disc. Faraday Soc.* **18**, 85 (1954).
Derjaguin and Landau, *Acta Phys.-chim. URSS.* **14**, 633 (1941) (in English); *J. exp. theor. Phys.* **11**, 802 (1941); **15**, 662 (1945).
90. Verwey and Overbeek, Ref. 47, pp. 61, 101, 103, 139, 150, 152, 160, 168, 171; *J. Colloid Sci.* **10**, 224 (1955).
91. Hoskin, *Trans. Faraday Soc.* **49**, 1471 (1953); *Phil. Trans.* **248**, 433 (1956); Hoskin and Levine, ibid. 449, (1956).
92. Hamaker, *Physica, 's Grav.* **4**, 1058 (1937).
93. Derjaguin and Abrikosova, *Disc. Faraday Soc.* **18**, 24 (1954);

Derjaguin, Abrikosova, and Lifshitz, *Quart. Rev. chem. Soc., Lond.* **10**, 295 (1956).
Kitchener and Prosser, *Nature, Lond.* **178**, 1339 (1956); *Proc. roy. Soc.* **A242**, 403 (1957).
94. Oster, *J. gen. Physiol.* **33**, 445 (1950).
94a. van Olphen, *J. Colloid Sci.* **17**, 660 (1962).
95. Eilers and Korff, *Trans. Faraday Soc.* **36**, 229 (1940).
96. Van der Minne and Hermanie, *J. Colloid Sci.* **7**, 600 (1952); **8**, 38 (1953).
Garner, Mohtadi, and Nutt, *J. Inst. Petrol.* **38**, 974, 986 (1952).
Koelmans and Overbeek, *Disc. Faraday Soc.* **18**, 52 (1954).
Koelmans, *Philips Res. Rep.* **10**, 161 (1955).
97. van der Waarden, *J. Colloid Sci.* **5**, 317 (1950); **6**, 443 (1951).
Mackor, *J. Colloid Sci.* **6**, 492 (1951).
Mackor and van der Waals, *J. Colloid Sci.* **7**, 535 (1952).
Cousens, *Disc. Faraday Soc.* **18**, 191 (1954).
Dintenfass, *Kolloidzschr.* **161**, 60, 70 (1958).
98. Auerbach, *Kolloidzschr.* **74**, 129 (1936); **77**, 161 (1936); **80**, 27 (1937);
Calderbank, *Trans. Inst. chem. Engrs* **36**, 443 (1958); **37**, 173 (1959); *Brit. chem. Engng* **1**, 267 (1956).
99. Dodd, *Proc. Phys. Soc. Lond.* **B68**, 686 (1955);
Glaser and Rahm, *Phys. Rev.* **97**, 474 (1955).
100. Westwater, *Advanc. Chem. Engng* **1**, 1 (1956); **2**, 1 (1958).
Newitt, Dombrowski, and Knelman, *Trans. Inst. chem. Engrs* **32**, 244 (1954).
101. Boys, "Soap Bubbles", Macmillan, London and New York (1924);
Mysels, Shinoda, and Frankel, "Soap Films", Pergamon Press, London (1959).
102. Peterson, Neill, and Jablouski, *Industr. Engng Chem.* (*Anal.*) **48**, 2031 (1956).
103. Ratzer, *Industr. Engng Chem.* (*Anal.*) **48**, 2013 (1956).
104. Metzner and Brown, L. F., *Industr. Engng Chem.* (*Anal.*) 48, 2040 (1956).
105. Ostwald and Siehr, *Kolloidzschr.* **76**, 33 (1937).
Peters, D., *Kolloidzschr.* **125**, 157 (1952).
106. Brady, *J. phys. Chem.* **53**, 56 (1949).
107. Perrin, *Ann. Phys., Paris* **10**, 180 (1918).
Raison, *C.R. Acad. sci., Paris* **232**, 1660 (1951).
107a. Sebba, *Nature, Lond.* **184**, 1062 (1959); **188**, 736 (1960).
108. Bikerman, "Foams" Reinhold, New York (1953).
109. Matalon in "Flow Properties of Disperse Systems" p. 323, (ed. Hermans) North Holland Publishing Co., Amsterdam (1953).
110. Ross, J. and Miles, *Oil & Soap* **18**, 99 (1941).
111. Brown, A. G., Thuman and McBain, *J. Colloid Sci.* **8**, 491, 508 (1953).
112. Brady and Ross, S., *J. Amer. chem. Soc.* **66**, 1348 (1944).
113. Jacobi, Woodcock, and Grove, *Industr. Engng Chem.* (*Anal.*) **48**, 2046 (1956).
114. Ross, S., *J. phys. Chem.* **47**, 266 (1943).
115. de Vries, "*Proc. 2nd Internat. Congr. Surface Activity*" **1**, 256, Butterworths, London (1957); *Rec. Trav. chim. Pays-Bas* **76**, 81, 209, 283, 383, 441 (1958).
116. Kitchener and Cooper, *Quart. Rev. chem. Soc., Lond.* **13**, 71 (1959).
117. Lawrence, "Soap Films", Bell, London (1929);
Dewar, *Proc. roy. Instn. G.B.* **22**, 192 (1917).
118. Davies, unpublished work.
119. Derjaguin and Titijevskaya, *Proc. 2nd Internat. Congr. Surface Activity* **1**, 211, Butterworths, London (1957), and quotations herein to other papers by Derjaguin.

120. Bikerman, *Proc. 2nd Internat. Congr. Surface Activity* **1**, 254, 255, Butterworths, London (1957); "Proceedings of 124th A.C.S. Meeting", 2I, 4, (1953).
121. Cumper, *Trans. Faraday Soc.* **49**, 1360 (1953).
121a. Goodman, *Chem. and Industry* (1963).
122. Ross, S. and Butler, J. N., *J. phys. Chem.* **60**, 1255 (1956).
123. Trapeznikov, *Proc. 2nd Int. Congr. Surf. Act.* **1**, 242, Butterworths, London (1957).
124. Marangoni, *Nuovo Cim.* **2**, 5–6, 239 (1871); **3** 97 (1878).
125. Gibbs (1878), see "Collected Works" **1**, 300-314, Longmans Green & Co. Ltd., London (1928).
126. Jones, T. G., Durham, Evans, W. P., and Camp, *Proc. 2nd Internat. Congr. Surface Activity.* **1**, 225, Butterworths, London (1957).
126a. Burcik and Newman, *J. Colloid Sci.* **15**, 383 (1960).
127. Hardy, *Proc. roy. Soc.* **A86**, 610 (1912); *J. chem. Soc.* **127**, 1207 (1925).
128. Talmud and Suchowolskaja, *J. phys. Chem., Moscou* **2**, 31 (1931); *Z. phys. Chem.* **154A**, 277 (1931).
129. Rehbinder and Trapeznikov, *Acta Phys.-chim. URSS.* **9**, 257 (1938);
Trapeznikov, *Acta Phys.-chim. URSS.* **13**, 265 (1940).
130. Kritchevsky and Sanders, World Congress on Surface-Active Agents, Summaries p. 21, Paris (1954).
131. Merrill and Moffett, *Oil & Soap* **21**, 170 (1941);
Camp and Durham, *J. phys. Chem.* **59**, 993 (1955).
132. Davies, *Proc. 2nd Internat. Congr. Surface Activity* **1**, 220, Butterworths, London (1957).
133. Ewers and Sutherland, *Aust. J. Sci. Res.* **5**, 697 (1952).
134. Gibbs, "Collected Works", **1**, 313, Yale University Press (1948);
135. Rose, *Nature, Lond.* **157**, 299 (1946).
Reidel, *Kolloidzschr.* **83**, 31 (1938).
Skogen, *Amer. J. Phys.* **24**, 239 (1956).
136. Baird, *Trans. Faraday Soc.* **56**, 213 (1960); *Nature, Lond.* (1960).
137. Danielli, *J. cell. comp. Physiol.* **7**, 393 (1936);
Sigwart and Nassenstein, *Naturwissenschaften* **42**, 458 (1955);
Z. Ver. dtsch. Ing. **98**, 453 (1956).
138. McBain, J. W., "Colloid Science", Boston (1950);
McBain, M. E. L. and Hutchinson, E., "Solubilization", Academic Press Inc., New York (1955);
Klevens, *Chem. Rev.* **47**, 1 (1950).
139. Stevenson, *J. Text. Inst.* **42**, T 194 (1951); **44**, T 12 (1953);
Adam and Stevenson, *Endeavour* **12**, 45 (1953).
Lawrence, *Proc. 2nd Internat. Congr. Surface Activity* **1**, 475, Butterworths, London (1957); *Nature, Lond.* **183**, 1491 (1959).
139a. Davies, *Proc. 2nd Congr. Int. Fed. Cosmetic Chemists*, Pergamon Press (1963).
140. Adam, *J. Soc. Dy. Col.* **53**, 121 (1937);
Palmer, *J. Soc. chem. Ind., Lond.* **60**, 59 (1941);
Adam and Stevenson, *Endeavour* **12**, 45 (1953).
141. Washburn, *Phys. Rev.* **17**, 273, 374 (1921);
Rideal, *Phil. Mag.* **44**, 1152 (1922).
142. Eley and Pepper, *Trans. Faraday Soc.* **42**, 697 (1946);
Peek and McLean, *Industr. Engng Chem. (Anal.)* **6**, 85 (1934);
Barrer, *Disc. Faraday Soc.* **3**, 61 (1948);
Carman, *Disc. Faraday Soc.* **3**, 72 (1948).

143. Durham and Camp, *Proc. 2nd Internat. Congr. Surface Activity* **4**, 3, Butterworths, London (1957).
144. Fowkes, *J. phys. Chem.* **57**, 98 (1953);
145. "Surface Activity and Detergency" (ed. Durham) MacMillan & Co., London (1960);
Wagner, *Proc. 2nd Internat. Congr. Surface Activity* **4**, 113, Butterworths, London (1957);
Durham, ibid. **4** 60 (1957).
146. Porter, *Proc. 2nd Internat. Congr. Surface Activity* **4**, 103, Butterworths, London (1957).
147. Wolfhoff and Overbeek, *Rec. Trav. chim. Pays-Bas*, **78**, 759 (1959).
148. Sven-Nilsson, *Kolloidzschr.* **69**, 230 (1934).
149. Sutherland and Wark, "Principles of Flotation", Melbourne (1955);
Also discussion in *Proc. 2nd Internat. Congr. Surface Activity* **3**, 202–385, Butterworths, London (1957); Sutherland, *J. phys. Chem.* **63**, 1717 (1959).
150. De Bruyn, *Min. Engng.* **7**, 291 (1955).
151. Milone and Cetini, *Proc. 2nd Internat. Congr. Surface Activity* **3**, 469, Butterworths, London (1957).
152. Pacter, *J. phys. Chem.* **59**, 1140 (1955).
153. "*Crystal Growth*", *Disc. Faraday Soc.* **5** (1949); *U.S. Pat. Syst. Leaflet.* 2,723,956;
Egli and Zerfoss, *Disc. Faraday Soc.* **5**, 61 (1949);
Buckley, "Crystal Growth" pp. 169–222 and 529–559. Wiley and Sons Inc., New York (1951);
McCartney and Alexander, A. E., *J. Colloid Sci.* **13**, 383 (1958).
153a. Bartell and Osterhof, *Industr. Engng Chem.* **19**, 1277 (1927);
Wolkova, *Kolloidzschr.* **67**, 280 (1934);
Studebaker and Snow, *J. phys. Chem.* **59**, 973 (1955);
Crowl and Wooldridge, *Res. Mem. Paint Res. Stn.*, Teddington, England, No. 283 (1960).
154. Dunning, Hsiao, and Johansen, *Oil Gas J.* **53**, 139 (1954);
Johansen, Dunning, and Beaty, *Soap Chem. Specialties* **31**, 41, 53 (1955); *Prod. Mon.* **20**, 26 (1956);
Dunning, Gustafsen, Johansen, *Industr. Engng Chem.* (*Anal.*) **46**, 591 (1954);
Guereca and Butler, H. S., *Prod. Mon.* **19**, 21 (1955);
Dunning et al., *Prod. Mon.* **20**, 29 (1956);
Johnson, C. E., *J. Amer. Oil Chem. Soc.* **34**, 209 (1957).
155. Johansen and Dunning, *J. Colloid Sci.* **12**, 68 (1957).
156. Rehbinder, *Z. Phys.* **72**, 191 (1931);
Rehbinder and Lichtman, *Proc. 2nd Internat. Congr. Surface Activity* **3**, 563, Butterworths, London (1957).
156a. Schnurmann, "Oil" (Manchester Oil Refinery) **2** (3) (1952);
Haworth, *Pr. & Wks. Engng*, May 1954.
157. Bowden and Tabor, "The Friction and Lubrication of Solids", Clarendon Press, Oxford (1954); *Proc. 2nd Internat. Congr. Surface Activity* **3**, 386, Butterworth, London (1957);
Bowden and Bastow, *Proc. roy. Soc.* **A134**, 404 (1931).
158. "Adhesion and Adhesives" (ed. De Bruyne and Houwink) Elsevier, Amsterdam (1957).
159. Jeffreys, *Proc. Camb. phil. Soc.* **26**, 204 (1930);
Derjaguin, *Acta. Phys. chim. URSS.* **20**, 349 (1945).

$\boldsymbol{a}$	amplitude
b	slope of correction curve (eq. (4.16))
b	constant of eq. (8.63)
c	concentration (*gm. ions l.*$^{-1}$ or *gm. moles l.*$^{-1}$ unless stated otherwise)
c	total ionic concentration of (uni-univalent) electrolyte (*gm. ions l.*$^{-1}$)
${}_s c$ & c_s	concentration at surface (*gm. moles l.*$^{-1}$ or *gm. ions l.*$^{-1}$)
c_∞	concentration of solute far from surface
$\mathbf{c}_h$	drag coefficient of gas stream on a liquid surface (dimensionless) (eq. (5.28))
c	drag coefficient of flowing film on underlying water eq. (5.22)
d	thickness of surface film.
f	farad unit of electrical capacity
μf	microfarad
f	function of $\varkappa a$ as in Fig. 3-11
f	force (*dynes* or *gm. cm. sec.*$^{-2}$)
$\boldsymbol{f}$	frequency (sec.$^{-1}$)
g	gravitational acceleration (about 981 *cm. sec.*$^{-2}$)
h	shortest distance between surfaces (e.g. of two spheres or gas bubbles)
h	Planck's constant ($=6.624 \times 10^{27}$ *erg sec.*)
i	electrical current
j	relative increase in area of monolayer due to disturbance
k	reaction velocity constant, or permeability constant
$\boldsymbol{k}$	Boltzmann constant ($=R/N=1.38 \times 10^{-16}$ *erg* (°*C*)$^{-1}$)
$\boldsymbol{k}T$	thermal energy unit ($=4.14 \times 10^{-14}$ erg at 25°C) (*gm. cm.*^{2} *sec.*$^{-2}$)
k'	constant of disaggregation of clumps (eq. (8.10), *sec.*$^{-1}$)
l	length
l_f	mean free path
l_+ & l_-	ionic conductances (*ohm*$^{-1}$ *cm.*^{2})
m	number of $-CH_2-$ groups involved in adsorption
m	mass of one molecule
mV.	millivolts
n	number of molecules or ions on 1 cm.^{2} of surface
n_0	number of molecules or ions in 1 cm.^{2} of close-packed monolayer
$n_1 - n_5$	defined as in eqs. (4.61) and (4.62)
n_c	number of counter-ions adsorbed into 1 cm.^{2} of Stern layer
n_e	equilibrium number of molecules adsorbed per cm.^{2} of surface
n_p	number of molecules of product per cm.^{2} of surface
n_r	number of molecules of reactant per cm.^{2} of surface
n_s	number of possible adsorption sites for counter-ions in Stern layer
n_w	number of waves visible on water surface
n	number of colloidal particles in 1 cm.^{3}

n	number of molecules in system
n_0	number of particles in unit volume
p	pressure, particularly vapour pressure (*dynes cm.*$^{-2}$ or *gm. cm.*$^{-1}$ *sec.*$^{-2}$)
pH_b	value of pH in bulk of solution
pH_s	value of pH in the surface
p	coefficient of eq. (5.17)
q	activation energy for molecular movement (*cal. mole*$^{-1}$)
q	moles of material transferring
r	radius (of an ion, or of wire ring)
r	roughness factor
s	fractional rate of turbulent replacement of liquid elements in a surface
t	time
t	tortuosity factor
$t_{1/2}$	time for number of dispersed particles to be reduced to half the original number (*sec.*)
$\bar{t}$	mean residence time of molecule in surface (*sec.*)
u	volume flow-rate (*cm.*^{3} *sec.*$^{-1}$)
u_1	cohesive energy of two molecules (*ergs*)
v	linear velocity (*cm. sec.*$^{-1}$ unless otherwise stated)
$\bar{v}$	mean linear velocity (*cm. sec.*$^{-1}$)
$\bar{v}_n$	mean velocity normal to surface (*cm. sec.*$^{-1}$)
v	molar or molecular volume
w	desorption energy of a $-CH_2-$ group
w	width
w	weight
x	distance normal to interface
δx	thickness of laminar sublayer
x	degree of polymerization
y	limit of ΠA as $\Pi \to 0$
z_1	valency of long-chain ions
z_2	valency of counter-ions
z	co-ordination number of monomer residue in surface

Δ	wave-damping coefficient (*cm.*$^{-1}$)
Π	surface pressure (*dynes cm.*$^{-1}$ or *gm. sec.*$^{-2}$)
Π_k	component of surface pressure originating in kinetic movements of long-chain molecules or ions
Π_r	component of surface pressure (repulsive) originating in the electrical energy of the double layer

Π_s	cohesive pressure within monolayer
Σ	foaminess (eq. (8.60))
Φ	total potential drop across interface ($=\psi+V$)

α	accommodation coefficient
$\boldsymbol{\alpha}$	$-(d\gamma/dc)$ (see eq. (4.8))
α	constant of adsorption equation (4.27)
β	correction factor in ring method of measuring surface tension
β	constant of Langmuir equation (4.28)
γ	surface or interfacial tension (*dynes cm.*$^{-1}$ or *gm. sec.*$^{-2}$)
γ_0	surface tension of plane surface of pure liquid
γ_i	interfacial tension of plane, clean interface
δ	mean distance of centres of ions held in Stern layer from plane of fixed charges
ε	electronic charge ($=4.774\times10^{-10}$ *e.s.u.*)
ζ	electrokinetic potential of interface
η	viscosity (*poises* or *gm. cm.*$^{-1}$ *sec.*$^{-1}$)
η_s	surface or interfacial viscosity (*surface poises* or *gm. sec.*$^{-1}$)
θ	fraction of surface or sites covered with adsorbed material
θ	contact angle
$\bar{\theta}$	mean contact angle of liquid on micro-rough surface
θ	angle
$\varkappa$	Debye-Huckel function: $1/\varkappa$ represents the mean thickness of the diffuse ionic double layer
λ	total free energy of desorption of a molecule or long-chain ion from the surface
λ_p	polarization energy of desorption of counter-ions or polar groups from interface
$\boldsymbol{\lambda}$	distance between successive equilibrium positions in liquid
λ	wavelength
μ	chemical potential (surfaces assumed planar)
μ°	standard chemical potential
μ_a	chemical potential of drop of radius a
μ_D	vertical component of net dipole moment of bonds in the head-groups of molecules adsorbed at the interface (*m.D.*)
μ_{ov}	overall dipole contribution from dipoles on surface and from counter-ions (*m.D.*)
μ_l	dipole moment of liquid
$\bar{\mu}$	electrochemical potential
μ	coefficient of friction

μf	microforad
ν	number of molecules per unit volume
ν	kinematic viscosity (=viscosity/density or *cm.*^{2} *sec.*$^{-1}$)
π	3.142
ρ	electrical charge volume density
ρ_l	density liquid (*gm. cm.*$^{-3}$)
σ	electrical charge density on surface
σ_c	surface charge density in Stern layer ($\sigma_c = z_2 \varepsilon nc$)
σ_D	surface charge density in diffuse double layer outside Stern layer (see p. 89)
σ	surface phase
$\boldsymbol{\sigma}$	see pp. 269–70
ϕ	volume fraction of dispersed phase
χ	contact potential (see p. 57)
ψ	distribution or outer potential (p. 57).
ψ_0	electrostatic potential in interface relative to adjacent conducting phase (mV.)
ψ_G	potential ψ_0 calculated by Gouy equation (2.27)
ψ_δ	potential in Stern phase (see p. 85)
$\psi_{Don.}$	potential ψ calculated from Donnan equation (2.37) or from (2.42)
ψ'	electrical field strength ($\psi = d\psi/dx$)
ω	overall flexibility of polymer molecule in surface
ω_0	flexibility of polymer molecule in surface in absence of internal cohesion
$\boldsymbol{\omega}$	angular velocity

Author Index

The bold figures are page numbers; the others are reference numbers under which each author is mentioned.

Subject Index